American Institute of Chemical Engineers

10th Topical Conference on Refinery Processing 2007

Held at the 2007 AIChE Spring National Meeting

April 22-27, 2007
Houston, Texas, USA

Printed from e-media with permission by:

Curran Associates, Inc.
57 Morehouse Lane
Red Hook, NY 12571
www.proceedings.com

ISBN: 978-1-60423-357-5

Some format issues inherent in the e-media version may also appear in this print version.

American Institute of Chemical Engineers

10th Topical Conference on Refinery Processing 2007

Held at the 2007 AIChE Spring National Meeting

AIChE

April 22-27, 2007
Houston, Texas, USA

ISBN: 978-1-60423-357-5

American Institute of Chemical Engineers
Proceedings
Three Park Avenue
New York, NY 10016-5991
Phone: 212-591-8100

www.aiche.org

American Institute of Chemical Engineers

10th Topical Conference on Refinery Processing
2007

TABLE OF CONTENTS

Session 3 – Advances in Hydroprocessing - I

Session 26 – Advances in Hydroprocessing - II

Session 45 – Advances in Petrochemical Production - I

Session 53 – FCC Tutorial and Advances

Session 87 – Bottom of the Barrel Tutorial

Session 105 – Bottom of the Barrel Processing

Session 114 – Oil Water Separation

Session 117 – Environmental Solutions - I

Session 126 – Environmental Solutions - II

Getting More with Less: Alon Big Spring Refinery ULSD Project
AIChE 2007 Spring National Meeting

Kirit Sanghavi, Alon USA, Big Spring, Texas
Sal Torrisi, Criterion Catalysts & Technologies, Houston, Texas

Abstract

Cost-effective solutions are the mantra of the capital projects group at ALON Big Springs Refinery, and the ULSD revamp project proved no exception. The refinery explored many options and exhausted all technological alternatives in converting their only DHT to ULSD service. This was particularly challenging given that its low operating pressure and feed difficulty (25% LCO), but creative technical solutions and close collaboration between refinery and technology suppliers yielded a solution with a capital cost of ~$850/B, much lower than the industry average cost. This paper will describe the trials and tribulations experienced along the road to the final solution, and the key decisions that lead to the successful startup in May 2005. The refinery will also describe how they have moved on the next phase of ULSD; getting the most out of the unit in an ever-changing economic environment.

Introduction

During the last six years, global ultra-low sulfur diesel (ULSD) legislation has promoted a steady improvement in diesel fuel quality. Adoption of Euro IV standards at the beginning of 2005 reduced sulfur levels in the European Union (EU) to <50 parts per million (ppm). More recently, many EU refiners have used tax incentives to adopt the Euro V standard of <10 ppm sulfur well in advance of the legislated 2009 date. In 2006, the majority of North American refiners made significant investments to upgrade their distillate hydrotreating capabilities to cut product sulfur by 97 percent in meeting the <15 ppm sulfur standards. The trend does not stop there. Monthly press announcements indicating other countries establishing timetables for ULSD include Mexico, Brazil, Russia, China and Australia.

Cost-effective solutions are the desire of refiners these days, and the Alon USA Big Spring Refinery (BSR) in West Texas proved no exception when it came to revamping their diesel hydrotreater (DHT) for ULSD service. The refinery explored many options and exhausted all alternatives in converting their only DHT to ULSD service. This was particularly challenging given its low operating pressure and feed difficulty due to 25 percent light cycle oil (LCO) in the feed blend, but creative solutions and close collaboration between refinery and technology suppliers yielded a solution with a capital cost of less than $900 per barrel (including the cost of catalyst), much lower than the industry average cost. This paper will describe the challenges and trials experienced along the road to the final solution, and the key decisions that lead to the successful startup in May 2006. The refinery has already moved forward on the next phase of ULSD: getting the most out of the unit in an ever-changing economic environment.

Background

In 1992 BSR went through a major revamp to produce low sulfur diesel (LSD) with less than 500 ppm sulfur. The 1992 DHT consisted of a single reactor with two sock loaded beds, a treat gas recycle compressor and an amine system to remove H2S from the recycle gas as well as sour makeup hydrogen. A hydrogen purification unit to clean the treat gas was operated for only a month at the start of the operation but was immediately shutdown due to marginal benefits.

The 1992 DHT was designed for a 20,000 barrels per day (BPD) feed rate but successfully processed rates up to 22,750 BPD at times. The unit operated with 20-26 percent LCO and at a reactor outlet pressure of 640 psig resulting in H2 partial pressure (H2 PP) in the range of 342-406 psia. The liquid hourly space velocity (LHSV ft^{-1}) was in the range of 1.5-1.75. Semi-regenerative reformer has been the sole source of hydrogen supply for the refinery and operated at 92-94 octane severity. The reformer was typically shutdown every January-February for regeneration, in preparation for gasoline season.

The only conversion unit at BSR is the fluid catalytic cracker unit (FCCU), and there is no hydrocracker or any outlet for No. 6 fuel oil for disposition of heavier sulfur laden streams. The bottom of the crude barrel is utilized for making regular and premium grades of asphalt. Stringent asphalt specifications preclude blending of any heavy sulfur laden streams such as LCO into asphalt.

The feed to the DHT consisted of straight run heavy naphtha (HNN), kerosene (Kero) and diesel (SRD) from the crude unit. The feed also contained LCO from FCCU. BSR has two separate LCO draws, namely Light LCO (LLCO) and Heavy LCO (HLCO). BSR has one of the shortest main columns in the world. The column, 12 ft 6 in diameter by 62 ft 6 in tall, consists of a total of 16 trays, a 2 ft section of structured packing and a 7 ft section of grid. The refinery pulled enough HLCO into diesel to maintain FCC LCO conversion in the range of 95-96 percent, with slurry API gravity commonly in the range of -2.5 to -6.0 and at times up to -8.0.

Table 1 shows analysis of heaviest streams, SRD and HLCO, in the DHT feed. By worldwide standards, these are aggressive blend stocks for a medium pressure LSD hydrotreater that's being considered for transformation to ULSD service. At first glance, BSR's HLCO would need to be undercut by more than five percent just to match average LCO 95 percent and end points in the United States. Table 1 also shows that greater quantity (lbs/hr) of sterically-hindered, thus difficult, sulfur species such as 4,6 dimethyl dibenzothiophene (DBZT) come in cracked stocks such as HLCO compared with SRD, despite flow rates that are nearly a third or less. Typical of a FCC main column, the HLCO tail (final boiling point minus 95 percent point) is longer compared with SRD, thus dragging in heavier and harder-to-treat sulfur species in the HLCO.

Table 1: Pre-ULSD Operation: Heaviest Streams in DHT Charge

		SRD Summer	SRD Winter	HLCO (1) Sample #3	HLCO (1) Sample #2
API		31.5	32.5	12.6	13.6
D-86-95%	F	664 (2)	621	703 (2)	705 (2)
D-86 FBP	F	678 (2)	641	722 (2)	726 (2)
Simdist 95%	F	714 (2)	673	729 (2)	732 (2)
Simdist FBP	F	**753** (2)	**710**	**795** (2)	**804** (2)
Tail: Simdist	F	**39**	**37**	**66**	**72**
Aromatics UV	wt%	12.1	11.5	43.7	42
Bromine Number	G-Br2	5	4.1	13	10.2 (to 13.8)
Nitrogen	ppm	**143**	**85**	**1005**	**775**
Sulfur	wt%	1.52	1.42	2.97	3.16
4 methyl DBZT	ppm	215	228	990	815
3 & 2 methyl DBZT	ppm	269	263	1190	1020
1 methyl DBZT	ppm	55	123	319	280
4,6 dimethyl DBZT	ppm	**227**	**189**	**940**	**800**
Typical flow	BPD	10940	9500	3500	3500
% DHT Charge	%	49.4	45.1	15.8	16.6
4,6 dimethyl DBZT	lb/hr	**31.5**	**22.6**	**47.1**	**39.8**
FCC slurry Gravity	API			-2.5 to -6	-2.5 to -6

(1) Excludes LLCO at approx 1700-2100 BPD. (2) Prevalent values

Exploratory Pilot Plant Tests

In early 2004, BSR conducted a series of exploratory pilot plant tests with the full range summer SRD and HLCO in the feed blend. The catalyst used in the pilot plant was identical to the catalyst in the DHT reactor at the time, which was deemed by the catalyst vendor to be very suitable for ULSD service and was recommended to other companies for ULSD units. The pilot plant was used to study the impact of varying reactor parameters such as treat gas rate (TGR, scf/b), H2 PP and LHSV on reactor operating temperatures to achieve less than 15 ppm product sulfur.

The pilot plant revealed that processing full range summer SRD and HLCO would result in very short run length of only a few months due to a high start of run (SOR) temperature. This too was at 500 psia H2 PP, a value much higher than 385 psia envisaged earlier for the BSR DHT under ULSD operation (Table 2). The exploratory pilot plant tests yielded the first wakeup call and sounded an alarm that there is substantial difference between ULSD operation (<15 ppm product sulfur) and LSD operation (at 300-450 ppm product sulfur). The test program exhibited the sulfur wall (sulfur equilibrium) phenomenon and showed that the wall can be avoided with lower LHSV and/or higher H2 PP. H2 PP could be impacted by reactor operating pressure, treat gas rate and/or hydrogen purity. Hydrogen purity in turn is impacted by makeup hydrogen flow rate and purity and also treat gas recycle compressor capacity.

Challenges

From the onset, it was clear that a number of challenges would need to be resolved before transforming the BSR DHT from LSD to ULSD service, as stated below.

Reactor/Catalyst Volume

LHSV would need to be reduced by approximately 50 percent by the addition of a new reactor.

Hydrogen Supply

The reformer would need to operate at a higher severity to supply additional hydrogen for ULSD service, reducing the reformer run length to potentially nine months.

Feed Blend

Exploratory pilot plant testing in 2004 with full range aggressive summer SRD and HLCO (Table 1) in the feed blend had already revealed that transformation from LSD to ULSD at the planned unit operating conditions which were based on minimum investment would be a major challenge.

H_2 Partial Pressure

For LSD service, the reactor operated with an outlet pressure of 640 psig. This would be reduced to 590-610 psig under the envisaged equipment arrangement and operating conditions for ULSD service including higher treat gas rate, an additional new reactor and a feed/effluent heat exchanger shell. Due to the combined effect of higher hydrogen consumption and lower reactor outlet pressure, with some offset from higher hydrogen make up flow rate, H2 PP at the reactor outlet would be reduced from 406 psia to 385 psi at the SOR. In accordance with Table 2, the unit that was classified as a medium pressure unit for LSD service would be re-classed as a low pressure unit for ULSD service. Typically, the low pressure unit does not process cracked stocks, particularly full-range material. However, there was never the slightest bit of consideration for not processing LCO at the DHT, so further pilot plant testing was required to determine the final solution.

BSR set a target of 12 months run length at the DHT under ULSD operation. The refinery was determined to tackle all the challenges methodically to achieve reliable ULSD operation and still stay with the goal of minimum investment.

Table 2: DHT H_2 Partial Pressures

	LSD	ULSD
H2 partial pressure, psia		
Low Pressure	<300	<450
Medium Pressure	300-500	450-650
High pressure	500+	650+
Big Spring Refinery		
Start of Run (SOR)	406	385
End of Run (EOR)	342	350

Methodical Approach

BSR formulated a plan to methodically progress the project. The plan included the following steps:

1. Perform detailed evaluation for selection of a catalyst vendor
2. Select best configuration for adding catalyst volume (parallel vs. series reactor arrangement)
3. Conduct additional pilot plant testing to establish SOR weighted bed average temperature (WABT) and H2 consumption as a function of 95 percent and final boiling point (FBP) for SRD and HLCO
4. Evaluate run length vs. investment, with involvement of the selected catalyst vendor
5. Carry out test runs on FCC main column to obtain pilot plant test samples and to identify unit constraints that would interfere with main column capacity and sustained reliable operation in a HLCO undercut mode
6. Explore all avenues to maximize the use of existing equipment to minimize new investment

Catalyst Vendor Selection

In late December 2004, BSR requested proposals for ULSD operation from five major catalyst vendors. On the basis of information in the proposals received, the refinery selected three vendors to evaluate their catalyst proposals in a pilot plant.

Testing was conducted at the three selected vendor's lab as well as an independent lab, using the same feedstock and at the ULSD operating conditions envisaged for the DHT. At the end of testing, Criterion Catalysts was selected on the basis of catalyst performance. BSR was particularly interested in the several benefits their reactor internals could provide, including increased catalyst volume and fouling protection.

Parallel vs. Series Reactor Arrangement

During the course of catalyst vendor selection, a quick decision was required regarding the size of the new reactor due to ever-increasing delivery times, which at that point were 12-15 months. An engineering firm instinctively recommended a parallel identical (9 ft 6 in diameter by 48 ft tall) reactor arrangement due to a desire to minimize reactor pressure drop (ΔP), resulting in higher reactor outlet pressure and thereby higher H2 PP. Two of the five vendors expressed concern with the parallel reactor arrangement because of the potential for flow maldistribution issues given the low liquid mass flux (approximately 1,300 $lb/hr{\cdot}ft^2$) at the design rate.

BSR weighed the pros and cons of parallel vs. series arrangement in light of the goal of achieving high reliability with minimum new investment and decided to go with series arrangement. By choosing this configuration, BSR eliminated the issues associated with parallel reactors such as difficulty in balancing flows and ΔP in both reactors, poorer catalyst utilization due to lower liquid and gas mass flux, and fouling of both reactors due to feed changes or upset conditions. Additionally, a series arrangement offers more flexibility to 1) stage reactor temperatures (via quenches), 2) use multiple catalysts optimized for the processing environment (Ni/Mo and/or Co/Mo) and 3) alter catalyst loading method (sock/dense) to accommodate pressure drop requirements. The decision was taken to order a new larger diameter 10 ft 6 in by 48 ft tall reactor to maximize catalyst volume while minimizing pressure drop for the series reactor configuration.

In hindsight, this choice has proven to be correct. With the limiting constraint being the operating pressure in high pressure hot separator (HPHS), the series arrangement will extend the run length by nearly 50 percent. This is due to increased catalyst volume in the new reactor with its larger diameter and dense packed bottom bed, higher operating pressures in the top three beds and more effective temperature distribution attainable in the four beds. The two reactor configuration options for BSR's DHT are shown in Figure 1.

Figure 1: Parallel vs Series Reactor Arrangement

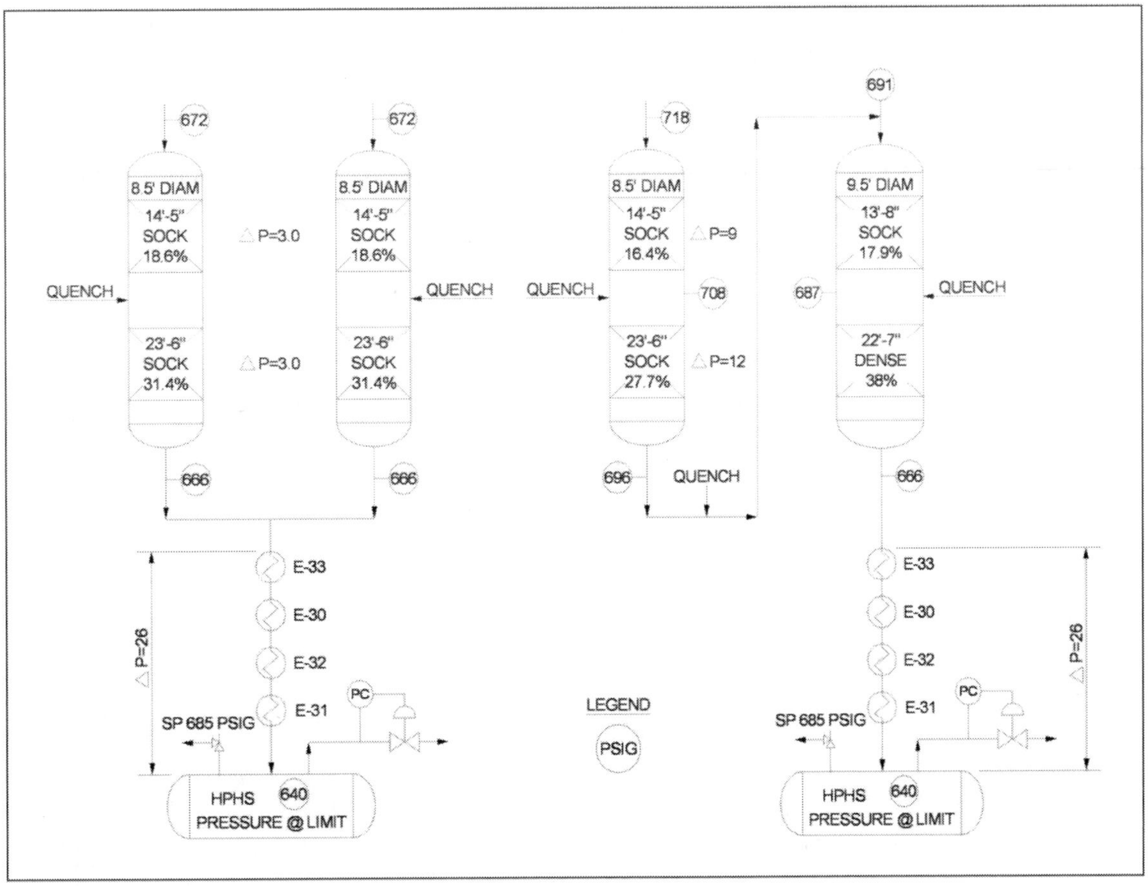

Additional Pilot Plant Testing

In an effort to better understand the impact of cut point on feed reactivity and subsequently the DHT cycle life, BSR conducted a series of pilot plant tests in an independent lab with progressively less difficult feeds. The cut points (defined as 95 percent point by distillation) for SRD and HLCO were evaluated to achieve a minimum 12-month run length at the expected unit operating conditions. At the same time, BSR continued to explore all avenues to maximize use of existing equipment, to minimize new investment and to reduce the extent of HLCO undercutting.

Pilot plant testing began with components having the full boiling range summer SRD and HLCO in the feed blend (Table 1). The first feed blend, Feed 5, contained 25.4 volume percent LCO. Test results for Feed 5 indicated that the heavy DBZT compounds in the tail end of SRD and HLCO are very difficult to treat and would require 703°F WABT at SOR to achieve 10 ppm product sulfur even at a higher H_2 PP (500 psia) than could be achieved in the unit. Cycle life for this option would be <6 months which did not meet the program objectives.

Next, plans were made to undercut only the HLCO and for more elaborate sulfur speciation testing of the feed and product to gain better process understanding. The newly created Feed 6 contained 21.7 volume percent LCO, reflecting the reduction in overall cracked feed volume due to undercutting. Despite operating the pilot plant at a higher pressure than expected in the unit, product sulfur climbed by 7.7 wppm over 10 days, and the operation was deemed too unstable. Further analysis of Feed 6 components (Table 3) revealed that both SRD and LCO streams still contained substantial quantities of difficult sulfur species, particularly the C2 to C5$^+$ DBZT's.

Table 3: Pilot Plant Data

	Pilot Plant Feed 6 with 21.7 total LCO's		Pilot Plant Feed 7 with 24.2 % total LCO's	
	HNN/Kero/SRD	LCO	HNN/Kero/SRD	LCO
API	34.60	16.45	35.30	18.00
S wt%	1.14	2.52	1.17	2.26
N2 ppm	90	656	60	485
Aromatics UV%	10.7	40.3	10.5	39.5
90% D-86/Simdist F	625/673	615/637	595/639	593/628
95% D-86/Simdist F	**646**/700	**635**/663	**611**/661	**624**/655
FBP D-86/Simdist F	669/745	658/722	631/701	650/721
C2 DBZT	588	**1248**	706	583
C3 DBZT	680	**1531**	787	**1444**
C4 DBZT	737	**1174**	736	**624**
C5$^+$ DBZT	**1864**	**1160**	**717**	**543**
	only SRD (summer)	only HLCO	only SRD (winter)	only HLCO
95% D-86 F	658	647	617	**640**
98% D-86 F	675	**666**	628	661
FBP D-86 F			635	**665**
	Conditioning with HNN/Kero/SRD		Conditioning with HNN/Kero/SRD	
LHSV	0.875		0.875	
Pilot Plant psig	570		570	
WABT degf	641		619	
Product S ppm	11		10.5	
	Pilot Plant Test with Feed 6		Test with Blend 7 @ normal conditions	
LHSV	0.875		0.7875	
Pilot Plant psig	570		488	
H2 PP sia	500		418	

Due to the short time remaining for completion of pilot plant testing prior to pre-ULSD turnarounds at the DHT and FCCU, further undercutting of both SRD and HLCO was evaluated in an effort to identify a feed blend that would be suitable for startup under ULSD planned operating conditions. The new feed blend, Feed 7, contained SRD and LCO that were cut such that just about 10 percent of the these streams (Table 3) contained the most difficult species with a boiling point >640°F (SimDist), and that is more typical of a U.S. refiner's ULSD cut points.

With these adjustments, the negative LCO impact on a low pressure DHT ULSD operation was reduced, and the SRD sulfur species became controlling as shown in Figure 2.

Figure 2: Feed 7 Sulfur Speciation

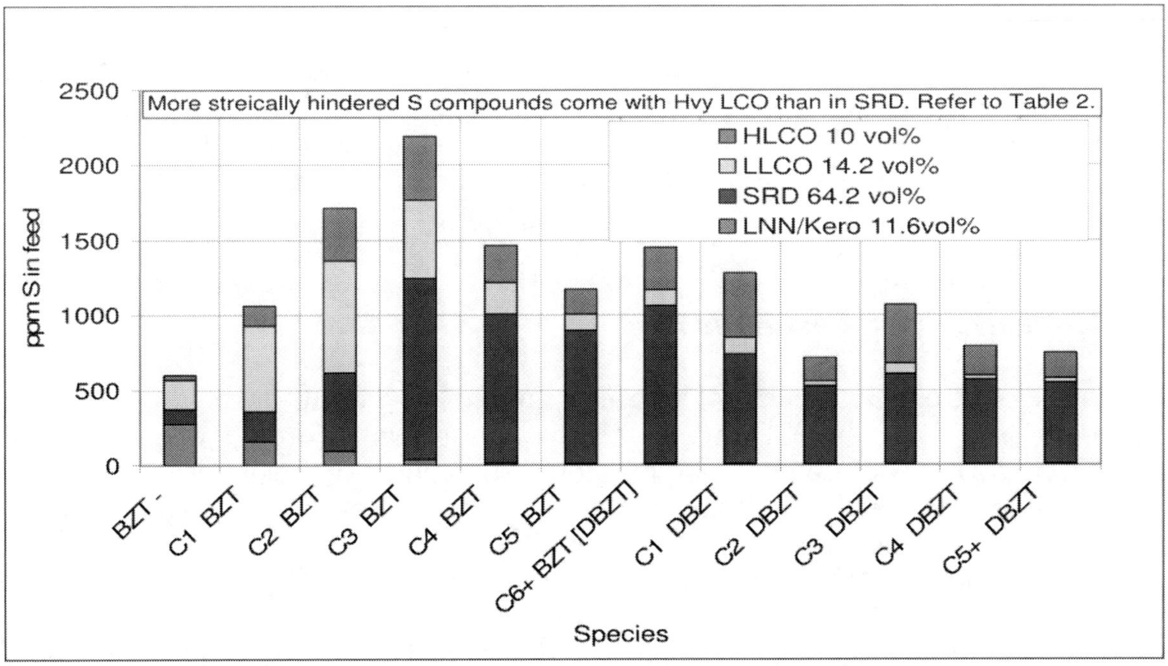

Pilot plant simulation with this feed indicated stable operations where product sulfur stayed at 6-8 wppm from day 35 to day 54 (Figure 3). Thus, a successful operating point was established for the start of ULSD operation.

Figure 3: Product Speciation When Processing Feed 7

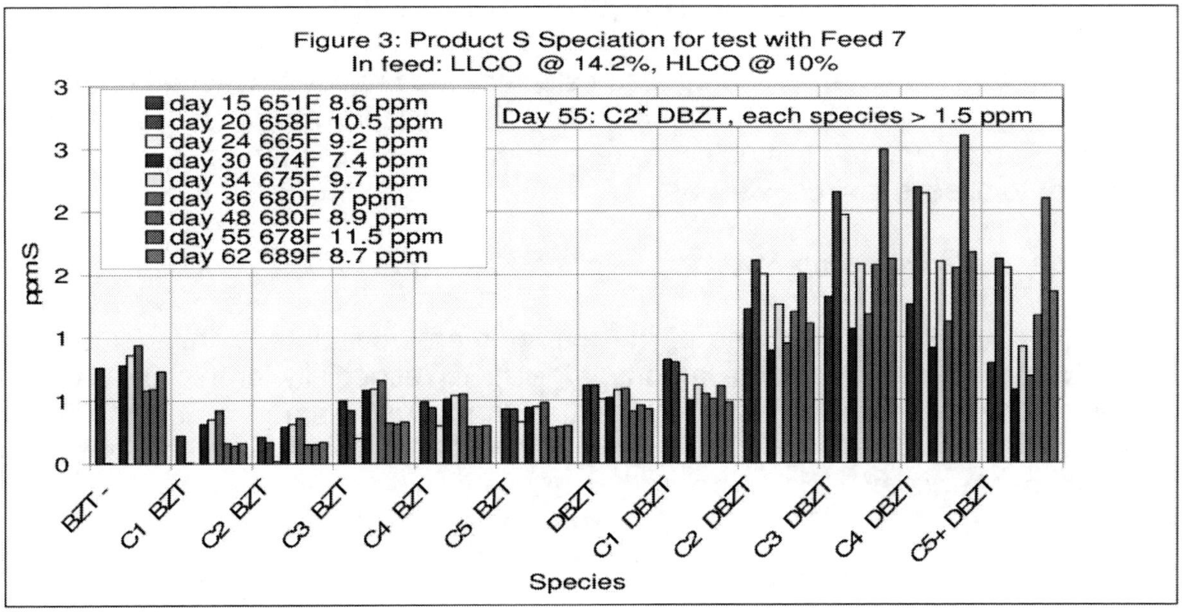

This last testing established a starting point for the 95 percent cut point limits for SRD and HLCO in the feed blend for ULSD operation. It showed the importance of controlling $C2^+$ DBZT's in HLCO due to its generally longer tail, in the feed to low pressure ULSD unit. The plan was to startup the DHT for ULSD service with feed qualities established herewith and then to gradually increase the cut points for SRD and HLCO, while checking the impact on reactor WABT and catalyst deactivation rate

Samples of more undercut HLCO for Feeds 6 and 7 were obtained from an actual test run at the FCCU main column. Each time, FCC main column was evaluated for constraints to be addressed for achieving sustained reliable operation in an HLCO undercutting mode without decrease in main column capacity.

FCC Main Column Test Runs

No retrofit ULSD project involving undercutting of LCO would be complete without test runs at FCC main column. Undercutting would decrease main column draws and bottoms temperature, reducing the column's pumparound heat removal capacity and internal refluxes thereby decreasing column capacity and quality of fractionation.

During the course of generating the undercut HLCO samples, BSR thoroughly studied the main column operation and identified a few modifications necessary to maintain FCCU main column capacity and quality of fractionation. The selected solution included use of an idle naphtha stripper for a new heavy cycle oil (HCO) draw and an addition of a new HCO product cooler. This was required to raise the main column bottoms and HCO draw temperatures. The top reflux was increased to maintain main column heat balance. Additional detail on this work is provided in reference 1.

Maximizing Use of Existing Equipment

Once a catalyst vendor was selected, the project proceeded on several fronts: the new reactor was ordered, pilot plant testing proceeded as described above, and parametric studies were carried out with input from the catalyst vendor to assess unit capabilities.

Options for Increasing Hydrogen Partial Pressure

Unit H_2 PP was one of the first elements evaluated with a parametric study, since it was viewed as important for ULSD, especially for this low pressure unit. Study results using Feed 5 indicated that a minimum of at least 400 psia and preferably 425 psia H2 PP was required at the reactor outlet to achieve a run length of 12 months.

Figure 4: Feed 5 – H_2 Partial Pressure vs Run Length

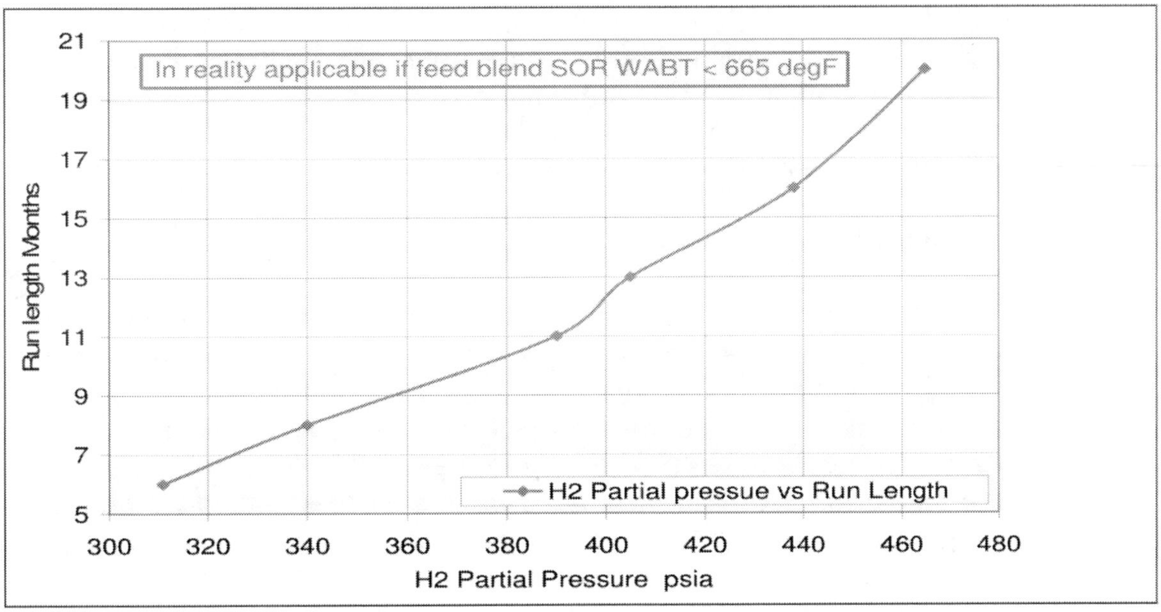

H_2 Plant Costly

The parametric study also revealed that a new hydrogen plant supplying 10 mmscfd of H_2 to the DHT could increase H2 PP to 450-465 psia and extend the run length to approximately 20 months. Investment for a new hydrogen plant was in the range of $20 million. This sizable investment could not be justified on the basis of a perceived need and BSR decided to wait for actual operating experience before making any decision on a H_2 plant. Simultaneously, the refinery was considering a proposal for converting semi-regenerative reformer to CCR, which gave more support to deferral of the H2 plant option until CCR option was fully explored. A PSA unit was explored to improve H2 purity of makeup gas but this benefit came at the expense of some of the hydrogen being sent to fuel, thereby limiting the already tight supply of hydrogen to the rest of the refinery hydrotreaters.

After evaluating the importance of H2 PP and the more expensive options to increase hydrogen production and purity, BSR started to examine lower cost options based on maximizing the use of existing equipment to achieve at least 400 psia H2 PP and a lower SOR WABT. Options considered included:

1. Increase purity of treat gas by commissioning an idled H_2 purification (permeate) unit
2. Increase reactor operating pressure
3. Improve pressure controls for the hydrogen system so that the reciprocating treat gas compressor can be operated at 100 percent load with increased compressor inlet pressure
4. Increase capacity of the amine system to better manage higher H_2S levels, thus improving desulfurization kinetics to achieve a lower SOR WABT.

Idled Permeate Unit Attractive

Evaluation of the idled permeate unit indicated that treat gas H_2 purity could be increased by 4-5 mole percent, resulting in an additional 25-30 psia H2 PP at the reactor outlet. A decision was made to equip all six separators in the permeate unit with state-of-the art membranes. Only three of the six units would be required for the service, with the rest on standby.

Check Equipment Ratings

A thorough check of the equipment in the reactor loop revealed that some of the equipment had higher maximum allowable working pressure (MAWP) on the equipment nameplate than the design pressures quoted on the P&ID's. Only one exchanger needed replacement for higher design pressure, and the existing reactor was re-rated taking into account portion of the allowable credit for the cladding. Pilot operated relief valves were installed as needed to allow a further increase in the operating pressure. With these changes, the reactor loop operating pressure at the outlet of the reactor was raised from the originally envisaged value (590-610 psig at the SOR) to 665 psig, the HPHS operating pressure was raised from approximately 614 psig pre-ULSD to 640 psig for ULSD operation and the compressor discharge pressure increased from 765-800 psig to 830 psig. These beneficial changes were all unexpected at the start of the project. Proper precaution was taken to prevent pluggage of the sensing lines for the pilot operated relief valve on the HPHS.

Heater Reconfiguration

BSR decided to add a third pass to the two-pass feed heater as well as a third feed/effluent exchanger shell. These changes reduced the heater feed side pressure drop from 50 psi to 24 psi, thereby permitting an increase in the reactor operating pressure by 16 psi without affecting upstream equipment. Since the heater had constrained the unit feed rate due to a permit limit on fuel firing, the third pass allowed higher unit feed rate to be attained.

Gas Circulation Benefits

Maximizing treat gas circulation rate was also important to ensure good H_2 PP and stable catalyst performance. The treat gas compressor is a positive displacement machine. Treat gas compressor suction pressure was controlled using makeup hydrogen. Before ULSD, the compressor supplied hydrogen to reactor inlet, gas quenches and purge, with all streams on flow controls. Thus, there was no clear bypass for compressor pressure control. This made compressor operation very difficult as manual control varied with each operator. During this period, the compressor was 75 percent loaded to avoid any overpressure that could lift a relief valve. As part of the ULSD project, a primary bypass around the DHT feed heater and a secondary bypass that routes excess hydrogen directly to reactor effluent product airfin cooler were installed. This permitted the compressor to operate at 100 percent load on automatic controls and increased treat gas rate by one-third.

In order to accommodate the higher operating pressure, the charge pumps required a larger impeller and new motors. A new relief valve was added on the discharge line to protect downstream equipment from overpressure.

Improved Gas Scrubbing

Lastly, DHT amine circulation was increased from 120 to 150 gpm with larger particulate filters and a carbon filter. New rich amine pumps were added to lower rich amine flash drum pressure from the envisaged value of 100 psig with a non-pumped system to 30 psig. This was to reduce the amount of hydrocarbons in the amine to the stripper and thereby reduce potential for foaming.

With the modifications described above, BSR was able to maximize use of existing equipment to increase H2 PP at reactor outlet from the originally envisaged value of 385 psia (Table 2) to the desired 425 psia at SOR.

Successful Startup

BSR had a very successful startup of the revamped unit for ULSD operation. Key factors behind the successful startup included:

1. Thorough equipment evaluation including a verification that existing towers, drums, exchangers, pumps and control valves would handle higher loads
2. Meticulous inspection of equipment during turnaround, especially new reactor internals for proper installation and levels
3. Comprehensive startup procedures including detailed procedure for presulfiding the catalyst and complete technical coverage during startup

Startup procedures for sulfiding the new catalyst were so complete and well-planned that the vendor's senior representative at the site stated, "This was the smoothest catalyst activation of all the ULSD units I have personally supported on the past 12 months."

Actual unit performance has exceeded the Feed 7 pilot plant results. The BSR ULSD unit achieved a stable SOR WABT of only 663°F compared with 678-680°F for Feed 7 tested in pilot plant and as measured by stable pilot plant performance over approximately 20 days.

Following this initial operation, BSR has increased feed rate and thereby LHSV, from 0.78 for Feed 7 pilot plant test to 0.84-0.87 hr^{-1}, while at the same time increasing cut point on both SRD and HLCO components. The SRD D-86 95 percent point was raised by approximately 20°F and the HLCO 95 percent point increased by 10°F. The test results for increasing HLCO D-86 95 percent point are shown in Figure 5, showing a reactor temperature increase of 1°F for each 1°F increase in HLCO 95 percent point.

For the time being, BSR is adhering to the current targets for LCO cut point (Figure 5) while continuing to move up SRD cut point, since there is a lot of value associated with maximizing crude unit diesel barrels. One way to measure the extent of LCO undercutting is to check the slurry gravity. At the current LCO cut points limits, the carbon black oil (CBO), now a blend of slurry oil and stripped HCO, has gravity is in the range of -1.9 to + 0.2 API, which is very much in line and within the range of -2 to +1 values that prevailed across the United States pre-ULSD. This is a very good achievement for a first benchmark ULSD run at BSR, with a low pressure ULSD DHT.

Figure 5: DHT Reactor WABT vs HLCO D-86 95% and FBP

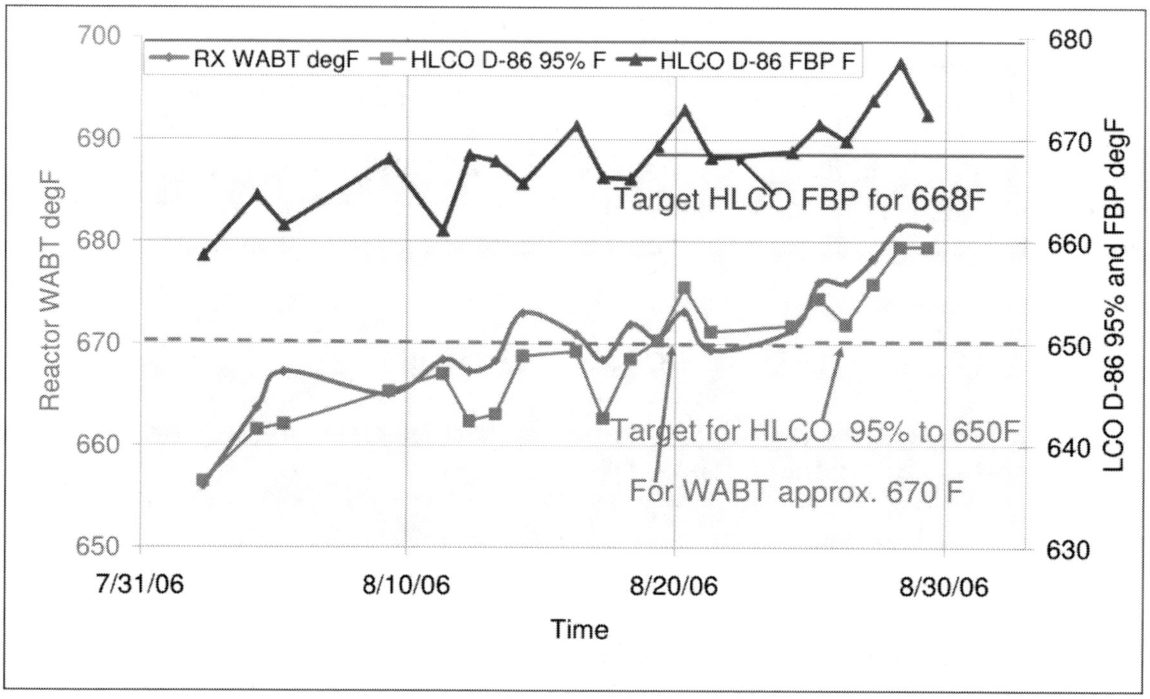

Full catalyst utilization is critical to achieving ULSD specifications, and good reactor internals performance is required to ensure that all catalyst particles are seeing uniform liquid flows and temperatures. Measuring axial and radial temperature spread (ΔT) is one of the simplest ways to evaluate performance of the internals (liquid/gas distributors and quench assemblies). A good benchmark performance for dense-loaded catalyst beds should show a maximum radial temperature spread of 20 percent of the axial temperature rise or 10°F (for example -5 to +5°F from the average), whichever is larger.

The data shown in Figure 6 show that this criteria is achieved in all four catalyst beds, even in the top three sock loaded beds, thus demonstrating good gas and liquid distribution with the Shell Global Solutions HD Distributor Tray and UFQ Quench tray. Figure 6 shows temperature spread from the average values for a set of six top TI's, six middle TI's and twelve bottom TI's for each bed. All TI's in each set are located at the same elevation. The last bed is the only dense loaded bed and has a maximum radial temperature difference of 4°F at the bottom of the bed and typically it is 3°F. The TI's below the feed distribution tray and quench tray has a radial difference of only 2°F and 3°F respectively. Depending on the EOR pressure drop experience from this first benchmark run, BSR could decide to dense load more beds knowing that a dense bed would have twice the pressure drop of a sock loaded bed.

Figure 6: Reactor Internals Perforamance – Radial Temperature Spread

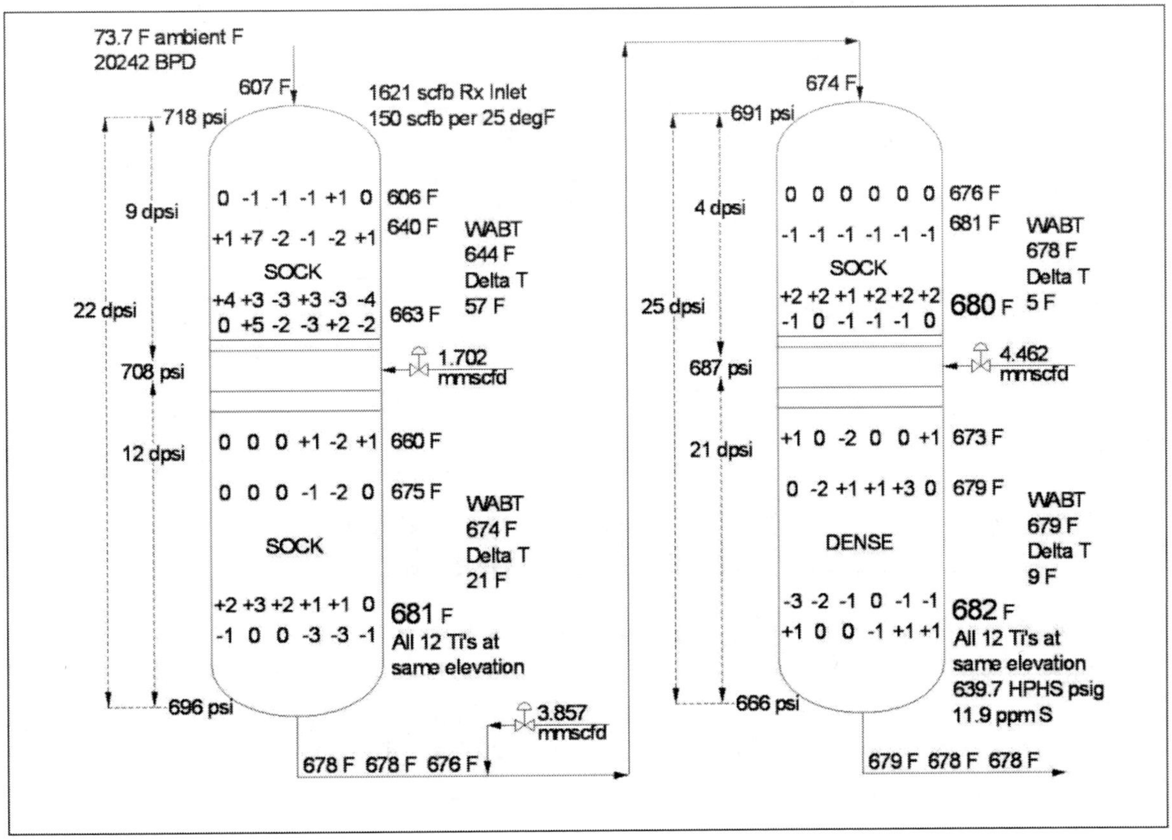

The addition of the third pass to the charge heater has been very successful. It has reduced the reactor loop pressure drop in line with expectations and increased heater efficiency as the stack temperature has been reduced from 500-540°F to 375-400°F with corresponding fuel savings in the range of $0.7 million per year.

Pre-ULSD the treat gas compressor ran at suction pressure anywhere in the range of 525-550 psig and up to 580 psig with discharge at approximately 765-800 psig and operated at only 75 percent load. Now the compressor runs at a steady 545 psig suction pressure, at a steady 830 psig discharge and at 100 percent load, thereby raising the compressor capacity by 33 percent (Figure 7).

Figure 7: Treat Gas Compressor Pressure Controls

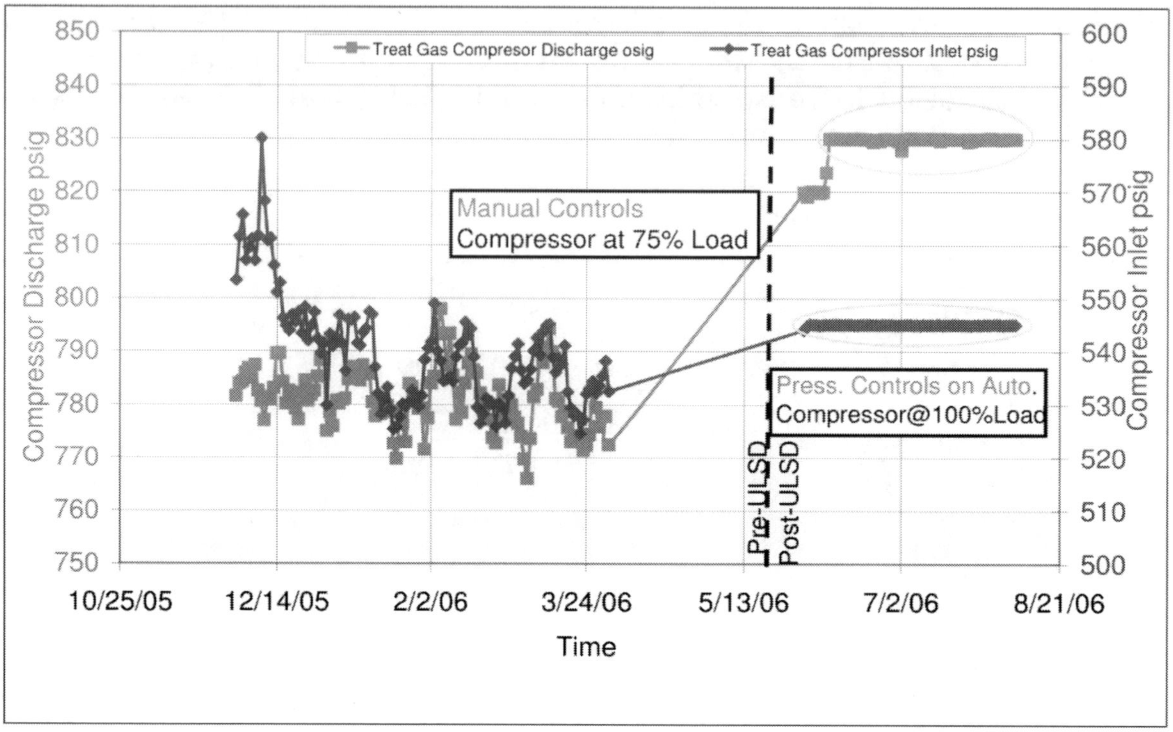

The amine system operates at 145 gpm and with lower frequency of filter change-outs and reduced incidents of flooding in the stripper.

The FCC main column has performed as expected in the undercut mode, and pleasantly CBO gravity, now slurry mixed with HCO product, has stayed in the range of +0.2 to -1.9 API compared with a customer specification of +2.0 or heavier. This has been achieved with one of the shortest main column in the world.

ULSD Transformation

In conclusion, BSR has transformed their only DHT from LSD to ULSD service at a very low investment cost, in the range of $900 per barrel. This also included the cost of the new catalyst, which has performed better than projections based on pilot plant testing. BSR continues to move forward with raising cut points for SRD and HLCO and is in the process of further evaluation of newer catalysts in the pilot plant that could lower SOR WABT by an additional 15-20°F. This would enable the refinery to further increase HLCO cut point. At the same time, BSR has kept its option open for a new CCR.

This success was achieved through an in-house think tank team working with a very methodical and sound engineering approach, without losing sight of the goal of achieving more with less investment. BSR is now focused on a clear direction for further strides in ever changing environment.

5 YEARS CYCLE EXPERIENCE FOR 10 PPM S DIESEL PRODUCTION

K. Vogt, Albemarle Catalysts, USA

ABSTRACT

On January 1, 2003, German legislation forced refiners to supply diesel with 10 ppm S, while in other area's there were also considerable tax incentives for such fuels. BP is in the forefront of obtaining experience with UD HDS and was able to produce the new specifications without revamping their unit in Lingen, Germany. It will be shown what improved catalyst was required and commercial results prove the viability of this approach.

Commercial data on required operating temperatures for producing ultra low sulfur diesel and related H_2 consumption will be shown. The importance of nitrogen inhibition will be demonstrated.

The Lingen case is an excellent commercial experience that will benefit North American refiners in their operational plans for structural production of ULSD. This paper will discuss the operation at Lingen and present the lessons learned from this long cycle of operation.

I. INTRODUCTION

New diesel specifications put a great pressure on oil refineries to increase the severity of their hydrotreating operations. New fuel specifications envisaged by the Auto Oil Program (AOP II) represented a challenge for European refiners. Driven by ambitious German environmental policy and by an agreement between the German car and oil industry on the one side and the government on the other side, more stringent low sulfur fuels were introduced in Germany already on January 1, 2003, and consequently German refineries were producing the 10 ppm sulfur products in 2002 to be able to deliver the product on spec at the pump.

In previous papers [1-6] it has been shown that for the conversion of sterically hindered sulfur species, more hydrogenation activity is required, having the following impact:

- H_2 consumption increases (figure 1)
- At low pressure and elevated temperatures, a sulfur wall will be met, at which sulfur cannot be further reduced (figure 2)
- To desulfurize difficult feeds, higher H_2 partial pressures are required. NiMo catalysts perform better at high pressure as they are capable to hydrogenate the sterically hindered sulfur species (figure 3)
- Nitrogen species inhibit HDS reactions. At the conditions at which nitrogen can be removed, NiMo catalysts have a higher HDS activity than CoMo catalysts (figure 4)

Albemarle introduced several catalyst technologies, which were tested extensively, such as:

- KF-757, a high activity CoMo STARS catalysts [7]
- KF-848, a high activity NiMo STARS catalysts [7]

Catalysts with a high hydrogenation activity, typically NiMo catalysts such as KF-848, consume more hydrogen at constant temperature. At constant product sulfur the difference in hydrogen consumption is much smaller, and usually not more than 10% difference between catalyst grades has been found. Operating at 350 or 10-ppm product sulfur makes a big difference though. Consequently sufficient hydrogen should be available to avoid hydrogen starvation. Maldistribution may result in local hydrogen starvation and coking, and consequently the distributors must be in an optimal condition.

Based on test results, a strategy was set to produce sulfur free diesel at Lingen. In this paper we will show you one result of their efforts. It will show you how at Erdöl-Raffinerie-Emsland (ERE) in Lingen (now a BP refinery) <10 ppm could be produced just by switching catalyst to KF-848 and installing an additional water injection in effluent pass. The impact of the selected strategy and the effect of catalyst quality, feed quality and nitrogen removal on the overall performance will be shown.

Figure 1: H_2 consumption with KF 756 & 757 (CoMo catalysts)

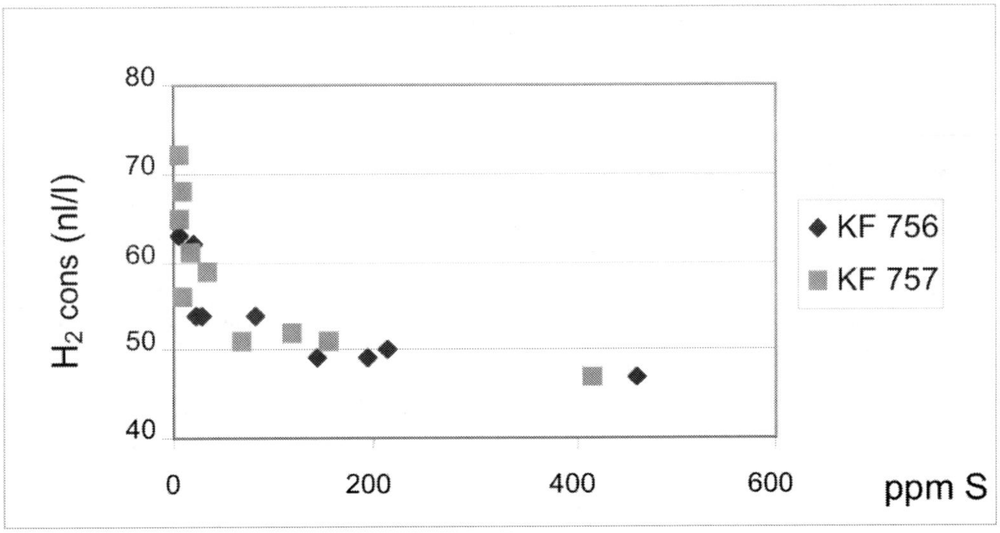

Figure 2: Wall effect: minimum sulfur reached at 653°F/345°C, 10 ppm S cannot be reached at low pressure

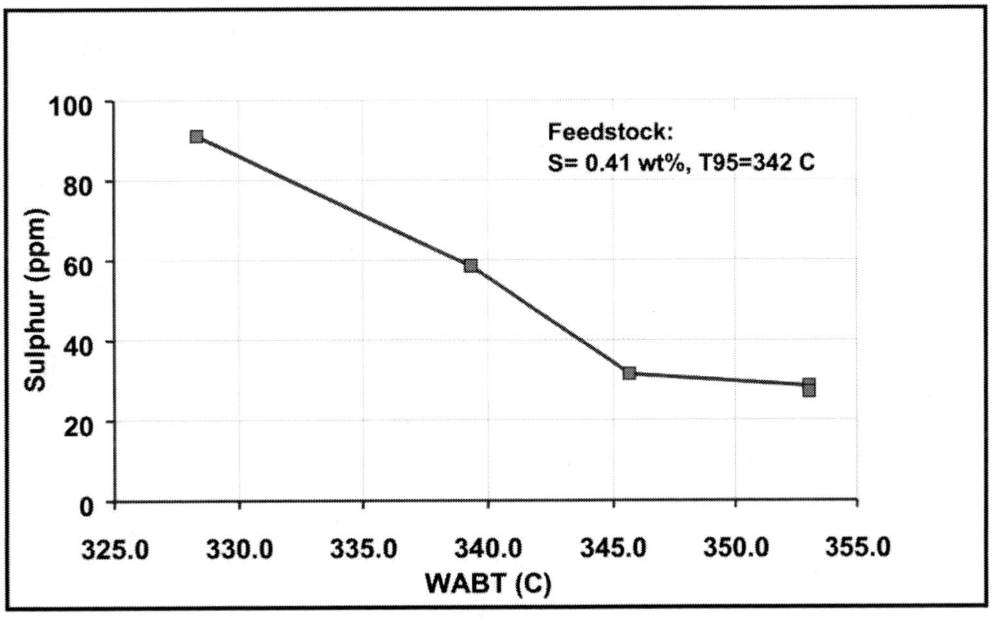

Figure 3: HDS activity of NiMo vs. CoMo catalyst at various pressures

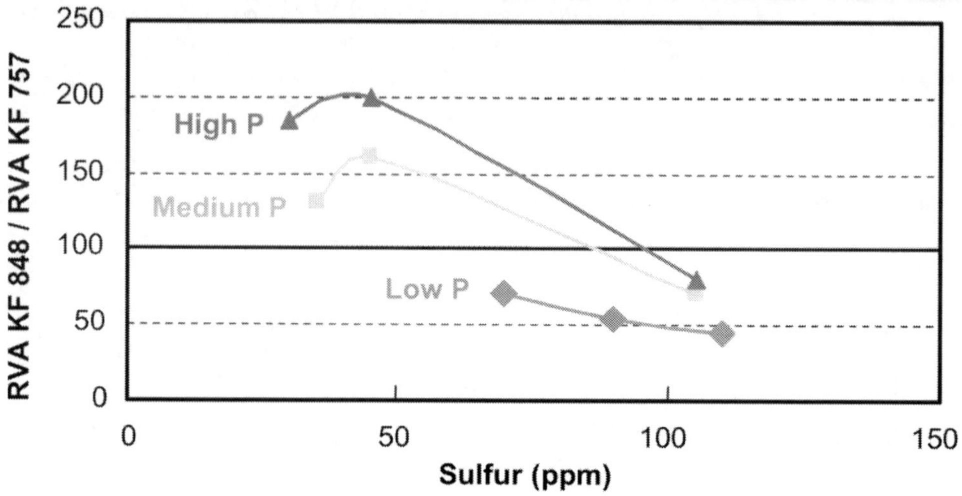

Figure 4: NiMo catalysts have a high activity at low product N

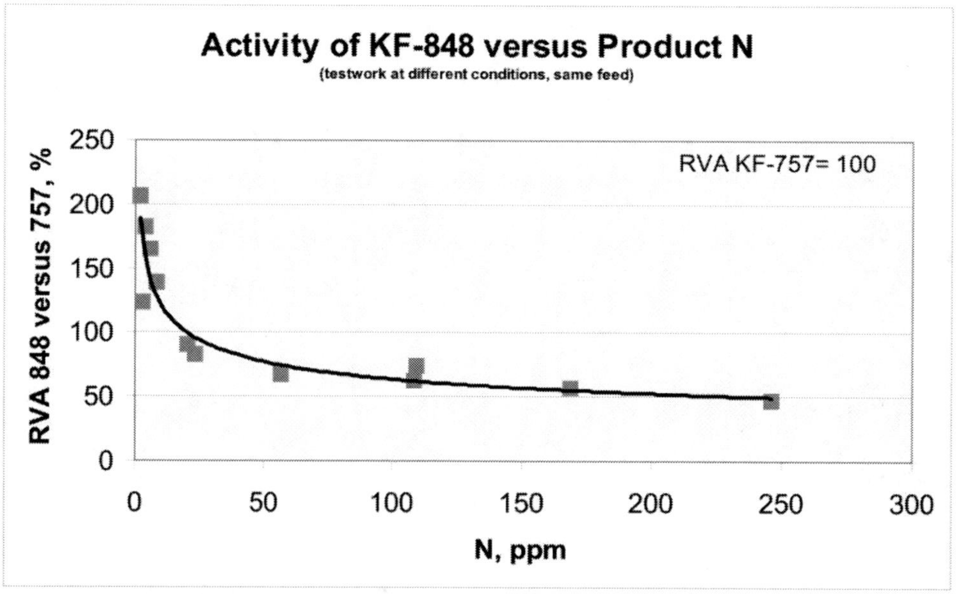

II. COMMERCIAL EXPERIENCE WITH KF-848

Albemarle recommends KF-848 for units operating at medium to high pressure, and treating their feeds to very low sulfur and nitrogen levels.

The diesel hydrotreater at ERE in Lingen was designed in 1995 for 500-ppm sulfur production. In August 2001 the first turn around took place. The unit processes on average 30 % coker gasoil. Coker gasoil is less aromatic than LCO, but contains more nitrogen. To be able to produce 10 ppm product sulfur in this high pressure unit, ERE decided to load KF-848.

By evaluating the results of several investigations and extensive test runs it was found that in this unit the energy efficiency will be a main challenge under the new conditions. In order to increase the heat recovery and to avoid ammonia salt formation, a new water injection system was installed upstream of the already existing one.

The averages of the hydrogenation conditions and the feed quality are given in table 1-2.

Table 1, Average operating conditions at ERE Lingen

Total pressure (max), psig	1000
H2 purity, %	93
LHSV, hr-1	up to 1.8
H2/oil, scfb	1500-2000

Table 2, Average feed quality at ERE Lingen

Density, API	34.7
Sulfur, wt%	0.36
Nitrogen, ppm	450
D86 IBP, 95%, FBP,°F	400, 690, 705
Coker gasoil content, wt%	30

From August 2001 till August 2006 ERE observed a very stable operation as shown in the next figures. The experience (figure 5) includes several months at <50 ppm sulfur during the first year after catalyst change out and a very successful phase-in of operation with a product sulfur content of approximately 10 ppm during the rest of the cycle.

Figure 5, Feed and Product Sulfur

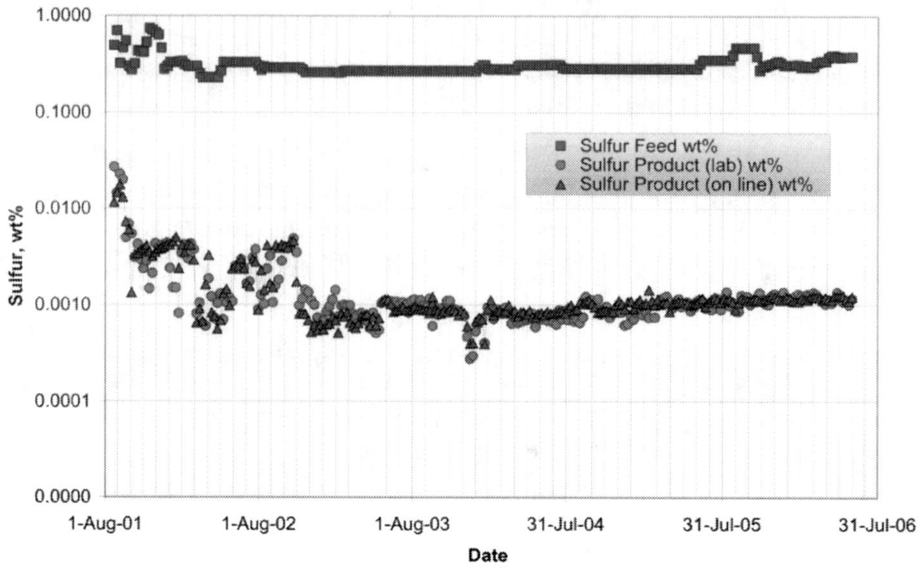

Since nitrogen inhibits HDS reactions and since cracked feedstocks usually contain a lot of nitrogen, it is of prime importance to investigate the impact of feed quality and operating conditions on the effectiveness of KF-848. In this high pressure unit with a feed blend of up to 50% coker gasoil material being processed the unit is operated to produce both 50 and 10-ppm product sulfur. When operating at 50-ppm product sulfur, product nitrogen is already less than 5 ppm and the inhibiting effect of nitrogen is not strong. As shown in figure 6, the high HDN activity of KF-848 reduces product nitrogen to less than 10 ppm.

Figure 6: WABT normalized for feed quality and conditions versus predicted product N

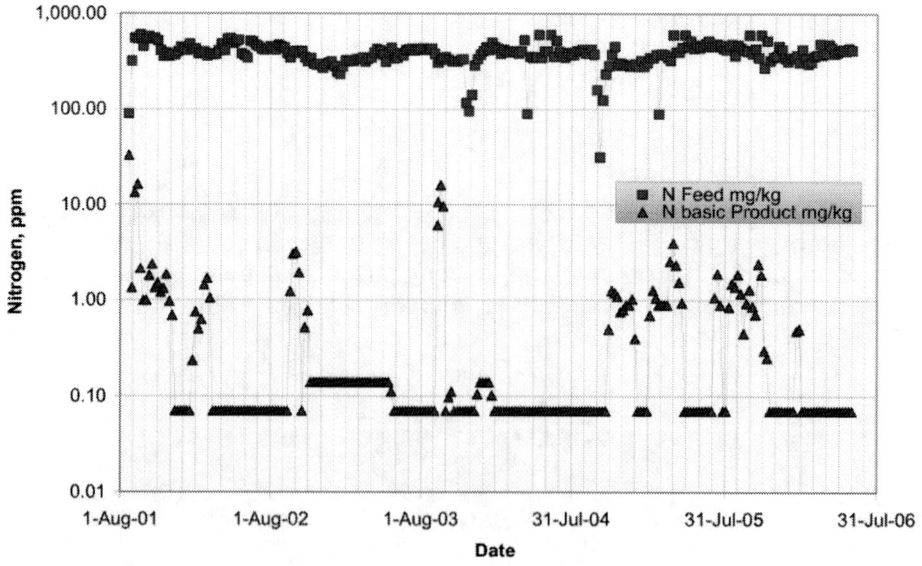

To reduce the product sulfur from 50 to 10 ppm, WABT needed to be increased by 10°C. Key to the success of the operation is the high HDN activity of KF-848, resulting in a reduction from 500-ppm nitrogen in the feed to <1 ppm in the product. The temperature variations on both levels are more related to the feed type and amount. As mentioned above, the WABT behavior is very predictable and varied seasonally with the need for heating oil production.

The H_2 consumption, as estimated using the reactor exotherm, increased by 10 to 15 % while desulfurization was driven from 50 to 10 ppm. A comparable result is observed by using the whole hydrogen consumption including solution losses.

In figure 7 we plot the operating temperatures of the diesel hydrotreating unit of the 5 years of uninterrupted cycle and without any off spec diesel.

Figure 7, Unit Temperature, °C

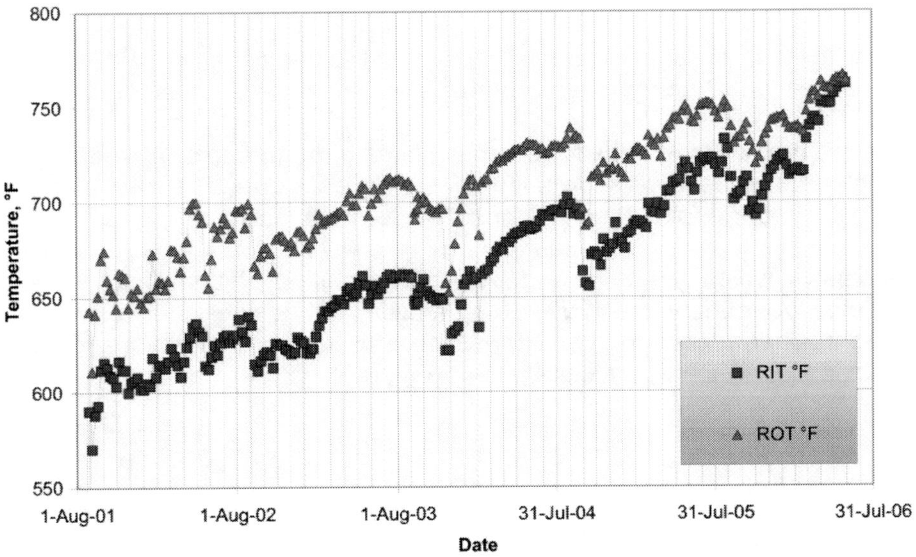

All in all, taking into account the unit design, catalyst activity and, of course, the experiences earned during the previous cycles, the summary of the foregoing provided a proven base to enable the hydrotreater to operate in the ULSD era with very little investment and still achieve a world class performance.

III. GENERAL OBSERVATIONS AND CONCLUSIONS

The drive for lower sulfur fuels has triggered an acceleration in catalyst development over the last few years. With STARS technology Albemarle Catalysts and Nippon Ketjen have provided breakthrough solutions for the refining industry to help reach the new targets. STARS catalysts have become the leading technology choice for production of ULSD and have been proven to be a "drop-in" solution to clean fuels production in some cases. As important as the catalyst developments themselves is understanding the application of the catalysts. Knowledge and understanding of the kinetics of ultra low sulfur diesel continues to improve.

Achieving <10 ppm sulfur by applying catalysts with a high denitrogenation and hydrogenation activity, such as NiMo catalysts, can be very successful. However, it should be taken into account that operating at ultra-low sulfur diesel conditions is more critical with respect to:

- Liquid distribution
- Nitrogen inhibition
- H_2 partial pressure and H_2/oil ratio

Especially the last factor is of prime importance for NiMo catalysts to avoid H_2 starvation and premature deactivation. Proper process control is essential to obtain stable operation.

BP has achieved breakthrough business performance through use of Albemarle's latest generation catalysts to make ultra low sulfur diesel. BP was one of the first major user of STARS technology for 50 and 10 ppm S diesel in Germany and the UK.

KF 848 is a catalyst which can have, under specific conditions, higher activity than KF 757, and which allows the refinery to achieve 10-ppm sulfur with less or no investments. At ERE Lingen this approach has proven to be very successful. Nitrogen removal and reduced nitrogen inhibition is key to this success.

REFERENCES

1. E.Brevoord, L.A.Gerritsen, F.L.Plantenga, Production of ultra-low Sulfur Diesel, Commercial experience with STARS Technology and the next step in ultra-deep HDS, ERTC 2000, Rome.
2. S.W.Mayo, E.Brevoord, L.A.Gerritsen, F.L.Plantenga, Process Ultra-Low Diesel, Hydrocarbon Processing, Feb 2001, Vol. 82, No.2.
3. S.W.Mayo, F.L.Plantenga, R.G.Leliveld, Elegant solutions for ultra-low sulfur diesel, o, F.L. Plantenga, R.G. Lelive, NPRA 2001.
4. R.G. Leliveld, S.E. Eijsbouts, S. Mayo, K. Fujita. E. Brevoord, Ultradeep HDS catalysis in production of clean diesel fuels, ACS Symposium, 'Emission control through Catalysis', San Diego, 2001.
5. T.Hashiguchi, Y.Miyauchi, K.Fujita, T.Funakoshi, STARS hydrotreating catalysts for ultra-low sulfur diesel production, Nippon Ketjen, Korean Seminar 2000.
6. J.Reinhardt, U.Balfanz, E.Brevoord, S.W.Mayo, F.L.Plantenga, Y.Miyauchi, K.Fujita, STARS technology for the production of ultra-low sulfur diesel, Akzo Nobel Catalyst Symposium 2001, Noordwijk
7. Eijsbouts S., Uragami Y., Inoue Y., Takahashi Y., Improved Hydrocracker Pretreat Catalysts, Akzo Nobel Catalysts Symposium, Noordwijk, July 1998.
8. F.L.Plantenga, R.Cerfontain, S.Eijsbouts, F.v.Houtert, K.Riley, S.Soled, S.Miseo, K.Fujita, "NEBULA" a breakthrough technology for hydroprocessing of Clean Fuels, Akzo Nobel Catalyst Symposium 2001, Noordwijk
9. S. Mayo, E. Brevoord, F. Plantenga, N. Gudde, G. Hoekstra, ULSD in real life: commercial performance of STARS and NEBULA technology, NPRA 2002
10. E. Brevoord, S. Mayo, F. Plantenga, J. Reinhardt, U. Balfanz, N. Ahr, N. Gudde, VEBA and BP are prepared for 10 ppm S diesel production, ERTC 2002.

Thermochemistry of Catalyst Deactivation II: Competition between Hydrogenation, Condensation and Hydrocracking

Paul R. Robinson, Chevron Lummus Global, 100 Chevron Way, 10-3210, Richmond, CA 94801

The first paper in this series presented the results of equilibrium calculations for the competition between the hydrogenation of naphthalene to form tetralin and decalin (Reactions 1-2) and the condensation of naphthalene with o-xylene to form chrysene (Reaction 3). This set of reactions was proposed as a simple model for the coke-induced deactivation of hydrotreating catalysts.

Saturation (Reversible)

1. Naphthalene ($C_{10}H_8$) + $2H_2$ $\leftrightarrow$ Tetralin ($C_{10}H_{12}$)

2. Tetralin + $3H_2$ $\leftrightarrow$ Decalin ($C_{10}H_{18}$)

Condensation (Irreversible)

3. Naphthalene + o-Xylene $\rightarrow$ Chrysene ($C_{18}H_{12}$) + $3H_2$

4. Chrysene $\rightarrow\rightarrow\rightarrow$ Coke + xH_2

This paper shows the effect of adding a hydrocracking reaction (Reaction 5).

Hydrocracking (Irreversible)

5. Tetralin + H_2 $\rightarrow$ 1-propyl-2-methyl benzene ($C_{10}H_{14}$)

In addition, for all reactions the thermochemical effects of hydrogen purity at constant total pressure are computed, along with the effects of excess hydrogen, i.e., "gas-to-oil ratio."

Molecular Weight and Its By-Boiling-Point Distribution of Middle Distillates for Hydroprocessing Modeling and Simulation

Jinwen Chen, Norma Mclean, Yevgenia Briker, Darcy Hager, and Zbigniew Ring
National Centre for Upgrading Technology
One Oil Patch Drove, Devon, AB T9G 1A8, Canada

Prepared for Presentation at the 2007 AIChE National Spring Meeting, TA001: Advances in Hydroprocessing I, April 22-27, 2007, Houston, TX, USA

March 2007

Abstract

Molecular weight (MW) and its by-boiling-point distribution are important physical properties of petroleum feedstocks. This paper presents a quick and convenient method to determine the MW and its by-boiling-point distribution using off-the-shelf GC-FIMS (Gas Chromatography-Field Ionization Mass Spectrometry) data. The predicted MW by-boiling-point distributions matched reasonably well with those measured by experiment.

1. Introduction

The implementation of ultra-low sulphur diesel (ULSD) specifications requires that refineries produce diesel fuels containing sulphur less than 15 wppm (Chen et al., 2003, 2004; Song, 2003). To achieve this target, refineries have to either use highly efficient hydrotreating catalysts and/or optimize hydrotreating process operations through mathematical modeling and simulation (Ito and van Veen, 2006; Babich and Moulijn, 2003; Song, 2003; Hu et al., 2001). To precisely establish the hydrotreating process model and to accurately evaluate the required model parameters, it is essential to have detailed physical properties and characterizations of the feedstocks and hydrotreated products. Among these are molecular weight (MW) and its by-boiling-point distribution, which is required in process and reactor modeling, reaction kinetics studies, vapor-liquid equilibrium analysis, etc.

Currently, there are no universally applicable methods or correlations to estimate the average MW for petroleum samples with wide boiling ranges, even for middle distillate samples, although some studies have been published (Retzekas et al., 2002; Sun and Yang, 2000; Khan and Kumar, 1989; Glavincevski and Gardner, 1986). An efficient method to determine the MW distribution by-boiling-point is simply unavailable in the open literature. Direct measurement is not easy for a sample with a wide boiling range. Established methods are either for samples with low (e.g. Freezing Point Depression, FBP<300°C) or high (Vapor Pressure Method: ASTM D2503, IBP>220°C) boiling points to provide an averaged MW. There is no suitable method for samples in the diesel boiling range without pre-separation. Measurement of MW distribution by-boiling-point requires distillation of the sample into a number of fractions, which is time and effort consuming.

This paper presents a quick and convenient method to determine the MW and its by-boiling-point distribution using off-the-shelf GC-FIMS (Gas Chromatography-Field Ionization Mass Spectrometry) data. This method is simple, requires a very small amount of sample, and does not need any distillation.

2. Experimental

To establish and validate the method, a total of 12 different middle distillate samples were chosen to undergo distillation to obtain 8 fractions for each sample. MW was then measured for each of the fractions by using either an in-house developed Freezing Point Depression method (using Cryette WR™ Cryoscope, Model 5009, Precision Systems Inc.) or ASTM D2503, depending on the boiling range of the fraction. At the same time, density and simulated distillation (SimDis) were also measured by using standard ASTM methods. From SimDis data, the average boiling point of each fraction of the sample was calculated, and, therefore, the by-boiling-point distributions of MW and density were obtained for each sample. These measured MW by-boiling-point distributions were then used to validate and calibrate the GC-FIMS predicting method.

The main properties of the 12 middle distillate samples are listed in Table 1. These samples have quite different boiling ranges, densities and hydrocarbon type compositions.

Table 1 Main properties of 12 middle distillate samples

	Density, g/ml	IBP, °C	FBP, °C	Paraffins, wt%	Cyclo-Paraffins, wt%	Aromatics, wt%
A-LT-Dist	0.8763	143	359.5	13.07	29.13	57.80
B-HY-Dist	0.8546	246.8	425	20.65	36.14	43.22
B-HY-Naphtha	0.7985	97	277.3	36.31	39.75	23.93
B-LT-Dist	0.8291	120.9	344.5	33.6	39.92	26.48
ColGO	0.9569	138.5	375.1	2.35	4.31	93.34
H-Gas Oil	0.8207	92.5	341.9	38.71	33.57	27.72
K-HY-Dist	0.8694	145	403.1	0.12	69.79	30.09
K-LT-Dist	0.8357	118	334.1	5.58	73.92	20.49
Nkht PC-Stove	0.8237	138.5	297.8	28.95	42.21	28.85
S-Bottom HGO	0.8694	204.7	371.4	13.05	33.29	53.66
S-Liquid T-Stove	0.8537	151.6	299.5	18.62	36.12	45.26
Stove	0.8184	142.2	312.8	24.55	53.92	21.53

Some of them have relatively high aromatics contents, such as GolGo, S-Bottom HGO and A-LT-Dist, while others have relatively high paraffin contents, such as K-LT-Dist, Stove, and B-HY-Naphtha. Therefore, these samples are a good representation of middle distillates.

The distillation of the 12 samples was conducted in a spinning band distillation system (Model 36 100A, B/R Instrument Corporation). Each sample was distillated into 8 fractions. Each fraction was then measured for MW, SimDis and density. All the fractions (total of 96) together with the12 original samples were submitted for GC-FIMS analysis. Typical measured molecular weight by-boiling-point distributions are shown in Figure 1.

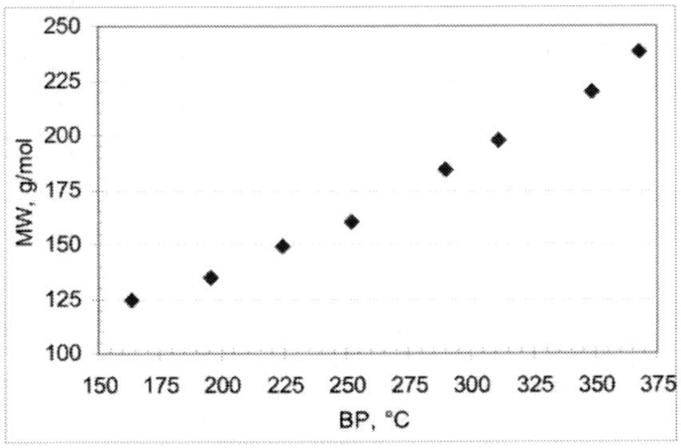

(a) ColGo

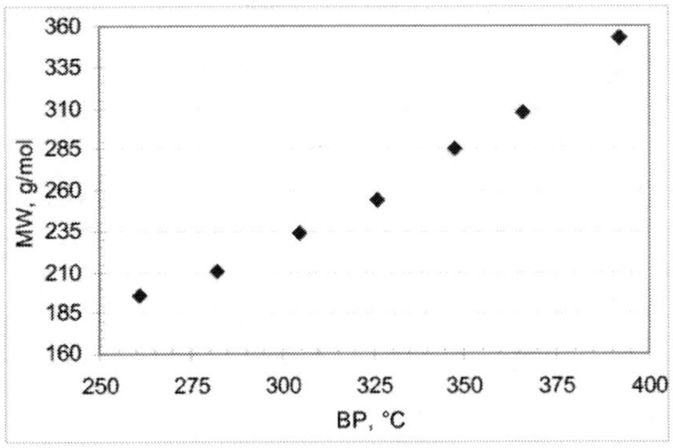

(b) B-HY-Dist

Figure 1. Measured molecular weight by-boiling-point distribution

The GC-FIMS analysis was performed on a Macromass GCT instrument with an orthogonal acceleration TOF (Time of Flight) MS (mass spectrometer) coupled with an HP 6890 GC (Gas Chromatograph). The GC column was 30-m Factor Four VF-5ht (fused silica in the range of -60°C to 400°C with 0.1-μm film thickness and 0.32mm i.d) from VARIAN. The oven temperature was ramped from room temperature to up to 370°C at the rate of 10°C /min and held for 15 min. The injection of 1μL sample was made with the program temperature vaporized (PTV) injector used in the splitless mode. The inlet was heated from room temperature 380°C at the rate of 100°C/min. Constant flow of 1mL/min was maintained throughout the run. The temperature on the GC-TOF interface was maintained at 300°C.

In order to ionize petroleum molecules and produce molecular ions, the FI emitter from CarboTeck was used. It consisted of 10-μm tungsten wire on which the carbon dendrites were generated. The emitter was placed in a high electric field (~10-7-10-8 V/cm) generated by a pair of extraction rods held at a high potential (-12Kv). Under these conditions it was believed that an electron could be removed from the molecule via quantum tunneling effects, generating radical molecular ions with minimal fragmentation. The emitter current was set at 0 mA, and the flash-off current was set at 30 mA with the scan time of 1.2 seconds and inter-scan delay of 0.2 second. Tris (trifluoromethyl)-1, 3, 5-triazine with the molecular weight of 284.9914 DA was used as the internal standard for mass calibration

3. Methodology Development

3.1 Determination of Response Factors

The raw data from GC-FIMS analysis are the mass/molecular weight, abundance and GC retention time for each detected compound/ion. The abundance represents the concentration of the compound in the oil sample. The abundance threshold of 4 was chosen to filter out the instrumentation noises. Normally there are about 5000 compounds detected in a typical sample. These compounds are identified with a mass window of 0.03 using a pre-established database containing 90 groups of hydrocarbons with carbon number raging from 1 to 50. To estimate the concentration of each compound, it is necessary to perform a pre-calibration using a standard sample with known hydrocarbon type distribution to establish the abundance-concentration correlation, which is defined here as response factor. In this study, a standard sample was analyzed using GC-MS to obtain hydrocarbon type distribution. A total of 10 hydrocarbon groups were identified and quantified. These 10 groups were: paraffins, cycloparaffins, dicycloparaffins, tricycloparaffins, naphthalenes, benzenes, pyrenes/ fluoranthenes, acenaphthalenes/fluorenes, benzothiophenes, phenanthrenes/anthracenes. At this stage, we assumed that all the compounds in the same group had the same response factor. Therefore, the response factor of each group was obtained by dividing the weight percentage of that group by the sum of the abundance of the compounds in the same group. For better data representation, relative response factors were defined and used by dividing the response factor of each group by that of paraffins. The other groups that are present in the sample but without response factors from GC-MS data were assumed to have response factors of 1. This simplification is reasonable since a) response factors are normally from 0.5 to 2; and b) the final result was not sensitive to the response factors of other groups since their contents in the sample were very low. Once the response factors were evaluated, they were included in the database for data processing of real samples. The determination flowsheet of response factors is illustrated in Figure 2.

3.2 Data processing of real samples

The raw data from GC-FIMS for real samples were also molecular weight, abundance and retention time of each compound (about 5000 in total). Data filtration was performed first to remove instrumentation noises. The compounds were then identified and assigned for response factors. A pre-established relationship between boiling point and GC retention time, obtained from standard paraffins, was used to determine the boiling point of each compound.

From the abundance value, the mass concentration of each compound was calculated using the corresponding factor. The entire boiling range was divided into a certain number of (normally 8 to 25) equal interval slices. The averaged molecular weight and the boiling point of

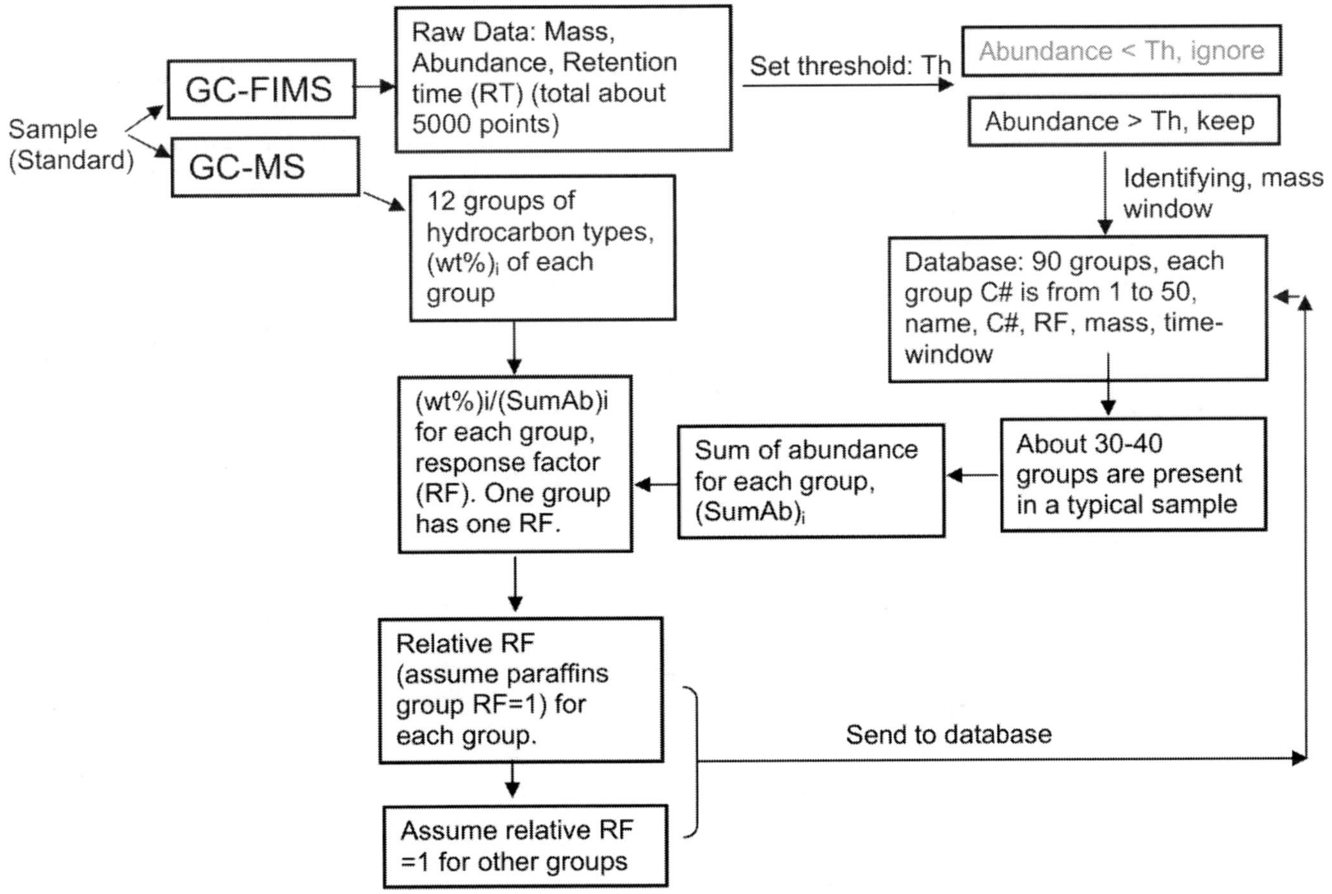

Figure 2. Response factor determination flowsheet

the compounds in each slice were calculated using weighted averaging and sent to Microsoft EXCEL® for data plotting and curve fitting. The calculation procedure is shown in Figure 3.

Figures 4 (a), (b), (c) and (d) show the predicted MW by-boiling-point distributions of selected middle distillate samples. For comparison, the corresponding experimentally measured ones are also plotted in these figures. Clearly the predictions obtained from GC-FIMS analysis are quite close to those measured. Note that the predictions and the experimental measurements covered different boiling ranges of the sample. The experimental measurements could not reach the lower boiling end but could reach the higher boiling end while the predictions had the exact opposite trend. Further studies are underway to improve the GC-FIMS analysis and prediction to reach the higher boiling end.

Since the 12 middle distillate samples used in this study have quite different boiling ranges, densities and hydrocarbon type distributions, together they are a good representation of middle distillates. Response factors based on these 12 samples should be universally applicable and should provide better predictions. Therefore, a procedure is being developed to optimize the response factors by fitting the predicted molecular weight with the experimentally measured ones for all the individual fractions obtained from the distillation of the 12 samples. The results will be reported in the future.

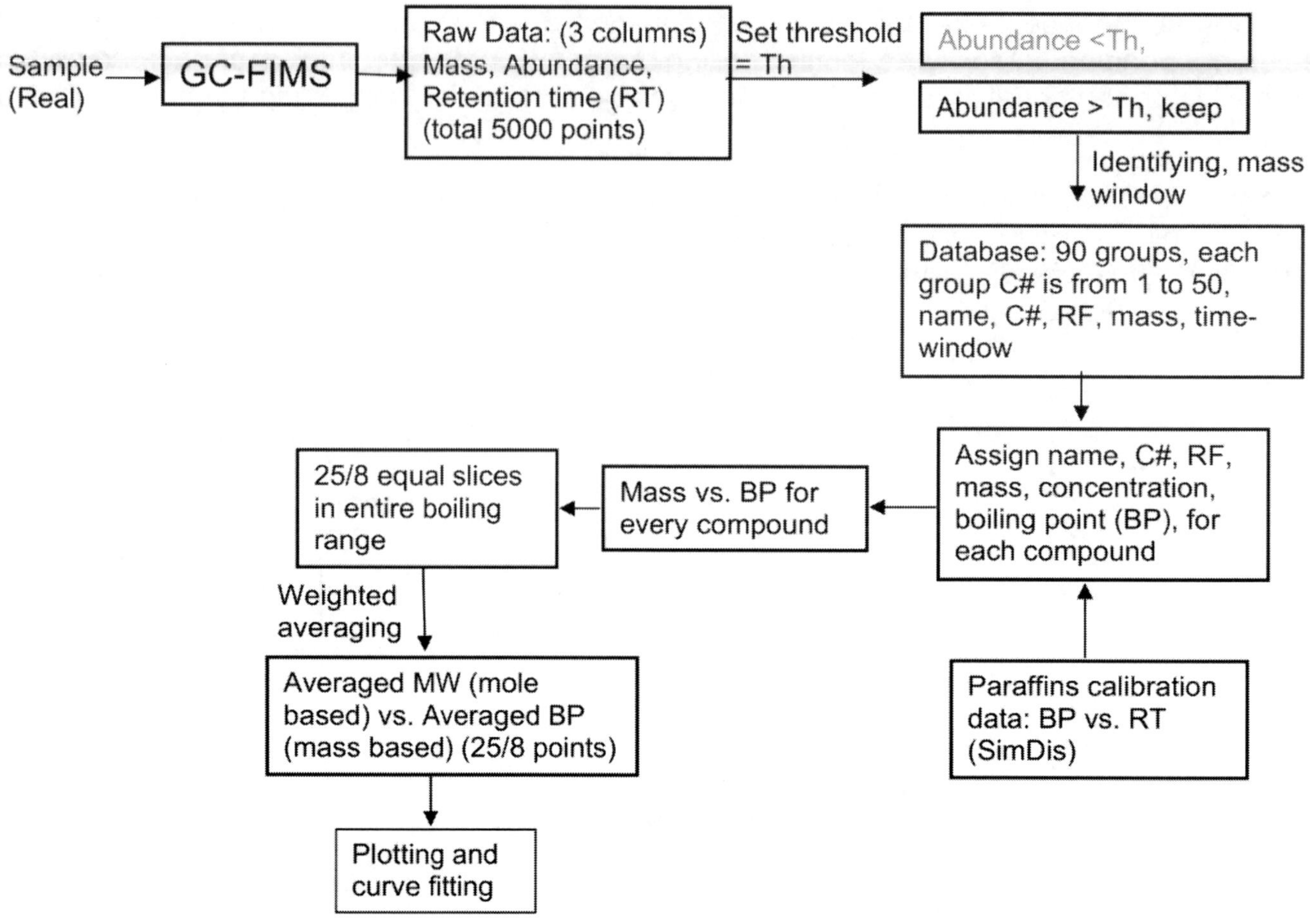

Figure 3. MW prediction flowsheet

4. Summary

A methodology based on GC-FIMS analysis has been developed to predict the molecular weight and its by-boiling-point distribution for petroleum middle distillates. 12 middle distillates samples, which have different boiling ranges and hydrocarbon type distributions, were used to validate the method. It has proved to be simple and convenient to use. The predicted MW by-boiling-point distributions were close to those obtained by experimentally measured ones. This method needs only a small sample for GC-FIMS analysis and does not need the time- and effort-consuming distillation otherwise required. This method overcomes the shortcomings of the established ASTM and in-house developed methods that can only be used for samples with either low boiling range or high boiling range.

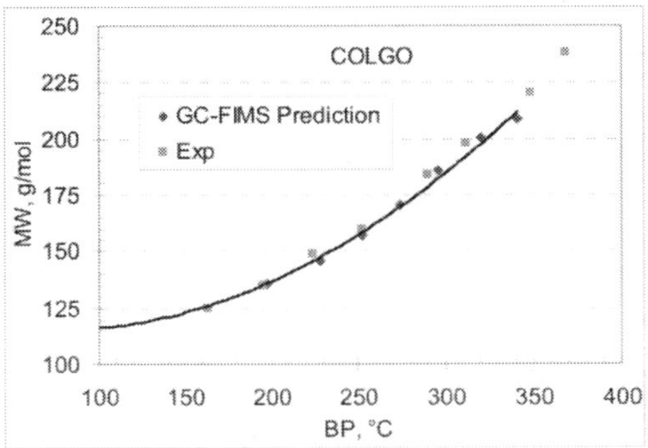

(a) ColGo

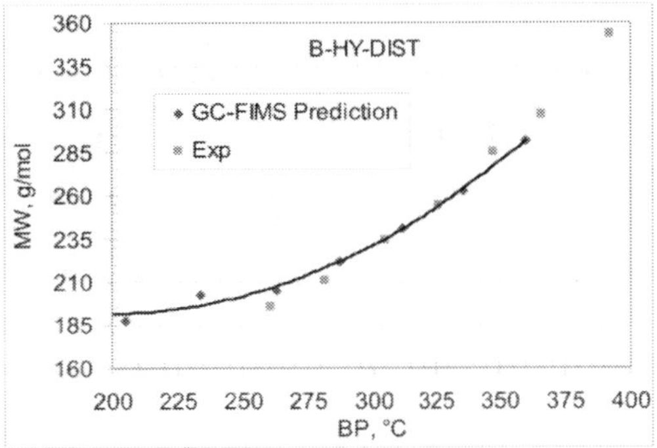

(b) B-HY-Dist

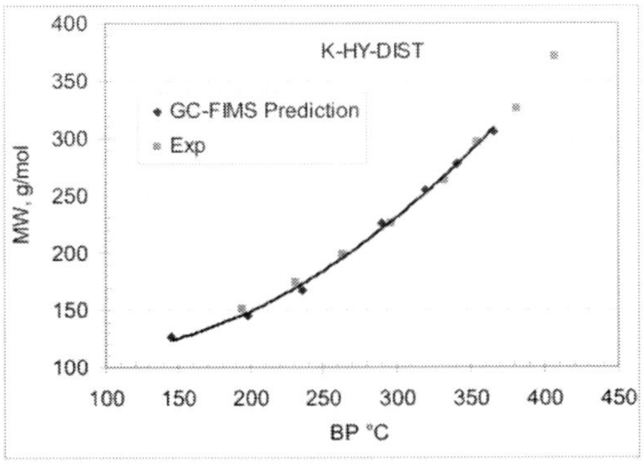

(c) K-HY-Dist

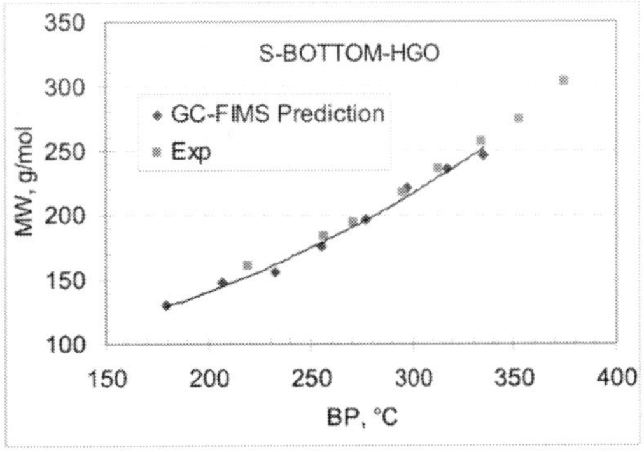

(d) S-Bottom-HGO

Figure 4. MW prediction from GC-FIMS

5. References

Babich, I. V. and Moulijn, J. A., "Science and Technology of Novel Processes for Deep Desulphurization of Oil Refinery Streams: a Review", Fuel, 82(6), 607-631, 2003.

Chen, J., Te, M., Yang, H., and Ring, R. "Hydrodesulphurization of Dibenzothiophenic Compounds in a Light Cycle Oil", Petrol. Sci. Tech., 21(5&6), 911-935, 2003.

Chen, J., Yang, H., and Ring, Z., "HDS Kinetics Study of Dibenzothiophenic Compounds in LCO" Catal. Today, 98, 227-233, 2004.

Glavincevski, B., and Gardner, L., "Expression for the Average Molecular Weight of Diesel Fuels, SAE Special Publications, 55-64, 1986.

Hu, M., Ring, Z., Briker, J., Te, M., "An Integrated Approach in Meeting the Challenges of Low Sulphur Diesel – Analytical Support, Process Research and Computer Simulation", NPRA Conference Proceedings, paper CC-01-161, Dallas, TX, USA, 2001.

Ito, E. and J. A. R. van Veen, "On Novel Processes for Removing Silphur from Refinery Streams", Catal. Today, 116, 446-460, 2006.

Khan, R., and Kumar, H., "Pseudocomponent Model for Predicting of Molecular Weight Distribution of Pyrolysis Liquids Generated at Slow and Rapid Heating Rate Reactors", Energy. And Fuels, 3(3), 312-315, 1989.

Retzekas, E., Voutsas, E., Magoulas, K., and Tassios, D., "Prediction of Physical Properties of Hydrocarbons, Petroleum and Coal Liquid Fractions", Ind. Engng. Chem. Res., 41(6), 1695-1702, 2002.

Song, C., "An Overview of New Approaches to Deep Desulfurization for Ultra-Clean Gasoline, Diesel Fuel and Jet Fuel", Catal. Today, 86, 211-263, 2003.

Sun, Y., and Yang, C., "New Method for Calculating the Molecular Weight of Petroleum Factions", J. Univ. Petrol. China, 24(3), 5-7, 2000.

Improving Vacuum Gas Oil HydroTreating Operation

via

a Lumped Dynamic Simulation Modeling Approach

By

Darius Remesat, University of Calgary

Presented in the Topical A: 10th Topical Conference on Refinery Processing

Advances in Hydroprocessing - I (TA001)

of the
2007 Spring AIChE Meeting,
Houston, TX,
April 22-26, 2007"

Unpublished
Excerpts from Thesis of same title used in this Paper

Improving Vacuum Gas Oil HydroTreating Operation

via

a Lumped Dynamic Simulation Modeling Approach

By

Darius Remesat, University of Calgary

Hydrotreating, a process of using hydrogen and a catalyst to remove sulfur from crude streams, has become a larger part of refining operations due to transportation regulations and the necessary processing of heavier, more sour crudes.

Refiners/Upgraders rarely achieve their run length and crude flow through objectives for hydrotreaters due to the occurrence of disturbances reducing the effectiveness of the catalysts. The disturbance, in the form of crude flow changes, feed compositional changes, contaminants such as sulfur, or hydrogen partial pressure (H_2 PP) changes, is not typically preventable but the impact can be mitigated during operation of the hydrotreater.

Most refiners struggle to make the necessary changes to mitigate the impact of the disturbance to recover lost run length or crude production since there is very little supporting evidence that they can rely on to provide mitigating steps. Over the author's ten years gathering first-hand information at over forty North American refineries/upgraders apprehension continues to exist to use hydrogen as a tool to improve refinery operations. Most operators surveyed agree in principle that improving

H_2 partial pressure will improve hydrotreater operation but can not make a justifiable, economic case to increase hydrogen use.

There are proprietary models that can provide some cursory indication of performance enhancement during operation but due to cost are not readily available in the public domain. Even if these models were available, it is not certain whether the model would be effective in simulating the process and any disturbances over the run length of the process.

A high fidelity dynamic model, using a lumped parameter approach, with both Excel and Hysys software, for industrial refinery/upgrader VGO hydrotreaters has been developed based on fourteen operating hydrotreaters. The model takes the best of proprietary and public steady state Hydrotreater models and successfully applies it to a commercial dynamic simulation package. The model tracks changes in intrinsic reaction rate based on catalyst deactivation, feed properties and operating conditions to provide useful process information, such as required operating temperature, optimal hydrogen partial pressure (function of operating pressure and hydrogen purity) and chemical hydrogen consumed. The model accurately simulates local disturbances, and performs well over the three distinct operating zones during the run length of the Hydrotreater (start, middle and end of run). This model can be applied to demonstrate the tangible economic benefit of increasing hydrogen use to improve the operation of a hydrotreater by increasing run length and/or improving crude processing.

It is the author's desire to have this model used to maximize refining assets to extend the productive life of our society's current main energy source.

Nearly all literature for HDS dynamic models (listed in Table 1) is based on validation with pilot plant data. This approach does not transfer well to industrial applications for the following reasons:

1. Dynamic models only run from approximately forty minutes to 3 months in length
2. Multiple variable disturbances and operator responses are not considered
3. Low conversion occurs since pilot plant reactor is 10-20 times shorter

Table 1: Summary on literature found and cited for modeling catalytic reactors

Reference	Model		Catalyst Deact.	Experiment P=Pilot I= Industrial	Application
	Steady State	Dynamic			
Levenspiel et al (1968)	X		X		Fixed bed reactor
Kodama et al (1980)	X		X		Residue desulfurization
Tamm Et al (1981)	X		X	P	Industrial Desulfurization, metals deposition
Baiker and Bergougnan (1985)		X			Fixed Bed – 1-d vs 2-d modeling
Döhler and Rupp (1987)	X			P/I	VGO-treatment
Chao et al. (1990)		X	X		Trickle Bed
Barto et al (1990)		X			Fixed bed reactor
Chao et al. (1991)	X		X		HDS
Barto et al. (1991)		X	X		Pellet Dynamics
Miyauchi and de-Wind (1994)	X				Residue Hydrotreating
Lucien et al (1993)	X			I/P	Shell HDS process
Froment et al (1994)	X			P	HDS, 1-d heterogeneous
Borio and Schbib (1995)		X	X		Four stages reactor
Koyama et al (1995)	X	X		I	Vacuum Residue HDS
Kulkarni and Dudukovic (1996)		X			Gas and solid phase fixed bed reactor
Warna and Salmi (1996)		X			3-phase trickle bed reactor hydrogenation of toluene
Korsten and Hoffmann (1996)	X				HDS, 3-phase, Trickle Bed
Recke and Jørgensen (1997)		X			Catalytic oxidation of hydrogen to form water
Iliuta et al. (1998)		X		P	Residence time distribution
Lababidi et al. (1998)	X		X		Desulfurization

Juarez et al (1999)	X		X	P	H2S impact on HDS
Kokayeff (1999)	X		X	P	HDS Aromatic Saturation
Ho (1999)	X				HDS kinetics
Oballa (1999)	X		X	P	Hydrocracking
Julcour et al (1999)		X		P	3-phase fixed bed
Matos and Guirardello (2000)	X		X		HDS + HDM
Hu et al (2001)	X			I	HDS on Industrial data
Chowdury et al (2002)	X			P	HDS of Diesel oil
Lababidi et al. (2004)		X	X	P	Advanced control
Mederos et al (2006)		X		P	HDS, Trickle Bed

All the dynamic models developed are run over such a short time, that the accuracy of any catalyst deactivation factors, if included, cannot be ascertained. In addition, the robustness of these models cannot be verified since the interaction of various variables (crude, temperature and hydrogen changes) is not demonstrated. Finally, non-validated correlations have been developed to factor in the incomplete catalyst wetting through contacting effectiveness that creates lower sulfur conversion in pilot plants. As a result, there is little or no confidence in scaling up a pilot plant validated dynamic model to industrial scale conditions. Any one of these factors show that there is a need to create a VGO HDS dynamically based model that is valid over an entire run length (at least 1.5 years).

There are a few papers (Table 1) that try to show compatibility with industrial data by adding 1 or 2 commercial reactor data points from other literature to show how well the model could be applied to industry. Any model can match up with one or two data points. Also, they exhibit one of the three noted deficiencies when trying to extend applicability to actual industrial conditions.

The literature review of dynamic modelling is still important for this body of work, since there are definitely valuable findings from pilot-plant validated dynamic model development that can be used for creating a novel industrial validated dynamically based Vacuum gas-oil hydro-treater model.

Data Acquisition

Over ten years, a database of 24 hydrotreaters (HDS) in various services (Distillate, Vacuum gas oil (VGO), naphtha, kerosene) was developed. Of the 24 in the database, 14 Hydrotreater units have the necessary process and equipment information to adequately build a representative model of the hydrotreater unit. For publishing purposes, confidentiality agreements with 6 of the refinery/upgrading operators (all in North America) have been obtained (labeled plants A-F, shown in table 4.1a,b). Multiple site visits to witness the data retrieval and operation for most of the units have been performed. The necessary data from the operating plants for model development is:

1. **Operating**- pressures, temperatures, (WABT = weighted average bed temperature), flows

2. **Frequent/reliable compositional laboratory data** - crude (including aromatics, cracked gas, olefins) off gas, hydrogen, contaminants (including sulfur, nitrogen, metals (vanadium, nickel)

3. **Equipment**- vessel size, pump sizes, piping sizes, control valves

4. **Catalyst**- k (activity/equilibrium) values, size, shape, expected performance processing schemes and requirements.

Table 4.2 shows a summary of catalyst/reactor information crucial for developing a hydrotreater model.

Table 4.1a Hydrotreating Operating Conditions				
	Industry Range of HDS Units	Plant A	Plant B	Plant C
Space Velocity, LHSV	.1 – 15	0.85	1.1	0.6
Reactor Pressure, PSIG	300 – 2200	675	1065	970
Temperature, °F average	500 – 800	630	645	666
Hydrogen to Oil, SCF/Bbl	300 – 5000	2500	840	3050
Product Sulfur	5 ppm-0.5%	<200 ppm	<100 ppm	<5ppm
Product Nitrogen	5 ppm-0.5%	<10ppm	<3 ppm	<0.3%

Table 4.1b Hydrotreating Operating Conditions				
	Industry Range of HDS Units	Plant D	Plant E	Plant F
Space Velocity, LHSV	.1 – 15	0.5	1.4	.7
Reactor Pressure, PSIG	300 – 2200	1550	460	475
Temperature, °F average	500 – 800	685	635	675
Hydrogen to Oil, SCF/Bbl	300 – 5000	4400	750	925
Product Sulfur	5 ppm-0.5%	<200ppm	<5 ppm	<5 ppm
Product Nitrogen	5 ppm-0.5%	<10ppm	<3 ppm	<0.1%

Table 4.2 – Catalyst/Reactor information

UNIT		Plant A	Plant B	Plant C	Plant D	Plant E	Plant F
REACTOR DIAMETER	ft	13	12	12.5	12	11	14
CATALYST TYPE		NiMo on Al	NiMo on Al	CoMo on Al	NiMo on Al	NiMo on Al	CoMo on A
CATALYST VOLUME	cuft	4424.0	2865.0	10370.0	6523.0	2415.0	4189.0
CATALYST WEIGHT	lbs	207928.0	151488.0	466650.0	316650.0	181678.0	163371.0
CATALYST DENSITY	lbs/cuft	47.000	38.400	45.000	48.544	53.200	39.000
CATALYST SIZE	mm	2.500	2.000	2.200	1.870	2.540	1.650
Cycle length	years	2.60	2.00	1.80	1.45	3.0*	2.28
REACTION TYPE (HDS,HDN)?		HDS	HDS	HDS	HDS	HDS	HDS
DESIRED PRODUCT S	%WT	<200wppm	<100 wppm	<5 wppm	<200 wppm	<5 wppm	<5 wppm
TYPICAL FEED S	%WT	1.70	0.80	1.50	2.20	1.40	0.90
SOR AVERAGE TEMP.	°F	630.0	645.0	660.0	685.0	635.0	675.0
TYPICAL LHSV	H-1	0.85	1.1	0.6	0.5	1.4	0.7
FEED TYPE (N,KERO,LGO,VGO)?		LGO	LGO	LGO	LGO	LGO	LGO
SUGGESTED REACTION ORDER	-	1.65	1.5	1.6	1.7	1.55	1.4
ACTIVATION ENERGY	CAL/MOLE	27000	29000	32000	24500	26500	27500

The model needs to be able to accurately calculate the outlet product sulfur. Key plant information from one plant is provided in this paper for illustrations purposes. Figure 4.5

provides an illustration of key disturbances on WABT for plant D, thus on the operation. It is these disturbances, with all the necessary information provided by the operator, that allows the development of a high fidelity dynamic model. The figure shows three disturbances impacting the overall run length and crude processing, and outlines what happens in the process as a result of the first sulfur disturbance (catalyst life is shortened, crude charge is reduced, and WABT is increased to increase catalyst activity).

The other two disturbances noted in Figure 4.5 are another sulfur distrubance and an increase in crude flow.

In all cases, the WABT increased in response. Figure 4.6 adds the crude flow to the WABT graph in 4.5. For the first disturbance, the crude flow is reduced since it is hypothesized that the rapid increase in temperature to increase catalyst activity severely impacted the performance of the reactor. As a result, the amount of crude that could be processed was reduced (on average 4000 BPD). This hypothesis was confirmed by the operator. If the reactor operation can mitigate the impact of various disturbances, for this case, a maximum savings of 4000 BPD could be obtained.

<u>Figure 4.5</u> Major disturbances and WABT for Plant D

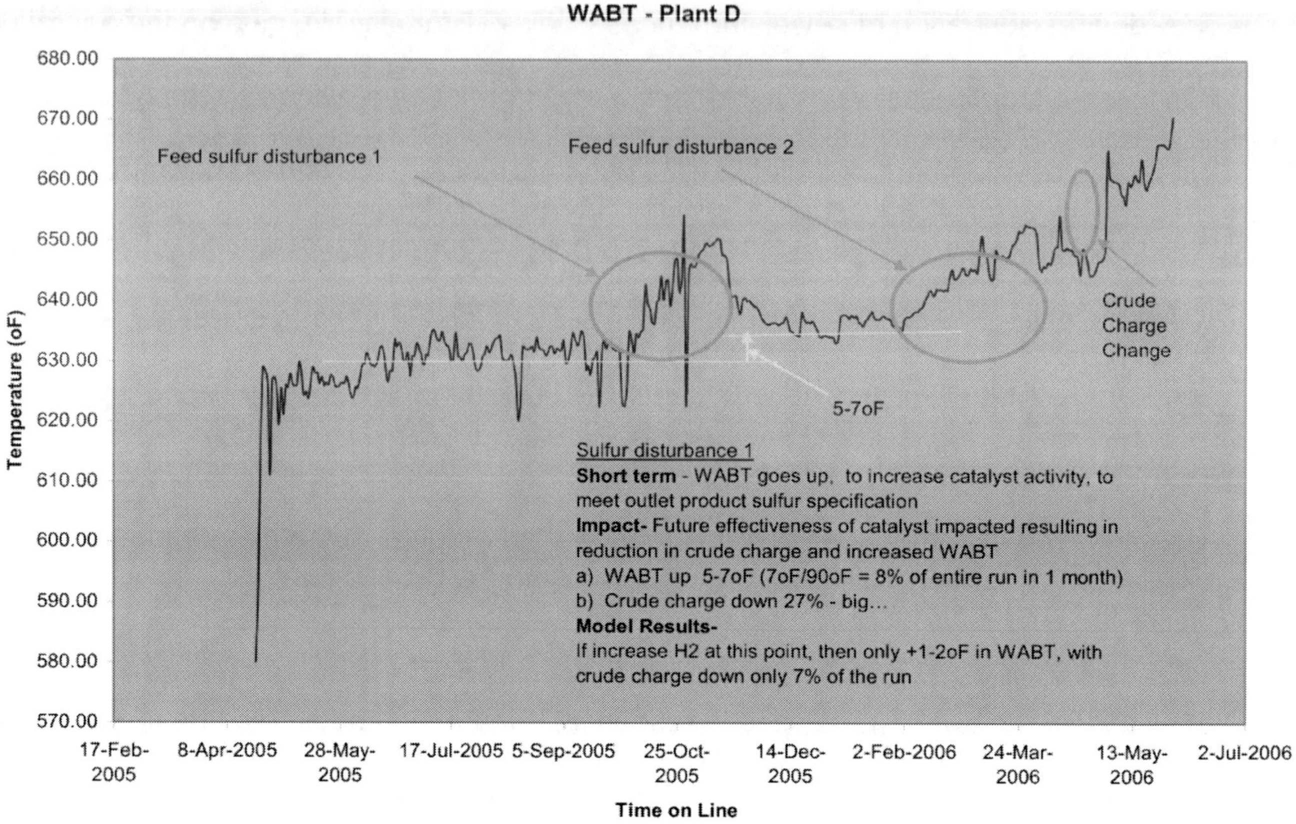

Figure 4.6 – Plant "D" Crude and WABT from Start of Run (SOR) to End of Run (EOR)

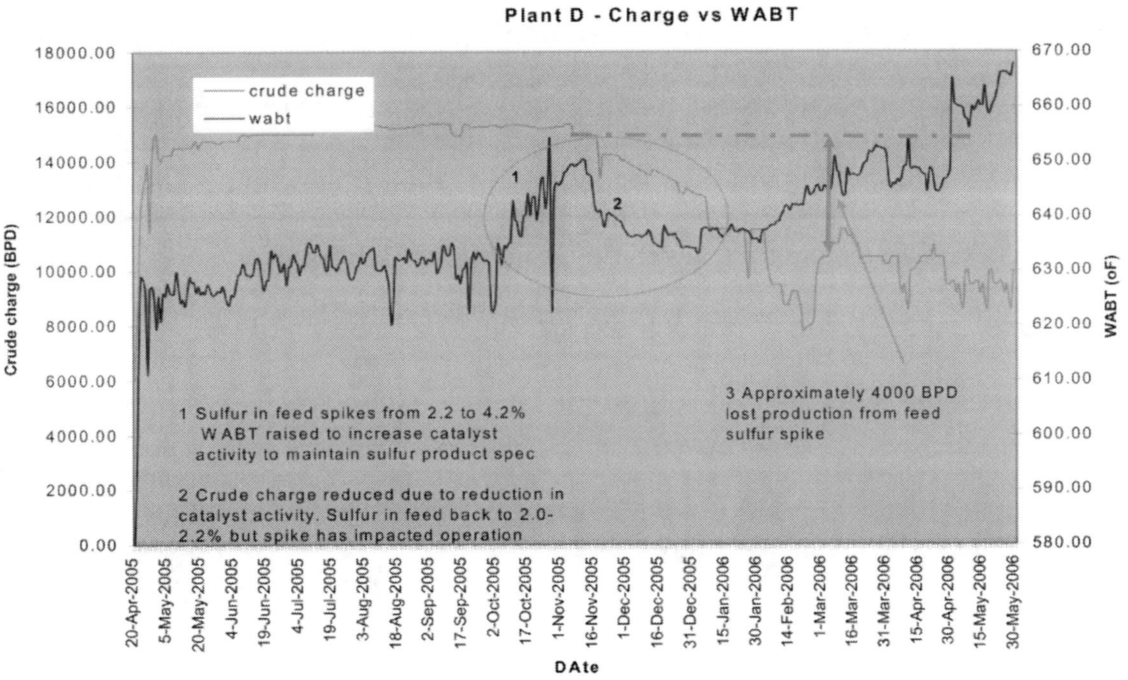

Model

A first-of-a-kind dynamically oriented lumped VGO hydrotreater model that incorporates key variables to provide an accurate representation for industrial conditions was developed. The model:

1) Is lumped to match data available from industrial operations.
2) Uses a mix of industrial correlations, kinetics theory, and academic research findings
3) Factors in changes in operating conditions in response to disturbances in operation.
4) Incorporates changing reaction rate, wetting efficiency, catalyst deactivation in three zones
5) Has factors for all key process variables
6) Is set up to run in dynamics mode to track the key variable product sulfur and representative value of performance at the plant site: WABT.
7) Uses familiar software (Excel® and HYSYS®) for easy translation into existing operations and acceptance by users.

The model was developed to meet the following objective: Create a dynamically functioning VGO hydrotreating model that can be used successfully in industrial settings. Process data, equipment specification, and catalyst information from actual commercial-scale VGO hydrotreaters will be used for building and

Since this model is developed for industrial settings, and to be used by refiners/operators, the software considered for the model should have the following features:

a) Should be readily available to the operators
b) Users should be familiar with tool(s)
c) Relatively inexpensive
d) User friendly

All refiners/operators use spreadsheet based software for various functions (ex. linear programming to optimize plant economic objective function, and evaluate operating data). Excel® (with Visual Basic®) was chosen as the primary tool to store, and calculate the various parameters of the VGO hydrotreater model and to display the output. For the dynamic portion of the model, HYSYS® was chosen. All operators[3] are familiar with HYSYS®, and use a simulator (ex. HYSYS®, Aspentech, PROII (SIMSCI), VMGSim) to provide plant/unit calculations and sensitivity analysis.

The calculation for the outlet sulfur product in the model is shown in the following equations:

$$S_{pi} = \left\{ \left[\left(\frac{1}{S_{p(i-1)}^{2\eta_e}} - \frac{1}{S_{f(i-1)}^{2\eta_e}} \right) \frac{LHSV_i}{LHSV_{i-1}} \left(\frac{P_i}{P_{i-1}} \right)^p \left(\frac{G_i}{G_{i-1}} \right)^g e^{\left(\frac{b}{T_i} - \frac{b}{T_{i-1}} \right)} + \frac{1}{S_{fi}^{2\eta_e}} \right]^{-1} \right\}^2 \quad \text{(a)}$$

or

$$S_{pi} = \left\{ \left[k_{\eta_d i} \frac{LHSV_i^2}{LHSV_{i-1}} \left(\frac{P_i}{P_{i-1}} \right)^p \left(\frac{G_i}{G_{i-1}} \right)^g e^{\left(\frac{b}{T_i} - \frac{b}{T_{i-1}} \right)} + \frac{1}{\sqrt{S_{fi}}} \right]^{-1} \right\}^2 \quad \text{(b)}$$

A crucial part in creating an accurate hydro-treater model is developing a realistic representation of catalyst deactivation and including it in the reaction rate expression as noted in this equation

$$k_i = k_{actual} = f(Catalyst\_activity) * k_{hds}$$

where

$$f(catalyst\_activity) = 1 - \frac{q}{q_o} = 1 - \phi$$

The variable q represents the concentration of blocked sites (or concentration of poisons of catalyst) from metals deposition and coking, and q_o represents the maximum concentration corresponding to completely deactivated catalyst[15]. Since it is difficult to determine q_o and q, in industrial and pilot plant settings, industrial operators and catalyst vendors have developed a series of correlations to determine the life left on a catalyst[3,7], (1-ϕ), or sites that are not poisoned. The correlations used for catalyst deactivation are found in the thesis recently published.

Results

Table 6.1 shows the comparison of plant WABT and outlet sulfur to the model developed. The R^2 for product sulfur ranged from 0.84 to 0.94, while R^2 for WABT (weighted average bed temperature) ranged from 0.89-0.93.

Table 6.1 – Statistical Results for each Plant – Model vs. Plant

	Plant A	Plant B	Plant C	Plant D	Plant E	Plant F
Sulfur R^2	0.84	0.90	0.85	0.917	0.94	0.84
WABT R^2	0.90	0.92	0.91	0.916	0.93	0.89

Plant Results – Entire Run Length

The output for the model has been set to reactor temperature and sulfur outlet product. The sulfur outlet product is the primary specification for the VGO hydrotreater, while the reactor temperature is directly related to the catalyst activity (as catalyst activity goes down, reactor temperature goes up). The results from comparing the six operating plants' data to the model are provided in the thesis, with one plant's results illustrated in this paper.

Plant D Results

Plant D operated its VGO hydrotreater for 1 year and 4.5months, well below the desired run length of 4 years. There were numerous major and minor (explained separately) disturbances that severely reduced the ability of the VGO hydrotreater to meet the crude production targets over the anticipated run length of four years.

The results of the model are shown in Figure 6.6 for the outlet sulfur concentration and figure 6.7 for the WABT (Weighted average bed reactor temperature). The model provides an R^2 of 0.917 for all the sulfur values with a high match (R^2 =0.95) to sulfur product values in the 20 to 130 ppm range. Above 130 ppm, the model provides an R^2 of 0.88, at times struggling to fully match the “spikes” in outlet sulfur results. Also, for the really low sulfur values of less than 20 ppm, the model only provided a match to within an R^2 of 0.81.

Issues with accuracy of lab measurements, quality of the lab samples can easily explain the differences between the model and the plant data.

Figure 6.7 shows how the model WABT compared to the plant data. Overall, the R^2 for the comparison between plant and model WABT was 0.916

Figure 6.7 – Plant D – WABT, Model vs. Plant Data

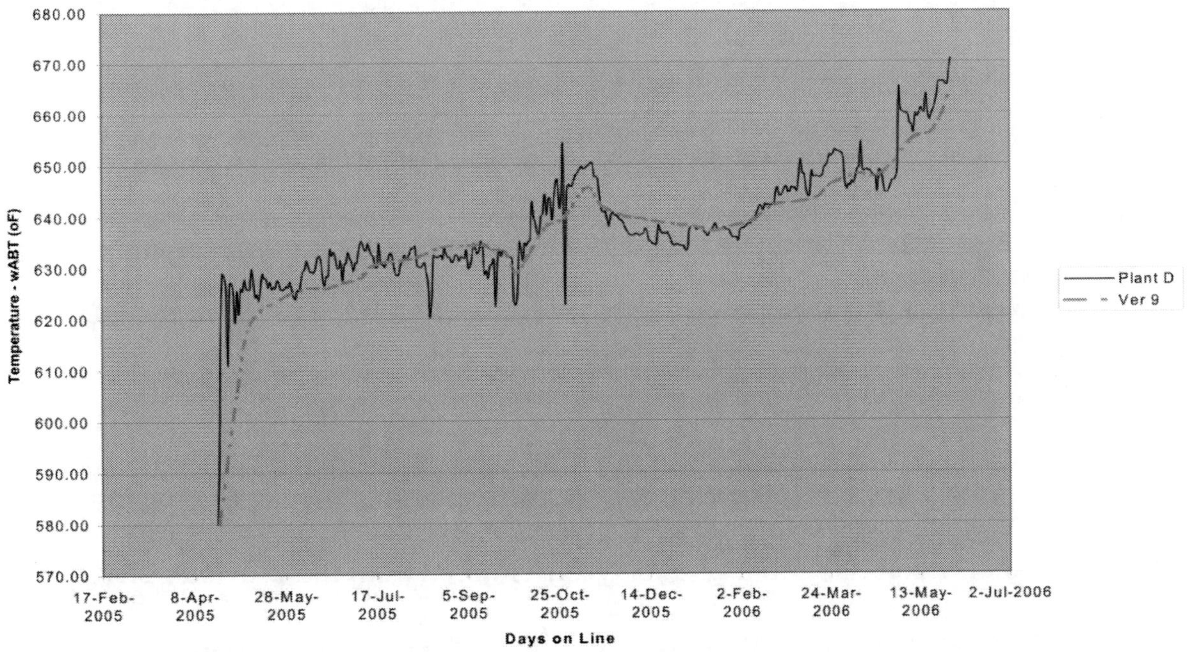

Figure 6.6 – Plant D – Outlet Sulfur Composition, Model vs. Plant Data

example is around March 24, 2006, where the WABT trends up 5-7oF and the model does not respond accordingly. None of the key process variables tend to indicate a direct reason for the discrepancy (i.e. changing crude flows, sulfur inlet concentration, hydrogen purity or flow). Just prior to, and just after this date two large disturbances are

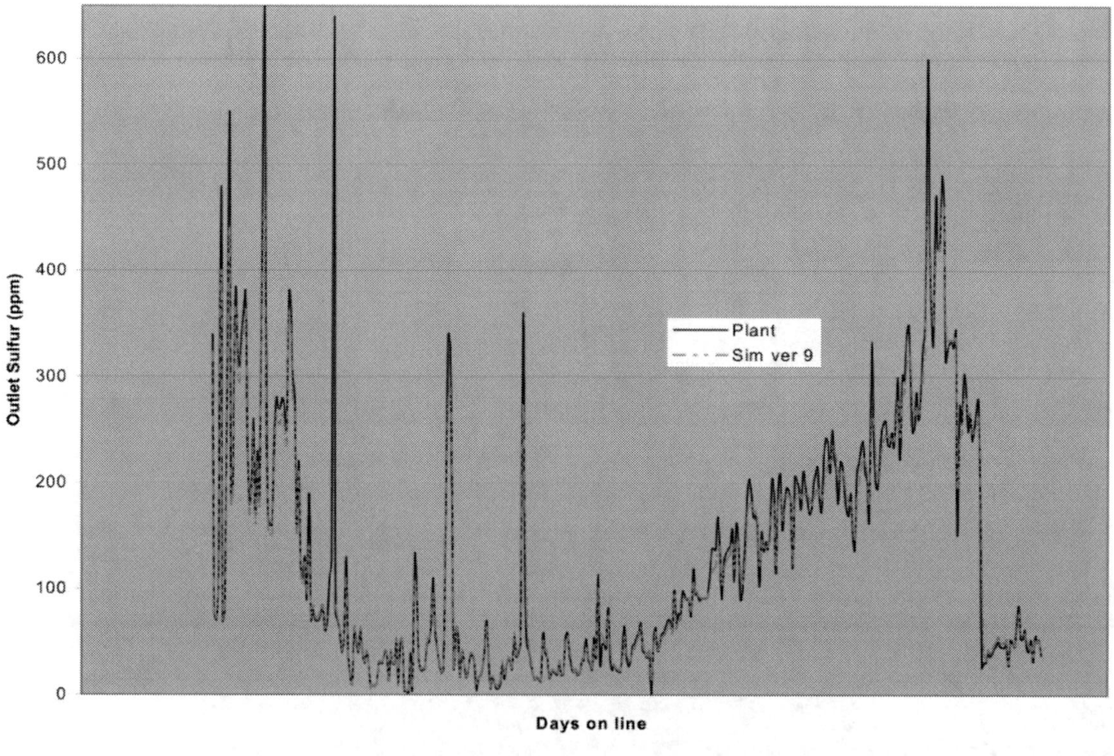

noted and accurately tracked by the model. A few explanations can be considered, like the catalyst deactivation had more of a long term impact from the previous sulfur inlet disturbance than what was considered in the model. However, the model did accurately predict the temperature rise at the time of the disturbances. Inquiries to the operator did not reveal anything specific regarding the events around March 24, 2006. Possibly, some human operating factor may have been the result of a deviation from normal temperature response logic.

Considering this plant had the most significant disturbances of any studied, with such an uncharacteristically short run time, having a model that can get an $R^2>0.9$ is a satisfactory result.

Application of the model-

Specific

With a good dynamic model of the actual process, various scenarios for Plant D were evaluated to determine what mitigating steps in the process could be used to improve the operation.

Hydrogen was used as the primary variable to impact the hydrogen partial pressure. For Plant D, Hydrogen purity was increased to 99.9% from an average of 90.5%. Access to 3rd party on-the-spot Hydrogen was assumed. Hydrogen partial pressure was increased from an average of 1900 to 2050 psig. The following scenarios were performed

1. Increase hydrogen partial pressure at point of 1st major disturbance and maintain same run length.

   Result: As stated previously, and shown in Figure 2 (lime line showing simulated crude charge), the crude charge could be increased to 14000 BPD, thereby saving an average of 4000 BPD of crude for 4.5 months. The value to a refiner based on $20/bbl upgrade is $8.1MM. The pre-first disturbance charge rate (15000 BPD) could not be fully recovered due to the impact of metals contamination on the catalyst. Hydrogen partial pressure positively impacts deactivation due to coke formation, but negatively impacts metals contamination (various studies illustrate this), and the model accounts for this catalyst phenomenon.

2. Increase hydrogen partial pressure for entire run length

Result-

a) The run length could be increased by 20% since the initial coke deposition on the catalyst is mitigated by the increased hydrogen partial pressure (WABT will not be as high) with the increased crude (15000 before 1st disturbance and 14000 BPD after).

b) If the crude charge is kept as actually happened during the run (and other variables extrapolated), then the run length could be increased by 45%.

However, with no real data after the actual run length the model showing an increase in crude production from the increased hydrogen partial pressure for the same run length is the best demonstration of the model. The trade off between crude charge and run length needs to be decided by the plant manager so both options were evaluated for consideration purposes.

Generic Application

Typically, for justification proposals on improving hydro-treater operations, increasing run length of a hydrotreater to match overall mechanical shut down schedules is a large enough benefit to refiners/upgraders to justify to management (versus disturbance mitigation, even though it can provide the largest economic benefit). The model developed in this thesis can provide tangible evidence to justify the benefit of using more hydrogen to improve hydro-treater operations. Using the model on plant B (data on shutdown and hydrogen costs were made available for evaluation), table 3 illustrates that run length can be increased by up to 60% by increasing the H2 partial pressure (via increased H2 purity). Incorporating all the costs, the value can be up to $24MM (NPV)

Table 3– Application of Model: Vary H2 purity and H2 recycle to increase Hydro-treater run length

Application of Dynamic model		Base	Purity 1a	Purity 1b	Purity 2a	Purity 2b
LGO Hydrotreater						
Charge Rate		14.4	14.4	14.4	14.4	14.4
Recycle Purity		76.8%	92.0%	92.0%	84.0%	84.0%
Recycle Rate MMSCFD		21	21	11.1	21.0	12.8
Average H2 Partial Pressure Full Feed Vaporization		560	675	578	631	571
% Change in Deactivation Rate		Base	~ 60% Decrease	~15% Decrease	~40% Decrease	~10% Decrease
Delta SOR Temp F		Base	Between -5 and -10	~ -1 and -2	~ -5	~ -1 and -2
Cost:	Increased H2 cost (2 yr run length)		$16.5MM	$7.2MM	$10.8MM	$4.8MM
Benefit:	Extend run length to x years		4	2.3	3.3	2.2
	S/D savings		Save 1 every 2yr	Save 1 every 16 yr	Save 1 every 6 yr	Save 1 every 22 yr
	$ Value (NPV -$40MM for S/D)		**$24 MM**	$-3MM	**$7MM**	$-3MM

Value of the model

- Using actual plant data to develop model
- No issues with scaling up model from laboratory plant scale data
- Model matches disturbances reasonably well, overall performance very encouraging
- Provide confidence to refiners/upgrader's to justify increasing Hydrogen use to improve operations.

Limitations of the model

- Model dependent on available process and laboratory data

- Need to modify model's parameters to address each refiner's/upgrader's specific crude and catalyst

Summary

A high fidelity dynamic model for industrial refinery/upgrader hydro-treaters has been developed based on 14 actual operating hydro-treaters. The model accurately simulates local disturbances, and performs well over the entire run length of the hydrotreater. This robust model can be applied to demonstrate the tangible economic benefit of increasing hydrogen use to improve the operation of a hydro-treater by increasing run length and/or improving crude processing.

It is the desire of the author to have this model used by refiners/operators to maximize their assets to extend the productive life of our society's current main energy source (crude oil).

SELECTIVITY OF RING-CONTRACTION PRODUCTS IN THE HYDROISOMERIZATION OF METHYLCYCLOHEXANE

Teerawit Prasomsri [1], Siriporn Jongpatiwut[1], Malee Santikunaporn[2], Daniel E. Resasco[2]
[1] The Petroleum and Petrochemical College, Chulalongkorn University, Bangkok, Thailand
[2] School of Chemical, Biological, and Materials Engineering, University of Oklahoma, Norman, OK, USA.

Abstract

The hydroisomerization of alicyclic compounds (e.g., methylcyclohexane) attracts the interests of researchers due to high demand for clean fuels, which can maintain high octane numbers while decreasing aromatic content. The composition of the products from skeletal isomerization of hydrocarbon is principally dependent on the shape selectivity and the acid strength of the acid catalyst, as well as the hydride transfer step. Experiments were conducted at 533 K in the presence of hydrogen at a total pressure of 2 MPa over bare and Pt-supported zeolitic and non-zeolitic solid acid supports. The products obtained under these conditions were mainly those from the methylcyclohexane ring-contraction (RC), that is, isomers consisting of 1,1-dimethylcyclopentane, cis-1,3-dimethylcyclopentane, trans-1,3-dimethylcyclopentanes, and trans-1,2-dimethylcyclopentane, and ethylcyclopentane. It was found that the acid strength plays an essential role in defining the product distribution due to the relative stability of the different carbenium ion intermediates. In addition to the acid strength, the shape selectivity of the catalysts was seen to be important in determining the RC product distribution. Interestingly, the high rate of hydride transfer was seen to greatly influence the selectivity of RC isomers over all zeolitic supports, but much less on non-zeolitic supports, such as tungstated zirconia.

Introduction

The conversion of naphthenic molecules has been extensively investigated on metal, acidic, and bifunctional (metal/acid) catalysts. Several reactions are known to occur, such as isomerization (including ring-contraction), ring-opening, dehydrogenation, and cracking, depending mainly on the reaction conditions, the feed, and the catalysts used. Methylcyclohexane (MCH) was considered as a suitable representative reactant for this catalytic test since it leads to the informative product mixture. In this work, the catalytic properties in the hydroisomerization of MCH in the presence of hydrogen over acidic and bifunctional catalysts were studied in order to elucidate the effects of shape selectivity and the acidic strength of catalysts on the selectivity of ring-contraction isomers, as well as the effect of hydride transfer.

Experimental

Catalyst Preparation

Both zeolitic (FAU, MFI, and BEA) and non-zeolitic materials, tungstated zirconia (WO_x/ZrO_2, WZ) were used as the catalysts and the supports. WZ samples were prepared by incipient wetness impregnation of zirconium oxyhydroxide solids with ammonium metatungstate solutions.

The Pt-containing catalysts were prepared by incipient wetness impregnation of both the proton form zeolites (FAU, BEA, and MFI) and WZ with 1 wt% Pt. An appropriate concentration was prepared from hexachloroplatinic acid solution. Before the impregnation, the supports were pretreated in air at 773 K for 3 h; and after the impregnation, the catalysts were dried at 383 K overnight and then calcined in air at 573 K for 3 h. Prior to each activity measurement, the sample was reduced in situ.

Catalyst Characterization

Temperature programmed desorption (TPD) of adsorbed ammonia was used to determined the acidity densities of catalysts. A 50 mg of catalyst was placed in the middle of a 1/4" quartz tube reactor. Before the adsorption of ammonia, the sample was pretreated at 773 K for 1 h in He flow. Then, the sample was cooled to room temperature. Prior to the measurement of the amount of adsorbed ammonia, the sample was purged in He flow at room temperature for 2 h in order to remove the excess and weakly adsorbed ammonia. Subsequently, the sample was linearly heated to 973 K in a flow of He at a heating rate of 10K/min while the evolution of ammonia was detected in a thermal conductivity detector.

Catalytic activity testing

The methylcyclohexane reaction was conducted in a 3/8" stainless steel reactor equipped with a thermowell and the thermocouple inside, which was placed in the center of the bed. Prior to each run, the designed amount of catalysts was diluted with inert alumina to make a volume of a catalyst bed of 2 ml and placed in the center of the reactor. The top part of the catalyst bed was filled with glass beads of 1 mm diameter, to ensure effective preheating and minimize heat loss. Before the activity measurement, the zeolitic catalysts were reduced in situ at 673 K for 2 h in hydrogen flow. The catalytic activity was measured at the temperature of 533 K, at the total pressure of 2 MPa and at a H_2/MCH molar ratio of 40. The space velocities were systematically varied by changing either the catalyst amount or the feed rate. In all experiments, the product gas mixtures were analyzed periodically on-line with HP5890 gas chromatograph equipped with HP-5 column and a FID detector.

Results and Discussion

Characterization of Catalysts

Table 1 summarizes the density of acid sites. As previously observed [1], the impregnation with Pt results in a loss of acid sites resulting from the interactions between platinum crystallites and acid sites.

Table 1. Acidity of catalysts

	Total acidity (μmol/g-cat)	
Sample	Bare catalysts	Pt supported catalysts
MFI	680	630
BEA	1617	1021
FAU	1109	795
WZ	158	48

The profiles of temperature-programmed desorption of adsorbed ammonia over various fresh catalysts are shown in Figure 1. Different desorption peaks were observed on the TPD profiles at different temperature ranges. Peaks shown at low and high temperatures can be attributed to ammonia desorbed from weak and strong acid sites, respectively [2]. The desorption peak at 470 K represents the very weak acid sites, which are due to an octahedral Al ion formed by dealumination [3]. The desorption peak at about 650 K represents the acid site formed by Al in the zeolite lattice structure.

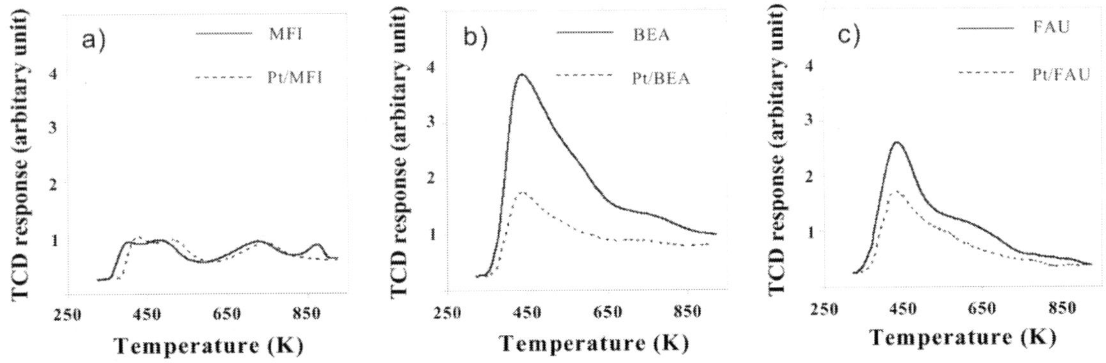

Figure 1. TPD of adsorbed ammonia on (a) MFI (b) BEA, and (c) FAU zeolites.

Among these catalysts, the acid strengths are in the order of WZ, MFI, BEA and FAU zeolites, while the number of acid sites are in the order of BEA, FAU and MFI zeolites and it is lowest in WZ.

Selectivity of MCH Ring-Contraction Isomers over Bare Acidic Catalysts

Under the investigated reaction conditions, the primary and major products are ring-contraction (RC) isomers. The equilibrium mixture of RC isomers is shown in Figure 2. Obviously, the selectivity to *cis*-1,2-dimethylcyclopentane was the lowest due to the most steric hindrance, the two methyl groups at the same side. The RC isomers found in this study consist of ethylcyclopentane, 1,1-dimethylcyclopentane, *cis*-1,3-dimethylcyclopentane, *trans*-1,3-dimethylcyclopentanes, and *trans*-1,2-dimethylcyclopentane.

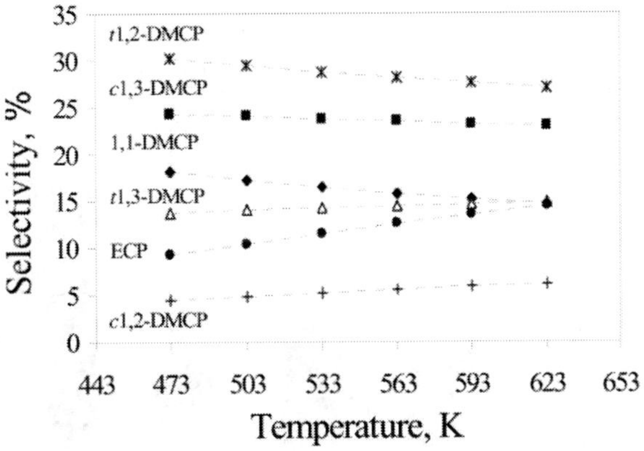

Figure 2. RC products selectivity as a function of temperature. The composition at thermodynamic equilibrium was simulated by using SimSci PRO/II PROVISION.
Reaction conditions: Total pressure = 2 MPa; H_2/feed molar ratio = 40.

The acid-catalyzed branching isomerization of alicyclic hydrocarbon compounds on solid acids is generally rationalized by the formation of protonated cyclopropane intermediates after the formation of a carbenium ion and each branching isomer (RC) requires the formation of carbocation at different positions as shown in Figure 3.

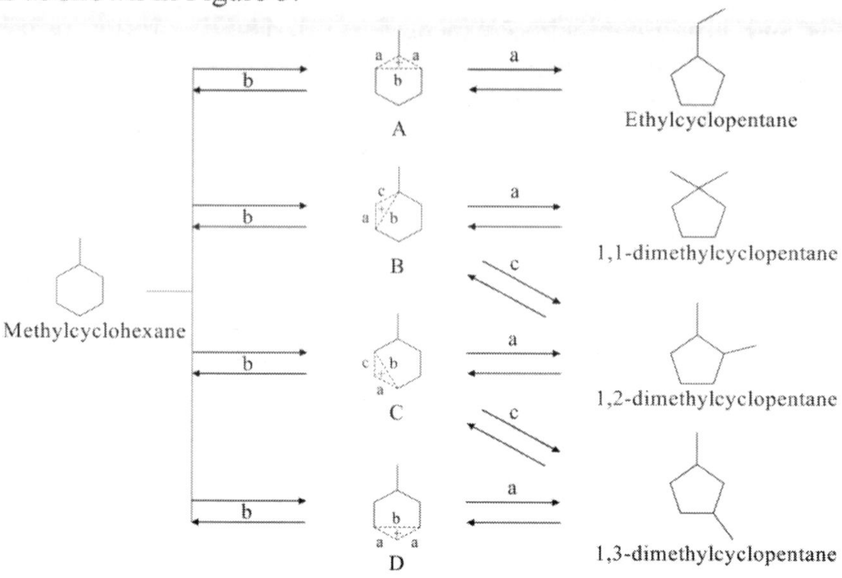

Figure 3. Carbocation chemistry for the ring-contraction reactions.

The monobranched ECP is formed via the rearrangement of a tertiary transition state (intermediate A), which is the most stable transition state, involving the internal alkyl shifts, in which the three C-atoms of the cycloproponium intermediates are involved [4]. High yield of ECP (high selectivity) is due to the easiest formation of this type of an intermediate on the weak acid sites. In addition, the shortening of the residence time will even increase the formation of ECP, resulting in high selectivity to ECP at low conversions (small reactor sizes). Unlike ECP, the dibranched DMCPs are preceded via corner-protonated cyclopropane (CPCP) intermediates (B, C, and D), which have a less stable transition state and require a strong Brönsted acid site [5]. These types of intermediates are least favored. Therefore, the formation of ECP is faster than the formation of DMCPs, and ECP can be produced on the weaker acid sites. In other words, the formation of DMCPs needs stronger acid sites or a longer surface residence time than an ECP, especially 1,3-DMCP isomers [6]. Therefore, correspondingly, the acid strength can be relatively compared by the DMCPs/ECP ratio.

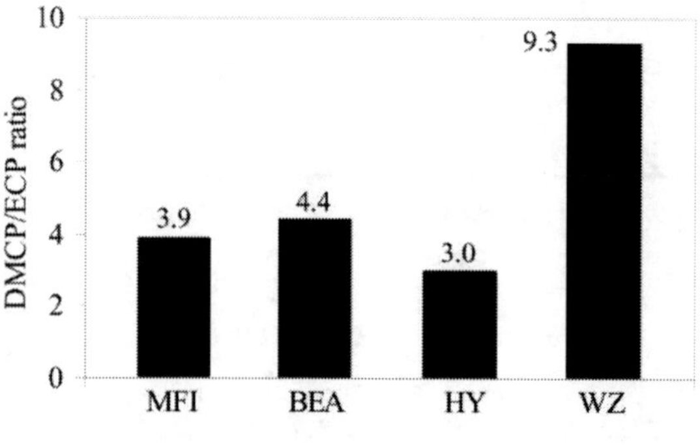

Figure 4. The DMCPs/ECP ratio over MFI, BEA, FAU zeolites and WZ at the same conversion (~3%).

As shown in Figure 4, the ratio on FAU zeolite (3.0) is less than that of the BEA (4.4) and MFI (3.9) zeolites resulting from the relatively weak acid sites of the FAU zeolite. The strongest acid sites, obviously, goes to WZ (9.3).

The selectivity to individual RC isomers as a function of the MCH conversion are plotted in Figures 5a (for MFI), 5b (for BEA), 5c (for FAU), and 5d (for WZ).

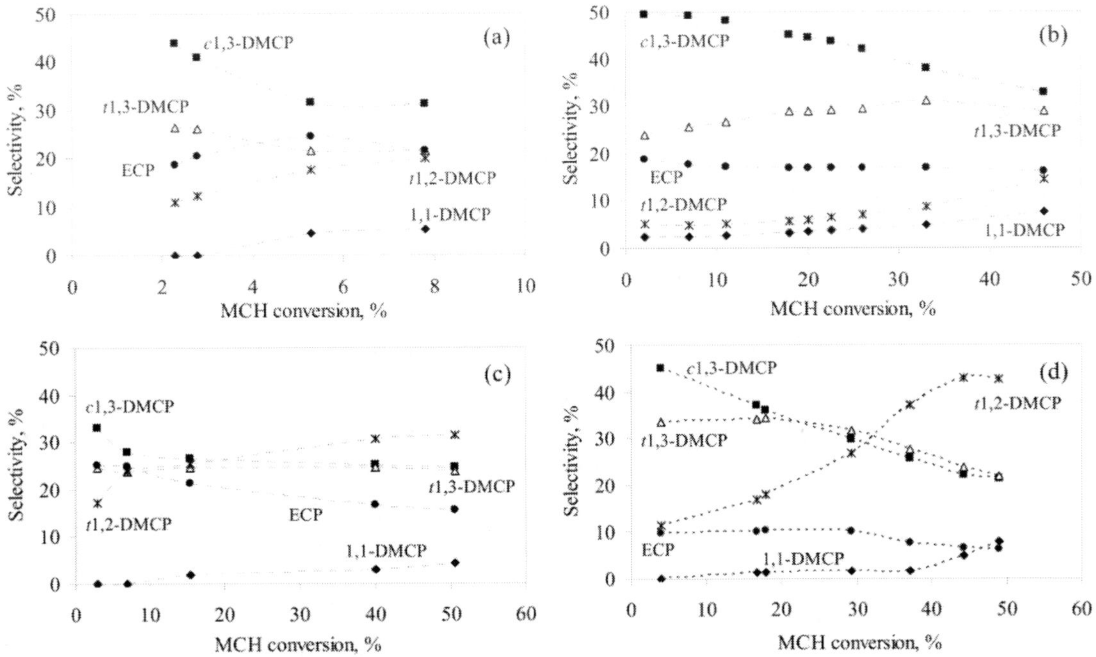

Figure 5. Selectivity to RC products vs. MCH conversion over (a) MFI, (b) BEA, (c) FAU, (d) WZ at different space velocities. Reaction conditions: Total pressure = 2 MPa; Temperature = 533 K; H_2/feed molar ratio = 40.

For strong acid strength zeolites (MFI and BEA), the selectivity to 1,3-DMCPs (*cis*-1,3-DMCP and *trans*-1,3-DMCP) are higher than that to *trans*-1,2-DMCP at all conversion ranges. But for WZ, the strongest acid catalyst, the selectivity to 1,3-DMCPs is high at low conversion, and at higher conversion (>30%), the selectivity is dominated by *trans*-1,2-DMCP which is more thermodynamically preferential (Figure 2).

Shape selectivity speaking, the bulkiest *trans*-1,2-DMCP exhibits the most difficulty for transport in the channels or pore mouth with a high constraint. When considering the selectivity of 1,2-DMCP over large pore size (BEA and FAU zeolites), the selectivity to *trans*-1,2-DMCP on the FAU zeolite is higher than that on the BEA zeolite. The smaller in the selectivity to *trans*-1,2-DMCP on the BEA zeolite was ascribed to the pore size restriction. Even though both FAU and BEA are 12-ring zeolites, the FAU zeolite contains the supercages. In addition, the selectivity to *trans*-1,2-DMCP on amorphous WZ which has a plenty of space for reactions can reach up to ~40%.

Interestingly, The formation of both *trans*-1,2-DMCP and 1,1-DMCP requires the same intermediate (B in Figure 3), but all of these acidic catalysts give the lowest selectivity toward 1,1-DMCP over all conversion ranges due to the high kinetic barrier associated with the intermolecular hydride transfer to the sterically-hindered intermediate (2,2-dimethylcyclopent-1-yl cation). This behavior was also observed on the isomerization of *n*-hexane which the 2,2-dimethylbutane isomer was suppressed [7].

In the case of *cis*- and *trans*-1,3-DMCPs, the selectivity to *cis*-1,3-DMCP is clearly higher than that to *trans*-1,3-DMCP over MFI and BEA zeolites, while they are close together over FAU zeolites and WZ. This can be explained by the means of transformation form cycloproponium intermediates to *cis*- and *trans*-1,3-DMCPs. To achieve *cis*-1,3-DMCP, flat surfaces (i.e., restricted areas) are required in order to allow both methyl groups faced away from the surface and shown up at the same side. In contrast, the possibilities to obtain *cis*- and *trans*-1,3-DMCPs are relatively equal to each other for the catalysts with huge reaction spaces (i.e., supercages and amorphous structures).

Effect of Metal (Pt) on the Selectivity to RC Isomers over Acidic Catalysts

As discussed above, the low amount of 1,1-DMCP found on bare acidic catalysts may be due to the limitation of the hydride/hydrogen transfer. Thus, to elucidate the effect of metal (Pt) on a hydride transfer step, the MCH reaction was performed over the Pt supported on the acidic catalysts (MFI, BEA, FAU zeolites, and WZ) under the same reaction conditions. The results suggest that Pt metals have an influence on the product distribution of RC isomers.

A significant difference was the selectivity to ECP. For all kinds of the Pt supported zeolites, higher selectivity to ECP was found at low conversion since ECP is easily formed via the internal alkyl shifts [4] and desorbs right away form the surface. Unlike zeolites, the selectivity to ECP is low on Pt/WZ. In addition, for all acidic catalysts, the increase in the selectivity to the thermodynamically favored DMCP isomers observed at higher space times may be due to a stepwise isomerization [8].

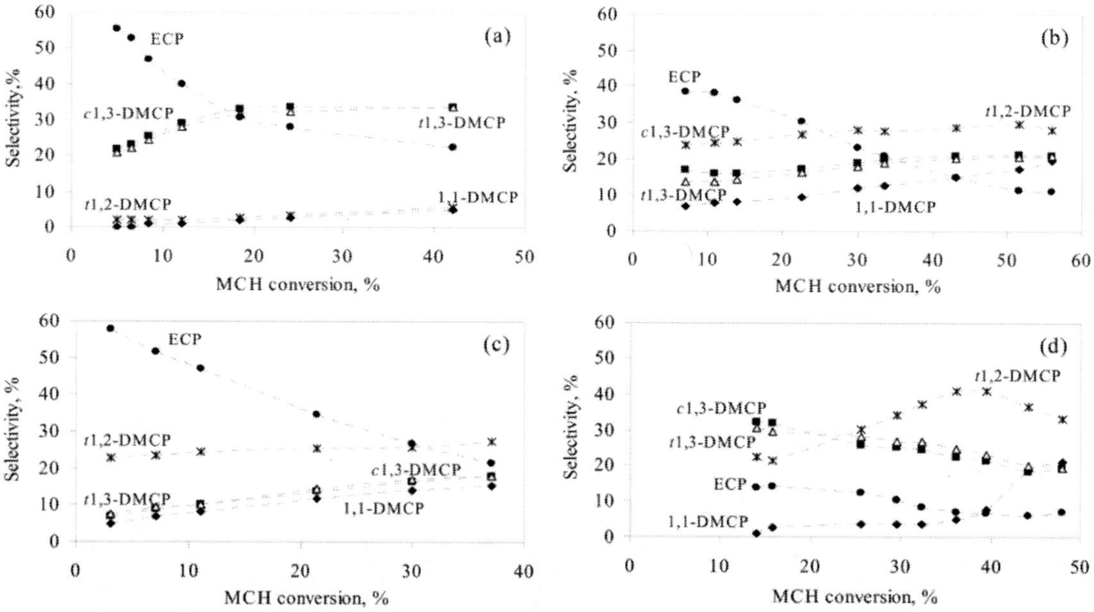

Figure 6. Selectivity to RC products vs. MCH conversion over (a) Pt/MFI, (b) Pt/BEA, (c) Pt/FAU, (d) Pt/WZ catalyst at different space velocities. Reaction conditions: Total pressure = 2 MPa; Temperature = 533 K; H_2/feed molar ratio = 40.

The enhancement of the formation of ECP, *trans*-1,2-DMCP, and 1,1-DMCP over the Pt supported catalysts should be derived from the enhancement of the hydrogen transfer step, which results in decreasing surface residence time of the intermediates [9]. On an acidic catalysts, the desorption of an intermediate (alkylcarbenium ions) on the surface to form RC isomers occurs by a bimolecular hydride transfer with the electroneutral molecules, mostly reactant. In this case, the hydride transfer step is slow, and it results in the slow desorption of intermediate transition states. On Pt catalysts, Pt metals are very active for the hydrogenation/dehydrogenation reaction that would accelerate the desorption rate, which decreases the lifetime for the intermediate. These lead to the difference in the distribution of RC isomers. However, unlike zeolites, the selectivity patterns of MCH isomers over bare and Pt promoted WZ are similar in the means of both were kinetically controlled at low conversion and thermodynamically controlled at high conversion. Therefore, it can be suggested that the hydride transfer step had less effect over WZ based catalysts.

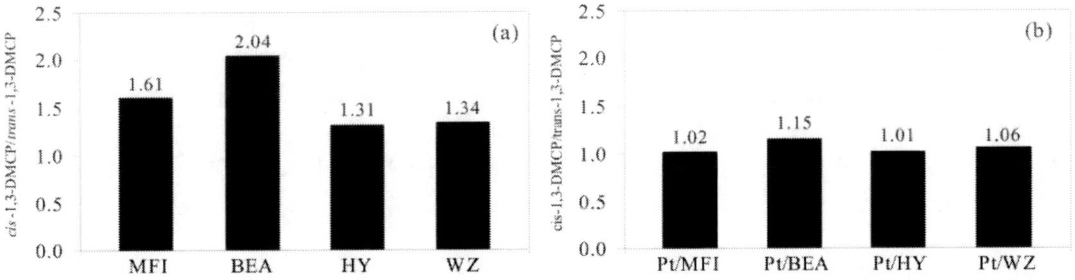

Figure 7. The *cis*-1,3-DMCP/*trans*-1,3-DMCP ratio over (a) bare catalysts and (b) Pt supported catalysts at the same conversion.

An interesting change in the *cis/trans*-1,3-DMCP ratio was observed in the product between bare (Figure 7a) and Pt supported (Figure 7b) acidic catalysts. The formation of *cis*-1,3-DMCP is easier than that of *trans*-1,3-DMCP due to its conformation. This ratio was significantly decreased on Pt supported catalysts due to the higher available hydrides.

Conclusion

The selectivity to RC isomers is highly dependent on the acidic strength but not the density of acid. The dibranching molecules (DMCPs) require a stronger acid strength than monobranching molecule (ECP). Therefore, the acid strength can be relative compared by the DMCPs/ECP ratio as follow WZ >> BEA > MFI > FAU. Over bare high acid strength zeolite (BEA, MFI), the product distribution is dominated by *cis*-1,3-DMCP. In contrast, HY and WZ having less pore constraint give rise to the higher selectivity to the bulky *trans*-1,2-DMCP than that over BEA and MFI. Moreover, the possibilities to form *cis*- and *trans*-1,3-DMCP are similar.

The presence of Pt enhances the formation of 1,1-DMCP, which is the most sterically hindered carbenium ion intermediate, due to providing surface hydrogen atoms that are readily used for the hydride transfer reaction. In addition, the *cis/trans*-1,3-DMCP are close to unity for all kinds of Pt supported catalysts.

The selectivity patterns over bare and Pt promoted WZ are similar. They were kinetically controlled at low conversion (i.e., favor *cis*-1,3-DMCP) and thermodynamically controlled at high conversion (i.e., favor *trans*-1,2-DMCP). Therefore, it can be suggested that the hydride transfer ability does not play an important role for MCH hydroisomerization over WZ based catalysts.

References

1. Kubicka, D., Kumar, N., Maeki-Arvela, P., Tiitta, M., Niemi, V., Karhu, H., Salmi, T. and Murzin, D.Y. (2004), Ring opening of decalin over zeolites. J. Catal., 227(2), pp. 313-327.
2. Lónyi, F. and Valyon, J. (2001), A TPD and IR study of the surface species formed from ammonia on zeolite H-ZSM-5, H-mordenite and H-beta. Thermo. Acta, 373(1), pp. 53-57.
3. Nakao, R., Kubota, Y., Katada, N., Nishiyama, N., Kunimori, K. and Tomishige K. (2004), Performance and characterization of BEA catalysts for catalytic cracking. App. Catal. A, 273(1-2), pp. 63-73.
4. Martens, J.A. and Jacobs, P.A. (2001), Introduction to acid catalysis with zeolites in hydrocarbon reactions. Stud. Surf. Sci. Catal., 137, pp. 633-671.
5. McVicker, G.B., Feeley, O.C., Ziemiak, J.J., Vaughan, D.E.W., Strohmaier, K.C., Kliewer, W.R., and Leta, D.P. (2005), Methylcyclohexane ring-contraction: a sensitive solid acidity and shape selectivity probe Reaction. J. Phys. Chem. B, 109, pp. 2222-2226.
6. Belatel, H., Al-Kandari, H., Al-Khorafi, F., Katrib, A. and Garin, F. (2004), Catalytic reactions of methylcyclohexane (MCH) on partially reduced MoO_3. App. Catal. A, 275(1-2), pp. 141-147.
7. Santiesteban, J.G., Calabro, D.C., Change, C.D., Vartuli, J.C., Fiebig, T.J., and Bastian, R.D. (2001), The role of platinum in hexane isomerization over $Pt/FeO_y/WO_x/ZrO_2$. J. Catal., 202(1), pp. 25-33.
8. Mignard, S., Caillette, Ph. and Marchal, N. (1994), Hydroconversion of methylcyclohexane on a bifunctional catalyst. Chem. Ind. (London), 58, pp. 447-58.
9. Iglesia, E., Soled, S.L. and Kramer, G.M. (1993), Isomerization of alkanes on sulfated zirconia: promotion by platinum and by adamantyl hydride transfer species. J. Catal., 144(1), pp. 238-253.

Green Diesel Production from Vegetable Oil

Extended Abstract 2007 AICHE

Jennifer Holmgren, Chris Gosling, Terry Marker, and Peter Kokayeff; UOP LLC
Giovanni Faraci, Carlo Perego, Eni S.p.A. Refining and Marketing Division

INTRODUCTION

The production of biofuels is expanding at a rapid pace worldwide as a result of increasing petroleum prices, government regulations, and commitments to greenhouse gas reduction. Future widespread use of biofuels depends on developing new technology to produce high quality transportation fuels from the highly oxygenated, biologically derived, feedstocks. These new biofuels need to be compatible with the existing fuel transportation and infrastructure.

Existing technology for producing diesel fuel from vegetable oil has largely centered on production of FAME or biodiesel[1,2]. While FAME has many desirable qualities, such as high cetane, there are other issues associated with its use such as poor stability and high solvency leading to filter plugging problems. UOP and Eni, S.p.A. recognized the need for different processing route to convert vegetable oils into a high quality diesel fuel or diesel blend stock that is fully compatible with petroleum derived diesel fuel. The two companies started a collaborative research effort in 2005 to develop such a process based on conventional hydroprocessing technology that is already widely deployed in refineries and utilizes the existing refinery infrastructure and fuels distribution system. The result of this effort is the UOP/Eni Green Diesel Process. This new technology utilizes widely available vegetable oil feedstocks to produce a high cetane, low gravity, aromatics and sulfur free diesel fuel. The cold flow properties of the fuel can be adjusted over a range of to meet various cloud point specifications in either the neat or blended fuel. The new process is targeted for commercialization in 2009 in an Eni refinery in Italy.

EXPERIMENTAL

Processing of biologically derived feedstocks is complicated by the fact that these materials contain a significant amount of oxygen. The feedstocks of primary interest in the Green Diesel Process are primarily vegetable oils such as soybean, palm, jatropha, or rapeseed oils. Other products such as animal fats and greases can also be used as a feedstock. Plant oils mainly consist of triglycerides with typically 1-2% free fatty acid content. The chemical structure of a triglyceride molecule is shown in Figure 1 (red=oxygen, white=hydrogen, grey= carbon atoms).

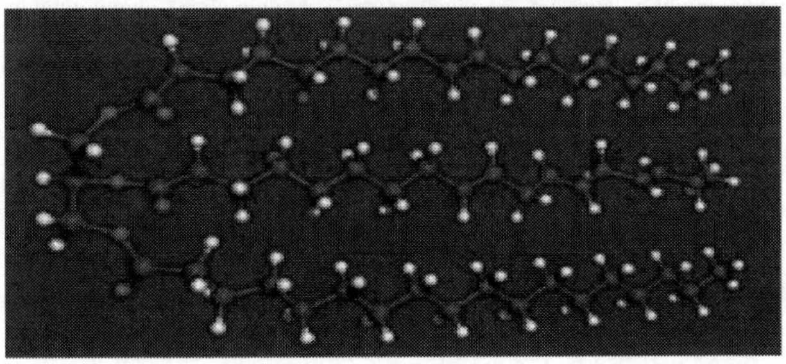

Figure 1: Chemical Structure of Triglycerides

Triglycerides and fatty free acids both contain relatively long, linear aliphatic hydrocarbon chains. The aliphatic hydrocarbon always contains an even number of carbon atoms, is generally unsaturated, and also corresponds to the carbon number range typically found in diesel fuels. There is also a three carbon "backbone" in the triglyceride molecule. The properties of various vegetable oils are compared to petroleum derived diesel fuel in Table 1. The volume yield of diesel fuel potential from vegetable oils is nearly 100% where diesel represents only about 20% of the volume of crude oil when distilled.

Table 1
Comparison of Vegetable Oil Feedstocks[3,4,5]

Bio -Feedstock	*Price ($/bbl)*	*Diesel Yield, v%*	*Carbon Number*	*Olefins, mol-%*
WTI Crude	68	20	11-22	0
Rapeseed oil	89	99	16-22	94
Soy oil	75	99	16-18	84
Palm oil	62	99	16-18	58
Jatropha oil	44	99	16-18	77

There are two processing routes for converting vegetable oil to diesel fuel as shown in Figure 2. The conventional processing route for diesel fuel production is via esterification with methanol in the presence of caustic to produce FAME with glycerol as a by-product. The glycerol is produced from the three carbon "backbone" and has a limited market, particularly in an unrefined form.

The hydroprocessing route uses hydrogen to remove oxygen from the triglyceride molecules, which is used in the Green Diesel Process. The oxygen is removed in two competing reactions: decarboxylation and hydrodeoxygenation. The extent of each reaction depends on the catalyst and process conditions used for the process. The three carbon "backbone" in the Green Diesel Process yields propane and can be recovered easily in when the process is integrated into a refinery. The oxygen contained in the feed is produced either as CO/CO_2 or water. In addition, all olefinic bonds are saturated, resulting in a product consisting of only paraffins.

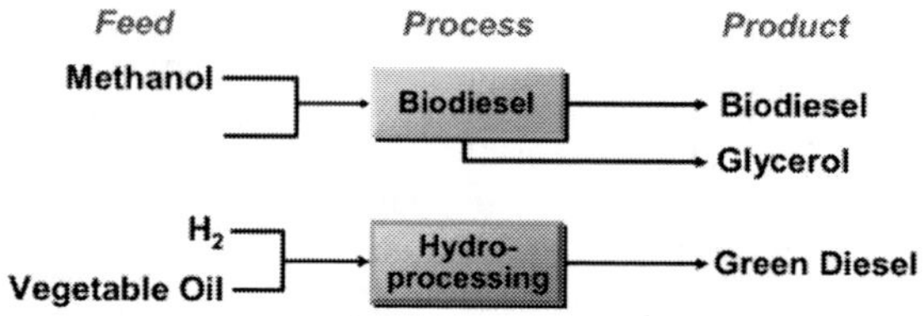

Figure 2: Vegetable Oil Processing Routes

The chemistry in the biodiesel or FAME route is problematic from a processing viewpoint. The reaction stoichiometry is as follows:

$$\begin{array}{lcl} \text{Triglycerides} + \text{MeOH} & \Longrightarrow & \text{FAME} + \text{Glycerol} \\ 100\ M^3 \quad\quad 13\ M^3\ \ \text{NaOH} & & 99\ M^3 \quad\quad 8\ M^3 \end{array}$$

About 8% of the product volume is glycerol, which is a low value product in unrefined form and has a limited market when refined. Methanol is required as a co-feed and feedstocks containing high concentrations of fatty acids can cause operational problems due to saponification reactions with the caustic used as a catalyst.

As noted previously, the Green Diesel Process removes the oxygen by reaction with hydrogen, producing a pure paraffinic product. The volume yield is 99% and the primary co-product is propane. By-products are water and carbon oxides. Feedstocks containing free fatty acids can be processed in the Green Diesel Process without the problems caused in FAME production.

$$\text{Triglycerides} + H_2 \Longrightarrow \text{Green} + H_2O/CO_2 + C_3H_8$$

Diesel.
100 M³ Catalyst 99 M³ 9 M³

The Green Diesel Process has several basic advantages over FAME production in a refinery setting. Hydrogen required for the reaction is readily available and the products are all ones that are already normal refinery products and do not require any special handling. The products can all be easily blended with conventional refinery products.

RESULTS AND DISCUSSION

While hydroprocessing was clearly the chemistry preferred for refinery applications, how to implement it in a process design was not obvious. There are two options that can be considered: co-processing in an existing distillate hydroprocessing unit or building a stand alone unit as shown in Figure 3.

The co-processing route was initially evaluated as attractive options, since existing equipment could be used, resulting in a lower cost implementation. Co-processing was found to be problematic after some initial evaluations were completed. Vegetable oils contain trace metal contaminants such as phosphorous, sodium, potassium, and calcium. These require addition of a pre-treating reactor for removal as the existing reactor will generally not have sufficient catalyst volume for the treating catalyst. The reactions involved in processing vegetable oil are fairly exothermic, and may require quench capabilities that may not necessarily be available in a given unit. The deoxygenation products (H_2O, CO, CO_2) will require revamping of the recycle gas system for removal, or use a substantial purge stream. The cold flow properties of the combined diesel product may limit the quantity of vegetable oil that can be processed, as normal paraffins are the primary product from vegetable oil and will impact the cloud point.

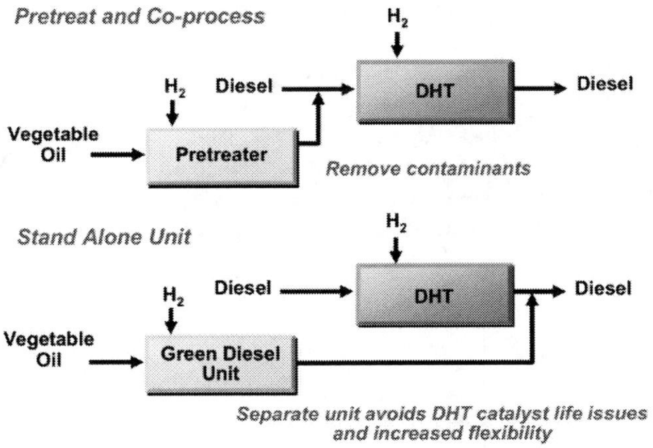

Figure 3: Vegetable Oil Hydroprocessing Alternatives

After considering all of the potential issues, UOP and Eni decided that it was much more cost effective to build a larger sized, dedicated unit that was optimized for vegetable oil processing due to the unique nature of these feedstocks. A flowscheme showing the process is shown in Figure 4.

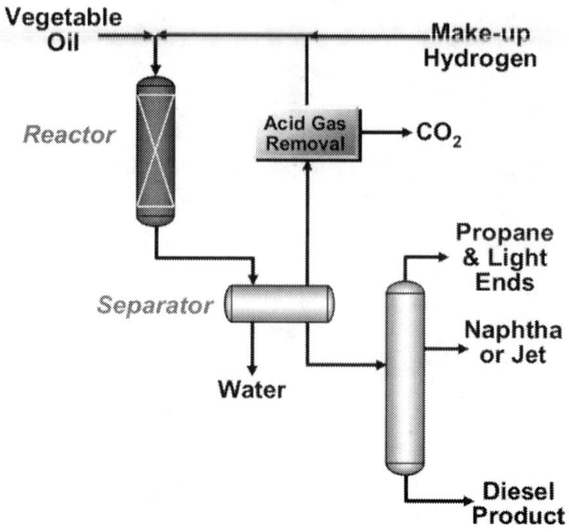

Figure 4: Simplified Green Diesel Flowscheme

In the process shown in Figure 4, vegetable oil is combined with hydrogen, brought to reaction temperature, and is then sent to the reactor where the vegetable oil is converted to the Green Diesel product. The product is separated from the recycle gas in the separator and the liquid product sent to a fractionation section. The design of the fractionation section can vary from a one column system producing on-specification diesel and unstabilized naphtha to a three column system producing propane, naphtha, and diesel products. It is envisaged that most installations will be a single column and the lighter products will be recovered in other existing refinery process units. The recycle gas is treated in an amine system to remove CO_2.

UOP and Eni have done extensive process performance testing to determine optimum process conditions, catalyst stability, and product properties. A range of vegetable oil feedstocks have been processed in pilot plants at both UOP and Eni facilities including soybean, rapeseed, and palm oil. An extensive program is underway to evaluate other potential feedstocks including tallow and greases derived from animals. We have done considerable testing to confirm stability of catalyst system employed and data from an on-going stability test evaluating rapeseed oil is shown in Figure 5. There is no measurable deactivation after over 2000 hours on stream.

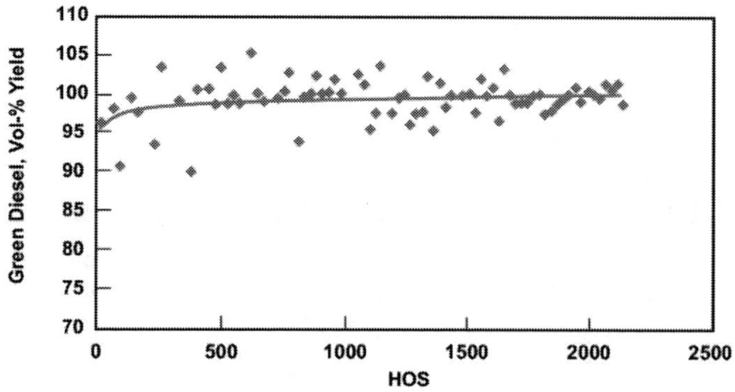

Figure 5: Catalyst Stability Testing

Product properties are similar for all the vegetable oils processed. Typical product yields are shown in Table 2. Cold flow properties can be controlled by the severity used in the process, so the diesel yield will vary from 99 vol-% to about 88 vol-% that is required to achieve a -10°C cloud point. The diesel yield loss is to kerosene and naphtha. In all cases, the product has a very high cetane number (>80) and contains no sulfur or aromatics. The range of hydrogen consumption shown in Table 2 is what is required for different feedstocks. Palm oil requires substantially less hydrogen than rapeseed oil due to the lower olefin content of that feedstock.

The Green Diesel product has excellent diesel fuel properties, and is very similar to diesel product that is produced via Fischer-Tropsch processes. The Green Diesel properties are

Table 2
Green Diesel Product Yields

Feed	
% Vegetable Oil	100
% Hydrogen	1.5 - 3.8
Products	
Vol% Naphtha	1 - 10
Vol% Diesel	88 - 98+
Cetane Number	>80
Sulfur, ppm	<1

compared to mineral diesel and biodiesel in Table 3. The diesel fuel properties for FAME are not ideal for diesel blending. The gravity is high, which can limit blending of lower value products like hydrotreated LCO into a refinery diesel pool. It also has a narrow boiling range and can have issues in meeting stability specifications. The heating value is about 12% lower than typical mineral diesel, which will reduce the fuel economy of FAME/diesel blends relative to mineral diesel or Green Diesel/diesel blends.

Table 3
Green Diesel Fuel Properties[5]

	Mineral ULSD	Biodiesel (FAME)	Green Diesel
% Oxygen	0	11	0
Specific Gravity	0.84	0.88	0.78
Sulfur, ppm	<10	<1	<1
Heating Value, MJ/kg	43	38	44
Cloud Point, °C	-5	-5 to +15	-10 to +20
Distillation, °C	200-350	340-355	265-320
Cetane	40	50-65	70-90
Stability	Good	Marginal	Good

Green Diesel, however, is a premium diesel blending component. The boiling range is comparable to typical diesel products, with substantially higher cetane and lower density. These are very valuable properties as it allows blending low value hydrotreated LCO into a typical refinery diesel pool and still meet EU diesel specifications. An example is shown in Table 4, where a straight-run diesel and kerosene blend can be produced to meet a 50 cetane ULSD product specification. No LCO can be blended without the use of cetane enhancing additives. If Green Diesel is added to the pool, the high cetane value and low density allow for blending a nearly equivalent volume of hydrotreated LCO into the ULSD product. Assuming a typical discount of $4.60/bbl for fuel oil relative to ULSD, this translates to a benefit of $9200 per day or $3 MM per year that could be realized through Green Diesel blending.

Table 4
Green Diesel Blending Benefits

Diesel Pool Components	Barrels in Pool	Cetane Index
Kerosene	500	41
Hydrotreated SR Diesel	7500	52
Hydrotreated LCO	2000	20
Green Diesel	2346	74
Blended Product Cetane		50

A Life-Cycle Analysis (LCA) of different diesel production routes was conducted using the Simapro™ LCA program. LCA is a method to determine and compare the environmental impact of alternative products or processes, including impacts of initial resource extraction to disposal of waste products. In this study, the scope of the analysis was from extraction of the crude oil through combustion of the refined diesel fuel in a vehicle. It was assumed that all fuels had the same performance in the vehicle for the purposes of this study. The primary focus of the analysis was on fossil energy use and emissions of greenhouse gases, although other impact categories are included.

The results of the analysis are shown in Figure 6. Both the FAME and Green Diesel products have a much lower total environmental impact scores than mineral diesel primarily because both have significantly lower production of climate-active CO_2. However, Green Diesel has both a lower environmental impact and lower climate active CO_2 production than FAME. The environmental impact of FAME production is higher due to the methanol feedstock requirement, which is produced from natural gas through a very energy-intensive process that has a high environmental burden.

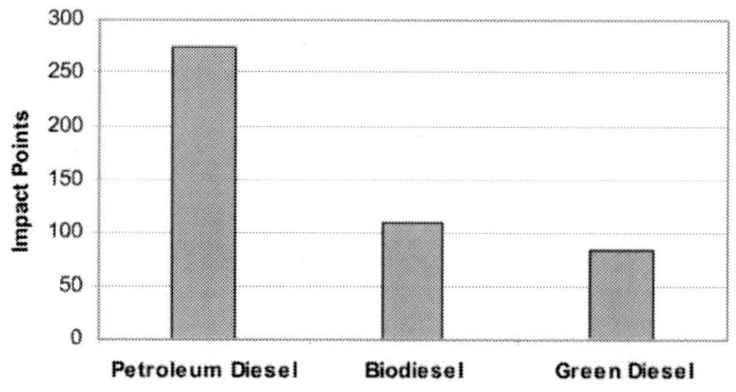

Figure 6: Life Cycle Analysis Results Impact Score

Conclusions

Refiners are well positioned to play a major role in renewable fuel production in the future. UOP and Eni have developed the Green Diesel Process, a new, sustainable, route for converting vegetable oil into premium quality diesel fuel as a first step in providing technology to enable refiners to participate in renewable fuels. The Green Diesel product is a superior alternative to FAME, with significantly better diesel product properties and is fully compatible with conventional mineral diesel products.

The UOP/Eni Green Diesel Process is fully developed and available for licensing from UOP. Eni is preceding with commercialization of the first unit in 2009 at one of the Eni refineries in Europe. UOP is currently working on developing projects with several other clients in North America and Asia.

References

(1) Alternative Fuels, National Renewable Energy Laboratory web site, http://www.eere.energy.gov/afdc/altfuel/whatis_biodiesel.html

(2) RADICH, A. Biodiesel Performance, Costs, and Use. Energy Information Administration, 2004. http://www.eia.doe.gov/oiaf/analysispaper/biodiesel/

(3) Worldbank Commodity Price Data, http://www.econ.worldbank.org , Jan-Aug 2006

(4) ICIS Chem Business (Chemical Market Reporter), October 2006

(5) Alternative Fuels Comparison Chart, National Renewable Energy Laboratory web site, http://www.eere.energy.gov/afdc/altfuel/fuel_comp.html

Hylube™ Process Commercialization: Recovering Value from Used Motor Oil

__T.N. Kalnes__, UOP LLC, 25 E. Algonquin Rd., Des Plaines, IL 60017
***Andreas Schuppel**, Puralube GmbH, Hauptstrasse 30, Gebaude 27, Elsteraue / OT Alttröglitz, D-06729, Germany*
***Laura Kadlec**, Technical Services, UOP LLC, 25 E. Algonquin Rd., Des Plaines, IL 60017*

ABSTRACT

Environmentally-conscious design of processes and products is increasingly viewed as an integral strategy in the sustainable development of new refining and chemical processes. Although re-refining of spent lubricating oils such as used motor oils has been practiced with varying technical and commercial success over the past 50 years, a sustainable processing technology has yet to become widely accepted. Poor on-stream efficiency, inconsistent product quality, and careless management of feedstock contaminants and byproducts have often resulted in widespread environmental problems and poor economics.

A novel catalytic process for regenerating spent lubricants has now been successfully commercialized. Puralube GmbH, a subsidiary of Puralube Inc., recently commercialized the UOP HyLube™ process. The first industrial scale plant, located in Elsteraue / Zeitz, Germany, converts used oil into lube base oils and green fuels in an environmentally conscious manner. Key process features include (1) continuous processing of the entire used oil charge stock in a hydrogen-rich environment without use of intermediate oil storage, (2) efficient removal of contaminants such as chlorinated, sulfurous, and oxygenated organic compounds and polyaromatic hydrocarbons, (3) management of all sulfurous and odorous compounds to eliminate malodorous and toxic emissions and (4) production of deeply desulfurized base oil and distillate fuel products of consistent high quality. Byproducts are managed to minimize environmental impact while maximizing energy recovery. This paper describes the HyLube process and reviews the performance of the commercial unit. It also compares the process life cycle of used oil re-refining to combustion of used oil as a secondary fuel.

INTRODUCTION

The HyLube Process is a patented catalytic process for re-refining used lubricating oils to produce saleable lube base stocks. This is the first re-refining process in which raw used oil is processed, without pretreatment, in a pressurized hydrogen environment. Fouling, coking, corrosion, and catalyst deactivation are minimized by the process design. The majority of the used oil is converted to high quality base oil products. Lesser quantities of hydrotreated low sulfur diesel, hydrotreated naphtha, heavy fuel oil and fuel gas are also produced. Environmentally undesirable compounds such as halogenated hydrocarbons and oxygenates are destroyed, and all water streams are treated.

A HyLube Process unit with a capacity of 220 metric tons/day (MTD) has been successfully commercialized by Puralube GmbH located in Elsteraue / Zeitz, Germany. The technology is licensed to Puralube by UOP LLC.

PROCESS DESCRIPTION

The HyLube process uses a unique feed preparation concept known as direct contact hydrogenation, which eliminates the need for much of the equipment used in other lube re-refining processes. The used lube oil feed is heated to reaction temperature by direct contact with a hot circulating hydrogen stream. As a result, the feed stream is maintained in a hydrogen environment which inhibits the formation of the polymeric and carbonaceous by-products which can cause equipment and catalyst fouling.

Figure 1 shows a flow diagram of the process. The first part of the process involves separation of the lube range and lighter components of the feed from the non-distillable residue portion. The used lube oil feed is mixed with heated hydrogen gas in a proprietary mixing device (R-115), vaporizing 50-70 percent of the feed. The hydrogen-rich vapor is separated from the remaining liquid in the Feed Flash Separator (B-115). The mixed hydrogen and vaporized lube stream is sent to the reactors (C-120 and C-121). Hot liquid from the Feed Flash Separator is charged to the Residue Stripper (K-150) where it is steam-stripped under vacuum to maximize heavy lube recovery overhead. The column vapors are condensed and the hydrocarbon phase is combined with the Feed Flash separator vapor and charged to the fixed bed reactors.

The HyLube process reactions are carried out at elevated temperatures and pressures in a hydrogen atmosphere using UOP proprietary catalysts. The Puralube unit operates with reactor section pressures of 870-1200 psig (60-80 barg) and reactor temperatures in the 570-660°F (300-350°C) range. As the feed is processed over guard and demetallization catalysts in C-120 and the proprietary hydrofinishing (conversion) catalyst in C-121, contaminants are removed and the quality of the lube base oil is rejuvenated and enhanced. In addition to converting hetero-atoms such as sulfur and nitrogen, the conversion catalyst increases viscosity index via saturation of multi-ring aromatic compounds.

After exiting the reactor, the purified products are separated from the reaction by-products and excess recycle gas. This is first done by staged cooling and flash separation, and then by fractionation. Acid gases such as H_2S and HCl are present in the reactor effluent. To prevent these components from becoming corrosive, mild neutralization chemicals are injected into the hot separator (B-125) vapor. The hydrogen rich vapor from the cold separator (B-140) is scrubbed, compressed, reheated and returned to the mixer. The hydrocarbon liquids collected in the separators are sent to the product stripper (K-160).

The product stripper separates acid gases and light fuel gas material from the lube oil product before the oil enters the main lube fractionation column. The product stripper normally operates at a pressure slightly above the plant fuel gas pressure. This allows the off gas to be used as fuel for the fired heater, increasing the energy efficiency of the HyLube process.

The stripped hydrocarbon liquid is sent to the lube fractionation column (K-200) which operates under vacuum. This column separates the hydrotreated products into various cuts to meet the desired lube oil viscosity grades. High quality naphtha or diesel cuts may also be recovered from this column. A photo of the Puralube unit, highlighting the lube fractionation column, is shown as Figure 2.

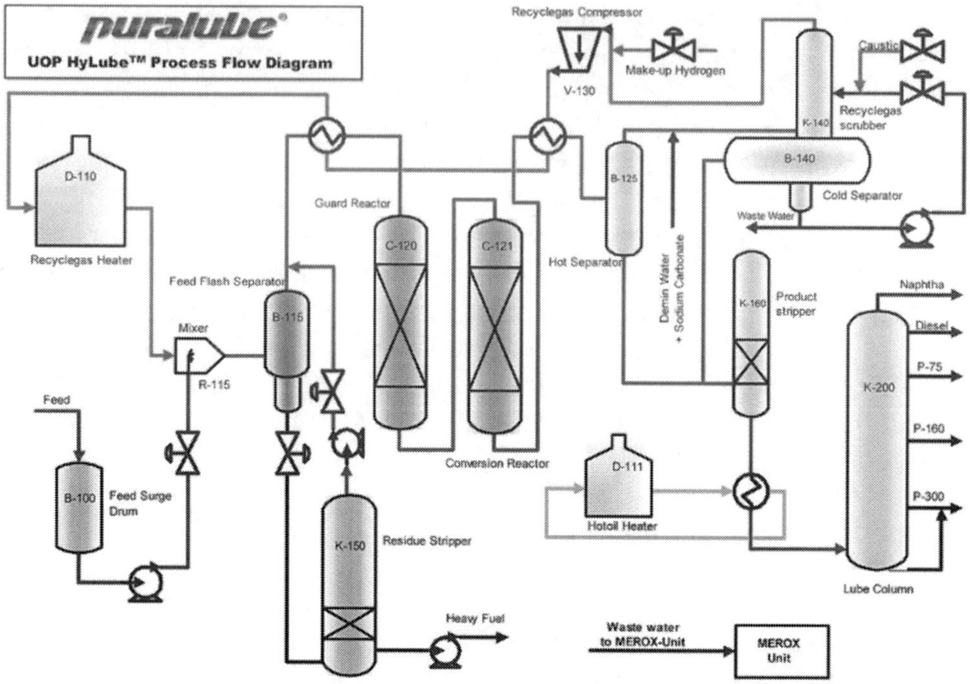

Figure 1: HyLube Process Flow Diagram

Because the Puralube unit is a stand-alone facility, all of the sulfur, chloride, nitrogen, and oxygen removed from used oil feed must be managed in an environmentally acceptable manner. The byproduct of acid gas neutralization is an aqueous stream containing sulfide and chloride salts as well as small amounts of ammonia and water produced in hydrotreating reactions. All of the aqueous streams leaving the HyLube unit are collected in the main column overhead receiver and then routed to the Sulfide Oxidation section of the unit.

The Sulfide Oxidation process is a patented extension of the widely used UOP Merox™ process. Here the sulfides are catalytically oxidized to sulfate salts with a resultant reduction in chemical oxygen demand. After sulfide oxidation, the process water is sent to a two-stage biological treatment facility prior to discharge.

Figure-2: HyLube Process Unit in Zeitz, Germany

COMMERCIAL OPERATION

Used oil feed to the Puralube HyLube unit includes a mixture of recovered motor oils, hydraulic oils, gear oils, and other paraffin-rich industrial oils. In addition to lube base oil and assorted additives, the used oil contains contaminants from the oil use and collection, including gasoline, diesel, combustion byproducts, and water. The feed also contains varying concentrations of particulate matter such as sludge, grit and corrosion metals.

Table 1 presents a partial characterization of the Puralube unit feed over a ten month period of operation. The feed typically contains less than 10% water, 5% light fuels and 1wt-% ash. The primary constituents of the ash are phosphorous, zinc, and calcium, all well-known components of motor oil additives. Sulfur content is typically less than 1wt-%. Day-to-day variations in feed composition are buffered by blending in a primary feed tank which holds approximately seven days of inventory.

Table-1: HyLube Process Unit Feed Characterization

Property	Units	Min	Max	Average
Viscosity @ 40C	cSt	46.8	70.6	56.2
Density	Kg/m3	881	907	888
Flash Point	°C	174	310	202
Total acid number	mg KOH/g	2.8	7.0	4.3
Water	wt%	3.1	10.7	6.0
Ash	wt%	0.33	1.11	0.72
Chloride	wt%	0.01	0.216	0.05
Sulfur	wt%	0.35	0.92	0.61
Nitrogen	wt%	0.00	1.60	0.12
Lead	wt ppm	6.9	72.7	29.4
Arsenic	wt ppm	0.02	3.8	0.81
Chromium	wt ppm	1	798	62
PCBs	wt ppm	0.1	1.2	0.3

This paper reviews the operation of the unit during the 10 month period from June 2005 to March 2006, during which detailed operating and analytical data were collected. Near the end of this period a ten day test run was conducted with the unit running at 110% of design capacity. All product quality, yield, and hydrogen consumption targets were achieved during this test run, as shown in Table 2.

Table-2: Summary of Commercial Unit Test Results

Parameter	Units	Design Target	Mar-06 Test Run
Fresh Feed (Used Oil)	m3/day	> 247	270
Hydrogen Consumption	Nm3/m3 FF	< 70.3	45
P-160 Product Quality			
Color	ASTM scale	1 nom	<0.5
Distillation 5% Point	°C	> 405	>413
Flash Point	°C	> 210	232
Sulfur Content	wt ppm	< 300	166
Saturates Content	Wt%	90	> 90 *
Total Lube Product Yield	% Lube range oil in FF	> 90	93

* Measured at a later date

The hydrogen consumption is calculated on the basis of total used oil charged to the unit. P-160 is the middle lube cut from the lube product fractionation column and is the primary lube product from in the unit. The total lube product also includes lighter (P-75) and heavier (P-300) base oils. Total lube yield is calculated as a percentage of the distillable lube base oil fraction present in the used oil feed.

A primary objective of the HyLube process is to produce consistently high quality base oil product. Table 3 summarizes average properties of the lube base oils and by-product oils from the commercial HyLube operation.

Table-3: HyLube Product Quality

		Lube Base Oils			Diesel	Heavy
Property	**Units**	**P-75**	**P-160**	**P-300**	**Product**	**Fuel**
API Gravity		34.6	34.0	33.2	37.6	20.4
Viscosity @ 40C	cSt	15	28	51	6	
Viscosity Index		105	117	119		
Sulfur	wt-ppm	115	117	114	111	5775
Color	ASTM	0.0	0.2	0.6		
Water	mg/kg				118	645
Heating Value	MJ/Kg				44	41

Table 4 compares the typical properties of the P-160 product to the specifications for different grades of virgin base oil. The P-160 quality exceeds the API Type II requirements.

Table-4: Lube Quality vs Typical for API Groups

		Puralube P-160	Group I	API Categories Group II	Group III
Property	**Units**				
API Gravity		**34.0**	32.5	32.6	35.3
Viscosity Index		**117**	94	104	118
Sulfur	wt-ppm	**<100**	484	<300	1
Nitrogen	wt-ppm		50	-	-
Color	ASTM	**<0.5**	< 0.5	< 0.5	< 0.5
Refractive Index					
Saturates	wt-%	**90-98**	89	>90	98
Aniline Point	°C	**109**	103	108	-
Copper Strip Corrosion	ASTM	**1a**	1a	1a	1a
Total Acid Number	mg/g	**0.0**	0	0	0
Flash Point	°C	**230**	211	224	-

During the period of operation discussed in this report, the viscosity indices of the medium and heavy lube products were consistently greater than 110, the sulfur content was usually less than 100 wt-ppm, and the color of the P-160 product was less than 0.5. Plots of these base oil quality measurements for the three lube products from the unit are shown in Figures 3-6.

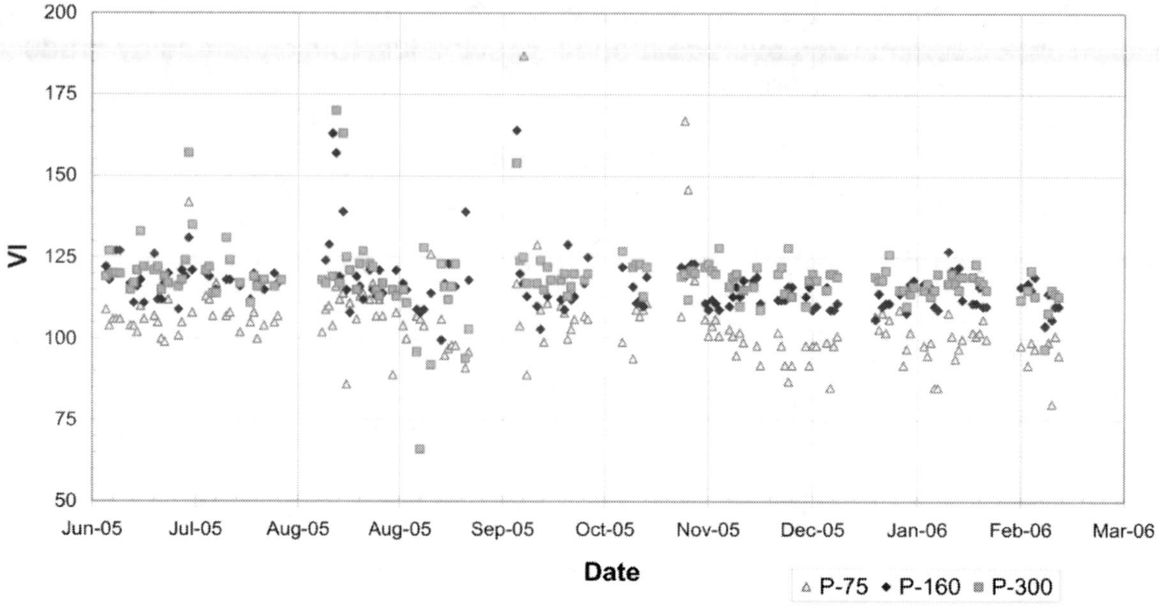

Figure-3: Product Viscosity Index

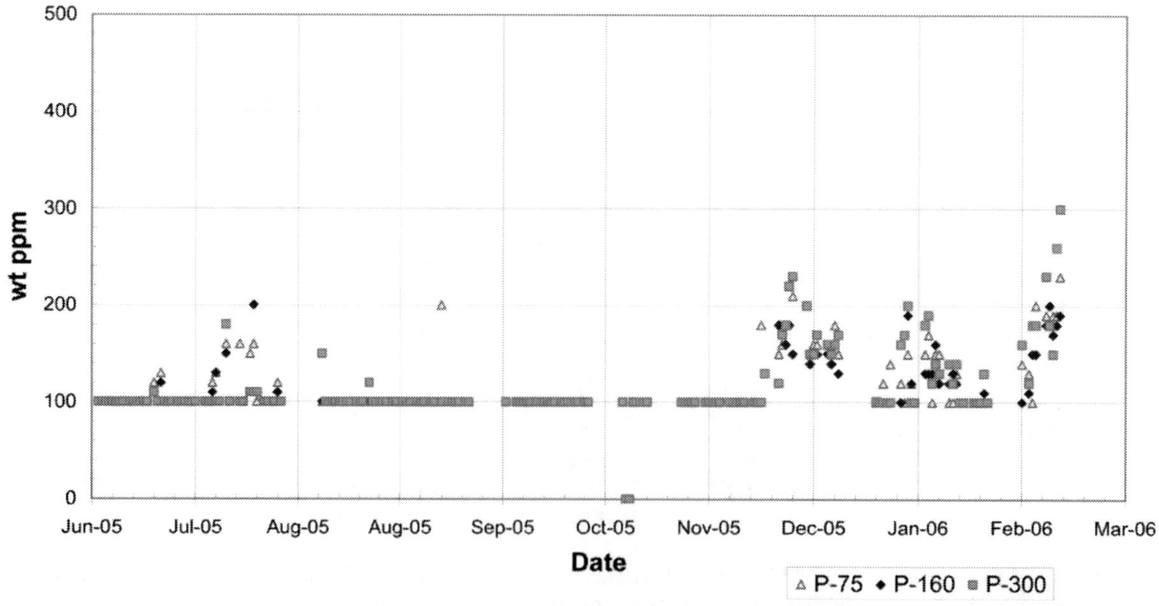

Figure-4: Product Sulfur Content

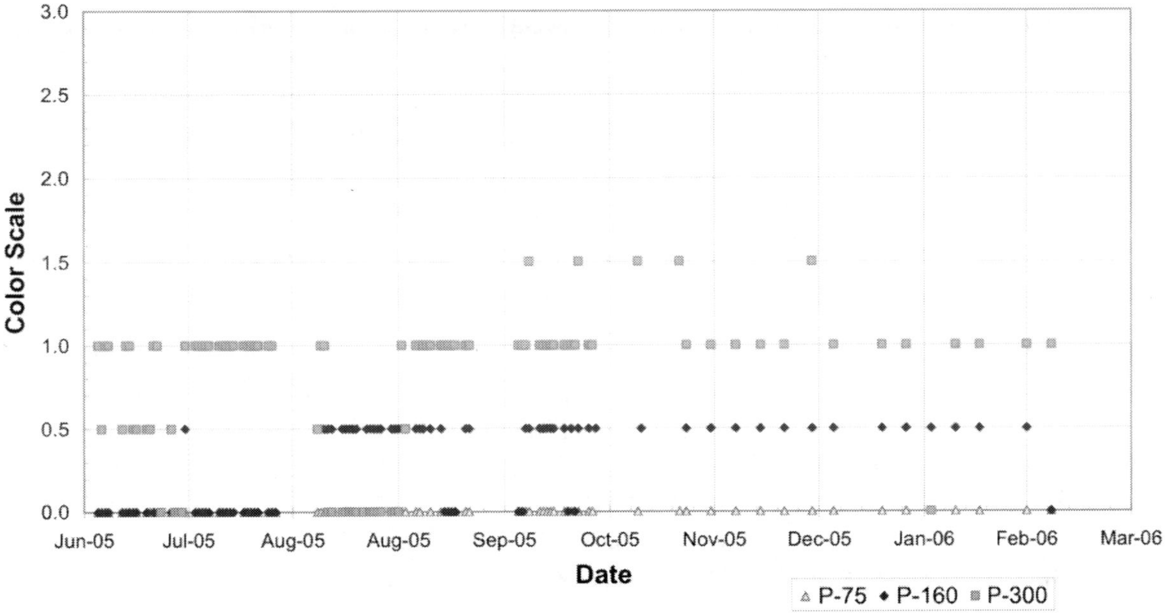

Figure-5: Product Color (ASTM D-1500)

During the same period of operation, the HyLube catalyst temperatures were stable and were increased only as needed to meet product specifications at higher feed rates. The C-121 conversion catalyst temperature is plotted in Figure 6.

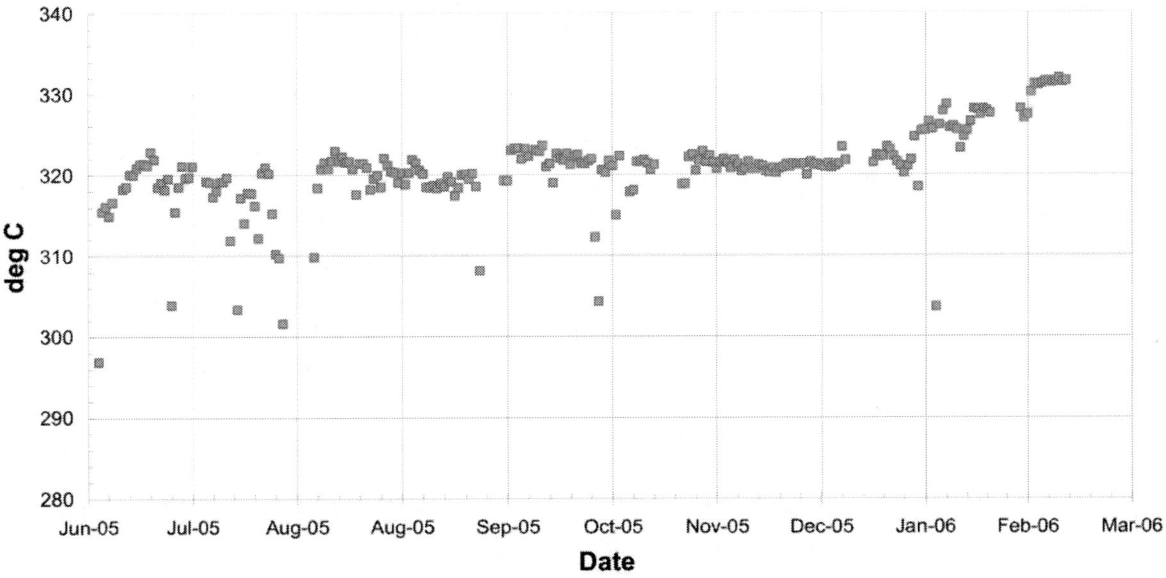

Figure-6: Average Catalyst Temperature

A key factor in maintaining stable HyLube catalyst activity is to minimize the level of inorganic contaminants in the reactor feed. The great majority of the inorganic components in the used oil feed leave the unit in the heavy fuel (residue) product. The guard reactor is designed to remove the remaining contaminants to prevent poisoning of the catalyst in C-121.

UOP has continued to work with Puralube to optimize the guard reactor catalyst loading in order to achieve maximum utilization of the proprietary conversion catalyst. Careful monitoring of the used oil feed for known catalyst poisons such as arsenic and silicon is ongoing. In addition, the latest generation of catalysts is being evaluated for use in the HyLube unit.

Another major objective of the HyLube process is to maximize unit on-stream efficiency. During the period of time discussed in this report, the Puralube unit was operating 88% of the time. One project goal is to achieve an on-stream factor for the HyLube process that is similar to what is currently achieved in conventional hydroprocessing units in modern petroleum refineries. As a reference point, a recent survey of UOP licensees operating hydrocracking units has indicated a typical on-stream factor of 95-97%, excluding planned major turnarounds. UOP and Puralube continue to work together to improve on-stream factor.

LIFE CYCLE ASSESSMENT

The purpose of Life Cycle Assessment (LCA) is to compare the effects of different processes or products on the environment, with the goal of determining which has the least negative environmental impact. Such studies were recently performed comparing used oil re-refining in the HyLube process to used oil combustion in cement kilns. (Frankl, 2005, Fehrenbach, 2005, Hartmann, 2005). The results indicate that the HyLube process is the more environmentally acceptable option. By displacing petroleum base oil and synthetic oil production, the re-refining process reduces the consumption of fossil fuels as well as the overall production of CO_2 emissions

The latest study (Kalnes, Shonnard, 2006) modeled the effects of re-refining used oil generated in the European market in the HyLube process. These effects were then compared to the impacts of burning the same oil in cement kilns, as a substitute for the primary energy sources of coal, heavy fuel oil (HFO) or natural gas (Nat Gas). ISO 14040 standards were followed in the study and SimaPro6.0 software was used for calculations.

Figure 7 shows the material flows for the utilization of 1 kg used oil (UO) by combustion in a cement kiln or as feed to the HyLube process. The impacts of UO transportation are included in both cases. For the combustion of UO, only the avoided flows of coal, heavy fuel oil, or natural gas were modeled. In the HyLube process, the avoided flows of virgin base oils, diesel and fuel oil were modeled. The base oil produced in the HyLube process has a high content of isoparaffins so that in addition to displacing petroleum base oil consumption, an estimated 0.2 kg synthetic oil can be avoided per kg UO.

This study emphasized the following categories of environmental impact: fossil fuels consumption, acidification / eutrification, and an aggregated single environmental score weighting all environmental categories according to the SimaPro Ecoindicator 99 (H) methodology.

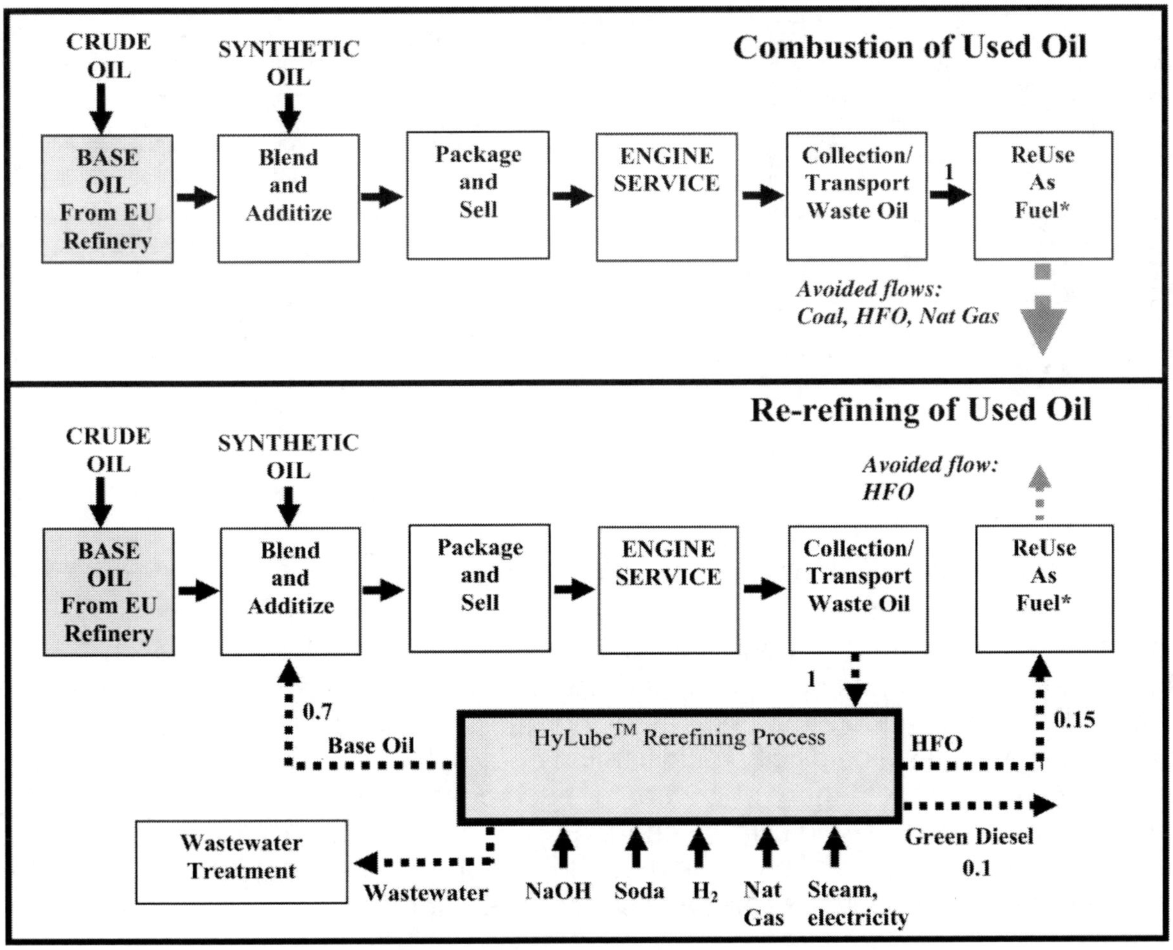

Figure-7: Material Flow Diagram for LCA

Figure 8 shows the cumulative energy demand for each of four alternatives: 1) used oil in cement kiln with displacement of coal, 2) used oil in cement kiln with displacement of fuel oil, 3) used oil in cement kiln with displacement of natural gas, and 4) re-refining of used oil in the HyLube process. Each of the alternatives exhibit a savings of energy compared to the reference flow described above.

The greatest savings of energy is achieved by re-refining of used oil in the HyLube process (-52.5 MJ/kg UO). This maximum benefit is realized by avoiding the production of virgin base oil, fuel oil and diesel in a refinery, and synthetic oil(s) in petrochemical plants. Producing base oils by conventional solvent refining methods and synthetic base oils by ethylene oligomerization to polyalfaolefin (PAO) require an estimated 58.2 and 95.6 MJ/kg respectively of cumulative energy demand. The next best alternative is combustion of used oil with displacement of fuel oil, followed by displacement alternatives of natural gas and then coal. The maximum difference in energy savings among these alternatives is only 16%, and therefore all alternatives exhibit a very positive benefit in "unweighted" energy savings compared to the reference flows.

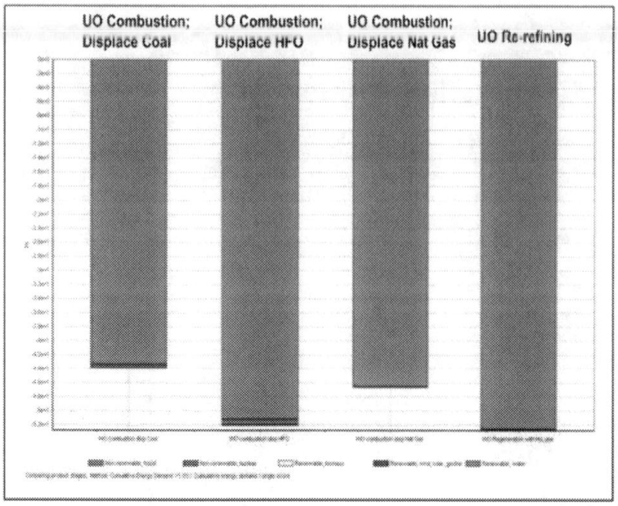

Figure 3. Comparison of used oil (UO) alternatives based on cumulative energy demand. HFO=heavy fuel oil, Nat Gas = natural gas

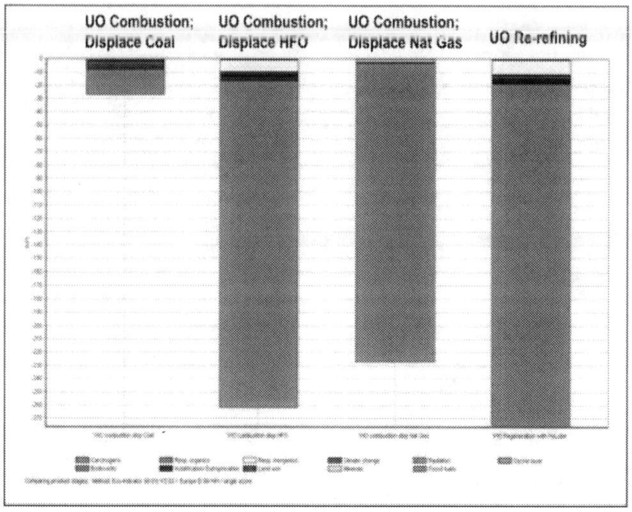

Figure 4. Comparison of UO alternatives based on single environmental score in SimaPro6.0 using Ecoindicator 99(H).

Figure-8: Comparison of LCA Results

To examine environmental impact, the EcoIndicator 99 (Hierarchist) method in SimaPro6.0 was used to aggregate eleven impact categories into one score. The impact categories in SimaPro are: carcinogens, respiratory inorganics, respiratory organics, climate change, radiation, ozone layer, ecotoxicity, acidification / eutrophication, land use, minerals, and fossil fuels. The dominant category of impact for each case is fossil fuel consumption, followed by respiratory inorganics and climate change, with minor contributions from the remaining categories. Coal consumption is weighted less than petroleum or natural gas in the EcoIndicator 99 method, as seen in the relatively small benefit for the case of UO combustion with coal displacement.

Figure 9 presents a comparison of environmental impacts for the four alternatives. Re-refining of used oil is the best case followed closely by used oil combustion with either fuel oil or natural gas displacement. Based on our data sources, it was believed that the highest accuracy predictions of impact are for climate change, acidification / eutrophication, and fossil fuel consumption.

The greatest benefit for climate change (primarlly CO_2 emissions) is for the case of UO combustion with coal displacement. This is logical because coal has the highest emission of CO_2 per unit of fuel energy for any of the fossil fuels in this study, and therefore when coal is displaced by a less carbon intensive fuel, lower greenhouse emissions occur. The next best case is re-refining, followed by UO combustion with fuel oil displacement. The least desirable alternative for climate change is UO combustion with natural gas displacement. When UO displaces natural gas in cement kiln combustion, a higher rate of CO_2 release occurs, but this increase is almost exactly compensated for by avoiding emissions from natural gas production. For the categories of acidification / eutrophication and fossil fuel consumption, re-refining of UO is the best alternative.

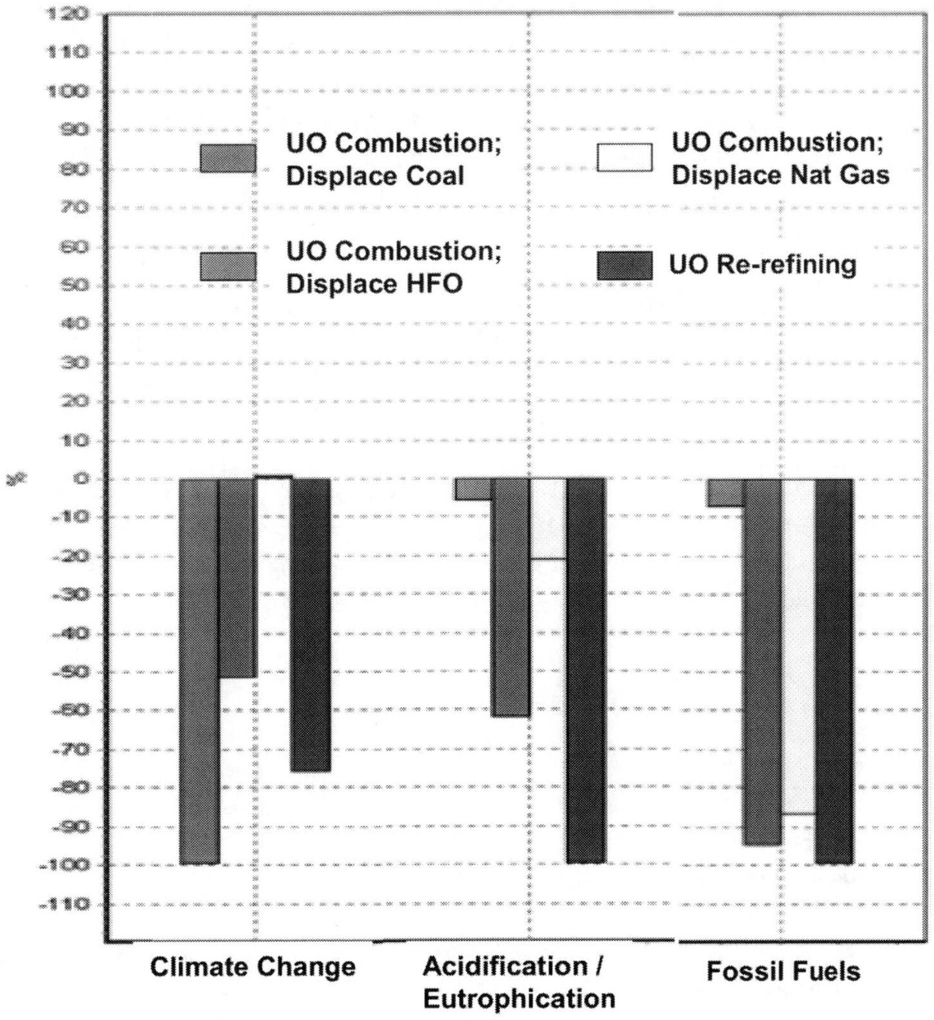

Figure 9: Environmental Impact Comparison

SUMMARY

Conventional base oil production methods are energy intensive, consume a diminishing fossil fuel resource, and place a large burden on the environment. The current trend of including higher percentages of hydrocracked base oils and/or ethylene-based synthetics in the lubricant blend further increases the overall life cycle burden of the finished product. Spent lube oils containing large quantities of high viscosity index and low pour point base oil represent a valuable resource which should be properly managed. While simple blending of used oils with low quality fuels will recover the energetic value of this material, the latent value of an engineered material containing very low aromatics and waxes is lost.

Historically, the great majority of used lube oil has been routed into low grade fuel. In order for that to change, a sustainable oil recycling process is needed. For such a process to be sustainable, the social and economic aspects of a given project must complement the environmental benefits. Factors that support re-refining as a more sustainable option for used oil management include contributions to (a) local job creation, (b) regional/national energy security, and (c) competition in the lubricant market by replacing foreign production with local production. However, the key enabler for re-refining is a cost-effective, eco-friendly technology which is capable of consistently producing valuable high quality products.

As automotive engine oil standards become more stringent and crude oil quality declines, production costs for base oil will increase. This trend coupled with more stringent environmental standards creates an opportunity to expand used oil recycling efforts worldwide. The successful commercialization of the first HyLube unit by Puralube GmbH represents an important step forward in the quest for a sustainable lube oil re-refining process.

Puralube's is now moving forward with a project to build a second HyLube unit in Germany for startup in 2009. Their longer term vision is to become the pre-eminent refiner of used oils in the world based on the UOP HyLube Process with a target to install 1.5 mm mta of processing capacity. Activities are already underway to expand the waste oil collection side of the business.

References

Frankl, P., Fullana, P., Baitz, M., 2005, Europe Life Cycle Considerations on Waste Oils and Implications for Public Policy - Ecobilancio Italia, ESCI

Fehrenbach, H. 2005, Ecological and energetic assessment of re-refining used oils to base oils: Substitution of primarily produced base oils including semi-synthetic and synthetic compounds, Institut für Energie- und Umweltforschung GmbH (IFEU), a study commissioned by GEIR - Groupement Européen de l'Industrie de la Régénération.

Hartmann, C., 2005, Clean, Competitive, and Sustainable: Why the Recycling of Waste Oils Must Remain an EU Policy Priority, GEIR (Groupement Européen de l'Industrie de la Régénération), position paper, Square Marie Lousie 49 – 1000 Brussels High Performance Automotive Lubricants, 1995, Chem Systems, May 1995, 93S4

Kalnes, T., Shonnard D., Schuppel, A., 2006, LCA of a Spent Lube Oil Re-refining Process, 9th International Symposium on Process Systems Engineering, (Marquardt et al editors) © 2006 Elsevier B.V./Ltd.

START-UP EXPERIENCE FOR VALUE RECOVERY FROM CCR NET GAS AT YANBU' REFINERY

Saleh M. Al-Shahrani, Saudi Aramco, Yanbu', Saudi Arabia
Yuv R. Mehra, Saudi Aramco, Dhahran, Saudi Arabia

Abstract

Yanbu' Refinery (YR) recently converted its Semi-Regenerative Platformer unit to the UOP-licensed Continuous Catalytic Regenerative (CCR) Platforming process. To recover value from CCR net gas, an LPG Recovery Unit, utilizing a new to Saudi Aramco technology, was installed to maximize LPG recovery and to increase the hydrogen purity of the net gas for use in the new Diesel Hydro-Treater unit. This paper will describe how the enhanced absorption process was incorporated at YR to complement the existing Saturate Gas Conditioning Unit, and share the experience with the start-up and operations of the LPG Recovery Unit. The recovered, higher-value LPG is separated in the existing debutanizer column. A new three-stage propane refrigeration system is used to chill the CCR net gas to maximize LPG recovery.

Introduction

The original fixed-bed platformer unit has been operational at Yanbu' Refinery (YR) since 1983. With a view to: (a) increase the unleaded gasoline production by 8 MBD; (b) upgrade YR product value by allowing light straight run naphtha and butanes to blend in the gasoline pool; (c) reduce requirement for expensive MTBE purchases; and (d) improve reliability, safety, and efficiency by reducing unit downtime, YR elected to transform the 38 MBD capacity UOP fixed-bed Platformer to a 40 MBD capacity UOP-licensed Continuous Catalytic Regenerative (CCR) Platformer system.

As part of the reforming process, the CCR system produces a net gas that is rich in C_3+ LPG and hydrogen. To maximize propane recovery from excess net gas and existing Sat Gas debutanizer overhead, the licensed LPG Recovery System, from Advanced Extraction Technologies, Inc. (AET), was selected because of its effectiveness in achieving very high propane recoveries (96+%) from about 10 kg/cm^{2}g *(151 psig)* off-gases without requiring feed gas compression. This paper presents the incorporation of new-to-Saudi Aramco technology and shares the start-up challenges and experiences with the AET LPG Recovery Unit at YR that started up during June-July 2006.

LPG Recovery Technology

The typical AET Process LPG Recovery Unit technology, as shown in Figure 1, utilizes a C_5+ solvent for absorption of C_3+ LPG components from a refinery off-gas stream. The selection of C_5+ solvent is based on the fact that it has low molecular weight, which reduces the solvent circulation, and it has a low vapor pressure, which reduces the solvent loss.

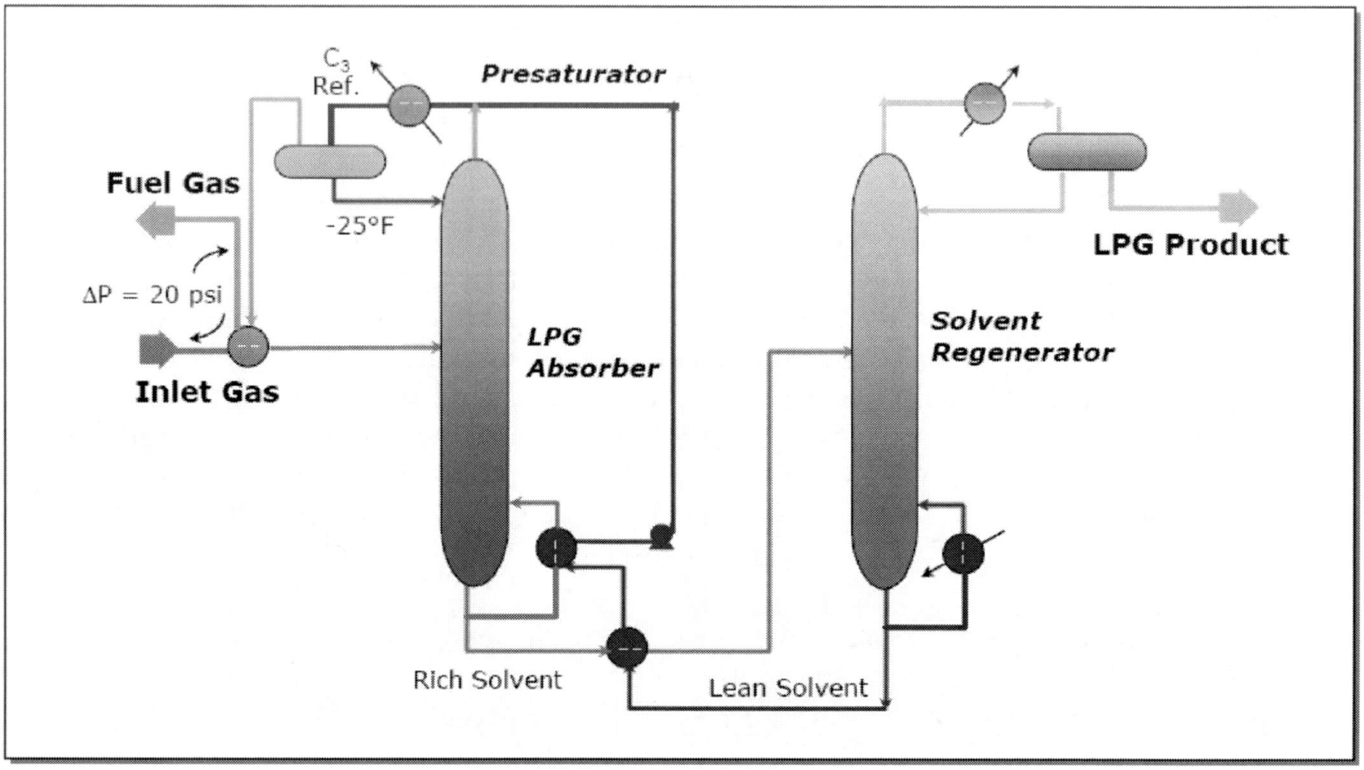

Figure 1 - AET Process LPG Recovery Unit [1]

In the AET Process, this absorption effectiveness is further enhanced by pre-saturating the lean C_5+ solvent with the "*undesired*" hydrogen and methane present in the inlet refinery off-gas that essentially forms the fuel gas, and is available at the overhead of the LPG absorber since the propane and butanes are removed from the off-gas inlet gas in the absorber column. The pre-saturated solvent at about -25°F, separated from the hydrogen/methane fuel gas, is pumped to the top of the absorber. This hydrogen/methane pre-saturated solvent upon entering the absorber focuses on absorbing only the propane and butanes.

As shown, the LPG absorber is reboiled at the bottom to ensure that the recovered propane-plus LPG meets the specification content for all light-ends present in the inlet refinery off-gas. The rich solvent from the bottom of the LPG absorber is fractionated in the solvent regenerator to separate the absorbed C_3/C_4 LPG as an overhead product and the bottoms being the C_5+ lean solvent product. By virtue of separation of C_3/C_4 LPG overhead, the solvent regenerator operation is similar to that of a debutanizer column. Therefore, any new AET process facility requires an LPG absorber column and a debutanizer-regenerator column.

[1] Adapted from AET web site: www.aet.com

LPG Recovery at Yanbu' Refinery

Since YR already had a debutanizer column as part of its Sat Gas Concentration Unit V14, alternatives were considered to eliminate the need for installing a new debutanizer column just for the new LPG absorber. Instead of securing a new segregated C_5+ stream for the lean solvent service, the stabilized platformate, which is a C_5+ product, available from the bottom of the existing debutanizer column was considered as the preferred lean solvent for the AET Process. Unfortunately, the solvent circulation rate of the stabilized platformate as lean solvent to achieve the 96+% propane recovery from the expected CCR net gas flow was large enough to require a larger diameter debutanizer column.

To utilize the existing debutanizer column in place, a part of the unstabilized platformate feed to the debutanizer was re-directed to the new LPG absorber to absorb some of the LPG from the CCR net gases. The remaining LPG present in CCR gases was absorbed by circulating a part of the stabilized platformate from the bottom of the existing debutanizer as the lean solvent for the LPG absorber. Doing so essentially splits the total C_5+ solvent requirement for the desired high propane recovery. The challenge of using the unstabilized platformate feed, which is essentially saturated with propane and butanes, as a bulk solvent for part of LPG absorption through the new LPG absorber, was overcome by chilling the unstabilized platformate to a significantly lower temperature of -29°C. The LPG absorber configuration at YR CCR system is shown in Figure 2.

To minimize platformate loss from the overhead of the absorber, and to increase the LPG absorption efficiency, the stabilized platformate solvent stream is pre-saturated with hydrogen, methane and ethane by mixing with the overhead gases from the LPG absorber and chilling to about -29°C *(-20°F)*. The chilled gases are separated from the chilled solvent stream in a presat separator. The pre-saturated lean solvent is pumped to the top of the LPG absorber column.

The unstabilized *(bulk solvent)* and stabilized debutanizer bottoms *(lean solvent)* Platformate streams are precooled by exchanging cold energy from the absorber bottoms stream and then chilled by propane refrigerant to -29°C *(-20°F)*. The CCR net gas is combined with the recycled overhead gases from the debutanizer and deethanizer columns to form the combined feed to the LPG absorber. The combined gas stream is first precooled by exchanging cold energy with the propane-free light-end gases *(hydrogen, methane and ethane)* and then chilled to -29°C *(-20°F)* before entering the LPG absorber column near the bottom. The upward flowing chilled gases are counter-currently contacted with the downward flowing pre-saturated lean solvent which absorbs the propane-plus components from the gases. The cold rich solvent from the bottom of the LPG absorber is pumped and warmed against the unstabilized and stabilized Platformate solvent streams and then flows to the unstabilized Platformate feed stream to the existing debutanizer joining at a point downstream of the take-off point for the unstabilized solvent.

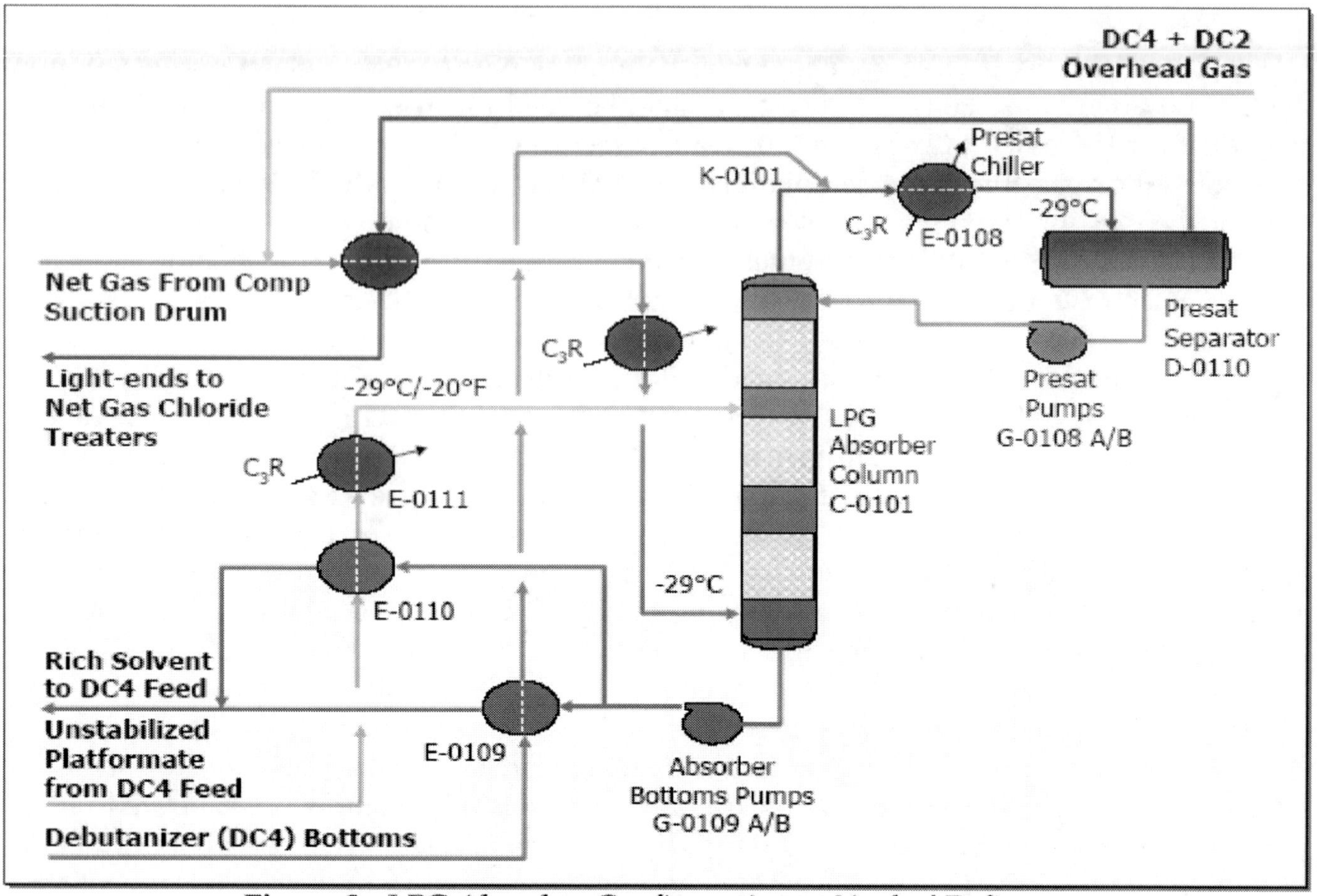

Figure 2 - LPG Absorber Configuration at Yanbu' Refinery

As may be noted by comparison of Figures 1 and 2 that the LPG absorber incorporated at YR does not have a reboiler. This simpler absorber design takes advantage of the availability of a deethanizer column as part of the Sat Gas Concentration Unit, which takes feed from the overhead of the debutanizer column to meet the ethane specification of the C_3/C_4 product.

As shown in Figure 3, at the debutanizer overhead the uncondensed light-ends are separated from the condensable C_3/C_4 mixture. The overhead net liquid C_3/C_4 product is further fractionated in deethanizer column followed by depropanizer column to produce propane and mixed butanes products.

The separated light-ends from the overhead of the debutanizer (DC4) column along with the separated ethane and lighter from the overhead of deethanizer (DC2) column are recycled back to the LPG absorber of Figure 2 to recover any contained propane and heavier hydrocarbons.

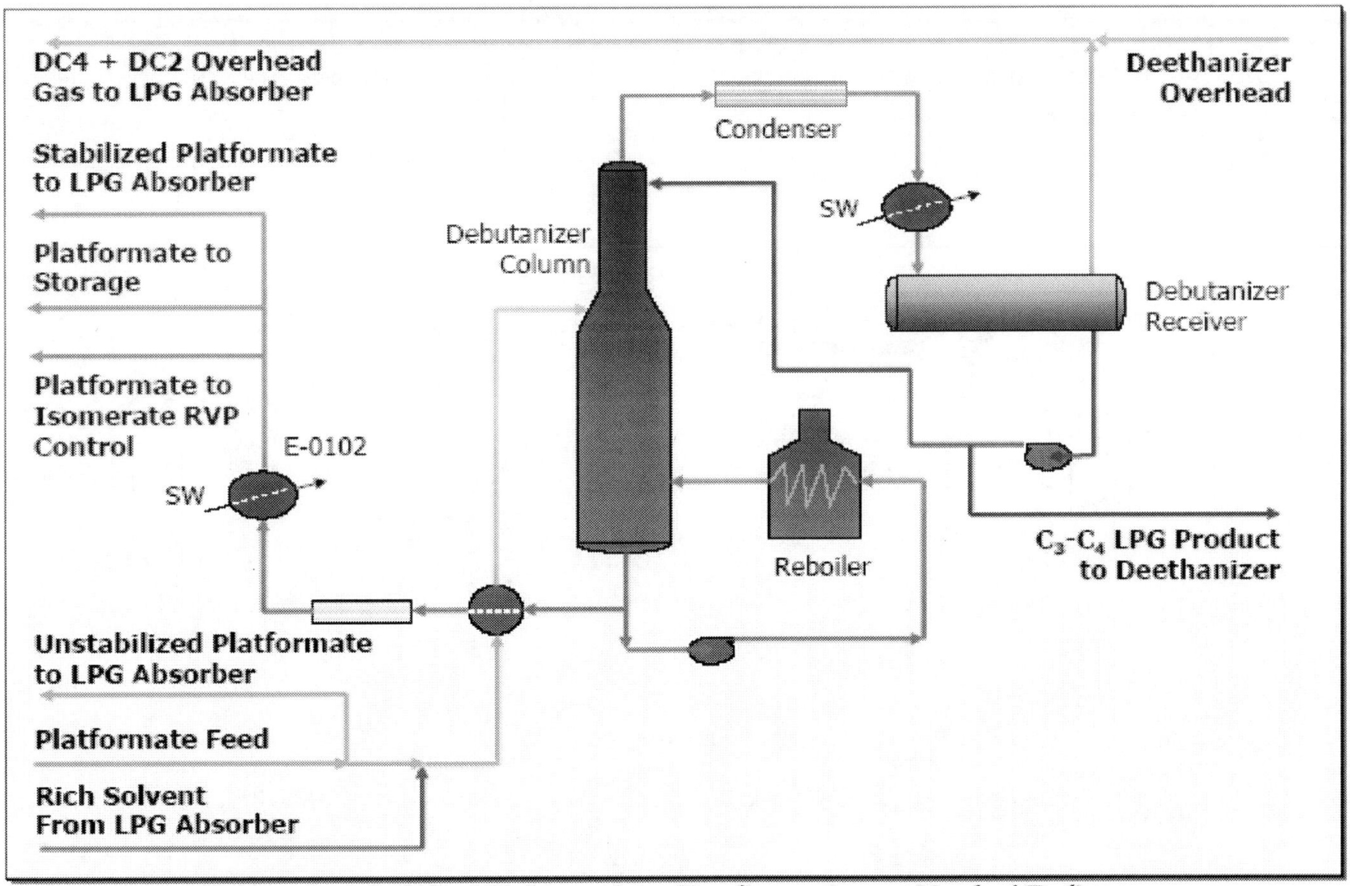

Figure 3 - Modified Debutanizer Configuration at Yanbu' Refinery

To handle increased processing load within the existing debutanizer column, modifications to existing equipment were as follows:

- Addition of series heat transfer surface to the feed/bottoms exchanger.
- Addition of parallel heat transfer surface to the bottoms product sea water cooler.
- Addition of new overhead sea water trim condenser.
- Replacement of overhead product and reflux pump.

Within the existing deethanizer system, the deethanizer feed pumps were replaced. Within the depropanizer system, the overhead air cooled condenser was replaced with a sea water shell and tube condenser, and the overhead net propane product cooler was replaced with a larger heat exchanger.

Propane Refrigeration Cycle

To provide the required chilling a propane refrigeration system is required. Figure 4 summarizes the closed loop three-stage propane refrigeration cycle to provide the necessary chilling for the unstabilized and stabilized Platformate solvent streams and the CCR net gas feed to the LPG absorber.

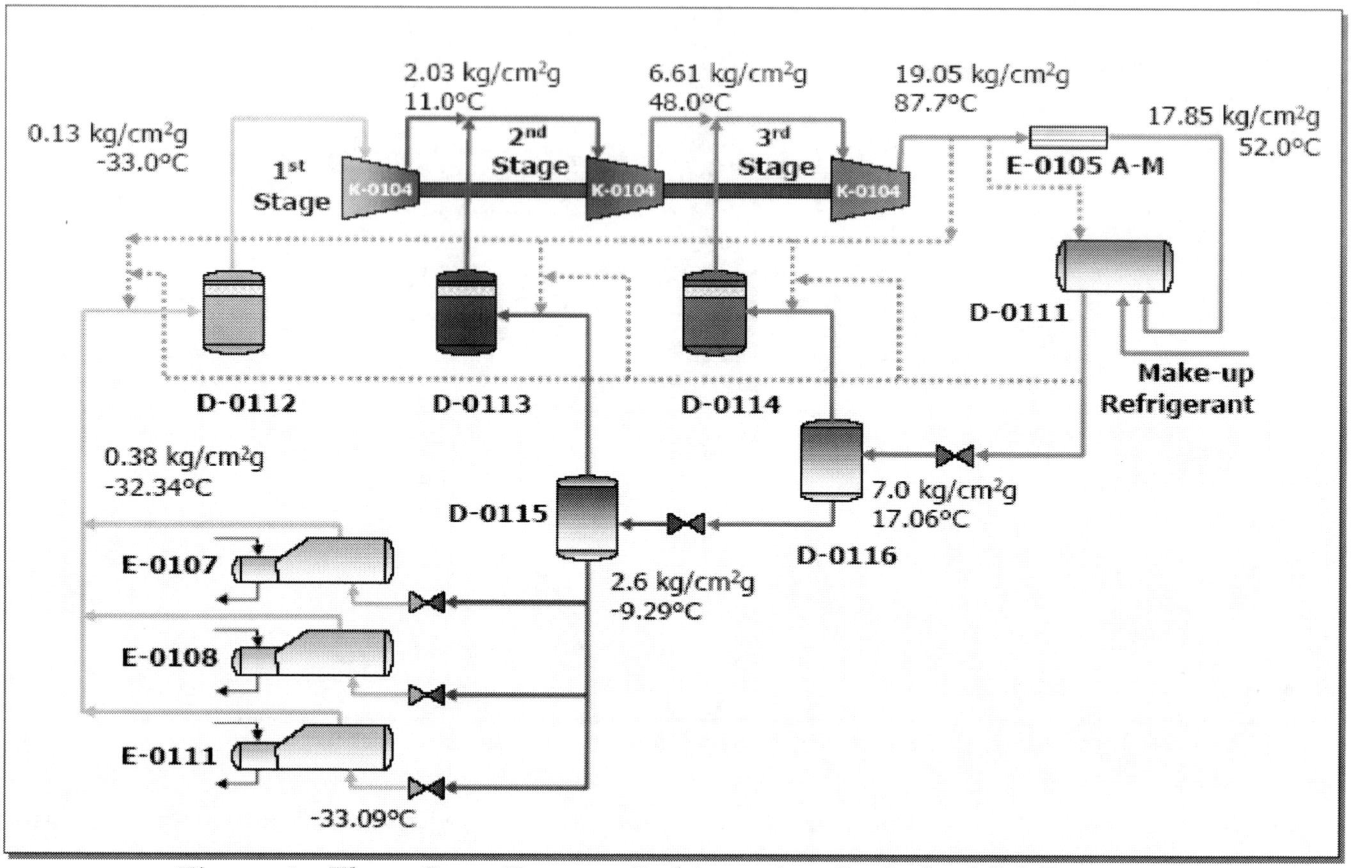

Figure 4 – Three-Stage Propane Refrigeration Cycle for LPG Recovery

All of the three process services, namely the Feed Gas Chiller E-0107, Platformate Chiller E-0111 and the Presat Chiller E-0108 chill the respective process streams to about -29°C *(-20°F)* using the -33°C *(-27°F)* propane refrigerant. The vaporized propane from the chillers enters the 1st stage of the Propane Compressor K-0104 through the 1st stage suction drum D-0112. The LP Economizer D-0115 and HP Economizer D-0116 reduce the refrigerant compressor horsepower by separating the 2nd stage side load flashed vapors at -9°C *(15°F)* and 3rd stage side load flashed vapors at 17°C *(63°F)*, respectively. The compressed propane refrigerant vapors are condensed by air cooled condenser E-0105 and returned to the propane accumulator D-0111.

As shown via the dotted line in Figure 4, the refrigerant compressor's anti-surge system controls the flow of the hot discharge propane and the quench liquid propane to the 1st, 2nd and 3rd Stage Suction Drums D-0112, D-0113 and D-0114, respectively.

Overall Design Performance

Because of the integration of the LPG absorber *Unit V11* with the debutanizer/deethanizer/depropanizer columns of the Sat Gas Concentration *Unit V14,* it is quite cumbersome to ascertain the material recovery performance of the LPG absorber system. By drawing an envelope around the integrated system, the LPG recovery is determined by analyzing the two gas streams summarized in Table 1.

Table 1 – Overall Material Performance

Component	Combined CCR Net & Unit V14 Gases		Product Gas to DHT/Fuel Gas		Products Yield	
					Gas Product	
	Kg-mol/h	*Mol%*	*Kg-mol/h*	*Mol%*	*Recovery, %*	*Flow, Nm³/h*
Hydrogen	2,170.20	73.62	2,168.31	88.63	99.9	48,601
Methane	107.15	3.63	104.21	4.26	**Liquids Products**	
Ethane	245.38	8.32	165.84	6.78	Recovery, %	Flow, US BPD
Propane	239.75	8.13	3.37	0.14	98.6	3,277
Iso-Butane	52.04	1.77	0.22	0.01	99.6	853
n-Butane	62.29	2.11	0.79	0.03	98.7	951
Iso-Pentane	22.91	0.78	2.29	0.09	90.0	380
n-Pentane	12.62	0.43	0.70	0.03	94.5	217
C_6+	35.41	1.20	0.64	0.03	98.2	719
TOTAL	2,947.75	100.00	2,446.37	100.00		6,398
Flow, Nm³/h	69,700		57,845			

Pre-Commissioning Activities

This unit was constructed as part of the YR platformer revamp project BI-3538. Upon completion of the mechanical construction of this unit, the following pre-commissioning activities had been undertaken:

1. A total of 120 hydro-tests were completed.
2. A total of 280 instrumentation loop were checked.
3. The eight (8) refrigeration loops were chemically cleaned.
4. Propane compressor lube oil system had been flushed.
5. A total of 31 equipment were boxed up.
6. A total of 22 motors were unclipped and run for 4 hours.
7. The instrumentation connection had been completed reinstated.

Subsequent to the completion of above pre-commissioning activities as part of mechanical completion, additional start-up pre-commissioning activities were as follows:

1. The refrigerant and lube oil circulation system was purged, evacuated and dried out with nitrogen to -40°C dew point.
2. The absorber system was purged and dried out by nitrogen to -40°C dew point.
3. All control valves where stroked and functionally tested.

4. All PZV were commissioned.
5. Refrigeration loop was filled with propane and samples collected to confirm the system dew point to be < -40°C.

LPG Recovery Process Start-up

Having completed all the commissioning activities as described above, and with the previously established operation of the debutanizer column which was producing a C_5+ Platformate product meeting the C_4 and lighter content specification, it was determined that the available initial CCR net gas flow on June 18, 2006 was about 33,000 Nm^3/h *(29.6 MMSCFD)*. Therefore, first the flow of unstabilized platformate *(bulk solvent)* from the feed to the debutanizer column was initiated at the rate of about 38 m^3/h *(167 US gpm)*. Once the level in the bottom of the LPG Absorber Column C-0101 was established at about 60% level transmitter (LT) range, the Absorber Bottoms Pump G-109B was started to circulate the bulk solvent through the two solvent cross-exchangers back to the debutanizer column feed.

After about an hour of the established circulation for the bulk solvent, flow of the debutanizer bottoms *(lean solvent)* stream was initiated at the rate of about 34 m^3/h *(150 US gpm)*. After filling the piping and exchangers, the lean solvent level began to increase in the Presat Separator D-0110. When the level was established at about 60% of the LT range, the Presat Pump G-0108A was started to transfer the solvent to the top of the absorber column. Within 20 minutes from the start of the G-0108A pump, level in the bottom of the absorber started to increase, thereby indicating that the two solvent circulations loops had been completed. The level transmitter controlling the flow from the discharge of the absorber bottoms pump G-0109B was switched to auto control.

With no gas flow through the absorber system, upon establishment of the lean solvent circulation, the column pressure started to decrease from an initial pressure of about 10 kg/cm^2g *(142 psig)* to about 8.6 kg/cm^2g *(122 psig)*. This indicated slight absorption of the hydrocarbons even at warm solvent temperatures on the order of 37°C *(98°F)*. To re-establish the pressure within the column to the CCR net gas compressor suction drum pressure without flowing gas through the LPG recovery unit, the inlet isolation valve was re-opened. The isolation valve blocking the flow from the LPG absorber overhead to the fuel gas system remained closed. With the circulating warm solvent the pressure increased and stabilized at about 9.8 kg/cm^2g *(139 psig)*.

While establishing the aforementioned two solvent circulation circuits several level and flow instrumentation calibration issues were identified.

- ***First Circuit***: between debutanizer feed → cross-exchanger E-0110 → platformate chiller E-0111 → LPG absorber C-0101 → absorber bottoms pump G-0109B → cross-exchangers E-0110 & E-0109 → debutanizer feed
- ***Second Circuit***: cool debutanizer bottoms from E-0102 → cross-exchanger E-0109 → mixer K-0101 → presat chiller E-0108 → presat separator D-0110 → presat pump G-0108A → LPG absorber C-0101 → absorber bottoms pump G-0109B → cross-exchangers E-0110 & E-0109 → debutanizer feed

Before attempting to start the propane refrigeration system, all instrumentation calibration issues were addressed over the following 24 hours.

Subsequently, the flow of CCR net gas through the LPG recovery process was initiated by opening the isolation valve at the outlet of the absorption system and closing the bypass isolation valve around the LPG recovery unit.

Propane Refrigeration System Start-up

With the established cooling load through the tube side of three chillers, start-up of the refrigeration system could now be commenced. The unfamiliarity of the refrigeration cycle within the refining industry as a whole was a significant challenge for the operations team in developing the feel for starting and controlling the three-stage, closed loop propane refrigeration system.

Table 2 summarizes the twelve compressor trips and the respective lessons learned that eventually resulted in a successful start-up of the three-stage centrifugal compressor and the closed loop refrigeration cycle at YR.

While the comparable quench-based suction drum configuration is quite extensively used at Saudi Aramco's gas processing facilities, one of the recurring problem in the start-up at YR had been managing and controlling the liquid levels in the various suction drums.

The hydrocarbon processing industry at large uses significantly sized refrigeration systems without the need to add quench liquid into the compressor suction drums by simply routing the anti-surge hot gas from the compressor discharge through a sparger line below the tube bundle in kettle chillers and in the economizer drums where liquid refrigerant is normally present. Doing so not only avoids the potential imbalance between the anti-surge gas flows and quench liquid temperature control valves, it also assures that the compressor suction drums remain dry and thus prevent compressor trips from high liquid level. Such a system at YR would have significantly simplified the start-up of the refrigerant compressor.

Throughout the compressor start-up attempts, starting flow of refrigerant vapors as quickly as possible into the refrigerant compressor to keep the discharge temperature within limits, and trying to close the anti-surge gas and liquid quench flows to reduce overloading the compressor motor was emphasized. A sudden realization after the twelfth trip that the inlet feed gas from the CCR unit was only about ⅓ the design gas flow and analysis of the compressor surge flows revealed that as long as the inlet gas flow from the CCR unit was less than 88% of the design flow for the LPG Recovery Unit, the installed refrigeration compressor will always have its anti-surge valve open.

This paved the way to change the start-up focus from trying to close the anti-surge valve to managing the introduction of available refrigerant vapors generated by the process streams through the chillers. This was the final straw that led to the successful start-up of the refrigeration system and within about five hours of continued integrated operation of the process-refrigeration systems, all temperatures reached their expected settings.

Table 2 – Propane Compressor K-0104 Start-up Trip History & Lessons Learned

TRIP NO.	TRIP TYPE OR SYMPTOM	LESSON LEARNED OR FOLLOW-UP ACTION TAKEN
1	High Vibration	Bypass trip setting during initial start-up.
2	High Amperage Load	Watch amperage indication to prevent overloading the compressor motor.
3	High-High Level 1st stage suction drum	Fix and calibrate LTs and FTs for accurate display on DCS screen.
4	High Discharge Temperature	1. Bypass current trip setting of 95°C and watch discharge temperature to not exceed vendor recommended limit of about 150°C. 2. Change start-up procedure to introduce chiller vapors into the compressor as quickly as possible.
5	Loss of Level Indication in Suction Drums and trip from High-High Level	Refill LTs with glycol and place plugs to prevent loss of fluid in LT legs.
6	Loss of suction pressure	DO NOT start compressor with isolation valves (ZV) OPEN on discharge and three suction stages.
7	High Discharge Pressure from cascading discharge temperature with discharge pressure	Avoid cascading discharge pressure and temperature until unit is stabilized and close in desired settings and the cascading configuration is checked and fixed.
8	Low suction pressure	Avoid imbalance from opening 5% of suction PV when compared to opening 30% discharge PV. Increase suction and discharge openings in equal increments.
9	Amperage overload from LP Economizer PV opening from 5% → 33%	Avoid rapid opening of LP and HP Economizer PVs.
10	3rd stage suction drum High-High level	Unreliable level transmitter operation. Avoid compressor operation until level indication is reliably displayed on DCS screen.
11	1st stage suction High-High level due to stuck 1st stage anti-surge valve and loss of suction pressure trip	1. Check, fix and recalibrate the 1st stage anti-surge valve and assure its operation. 2. Do not open drain valve too quickly. 3. Assure flow from 2" hot sparger gas line in suction drums to assure imbalanced quench liquid is vaporized.
12	1. 3rd stage anti-surge valve stuck…3rd stage suction High-High level 2. 1st stage anti-surge valve stayed open throughout	1. Fix solenoid and recalibrate the 3rd stage anti-surge valve functionality. 2. Recognize that the anti-surge valves will remain open at inlet gas flow from CCR < 88% of design flow to the LPG Recovery unit.
SUCCESSFUL START-UP		

A representative DCS display snapshot of the process flow is shown in Figure 5 and that of the refrigeration system in Figure 6.

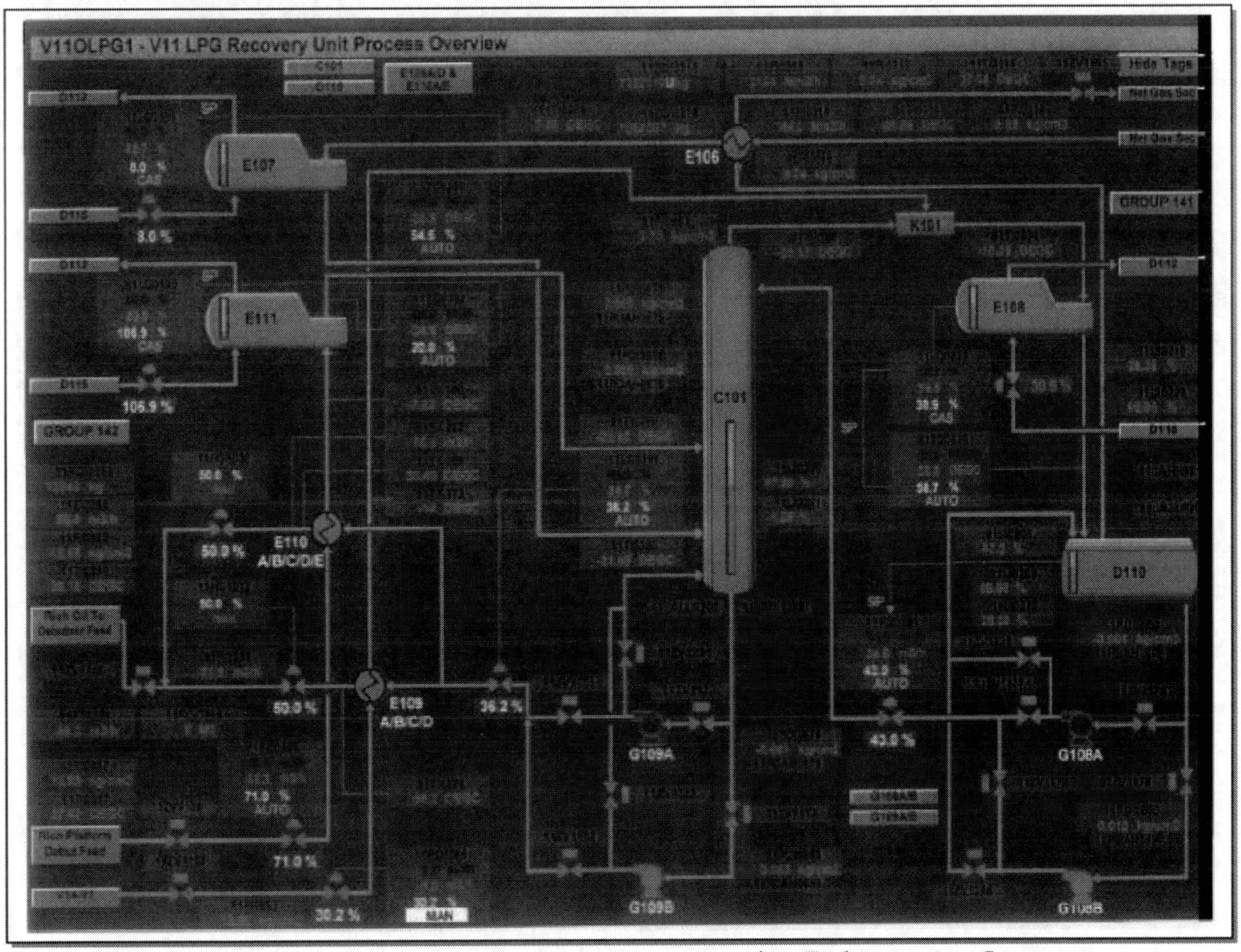

Figure 5 – LPG Recovery Unit Process Overview after Refrigeration Start-up

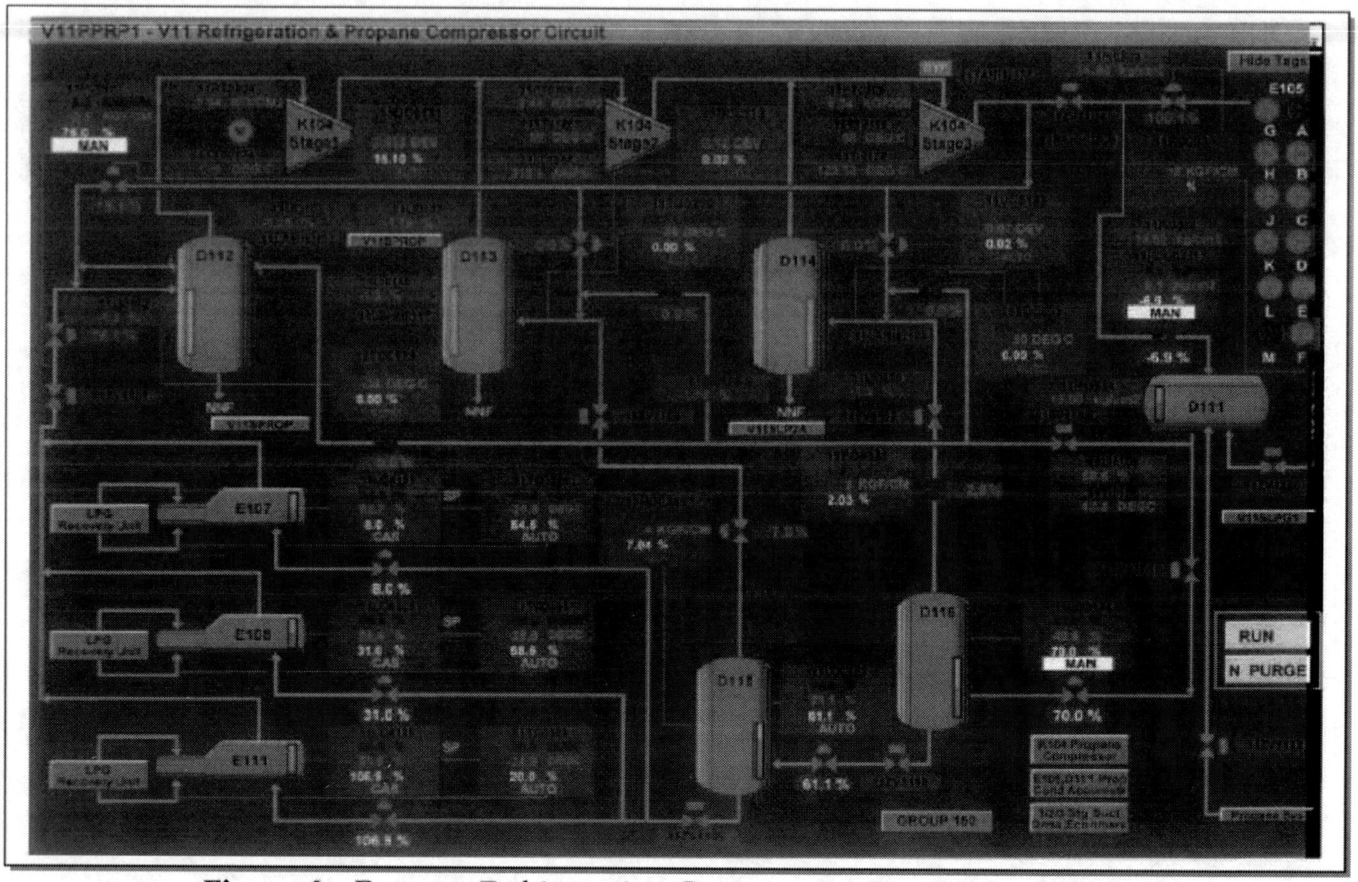

Figure 6 – Propane Refrigeration Compressor Circuit after Start-up

Conclusions

The new-to Saudi Aramco AET LPG Recovery Process was successfully incorporated at Yanbu' Refinery (YR) to utilize the Gas Concentration Unit's existing debutanizer and deethanizer columns in place to maximize LPG recovery from CCR net gas. Due to unfamiliarity of the refinery personnel with the closed loop, multistage propane refrigeration cycle, steady learning from each compressor trip led to eventual successful start-up of not only the propane refrigeration compressor but also the integrated process-refrigeration system. Most of the compressor trips were related to high liquid level in compressor suction drums caused by the imbalance of the anti-surge gas flow and its associated liquid propane quench flow. The operation of the integrated LPG recovery facility is very easy, stable, and flexible to gas flow and composition changes. Should the refrigeration compressor accidentally trip, the LPG recovery process can continue to operate stably and await restart of the refrigeration compressor. The operation of the YR unit has successfully demonstrated turndown to 30% of the design inlet gas flow capacity. With this start-up, YR has successfully pioneered the incorporation of colder temperatures within a refining environment to maximize recovery of value from its hydrocarbon resources.

OPTIMIZING CRACKER OPERATIONS: Selection of Downstream Configuration

BY

Osman Aboul-Nasr
Chemical Technology Center
Fluor
Greenville, SC 29607-2762

Prepared for Presentation at AIChE Annual Spring Meeting
Houston, TX
April, 2007

Session TA006
Advances in Petrochemical Production - I

"Unpublished"

Optimizing Cracker Operations:

Selection of Downstream Configuration.

Extended ABSTRACT

The increase in the price of crude oil over the past couple of years has led to a rapid increase in the investment in refineries, crackers and downstream petrochemicals, both in the Middle East and China. In the Middle East, governments have been encouraging private investors to take advantage of the cheap crude and natural gas to improve national economy. In the meantime, China continues to provide a cheap source of labor, as well as a market for the petrochemical intermediates and products.

The investment regulations in the Middle East stipulate, amongst other rules, that the products must have high value and lead to significant impact on employment in the country. As an example of the first instance, commodity polyethylene is not a product of first choice, while specialty grades are. As far as the national impact, the product slate should lead to secondary products. As an example, it is not acceptable to produce butadiene, but a significant portion be directed at SBR and butadiene rubber, as a prelude to a local tire industry. These, and other criteria, are used to screen applicant for a limited amounted of feedstock.

For an applicant to secure the feedstock, they must prove the economic viability as well as the superiority of the product slate. An applicant must therefore consider a large number of configurations and product selection.

Fluor developed proprietary software to assist the client in optimizing the selection of the petrochemical units, downstream of a cracker by quickly providing the investment and economic parameters for the product slate. It is a powerful pre-screening tool that was successfully used in several cases.

The software uses non-confidential data on the material balance and utility consumption for most cracker technologies, of sufficient accuracy to provide a basis for downstream processing. Its database contains standard, non-confidential conversion and utility consumption for over 500 product/licensor combination. In addition, it has investment data based on non-confidential data of the ISBL an OSBL for each unit.

The basic information about the project includes the following:

1. Location, necessary to estimate the investment cost
2. Project schedule, necessary to determine the escalation and interest during construction
3. Preliminary product slate, wants, needs and musts
4. Feed and product prices.

Other information can be developed in cooperation with the client, or using non-confidential information. Fluor compiles its own pricing information on feedstocks, intermediates and commodities from open literature

Confidentiality:

In the pre-screening stage, the client will have not executed confidentiality agreements with licensors or technology suppliers. Fluor will use only non-confidential information in its databases, as well as open-literature data. In addition, Fluor will use technology neutral estimates in the pre-screening stages. This approach ensures that all doors remain open for the clients to explore as many configurations as possible.

Approach

The approach used is fairly straightforward. First, the feedstock is defined. The state of the feedstock composition is more important than its chemical composition. At this stage, the client would know only whether the feed is gas, naphtha, propane, etc., which is enough for the accuracy at this stage. It is possible to fine-tune the cracker product by adjusting the severity of cracking, for instance. For an Aromatics Complex, the client will not know the exact feed composition, since it will come from a refinery that is yet to be designed. Should the client have a well-defined cracker feedstock, the product will be better defined, but still allows for some fine tuning by selecting the proper severity, technology, recycles, etc.

Once the cracker output is determined, the disposition of the product streams will be determined. For instance, ethylene can be allocated between various product chains: PE, EG, EO derivatives, PVC, styrenics, etc. The overall choice of these products will be dictated by the client's business strategy. The software will provide different options within these broad requirements, that is, how much ethylene would go into each chain, and the economic parameters associated, investment (+/- 20%), and IRR (+/- 3%).

The capacity of each product line is then determined. This determines the investment and the operating cost associated with this product. The software will recommend the number of lines required to meet that demand. In certain instances, specific technologies have different capacity limitations. This is one of the few instances where a specific technology or licensor are recommended. Another instance of specific licensor recommendation is when the client is committed to a particular product.

Special Features

Should multiple lines be needed, the potential for capital cost savings will exist. These savings can be achieved in engineering and bulk procurement. The software has the capability to include such capital cost saving allowances. Similar savings will be possible in the OSBL for multiple lines.

The initial estimate Infrastructure and Utilities for each unit are estimated on a stand-alone basis. The software has the capacity to include savings for the scale factor and synergy for the whole complex.

The Owner's Cost is a significant part of the investment. It is estimated as a fraction of the TIC.

The project duration is critical in estimating the interest during construction, which can be as much as 5-15% of the total capital invested. The software can accommodate project schedules of 2, 3 and 4 years, with standard drawdown schedules.

Output

The software will provide the following information

1. Cracker output.
2. Annual capacity of each product.
3. Number of lines, and technology, if product specific
4. The investment details: ISBL and OSBL for each product, I,U&O for the complex, and Owner's Cost.
5. Working capital
6. NPV and IRR

Example

The Presentation will include a demonstration of the software on a typical pre-feasibility study.

Please contact the author for further details.

Contact Information

For information about the software, please contact
Dr. Osman Aboul-Nasr
Chemical Technology Center
100 Fluor Daniel Drive
Greenville, SC 29607
USA
1.864.281-5167

at osman.aboul-nasr@fluor.com

AICHE

2007 SPRING NATIONAL MEETING

CONDENSING AT VERY LOW PRESSURE WITH VERY LOW ALLOWABLE PRESSURE DROP, LOW LMTD AND LARGE DUTY REQUIREMENTS. A SUCCESSFUL SOLUTION IN VACUUM DISTILLATION COLUMNS FOR PETROCHEMICAL PRODUCTION.

BY
SYLVAIN BENEZECH
PHILIPPE HEBERT
FRANCOIS REVERDY
(ALFA LAVAL)

Why ZIEPACK?

The development of ZIEPACK heat exchangers was meant to fill a technology gap on the market place, namely the lack of large duty, ultra low pressure drop evaporators, condensers and reboilers that would be compact enough to fit directly inside distillation columns. Distillation towers operating under vacuum conditions in the petrochemical industry were the primary target.

The economic advantage of ZIEPACK's concept seemed obvious at the time:

- **Reduction of capital cost, plot space, and plant complexity via elimination of external components like reboilers, condensers, reflux separators, reflux pumps, piping, control valves and instrumentation.**

- **Reduction of column head (capital cost)**

- **Possibly reduction of liquid inventory in the reboiler section (Operating cost)**

- **Increased tower capacity via increased thermal duty and lower pressure drop**

- **Ideal tool for low-cost plant debottleneck, specially when space is a premium**

The target compactness was quite aggressive: Offer the same thermal and pressure drop duty as traditional S&T solution but using only 25% to 30% of the overall S&T volume. (up to 250 m^2/m3)

Pressure drop was not to be sacrificed to the need for compactness, in fact it was supposed to offer ultra low pressure drop when advantageous to the process.

Other design criteria included:

- **Heat duty up to about 25 MW**
- **Design temperature between -50°C and 500°C**
- **Differential pressure up to 30 bars**
- **Surface area 250 m^2 to 2500 m^2**
- **Materials ranging from CS to SS to SMO to austeno-ferritic alloys**

Only a welded plate type design could meet all these criteria.

MECHANICAL DESIGN

The base element: A "Double-Plate" Panel:

The double plate panel is made up of two flat plates laser welded together by a computer driven machine to a predetermined "tube layout".
Process flow passages are then formed between the two plates by inflation using high pressure water or air. The pattern of laser welds ensures that the inflation creates the desired layout of tubes.
This layout is typically a series of either straight or U-shaped independent circuits optimized for a high heat transfer coefficient.
The double plate panel is pressure resistant up to 30 bars (internal pressure). It is used to carry the cooling or heating medium of the heat exchanger.

The "Inter-Plate" Channel:

The passage between two panels forms a boiling or condensation channel. Its wavy surface optimizes heat transfer, hydraulics of boiling and condensation, and can offer ultra low pressure drop by playing with the gap between panels.

The Plate Bundle:

Hundreds of panels are stacked together to form a plate bundle. Header boxes are welded to the ends of the bundle to distribute and collect the cooling or heating medium (which could be also phase changing)
The whole bundle is inserted in a light frame which allows assembly, handling, as well as supporting inside the column.
On the shell side, a thin casing and baffle arrangement prevents any process flow by-pass around the plate bundle that would adversely affect heat transfer efficiency.

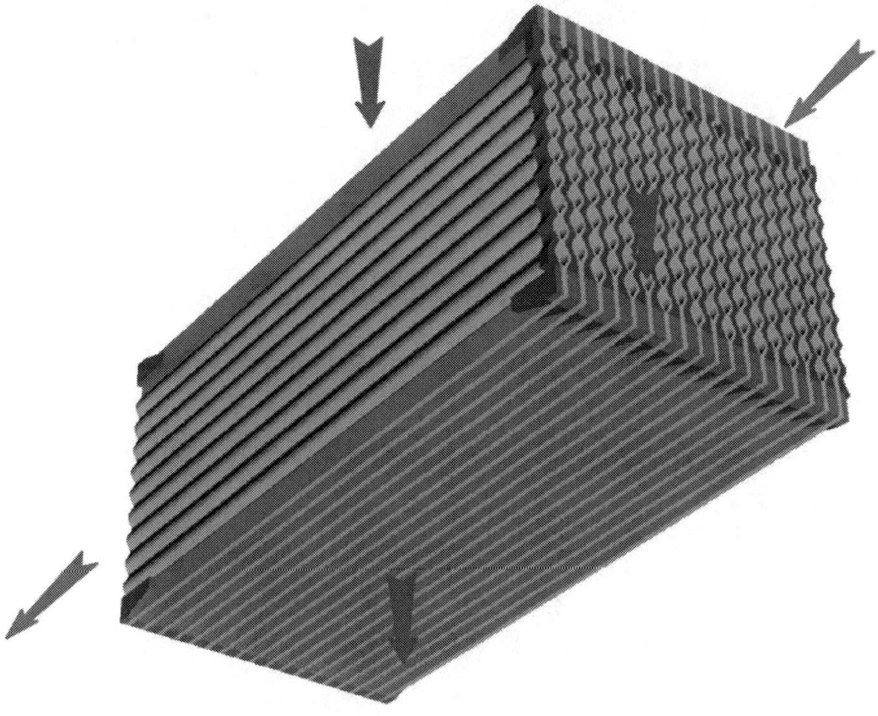

The plate bundle's simple and flexible structure provides good resistance to the thermo-mechanical stress that may occur during start up, shut down or upset conditions.

PROCESS DESIGN

Validation of the Thermal and Hydraulic Capabilities:

The theoretical thermal and hydraulic design of this unique design was ultimately verified by extensive R&D.

A full scale model was built for testing at the GRETH in Grenoble, France.

Bundle in test in GRETH Workshop

Among other things, the modeling tests allowed to verify the hydraulic limits for proper operability.
It also confirmed the expected excellent behavior of the design when the process is performed in a vacuum or requires particularly low pressure drop.

RESULTING DESIGN

The resulting design met all the initial targets (compactness, process capabilities), and the range of possible equipment sizes grew a bit:

- **Heat duty up to about 30 MW**
- **Surface area 300 m^2 to 4000 m^2**

Not all the materials of construction initially short listed are available today, but will be in the future

COMMERCIAL REFERENCES

At press time, there were 16 ZIEPACK exchangers in operation around the World, designed typically to the ASME code, with design temperature ranging from 45°C to 450°C and design pressure from 1 bar to 10.5 bar

a) SAMPLE REFERENCE "A"

YEARS IN OPERATION: 2.5

SPECIFICATION REQUIREMENTS:

DUTY SPEC (MW)	COLUMN DIAM. (m)	EXCHANGE AREA (m^2)
14.96	3.5	3 400

	Fluid (*), (**)	Flow rate	Temp. in	Temp. out	Pressure in	Pressure drop
Unit	-----	Kg/h	°C	°C	Bar (kg/cm^2 g)	mbar
Cold side	HP CW	2 662 464	33	38	0.3	700
Hot side	"Process"	100 781	44	38	-0.4	20

(*) HP CW = High pressure cooling water / Process = HC + water + inerts
(**) hot side is condensing

ON SITE MEASUREMENTS DURING ZIEPACK VISIT:

% OF SPEC RATE	PREDICTED DUTY	MEASURED DUTY
---	MW	MW
66	10.05	10.00

b) SAMPLE REFERENCE "B"

YEARS IN OPERATION: 2.2

SPECIFICATION REQUIREMENTS:

DUTY SPEC (MW)	COLUMN DIAM. (m)	EXCHANGE AREA (m^2)
12.63	3.3	900

	Fluid (*), (**)	Flow rate	Temp. in	Temp. out	Pressure in	Pressure drop
Unit	-----	Kg/h	°C	°C	Bar (kg/cm^2 g)	mbar
Cold side	BFW	150 783	168	168	7.7	700
Hot side	Hydrocarbon	153 250	185	184	1.2	210

(*) BFW = Boiled feed water
(**) hot side is condensing / cold side is vaporizing

ON SITE MEASUREMENTS DURING ZIEPACK VISIT:

% OF SPEC RATE	PREDICTED DUTY	MEASURED DUTY
---	MW	MW
100	11.32	12.52

c) SAMPLE REFERENCE "C"

YEARS IN OPERATION: 2.2

SPECIFICATION REQUIREMENTS:

DUTY SPEC (MW)	COLUMN DIAM. (m)	EXCHANGE AREA (m^2)
3.19	3.3	800

	Fluid (*), (**)	Flow rate	Temp. in	Temp. out	Pressure in	Pressure drop
Unit	-----	Kg/h	°C	°C	Bar (kg/cm^2 g)	mbar
Cold side	CW	542 000	31	36	1.03	700
Hot side	Hydrocarbon	29 950	54	37	-1.01	6.7

(*) CW = cooling water / Hydrocarbon = Hydrocarbon + inerts
(**) hot side is condensing

ON SITE MEASUREMENTS DURING ZIEPACK VISIT :

% OF SPEC RATE	PREDICTED DUTY	MEASURED DUTY
---	MW	MW
96	3.09	3.04

WHO IS ZIEPACK?

ZIEPACK was originally formed in 1997 by ZIEMANN GmbH (49%) and PACKINOX SA (51%) to design and manufacture welded plate heat exchangers for the hydrocarbon industry.

PACKINOX, a world leader in heat transfer, has extensive know-how and experience in large heat exchangers for refining, petrochemicals and gas processing. PACKINOX is recognized in a variety of other applications including Catalytic Reforming, isomerization, Paraxylene production and hydrotreating.

PACKINOX became part of the ALFA LAVAL GROUP in 2005.

ZIEMANN's dedicated technology, incorporating computer-driven laser welding and hydroforming, is applied to the production of large panel coils used in cooling vats. ZIEMANN's years of proven expertise in this field have resulted in the fabrication of tens of thousands of square meters of heat transfer area.

Investigations into aromatic thiol production: A joint theoretical, experimental and economic approach.

Bradley R. Atkinson Nicolas Rouckout and Rayford G. Anthony
Kinetics, Catalysis and Reaction Engineering Laboratory, Artie McFerrin Department of Chemical Engineering, Texas A&M University, College Station, TX 77843.

Aromatic thiols are of industrial importance due to their use in the manufacture of a wide variety of specialty chemicals. This list includes, but is not limited to, agrochemicals, polymers, pigments, dyes and pharmaceuticals [Roy, 2003; Roberts, 1997]. The thiol production reaction that involves the hydrogenation of 2, 5-dimethyl benzene sulfonyl chloride to 2, 5-dimethyl benzene thiol is focused on here. A new process has been patented for the hydrogenation reaction [Jacobsen, 2001] that eliminates the typical heavy metal salt by-product, which is an environmental, ecological and biological contaminant [Poland, et al., 2001; Tsai, et al., 2003; Piao, et al., 2003; Zerahn, et al., 1999; Bruce, et al., 2003].

As the first step in an investigation to find the transition state of the rate limiting step in the reaction system, adsorption of the 2, 5-dimethyl benzene sulfonyl chloride to a series of 4 (100) palladium matrices is studied by quantum mechanical simulation. These investigations utilized the Gaussian03 software simulation package. Density functional theory (DFT) was used for all calculations at the B3LYP level with a mixed basis set. H, C, and O utilized the 6-31G(d') basis set while the S, Cl and Pd molecules all were based with LANL2DZ sets. S and Cl molecules each took advantage of additional d-shell mathematical representations while Pd used the basic LANL2DZ basis set. Additionally, S, Cl and Pd all were restricted with effective core potentials (ECP) to ease the computational load on the computing cluster. The calculations were spin restricted to a neutral charge and singlet spins.

The creation of the palladium matrices for Gaussian03 simulation was the first step. How can the delicate balance of realistic accuracy and computational ease be reconciled on a molecular level? In testing their Pd-based reaction system for the effects of particle size and charge distribution throughout the catalyst surface, Duca and Barone simulated adsorption of various amino-nitro-benzene molecules over a layer of six Pd molecules arranged in a simple, rectangular plane [Barone and Duca, 2002]. Their further work shows a scheme for simulating a three-layered Pd matrix with only 22 atoms [Duca, et al., 2001]. The matrices used and designed in this study are shown in figure 1.

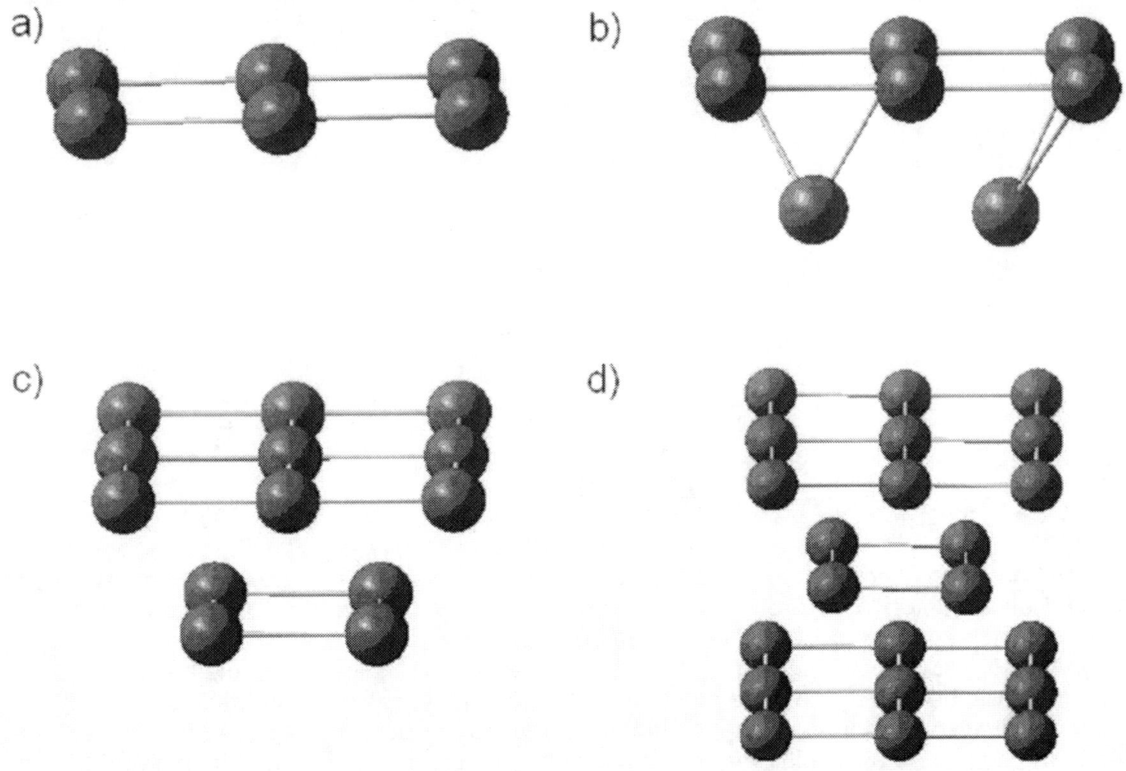

Figure 1 – A diagram of each of the four palladium structures. Each are described briefly below by letter, the enthalpies of formation shown are calculated single point energies of the locked-in structures. a) Pd6 – a single layered six atom structure, ΔH_f = -4.772 x 10^5 kcal/mol. b) Pd8 – a two layered, 8 atom structure in a dual inverted pyramid scheme 6 atoms over 2 atoms, ΔH_f = -6.363 x 10^5 kcal/mol. c) Pd13 – a two layered, 13 atom structure in an inverted pyramid scheme 9 atoms over 4 atoms, like two Pd8s side by side, ΔH_f = -1.034 x 10^6 kcal/mol. d) Pd22 – a three layered, 22 atom structure in an hourglass scheme 9 atoms over 4 atoms over 9 more atoms, like Pd13 with a extra layer on the bottom, ΔH_f = -1.75 x 10^6 kcal/mol.

Possible adsorptive modes were investigated over the simplest palladium matrix for ease of calculation. All of the aromatic molecules showed approximately the same behavior: a backing away from the palladium matrix and a rotation into a "sulfonyl chloride down" position. These rotations from the original adsorption modes are shown in figure 2. These studies showed that the adsorption modes for 2b) and 2c) (explained in the figure caption) are the main adsorption modes.

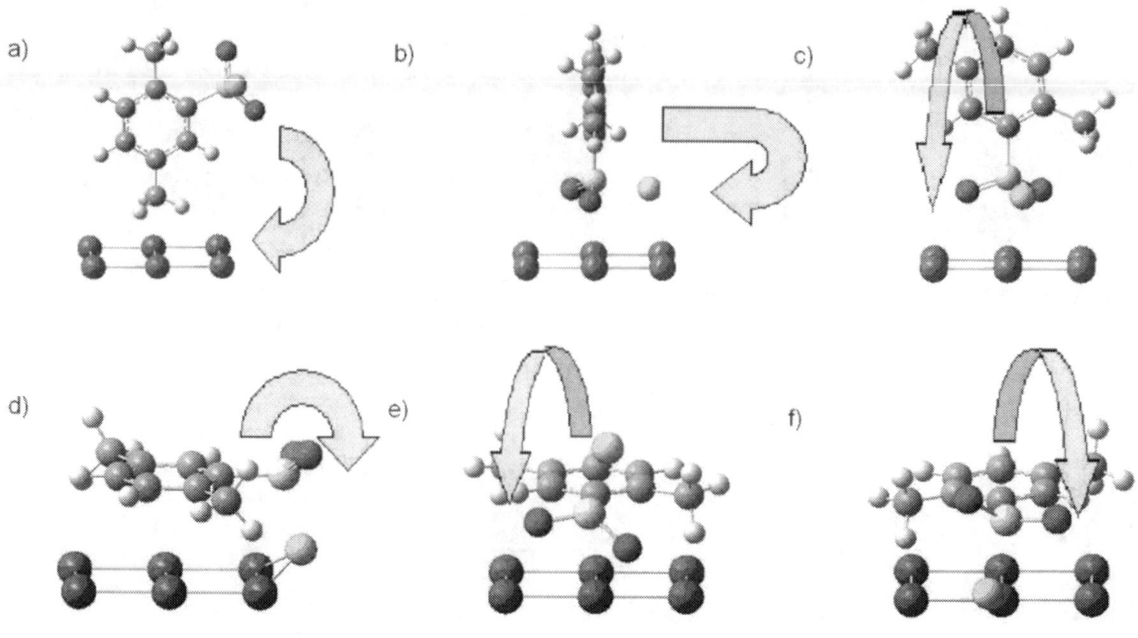

Figure 2. The adsorption modes considered during the course of the investigation. Sclads is an abbreviated filename used for the sulfonyl chloride adsorption calculations. a) a representative structure for the adsorption of the molecule via one of the methyl groups in the two orientations possible on the non-symmetrical Pd6 matrix (others not shown). All of the structures represented in this single example showed the same behavior. b) a vertical adsorption mode via the sulfonyl chloride functional group. c) same as 2b) but in the opposite Pd6 orientation. d) a horizontal adsorption mode with the chlorine molecule interacting with the surface immediately, similar to another adsorption mode by flipping the molecule so that the chlorine is farthest from the Pd 6 matrix (not shown) and exhibiting similar behavior. e) a horizontal adsorption that is simply 2d) flipped and in the opposite Pd6 orientation. f) a horizontal adsorption that is simply 2d) in the opposite Pd6 orientation. The movements from the first few steps of the DFT minimizations are represented by the arrows. In every case, the driving force was the need for one or both of the oxygen atoms bonded to the sulfur to interact directly with one or more of the palladium atoms.

The orientations from 2b) and 2c) were setup on the other, more complicated palladium matrices as well. Each step in the energy minimization calculation brings a lower energy and thus a more stable state. How does this simulation represent what is happening in the reactor? The turning of the molecules into the "sulfonyl chloride down" position shows that this is the most energetically stable of the postulated adsorption conditions. . The further minimizations are a picture of what the aromatic molecules would be doing on the surface of the catalyst without the presence of hydrogen. The results of the full minimizations over the 6 and 8 palladium matrices are shown below. The 22 and 13 atom matrices have proven beyond the current hardware/software configuration. Calculations with newly implemented convergence criterion and step processes are being run currently for these matrices.

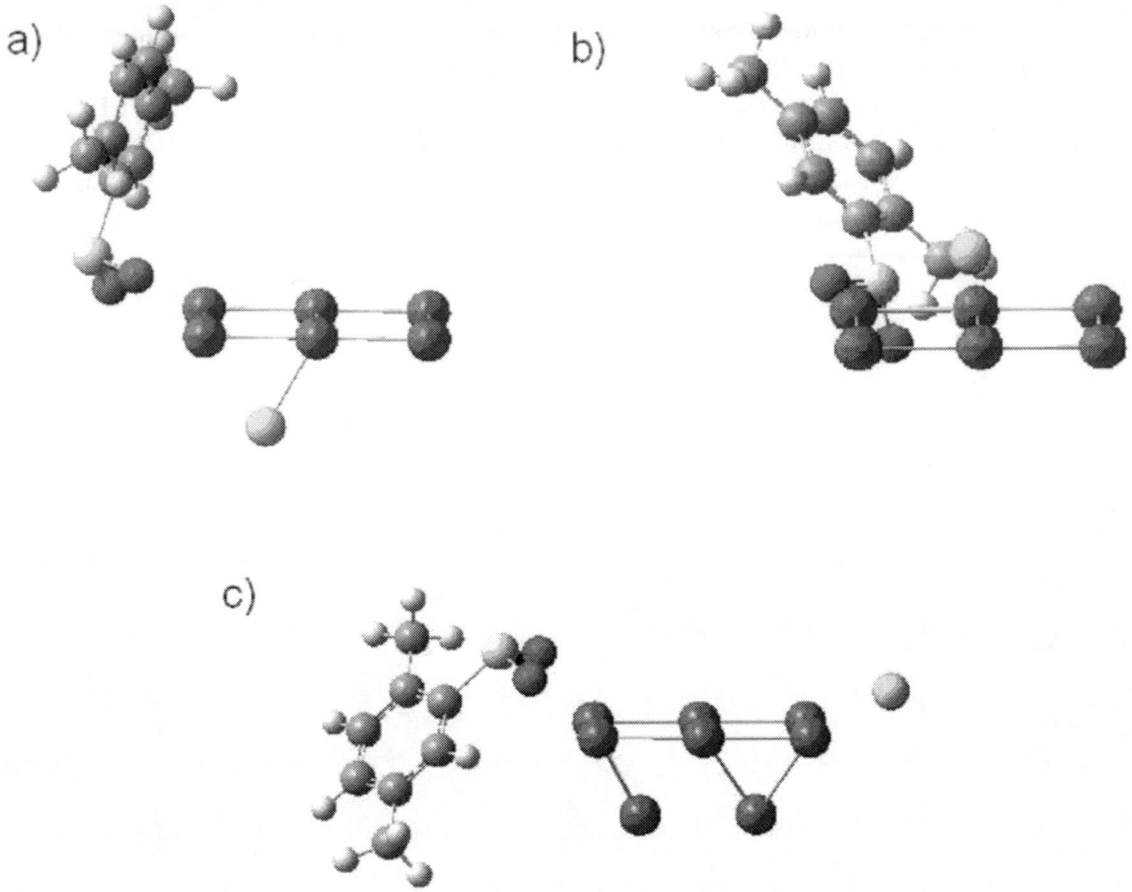

Figure 3. The results of the DFT optimization calculations on 2b), 2c) and the equivalent 2b) adsorption mode over Pd8 are shown here. These represent the minimums found by the Gaussian03 package. a) The endpoint structure of the 2b) calculation, ΔH_{ads} = -51.965 kcal/mol. b) The endpoint of the 2c) calculation, ΔH_{ads} = -12.995 kcal/mol c) The endpoint of the 2b) mode calculation over the Pd8 palladium matrix, ΔH_{ads} = -56.2816 kcal/mol.

These simulated adsorption energies for a) and c) are reasonable when compared to the individual adsorption energies of the atoms of the sulfonyl chloride group, but b) has a marked difference in energy from the other two. This has to do with the energetic differences between certain palladium atoms as compared to others.

In parallel with this theoretical work, an experimental set-up has been built to investigate the catalytic hydrogenation of an aromatic sulfonyl chloride into an aromatic thiol. A continuous flow Robinson-Mahoney reactor system has been designed and built from scratch in the laboratory. It has been used to obtain data on the intrinsic reaction rates at the surface of the catalyst. Hydrogenation reactions were performed at a reaction pressure of 364.7 psia and at three different reaction temperatures: 110 °C, 97 °C and 85 °C. The data was used to begin development of a kinetic model that includes adsorption of individual components and surface reactions. This combination of experimental and theoretical data has been used to create a set parameterized rate equations of the Hougen-Watson type. The parameters of the kinetic model were estimated using numerical methods. Though the work is still in its preliminary stages, some of the results will be presented.

On top of this combined theoretical/experimental approach, a rudimentary capital cost analysis through AspenTech and ICARUS software has been performed on a system to produce benzenethiol (thiophenol) from benzene sulfonyl chloride. This study involves a batch reactor of 5000 gallon total capacity, the post-reaction processing and recovery train. Initial results indicate a plant that can produce 1 million lbs. of benzenethiol in just over 310 batches. If run on a continuous plant basis (24 hours a day, 7 days a week, 300 operating days a year) it can produce near to 3 million lbs. benzenethiol per year. Some additional analysis must be performed to get more accurate capital cost numbers, but preliminary analysis shows an initial capital outlay from ICARUS of $4 million and a yearly operating cost of near to $2 million for this commercially available (from the Texas A&M University Licensing Office) process.

References

[1] Kathrin-Maria Roy, Ullman's, Encyclopedia of Industrial Chemistry, Wiley-VCH, New York, 2003, pp.625-652.

[2] John S. Roberts, Encyclopedia of Chemical Technology, Vol. 24, John Wiley and Sons, New York, 1997, pp. 19-34.

[3] Stephen E. Jacobsen. (13 September 2001) Hydrogen Reduction of Sulfonyl Chlorides to Thiols. International Patent Classification: C07C 319/02, 309/86. International application number: PCT/SU01/06129. International Publication Number: WO 01/66517 A1. Patent Assigned to E.I. Du Pont De Nemours and Company.

[4] John S. Poland, Scott Mitchell, Allison Rutter, Cold Regions Science and Technology 32 (2001) 93-105.

[5] W. T. Tsai, H. P. Chen, W. Y. Hsien, C. W. Lai, M. S. Lee, Resources, Conservation and Recycling 39 (2003) 65-77.

[6] Fengyuan Piao, Kazuhito Yokoyama, Ning Ma, Toru Yamauchi, Toxicology Letters 145 (2003) 28-35.

[7] B. Zerahn, A. Kofoed-Enevoldsen, B. V. Jensen, J. Mølvig, N. Ebbehøj, J. S. Johansen, I. –L. Kanstrup, Respiratory Medicine 93 (1999) 885-890.

[8] S. L. Bruce, B. N. Noller, A. H. Grigg, B. F. Mullen, D. R. Mulligan, P. J. Ritchie, N. Currey, J. C. Ng, Toxicology Letters 137 (2003) 23-34.

[9] Giampaolo Barone, Dario Duca, Journal of Molecular Structure (Theochem) 584 (2002) 211-220.

[10] D. Duca, G. Barone, Z. Varga, G. La Manna, Journal of Molecular Structure (Theochem) 542 (2001) 207-214.

[11] M. J. Frisch, G. W. Trucks, H. B. Schlegel, G. E. Scuseria, M. A. Robb, J. R. Cheeseman, J. A. Montgomery, Jr., T. Vreven, K. N. Kudin, J. C. Burant, J. M. Millam, S. S. Iyengar, J. Tomasi, V. Barone, B. Mennucci, M. Cossi, G. Scalmani, N. Rega, G. A. Petersson, H. Nakatsuji, M. Hada, M. Ehara, K. Toyota, R. Fukuda, J. Hasegawa, M. Ishida, T. Nakajima, Y. Honda, O. Kitao, H. Nakai, M. Klene, X. Li, J. E. Knox, H. P. Hratchian, J. B. Cross, C. Adamo, J. Jaramillo, R. Gomperts, R. E. Stratmann, O. Yazyev, A. J. Austin, R. Cammi, C. Pomelli, J. W. Ochterski, P. Y. Ayala, K. Morokuma, G. A. Voth, P. Salvador, J. J. Dannenberg, V. G. Zakrzewski, S. Dapprich, A. D. Daniels, M. C. Strain, O. Farkas, D. K. Malick, A. D. Rabuck, K. Raghavachari, J. B. Foresman, J. V. Ortiz, Q. Cui, A. G. Baboul, S. Clifford, J. Cioslowski, B. B. Stefanov, G. Liu, A. Liashenko, P. Piskorz, I. Komaromi, R. L. Martin, D. J. Fox, T. Keith, M. A. Al-Laham, C. Y. Peng, A. Nanayakkara, M. Challacombe, P. M. W. Gill, B. Johnson, W. Chen, M. W. Wong, C. Gonzalez, and J. A. Pople, Gaussian 03, Revision C.02, Gaussian, Inc., Wallingford CT, 2004.

FCC Operations

Pravin Sheth, ExxonMobil Research and Engineering Company, 3225 Gallows Rd., Room 8 A 0208, Fairfax, VA 22037

FCC Reactor-Regenerator Operations

The objective of this paper is to provide overview on the subject of operations and troubleshooting techniques used in FCC Units. Focus will be on Reactor-Regenerator section. Common FCC Operation problems such as Catalyst loses, Cyclones failures and cat circulation problems will be discussed.

Impact of Catalyst, Additives and Feedstock on FCC Unit Performance

E. Thomas Habib Jr., Grace Davison Refining Technologies, W. R. Grace & Co., 7500 Grace Drive, Columbia, MD 21044

Fluid Catalytic Cracking (FCC) units are the most prevalent molecular weight reduction process in petroleum refineries. Traditionally, FCC units are used to convert heavy vacuum gas oil (the overhead product from vacuum distillation) to lighter products, principally gasoline. However, the modern FCC unit may often process feeds containing light atmospheric gas oils to those containing significant portions of high boiling residual fractions. Catalysts employed vary greatly, although all commercial FCC units today utilize catalysts containing a form of the zeolite faujasite as their primary active ingredient. However, smaller amounts of non-faujasitic materials, referred to as catalyst additives, may be blended with the base catalyst in order to impart special catalytic functionality or selectivity. This tutorial presentation reviews the different types of catalysts and catalyst additives commonly utilized in FCC units. Feed effects and feed interactions with the catalyst system are also discussed, including the impact of feed impurities such as nitrogen, sulfur, and metals.

Effect of Vanadium Deposited Locations in FCC Catalyst on Naphtha Sulfur Content

Peng Wang, Huiping Tian, and Jun Long. FCC catalyst, RIPP,Sinopec, No.18 Xueyuan Road, Beijing, China

Recently, the positive effect of deposited vanadium on reducing FCC naphtha sulfur content has been gradually realized and come to be affirmed during FCC process. So the appropriate method by adding vanadium oxide in FCC catalyst was investigated in this work. Through the comparing of different vanadium deposited locations on FCC catalyst such as pseudoboehmite (converted to γ-Al_2O_3 after calcinated), clay and Y zeolite, it was found that the interaction between vanadium oxide and its supports can induce new active sites such as Lewis acid sites with proper acid strength and V-O-M (support) bond with active oxygen. The former benefits the selective adsorption of thiophene in the presence of cyclohexane while the latter benefits the oxidation of thiophene to SO_2. Through the evaluation by MAT test using sulfur contained VGO feedstock, it has shown that deposited 1000 μg/g V^{5+} on different component of FCC catalyst will achieve sulfur reduction effect in different extent comparing at the same heavy oil conversion. Vanadium oxide supported on zeolite is the best while on γ-Al_2O_3 is inferior.

AIChE 2007 Spring National Meeting

Dynamic Simulation of a Fluid Catalytic Cracking Unit.
Predicting the Operation Using a Hydrotreated Feed.

Yiannis Bessiris and Vassilis Harismiadis
Hyperion Systems Engineering, Modeling and Simulation Solutions, Pune, India

Subject area: Topical A: 10th Topical Conference on Refinery Processing - A007 Advances in FCC.

Keywords: Fluid catalytic cracking, simulation, dynamic modeling, operator training, yield shift calculations.

Fluid Catalytic Cracking (FCC) has been used for more than 60 years in order to convert straight-run atmospheric gas oils, vacuum gas oils, atmospheric residues, and heavy stocks recovered from other refinery operations into high-octane gasoline, light fuel oils and olefin-rich light gases. Due to the complexity of the process itself, and the need for efficient ways of handling various feedstock, a rigorous dynamic model that predicts the reactor yields and the unit dynamics is becoming a crucial tool in determining and achieving the optimum operating conditions. This includes determining potential bottlenecks in the design and "what-if" scenarios evaluation.

The simulation of the FCC unit provides an additional tool with such precision, that this can be used not only for the optimization of the process, but also as the Operator Training Simulator (OTS), for existing operators to learn the process, and to be used throughout the plant lifecycle to train new operating crews. OTS systems can result in fewer operation operating errors and quicker and safer start-ups and shutdowns; improve plant reliability due to improved response to abnormal situations; and can help optimize throughput and product quality.

In the specific plant that we modeled, the feed of the existing FCC would be altered from normal to hydrotreated one, after a new installation of a mild Hydrocracker unit. The potential control issues that would arise from such an alteration in the feed would have to be evaluated in the model before the actual change, and solutions to any potential problems would have to be tested and implemented in the simulator beforehand.

DESCRIPTION OF FCC PROCESS.

The process we modeled includes a FCC reactor of UOP design. UOP's process uses a side-by-side reactor/regenerator configuration. Feed distributors inject the feed into the riser, where it contacts the catalyst and reacts. A high efficiency stripper then separates the remaining hydrocarbons from the catalyst, which is then reactivated in a combustor-style regenerator and flows back to the mixing chamber to react again with the feed. The reactor zone features a short-contact riser and a termination device for quick separation of catalyst and vapors. The combustor-style regenerator burns coke in a fast-fluidized environment completely to CO2 with very low levels of CO. The circulation of hot catalyst from the upper section to the combustor provides added control over the burn-zone temperature and kinetics.

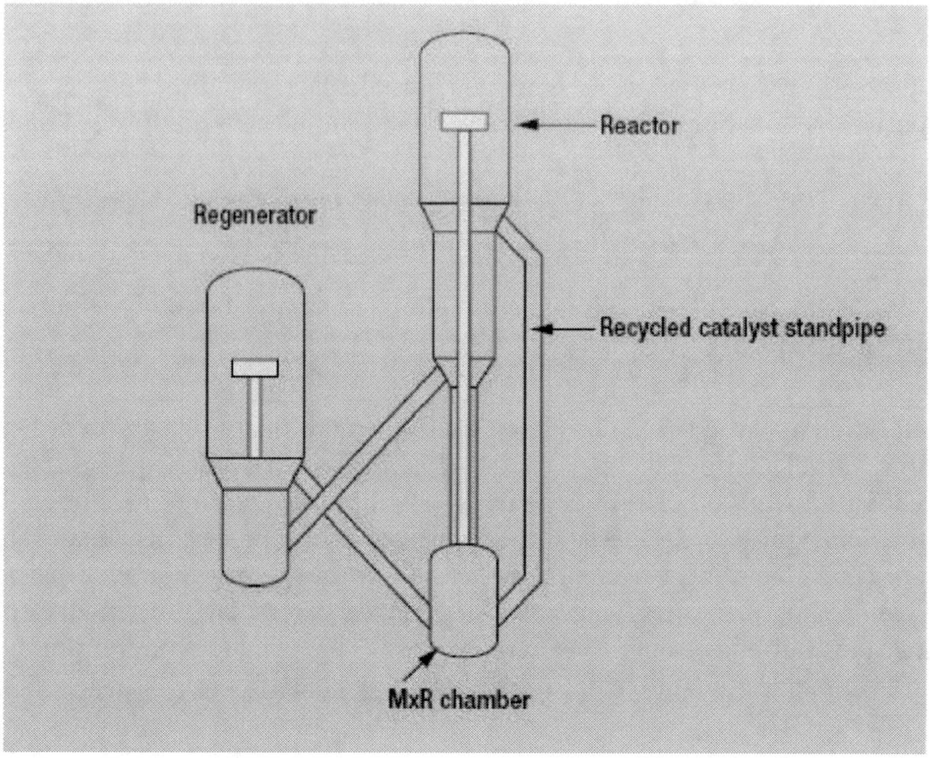

Figure 1. Typical Fluid Catalytic Cracking process. Source: Refining Process 2006 – Hydrocarbon Processing Magazine.

IMPLEMENTATION OF DYNAMIC SIMULATION – PROBLEMS AND SOLUTIONS.

The challenges in the modeling of an FCC reactor are due to the fact that:

- The reactions taking place are extremely complicated
- The mathematical systems of equations have to be resolved in real time with as much precision as possible so the simulator can also be used as operating training tool and not only for engineering studies.

The existing commercially available simulation software for the FCC processes, require significant computational time to compute the exact products and the reactor outlet conditions. Thus they could not be integrated to the dynamic model of the full refinery process while achieving real time step. The solution applied finally consists of the following steps:

a) Modelling of FCC reactor by using the high fidelity model of the AspenTech AspenFCC^{TM} software. Operating parameters that have been chosen are:
    - Riser outlet temperature
    - Reactor Pressure
    - Riser residence time
    - Catalyst / Oil ratio
    - Type of feed characterized by the following parameters
        - API SG (Specific Gravity)
        - Percentage of Sulphur content
        - Percentage of Nitrogen content
        - Boiling point
        - Characterization of the feed as paraffin, aromatic, hydrotreated e.t.c
    - Catalyst
        - MAT of the Catalyst
        - Catalyst's activity vs. the duration of its use (Start of cycle , end of cycle)
        - Catalyst's poisoning from metals and carbon monoxide (CO).

b) A base case (design case) was modeled and the yields of all the components in the feed were calculated so they would match the design of the FCC unit.

c) Step changes were carried out in each of the aforementioned operating parameters (a step), while keeping the rest of the parameters constant. The step changes were confined within the operating limits of the specific unit.

d) For all reactions taking place, yield shift variations for each component were calculated using AspenFCC^{TM} modeling environment and grouped into tables. In this way, we quantify the effect of a small change in a specific process condition to the yields of all components.

e) These results were imported into the modeling software and used for the calculation of the total yield for each component. This is done by accepting that each process condition is a series of changes from the base case. Thus, the total yield for each component was calculated by summing the yield shifts due to the differences from the base case.

In this way the time consuming calculations have been executed beforehand and the simulator can uses their results straightforwardly and in real time.

The dynamic simulator was completed by modeling, in detail, the mechanical equipment and the control systems surrounding the reactor of the unit. We had to create customized modeling routines in order to simulate the exact behavior of each component of the FCC system. New blocks to simulate the reactor, riser, cyclones, stripper, regeneration, standpipe and slide valves had to be created and modeled to interact with the catalyst flow and pressure profile, taking into consideration fluidizing effects in the catalyst itself. The coke deposition in the catalyst was also an important parameter to simulate as the

regeneration process has to be modeled in detail. Afterburner effects due to coke burning high in the regenerator were also a custom malfunction that we had to take into consideration.
The reliability and fidelity of the simulator can be proven through the comparison of our steady state and the actual unit operation.

The result was a dynamic model of the FCC unit with the ability to simulate accurately the start-up, shutdown and operation of the process and predict the reactor yields for different feedstocks, like Arabian light/medium oil, hydrotreated feed etc. The process bottlenecks were identified and resolved before the change in the feed of the unit, resulting in a smoother changeover and significant financial savings. Problems of low coke deposition on the catalyst due to the hydrotreated feed were observed in the simulator and corrective actions were taken to prevent this from happening in the actual plant.

RESULTS

Apart from the process simulation, the Emergency Shut Down system (ESD) of the operation unit has also been completely modeled. The final model can achieve up to five times the real time speed. The communication of the model with the DCS (Distributed Control System) of the simulator, which is a copy of the real DCS of the unit, is bi-directional. Thereby, the results of the computations of the simulator appear as indications on DCS consoles and all the manipulations of the operators influence the model exactly the same as in the real plant.
The issue of low coke deposition in the catalyst that was observed in both the simulator and the actual plant was corrected by recycling heavy oil in the FCC reactor, a solution that was first tested in the simulator.
The operators are already being trained on the following entities:

- Familiarization with the normal operation of the unit.
- Response of the unit to disturbances.
- Variation of the products' properties with respect to the operational parameters.
- Application of the normal start-up and shutdown operations.
- Application of emergency shutdown operation of the unit.
- Dealing with equipment malfunctions.

BENEFITS

The simulator has been widely accepted by the operating staff of the unit mainly thanks to the model's fidelity. The specific method used to model the system's reactors offers the required fidelity via analytical and detailed solution methods in real time, so that the results can be used not only for analysis and process optimization, but also as a tool of learning the operation of this new process.

The expected benefits are:

- Control of the normal process of start-up and shutdown of the unit.
- Reduction of the required time for normal start-up and shutdown of the unit.
- Increase in the continuous operation time.
- Increase in the operational time between sequential catalyst changes, by avoiding unsafe operational conditions.

- Continuous improvement of the real-time control system of the unit. The desired modifications can initially be tested on the simulator in different operating conditions, before being implemented operationally.
- The probability of a severe incident occurring during the unit operation is reduced.
- The impacts on the environment are reduced by avoiding an abnormal operation of the unit.

Biography

Vassilis Harismiadis has 8 years experience in the oil & gas industry with particular emphasis on the use of latest technology to improve plant effectiveness. His special expertise is in operator training simulators using virtual plant technology. He has led a number of training simulator projects which required the connection of dynamic models with the real plant's Distributed Control System and Emergency Shutdown System, thus providing significant operational benefits to clients. Vassilis holds a doctorate from NTU Athens, Greece in the thermodynamic modelling of complex systems and has published extensively in peer-reviewed journals.

Yiannis Bessiris has 7 years experience in development of large scale and full scope high fidelity process simulations for the oil and Gas and Power business. Project experience includes modeling for hydrocracking, lower olefins, refrigeration, sulphur recovery and fluid catalytic cracking units. He has presented his work in various Chemical engineering conferences. He is currently a business development manager for the Modelling and Simulation Solutions (MaSS) department of Hyperion systems engineering. He holds a MSc. in Chemical Engineering from the University of Patras, Greece.

Hyperion Systems Engineering Ltd. provides world-class professional consulting services and solutions in the areas of Supply Chain Management, Advanced Planning and Scheduling, Dynamic Simulation and Modeling, Advanced Process Control & Optimization and Plant Information Management. Hyperion enables Process Manufacturing companies to increase efficiency and profitability.
With headquarters in Cyprus and operating subsidiaries in the UK, Russia, Greece, India and Bahrain, Hyperion employs over 150 engineering professionals who have implemented over 200 projects worldwide. Hyperion has one of the world's most experienced teams of resource in the field of Dynamic Simulation and Operator Training, with in excess of 50 simulation and modeling professionals who have implemented projects for companies such as BP, Sasol, Motor Oil Hellas, Exxon, Texaco, Chevron, Shell, Saudi Aramco, Ferrostaal, in the Oil & Gas, Petro-Chemical, Chemical and Power generation industries.

Contact Details

HYPERION SYSTEMS ENGINEERING LTD
Modeling and Simulation Solutions
Gandhi Empire,Plot No 2, S. No 595/1
5th Floor, Sareen Estate
Kondawa Rd, Mouji Munjeri
Bibewadi, Pune-411040, India
E-mail: y.bessiris@hyperion.com.cy
Web: http://www.hyperion.com.cy

Optimization Approach for Production Planning in Olefins and Aromatic Plant, Cracking Vs. Reforming

Bhieng Tjoa[1], A. Cervantes[1], J.A. DeFazio[1], A. Fraga[2], and E. Terra[2]. (1) Optience Corporation, 1999 S. Bascom Ave., Suite 510, Campbell, CA 95008, (2) BRASKEM, Salvador, Brazil

In the cyclical petrochemical industry, there is a more competitive and globalized market than ever. Feedstock can come from any place in the world and products are also available in any part of the world. Associated with Globalization, geopolitical tensions and speculation make the oil market extremely volatile, directly influencing the petrochemical market.

On the eve of another "low cycle" in the petrochemical industry, with large capacities and high technology projects, the result is an offer surplus and a reduction in the profitability of the companies.

Competitiveness is a *key success factor* for the companies that are preparing themselves for this challenge. Among other actions, an efficient choice of feedstock, the best use of its assets and the "exploration" of the most profitable markets are the best answers to boost results.

The volatility of the prices, sometimes benefiting the olefins, sometimes the aromatics products and the complexity and diversity of Braskem's processes which include Reforming and Cracking units, strengthen the necessity of searching the best point between feedstock, market and operations.

To solve this highly complex problem, the acquisition of an advanced optimization application gives companies a competitive advantage. A year and a half ago Braskem, the largest petrochemical company in Latin America, acquired Optience SCMart Suite as the optimization application platform to help their planning and scheduling process.

The main model, developed in SCMart, is a nonlinear multiperiod monthly planning model which includes all plant constraints and uses a nonlinear correlation yield model based on SPYRO simulator for the cracking furnaces. It also includes a nonlinear correlation to estimate the reformer yields as well as a complex model for a naphtha fractionator. The planning model is supported by a scheduling model to achieve optimal results at the daily level. The system also uses a feedstock database library that tracks the available feedstock qualities and properties.

In this work, we discuss the planning module that is used to support the technical and economical decisions at Braskem. The nonlinear planning model consists on 5,000 variables per period including flows, inventories, properties and other process variables. It covers two ethylene plants, one reformer plant, plus other C4s, C5s and aromatics plants. The main products are ethylene, propylene, butadiene, benzene, para-xylene and other aromatics. These decision process variables that are optimized include COTs (Coil Outlet Temperatures), reforming and cut temperatures, possible co-cracks, furnace and reformer utilization, ethane recycling flows and loads to other downstream units.

We present different case studies and scenarios which have been running during these last months that demonstrate how the tool has been helping us in the decisions as the amount and quality of the Naphtha to be bought and processed in each unit. It has also helped us to identify the ideal markets and in the exploration of trade offs and ideal yields, as well as the optimal use of the assets, through concepts of shadow price and profit variability.

Higher productivity by using the new grade Sandvik APMT as radiant tube in ethylene cracking furnaces

Mats Lundberg
Dr.
Sandvik Materials Technology

Jan Samuelsson
M.Sc.
Kanthal AB

Claes Öhngren
M.Sc.
Sandvik Materials Technology

Johan Josefsson
M.Sc.
Sandvik Materials Technology

Bo Jönsson
Lic.Tech.
Kanthal AB

Roger Berglund
M.Sc.
Kanthal AB

Eduardo Gomes
M.Sc.
Sandvik Materials Technology

Prepared for Presentation at the Year 2007 Spring National Meeting
Houston, Texas, April 20-27, 2007

Higher productivity by using the new grade Sandvik APMT as radiant tube in ethylene cracking furnaces

Mats Lundberg, PhD
Sandvik Materials Technology
Jan Samuelsson, M.Sc.
Kanthal AB
Eduardo Gomes, M.Sc.
Sandvik Materials Technology
Roger Berglund, M.Sc.
Kanthal AB,
Bo Jönsson
Kanthal AB
Johan Josefsson, M.Sc.
Sandvik Materials Technology
Claes Öhngren, M.Sc.
Sandvik Materials Technology

Abstract
Selection of material for radiant tubes in ethylene furnaces is a key factor for increased productivity and reduced maintenance. Besides oxidation, carburization resistance, high temperature strength and tendency to catalyze coke formation are all material properties that have to be considered.

Conventional FeCrAl-alloys are known for their outstanding oxidation properties, but due to relatively poor mechanical strength at elevated temperatures, their use in highly stressed components are limited. However, this weakness can be bypassed by utilizing powder metallurgy processes. The new grade Sandvik APMT is a FeCrAl-alloy produced by a rapid solidification process. The resulting dispersion strengthened structure extends the use of this material towards application with higher stress levels at high temperatures.

Sandvik APMT is an alumina forming alloy that shows excellent properties when it comes to prevent coke formation and carburization. Results from computer simulations and field tests indicate that these properties provide great potential values for ethylene producers.

In this article high temperature mechanical data, coke formation and carburization properties for Sandvik APMT are presented and discussed in relation to other alloys. Results from test installations of Sandvik APMT tubes in ethylene furnaces are reviewed.

Introduction

FeCrAl-alloys form a protective alumina (Al_2O_3) scale that provides excellent corrosion properties, and maximum working temperatures which greatly exceeds any chromia forming alloy. One drawback of ferritic FeCrAl-alloys produced by conventionally melting processes has been the relatively low creep strength offered at high temperatures. However, the high temperature mechanical strength can be increased by introducing dispersion strengthening and grain size control. Both methods are applied since several years in commercial materials produced using Powder Metallurgic (PM) routes.

The first technique for production of Oxide Dispersion Strengthened (ODS) [1] materials is known as Mechanical Alloying (MA). Despite excellent creep properties developed in these materials, the use of MA-materials in large scale has been limited due to the high costs for production.

A second and more recent technique utilizes rapid solidification through gas atomization, these alloys are often referred to as Rapidly Solidified Powder metallurgical (RSP) materials [2, 3]. In RSP-materials the finely dispersed particles are not only oxides, but also nitrides and carbides. The resulting creep strength of RSP-materials typically falls in between that of conventionally produced and MA-materials.

The new grade, Sandvik APMT, is a new member in the group of commercially available RSP-alloys. In this paper some relevant properties of Sandvik APMT are presented together with some experiences from test installations in commercial ethylene furnaces.

Manufacturing

The production route of Sandvik APMT tubes is briefly outlined in Figure 1. Sandvik APMT is available as hot extruded tubes in outer diameters between 26 and 260 mm (1.02-10.24") and wall thicknesses between 3 and 12 mm (0.12-0.47").

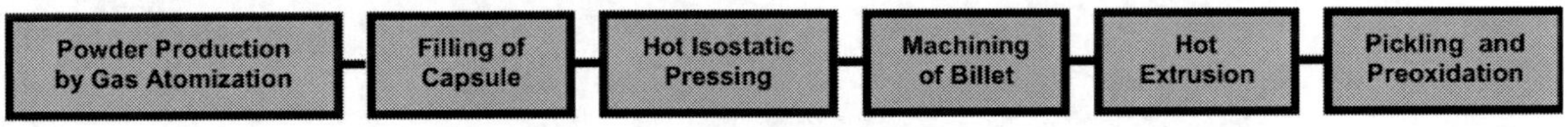

Figure 1: Schematic of the APMT production route.

Properties of Sandvik APMT

The nominal compositions of Sandvik APMT and Alloy 800HT are given in Table 1.

Table 1. Nominal chemical compositions of APMT and Alloy 800HT given in weight-%.

		Cr	Al	Mo	C	Mn	Si	Fe	Ni	Minor Add.
APMT	Min	22	5.0	3.0				Bal.	--	Present
	Max				0.05	0.4	0.7			
Alloy 800HT	Min.	19	0.15	--	0.06			Bal.	30	Ti
	Max.	23	0.60		0.10	1.5	1.0		35	

The density of Sandvik APMT is about 7.2 g cm^{-2}, which is approximately 10 % lower than that of most austenitic FeNiCr-alloys. Figure 2 presents the thermal expansion of Sandvik APMT compared to 25Cr-35Ni. From this it can be seen that Sandvik APMT has 10-15% lower thermal expansion.

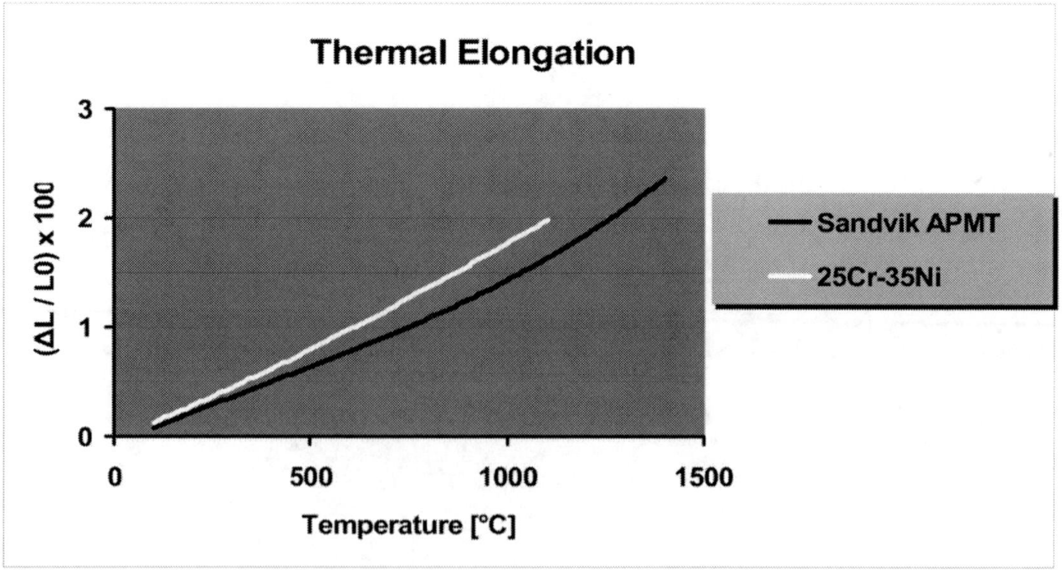

Figure 2: Thermal expansion as a function of temperature.

Creep Strength

Sandvik APMT is being evaluated regarding the creep properties at temperatures between 800°C and 1300°C. Figure 3 presents data of "time to rupture" for Sandvik APMT and Alloy 800HT, in a Larson-Miller plot.

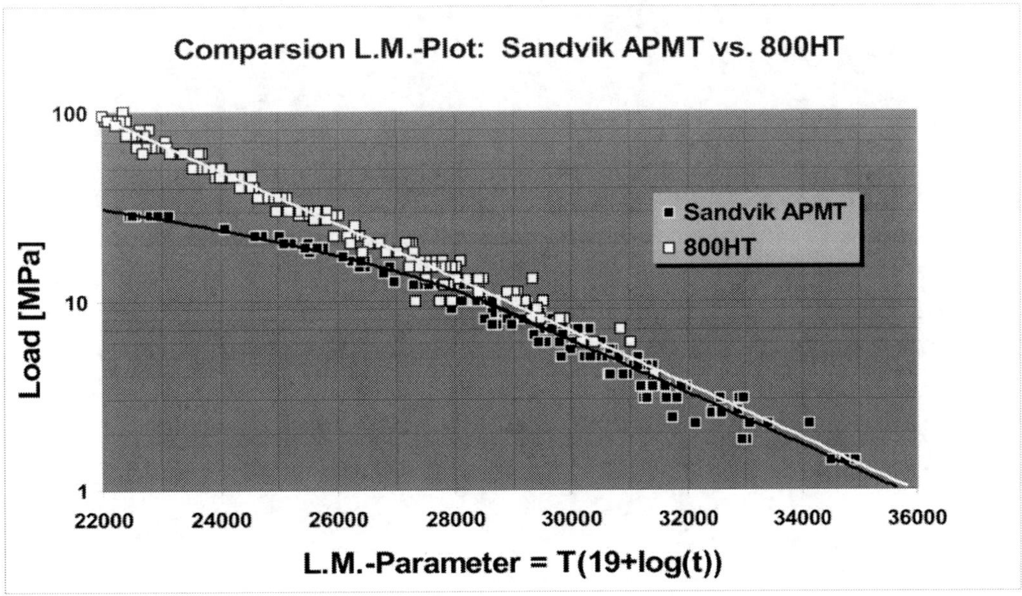

Figure 3: Creep strength of Sandvik APMT compared to Alloy 800HT material. Temperatures in Kelvin and times in hours.

Oxidation

Sandvik APMT has excellent corrosion resistance in most atmospheres due to the formation of a dense alumina scale. The protective properties of alumina scales compared to chromia scales have been shown in several studies [4, 5]. This scale is an effective barrier against diffusion of carbon and oxygen from the gas phase into the metal matrix, as well as for blocking the catalytically active alloying elements, like Fe, inside the alloy from reaching the oxide surface. In Figure 4 a cross section of an alumina scale formed at 1200°C during 3000 hours is presented. The thickness of the scale is less than 30 μm and no apparent spalling has occurred. This means that the aluminum reservoir in the alloy can sustain the scale for extended operating periods. As the structure of Sandvik APMT is ferritic at all temperatures, the diffusion rate of aluminum inside metal matrix is very fast. This is beneficial when it comes to rapid healing of a damaged scale. All these factors result in an extremely good resistance to high temperature oxidation.

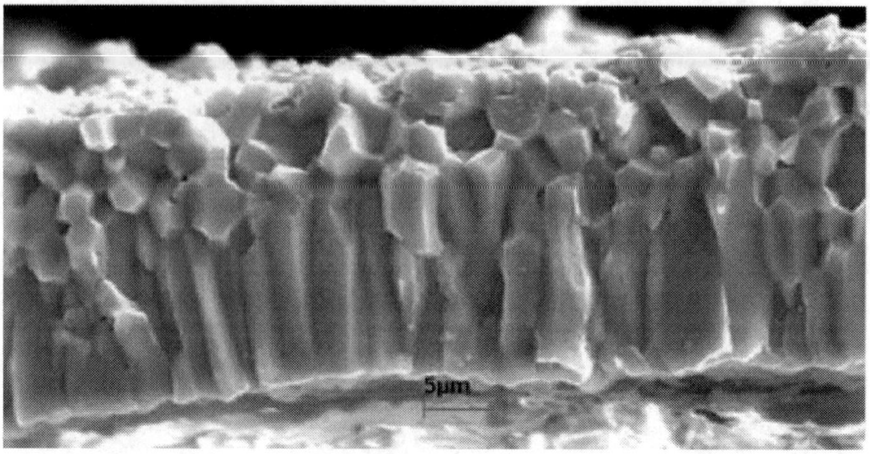

Figure 4: Cross section of alumina scale formed in air after 3000 hours at 1200°C (SEM).

Corrosion resistance of most materials conventionally used in steam cracking industry relies on formation of a protective chromia (Cr_2O_3) layer. In Figure 5 results from isothermal oxidation of a chromia forming alloy, 25Cr-35Ni, is compared with Sandvik APMT. The test was carried out at 1100°C in air. Oxidation rates are presented as weight change as function of time.

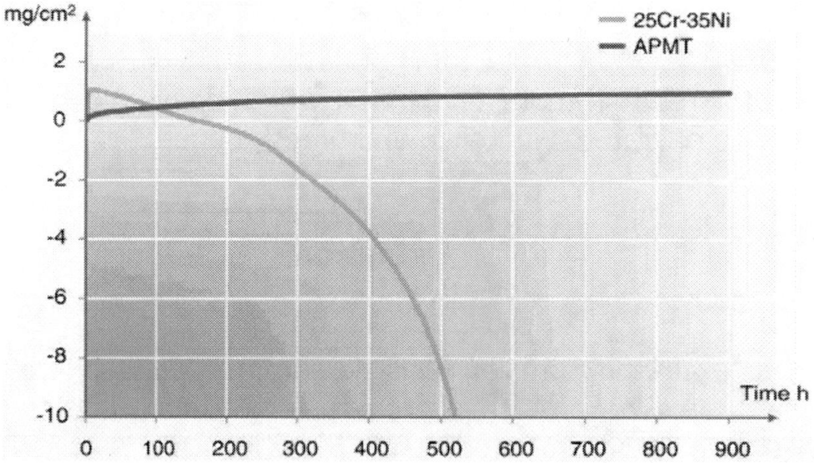

Figure 5: Mass change during intermittent oxidation in air at 1100°C. After initial period, cycle times were approximately 100 hours. Sandvik APMT shows a continuously parabolic weight gain, while the 25Cr-35Ni alloy starts to lose oxide scale, spallation, after less than 24 hours.

It can be noticed that the chromia scale formed on 25Cr-35Ni has a rapid initial growth rate, but at a certain point intrinsic stresses causes the scale to spall off. New oxide is formed on the spalled areas and chromium from the alloy is rapidly consumed. This process continues until the metal matrix is so depleted in chromium that it can not reform or heal the chromia scale. Sandvik APMT shows a slow, parabolic growth rate with no signs of spallation. The alumina scale of Sandvik APMT is protective up to 1250°C in continuous operation, but can handle temperature excursions for shorter period of times. In Figure 6 this is illustrated by a joint between Sandvik APMT and a 25Cr-35Ni. The material was exposed to a temperature close to 1400°C for 24 hours. As can be seen the upper part of Sandvik APMT is still intact, while the 25Cr-35Ni-alloy was fully oxidized.

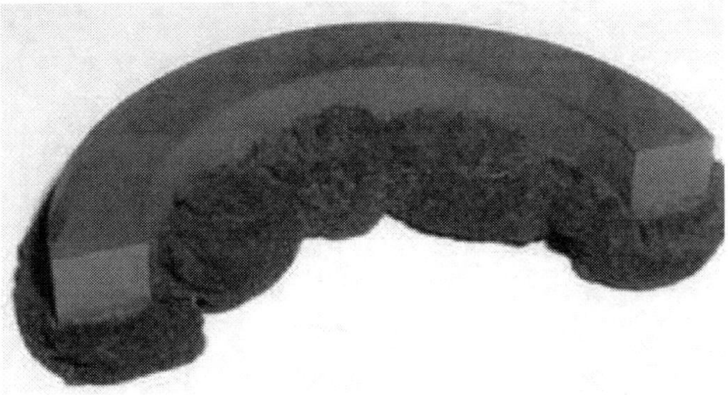

Figure 6: Photo of joint between Sandvik APMT and a 25Cr-35Ni material exposed at ~1400°C for 24 hours.

Carburization

Alumina forming alloys have been compared to chromia forming alloys with about 20 % chromium and a nickel content varying between 20-80 %. The experiments were carried out in a electrically heated furnace at 1050°C and 1150°C in a gas with a carbon activity >1. The results from this study showed that the corrosion rate for chromia forming alloys is strongly dependent on the nickel content and that alumina forming alloys are much less sensitive to corrosion in this environment. The corrosion rates for the alumina forming alloys were 5-20 times lower than that of the FeCrNi-alloys tested [6].

FeCrAl-alloys have also been submitted to carburization tests in packed graphite. Samples were packed in graphite, inside hermetically sealed containers and heated up to 1100°C for 96 to 192 hours. The results showed that no carbon up-take could be detected in the Sandvik APMT material, indicating that the alumina scale provides a very efficient barrier against carbon up-take.

Coking

It is well know that coke is formed by two different mechanisms, and that their relative importance depends mainly on the type of feedstock used. Catalytic coking is believed to be dominant when light feedstocks, like ethane are used, while heavier feedstocks like naphtha, tend to generate pyrolytically formed coke. It is also well documented that certain elements, such as Fe, Ni, Co and some of their oxides, are catalytically active in coke formation, while oxides of Al, Cr and Si are virtually inert [7, 8, 9].

Figure 7, presents a plot, showing the coke build-up as pressure drop vs. time for alumina and chromia forming alloys. The chromia forming alloy starts to build up a pressure drop after approximately 8 hours, while this pressure drop starts after 24 hours for the alumina forming alloys. All tube samples were pre-oxidized before they were exposed to the test environment.

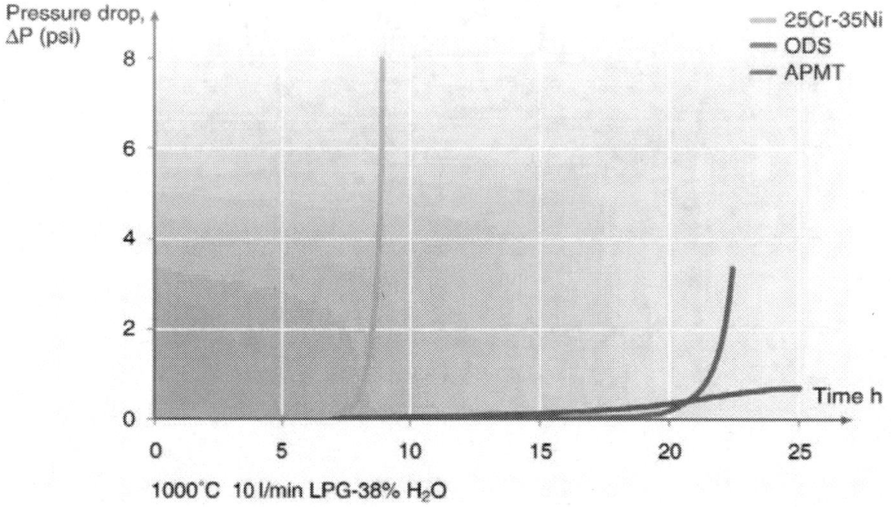

Figure 7: Coking resistance measured by Pressure drop as function of time.

The test was carried out in a laboratory gas cracker where LPG and 38 % H_2O were fed into the tubes at 1000°C.

Simulation of Sandvik APMT in Cracking Service

In order to estimate potential benefits in cracking service due to the properties offered by Sandvik APMT a simulation using PYCOS software has been made. The furnace design used was a 6-leg coil. Comparing the Sandvik APMT material versus a reference, cast 35Cr-45Ni material, differences covering run length, ethylene yield, ethane conversion and fuel gas were calculated. The feedstock used in the simulation was ethane with conversion of 68% and contact time of 0.46 seconds. All the tubes were smooth, straight and with the adiabatic volume at the coil outlet of 14 dm^3 per coil. The coke formation rate of Sandvik APMT was assumed to be 1/3 of the coking rate of the cast 35Cr-45Ni tubes, which is a realistic assumption based on experimental experiences of pressure drops in the test earlier described. All other input data were identical in the simulations. N.B. the data and the design are not taken from the real test installation ongoing. The results of the simulation are displayed in Table 2.

Table 2: Results from simulation of Sandvik APMT in ethane cracking service.

	APMT	35Cr-45Ni
Conversion	75%	68%
Ethylene yield	57%	53%
Run lengths	48 days	40 days

It can be concluded that Sandvik APMT tubes, due to the lower coking rate, facilitate higher conversion of ethane, increased yield of ethylene and longer run lengths compared to a cast 35Cr-45Ni material. Higher conversion is of course desirable because the need for recycling of feedstock is decreased. Increased yield of ethylene decreases necessary downstream capacity for processing of by-products. Longer run length means that the productivity increases as less time is spent on decoking the tubes.

Test Installation

In October 2004 Sandvik APMT tubes were installed in an ethylene furnace operated by a major international petrochemical company. Normal cracking conditions have been used during the operation. During operation, the performances of the tubes as well as operating conditions have been measured. In Figures 8 and 9 the collected data from two runs with ethane and naphtha, respectively, are plotted and compared to the performance of control tubes operating in the same furnace.

Coil Outlet Temperature (COT) is the temperature of the gas as it leaves the firebox. Tube Metal Temperature (TMT) is the temperature of the outer surface of the tube walls measured in the firebox, at a location not far from the coil outlet. A high COT and a low TMT indicates effective heat transfer into the process gas.

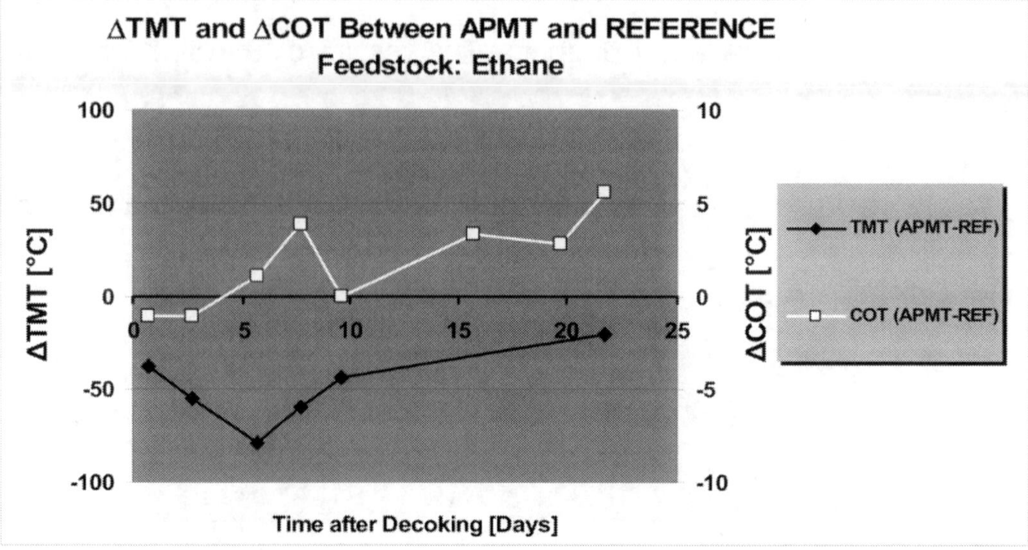

Figure 8: Run with ethane after totally 167 days of operation. The differences between Sandvik APMT and reference tubes, regarding TMT and COT respectively, are plotted as function of days after decoking.

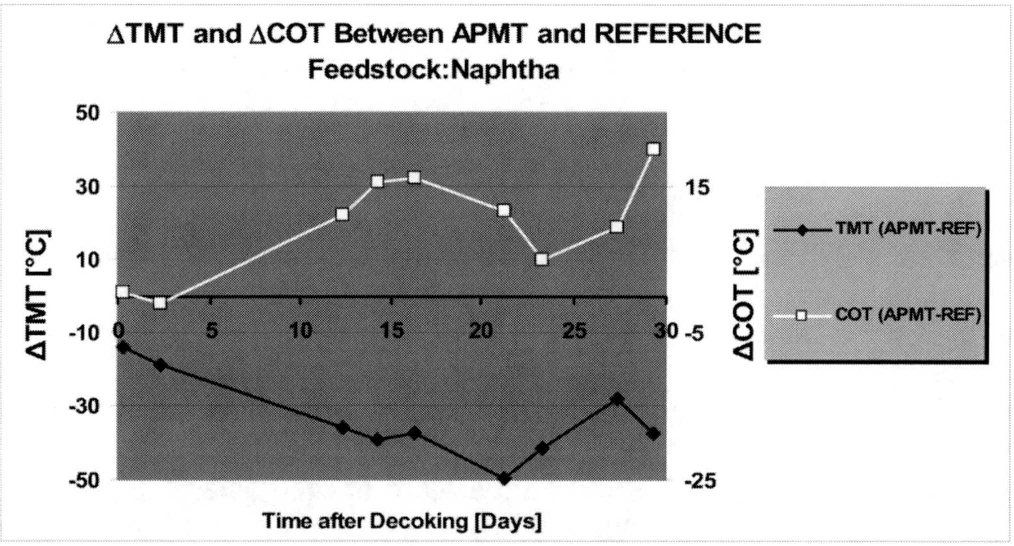

Figure 9: Run with naphtha after totally 400 days of operation. The differences between Sandvik APMT and reference tubes, regarding TMT and COT respectively, are plotted as function of days after decoking.

According to the results, the TMT measured on the Sandvik APMT tubes are ~10-30°C lower than that of the control tubes when ethane is used for feedstock. When naphtha was used as feedstock the difference was even bigger, ~20-60°C. In both cases the coil outlet temperature (COT) was measured to be slightly higher in the Sandvik APMT tubes. According to operators at the plant the difference in TMT can also be clearly seen during visual inspection. The difference in TMT is a strong indication that the coke layer formed inside the Sandvik APMT tubes is much thinner than that formed in the control tubes.

Conclusions

The substantial improvement of the creep properties to a higher level, well exceeding that of existing RSP FeCrAl-alloys, makes Sandvik APMT interesting for petrochemical processes where excellent oxidation, carburization and coking properties up to 1250°C are beneficial.

Laboratory experiments, as well as field installations, indicate that Sandvik APMT have excellent coking properties.

Measurements from test installations of Sandvik APMT tubes in ethylene furnaces, run at normal conditions, confirms that Sandvik APMT gives a significantly lower TMT than the control tubes in combination with a slightly higher COT. The reason for the differences in heat transfer efficiency has to be found in lower coking rate of Sandvik APMT.

Simulations performed of steam cracking processes support the conclusions that Sandvik APMT due to a low coking rate facilitates higher conversion, increased yield and longer run lengths compared to a conventional cast 35Cr-45Ni material.

References

[1] Czyrska-Filemonowicz, D. Clemens, W.J: Quaddakers*: The Effect of High Temperature Exposure on the Structure and Oxidation Behaviour of Mechanically Alloyed Ferritic ODS Alloys,* J. of Mat. Processing Technology 53 (1995), p. 93- 100

[2] B. Jönsson, R. Berglund, J. Magnusson, P. Henning and M. Hättestrand: *High Temperature Properties of a New Powder Metallurgical FeCrAl Alloy,* Materials Science Forum, Vols. 461-464 (2004) pp. 455-462

[3] J. Jedlinski: *The oxidation Behavior of FeCrAl alumina forming alloys at high temperatures,* Solid State Ionics 101 - 103, (1997), p. 1033-1040

[4] B.A. Pint, A.J. Garratt-Reed and L.W. Hobbs: *The reactive element effect in commercial ODS FeCrAl Alloys,* Materials at High Temperatures, Vol 13, (1995),

[5] I.G. Wright, B.A. Pint and P.F. Tortorelli: *High-Temperature Oxidation Behavior of ODS-Fe3Al,* Oxidation of Metals, Vol 55, Nos 3/4, (2001), p. 333-357

[6] M.B. Gutman, I.V. Voronkin, B. Magnusson and N. Lindskog: Untersucung der Wärmekorrosionsbeständigkeit von elektrischen Heizleiterlegierungen in kohlenstoffhaltigen kontrollierten Atmosphären, Elektrowärme international, 40. Jahrgang, Heft B 2/1990, S. B 85-B 90

[7] M.A. Harper, M. Ducasse, and D.J Young: *Cyclic Oxidation Plus Carburization of Heat-Resistant Alloys,* Corrosion Science, Corrosion Vol 51, No 3 (1995), p. 191-200

[8] E.M. Müller-Lorenz and H.J. Grabke: *Coking by Metal Dusting of Steels,* Materials and Corrosion 50, (1999), p. 614-621

[9] Shigeru Ando, Tomotaka Nagashima and Hiroshi Kimura: *Localized Coking on Austenitic Heat-Resisting Steels in CO-H2 Mixtures at 923°K,* ISIJ International, Vol 31 (1991), No 7 pp 744-749

IMPROVEMENT OF ETHYLENE PLANT PERFORMANCE USING HIGH-Cr, HIGH-Ni WELD OVERLAID FURNACE TUBES

Yutaka Yoshida, Kouji Ohya, Frans X. Terwijn[+], and Toshihide Maeda**
Daido Steel (America) Inc., Victoria, BC, Canada.
**Daido Steel Co., Ltd., Nagoya, Japan,*
[+] TACT / Daido Steel Co., Ltd., Rotterdam, The Netherlands

Abstract

High-Cr, high-Ni PPW (Plasma Powder Weld) overlaid furnace tubes have been evaluated in several commercial ethylene pyrolysis furnaces. In addition to extremely high carburization resistance due to the "dam effect" of the PPW layer, which leads to longer tube life, these tubes demonstrated significant improvements in the operational performance of the pyrolysis furnaces, especially when ethane feed was used. The lower TMT (Tube Metal Temperature) and lower rise of coil pressure indicate a reduction or slowdown of coke formation and as a result, run length was extended to twice or longer in the case of ethane feed and approximately 30% longer for naphtha feed. These effects are brought about by the chemistry of the internal surface of the tube. A high chromium weld overlay creates and maintains a tight chromium oxide layer without iron, nickel, or their oxides which act as catalysts for the coke formation. It was also found that conversion and ethylene yield were improved by 1 - 3%. Combined with higher furnace availability and higher robustness against unplanned shutdowns, the benefits of ROA (Return of Asset) improvement can far outweigh the initial cost penalty of using these tubes.

Introduction

Ethylene is one of the most widely produced and consumed chemical commodities and its derivatives are being used in a wide range of applications such as packaging, medical, food, electronics, automotive, and building industries. At the same time the production of ethylene is one of the most energy intensive processes (for example, it is responsible for one third of the energy consumed in the petrochemical industry in Japan) since it involves steam cracking of feedstocks, such as naphtha and ethane, in high temperature pyrolysis furnaces. It is also a very severe environment for the furnace tubes due to the high temperature, high pressure, carburising and reducing atmosphere in the cracking cycle and oxidizing atmosphere in the decoking cycle. Especially, coke formed in the tube inner surface is problematic since it causes a reduction of process efficiency (reduced conversion and yield), poor heat transfer and thus the increase of TMT (tube metal temperature), reduction of tube inner diameter and thus the increase of pressure drop, carburization and eventual tube plugging and failure.

In order to increase productivity and extend the tube life, over the years tube materials have evolved from 304 and 310 stainless steels in the 60's to HK, HP mod or HP micro alloy in recent years.[1] By increasing nickel and chromium content, these tubes have enabled higher temperature pyrolysis which, combined with shorter residence time, gave rise to higher conversion and ethylene yield while extending or at least keeping the tube life to an acceptable level. As this trend continues, 35%Cr-45%Ni alloy has been gaining popularity in the industry. However, this alloy tends to have weldability and creep failure issues and further improvement of reliability and tube life may not be feasible by constructing the whole tube with higher nickel and chromium alloys. In addition to the increase of Cr and Ni, there have been numerous attempts to improve the furnace tubes by additions of Nb, W, Mo, Si, and Al[2] and most of the commercial tubes have at least some of these additive elements. Another approach is to apply coatings that reduce coking, such as SiC or Al_2O_3, on the inner surface of the tubes using various methods, such as PVD, CVD, PLD, thermal spray, and diffusion coating, etc.[3,4] While there seems to have been some success in this approach, these coatings are relatively thin and would be lost over time through oxidation, reduction, spalling, and erosion.

Our approach is to apply a thick Cr-rich metal layer on the entire inner surface of conventional tubes by welding. Combined with our expertise in the PPW process, we have developed PPW overlaid furnace tubes ("EST", Ethylene Super Tube). The most obvious advantage of PPW overlaid tubes is its extremely high carburization resistance into the base tube ("dam-effect") which extend the tube life by a factor of two to four compared with conventional HP tubes.[5,6] Additionally, it has been demonstrated in the commercial furnaces that these tubes can contribute to ROA (Return of Asset) improvement through longer run length and increased ethylene yield, which will be described in this paper.

Construction of Tubes

Our radiant tubes are constructed by weld overlay of 45%Cr-50%Ni-1%Mo alloy on the inner surface of conventional base tubes using the PPW technique **(Fig. 1)**. HP-Nb (25%Cr-35%Ni) was selected for the base tube for its availability, low cost, and high creep strength. The thickness of the overlay is 2 - 3 mm and this differentiates PPW overlay from a number of other coating technologies where much thinner coatings suffer durability problems. The overlay composition was carefully selected to minimize carburization and catalytic coking **(Fig. 2)**, and to ensure high reliability through matching of thermal and mechanical properties to the base tube.[7]

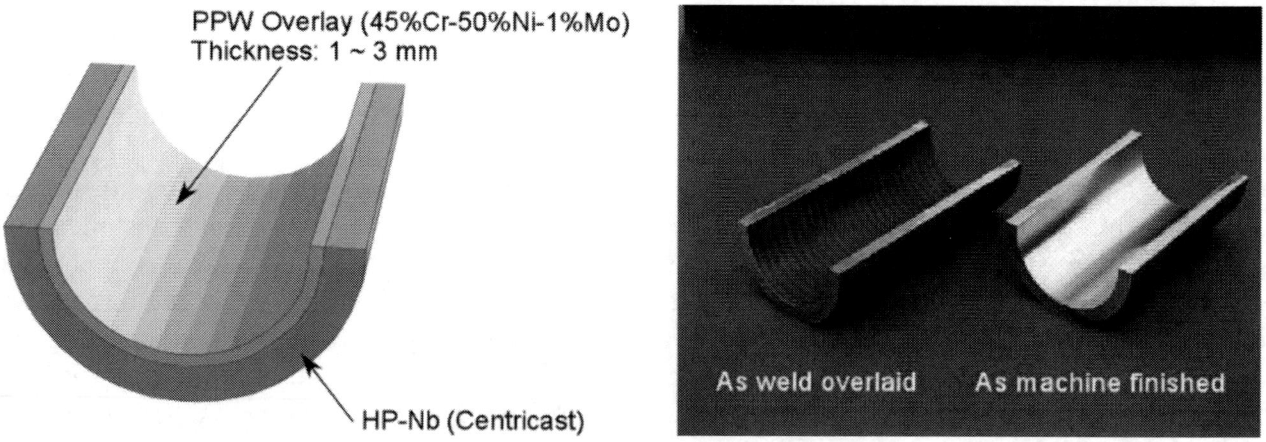

Fig. 1 PPW overlaid tube construction

Radiant coils were constructed with PPW overlaid straight tubes and conventional static cast fittings without PPW overlay. Although welding of PPW overlaid tubes together requires a special procedure (**Fig. 3**) to ensure PPW protection throughout the length of the straight tube, welding the complete straight tubes to fittings without PPW overlay can be done with standard welding procedures of HP-Nb.

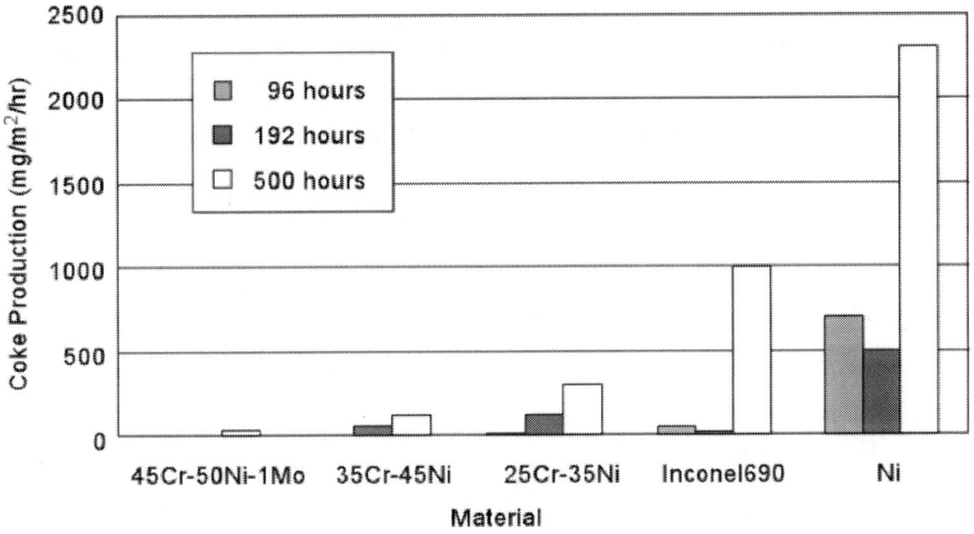

Fig. 2 Coke production on the surface of various materials (laboratory data)

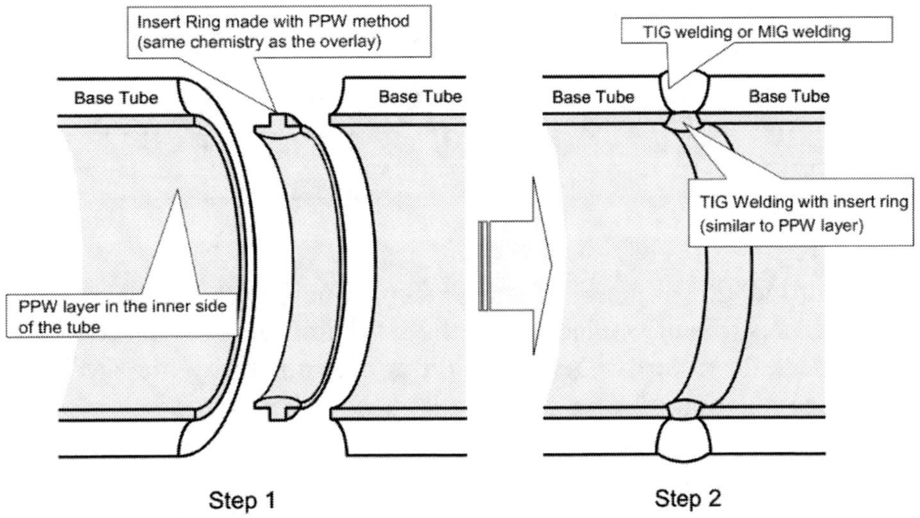

Fig. 3 Connecting PPW overlaid tubes using a two-step welding procedure

Performance Data from Commercial Furnaces

Several companies have installed PPW overlaid tubes over the years either for a full furnace, full coil or for partial application of coil outlet area. Some are being operated with naphtha feed and others are with ethane or ethane/propane feeds. In this paper we will mainly discuss the performance data, such as run length, conversion and ethylene yield, from two of the ethane furnaces with a special emphasis on the reduction of coking. As is well known in the industry, there are two mechanisms of coking: catalytic and pyrolytic (also called thermal, asymptotic, or amorphous). The former mechanism plays an important role in ethane furnaces and could be suppressed by assuring the process surfaces are free from any substances with catalytic effects (typically, iron, nickel, and their oxides) while the latter mechanism is dominant in naphtha furnaces and would be more difficult to control by the tube material.

Table 1 shows the configurations of the furnaces and evaluation conditions. In Furnace A, PPW overlaid tubes were installed in the last two legs (20% application) of two out of four coils in the furnace and comparison was made against the other two coils with conventional tubes (35%Cr-45%Ni). In Furnace B, PPW overlaid tubes were installed in one half of the furnace (50% application) and comparison was made against an identical furnace (Furnace C) with conventional tubes.

Table 1. Furnace configurations and process conditions for evaluation of PPW overlaid tubes

	Furnace A	Furnace B	Furnace C
Type	Lummas SRT 1.5	Stone & Webster WV coil	
Tube Materials - — 35Cr-45Ni ···· HK40 ▬ EST — Other	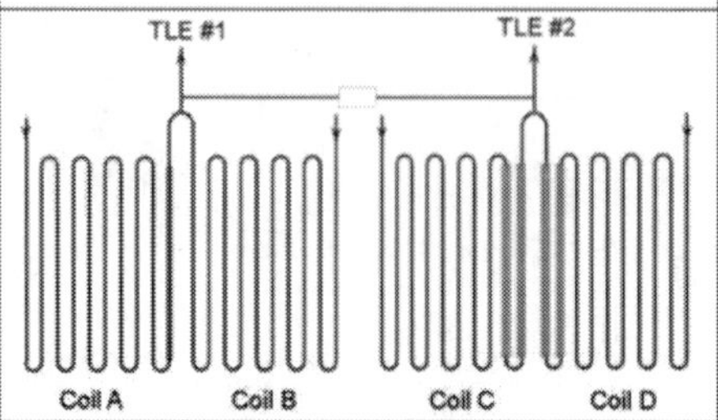	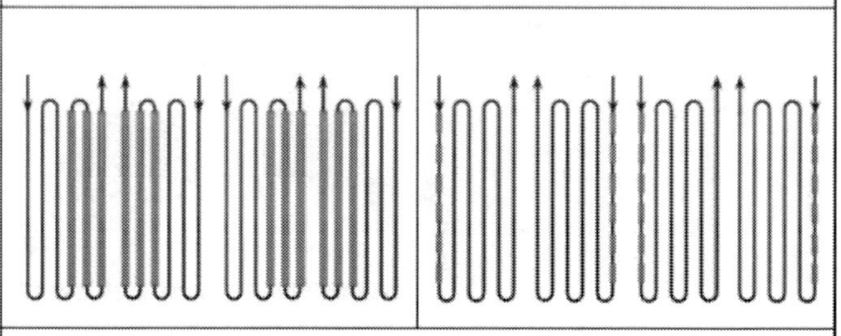	
Tube ID (where EST is applied)	139.2mm (5.48")	95.3 mm (3.75")	
Feed	ethane, 11.0 - 12.0 ton/hr	ethane, 5600 sm^3/hr	
COT	825 - 830 deg.C (1517 - 1526 F)	840 deg.C (1544 F)	
Conversion	~ 60%	~ 60%	
R/L Limit	TMT 1080 deg.C (1976 F), QOT 380 deg.C (716 F)	ΔP 0.15 MPa	
Typical R/L	80 days	30 days	
EST Application	2/10 pass (20%)	3/6 pass (50%)	None (reference)

Run Length

On-stream run length is one of the most important performance criteria for ethylene manufacturers since this is directly connected to the profitability of ethylene pyrolysis, especially if the plant is furnace limited. In most cases, run length is determined by the TMT which rises as the run progresses due to accumulation of coke and resultant poor heat transfer, or pressure drop across the radiant coil which increases, again, due to coking. In our case, run length of Furnace A is TMT limited and that of Furnace B and C is pressure drop limited.

Fig. 4 shows the TMT trends in Furnace A over the first three runs after the installation of PPW overlaid tubes. It is clear that TMT of PPW overlaid tubes rise slower than the conventional tubes and tend to stay at lower levels. Furnaces B and C have limited TMT data but PPW overlaid tubes in Furnace B showed much lower TMT than the same position in Furnace B (by about 60 deg. C). Although Furnace A is TMT limited and TMT of coil C stays low, it is not possible to extend the run length because other coils in the same furnace were conventional tubes and TMT was reaching the limit.

Fig. 5 shows the pressure drop across the coils in Furnaces B and C for the first several runs after the installation of PPW overlaid tubes. In these furnaces, runs are terminated when the pressure drop reaches around 0.15 MPa. Again the effect of PPW overlaid tubes is significant and in this case run length was extended to twice or longer than the furnace with conventional 35%Cr-45%Cr tubes even though this application was only for the 50% of the coils. Also noticeable is that this effect seems to be becoming more prominent after each decoking cycle. This means that the PPW overlay can re-make or fortify the pure surface chromium oxide by oxidization during the decoking process between runs ("self-healing"). Both data above indicate that there is less coking in PPW overlaid tubes, giving rise to better heat transfer and less decrease in effective tube inner diameter.

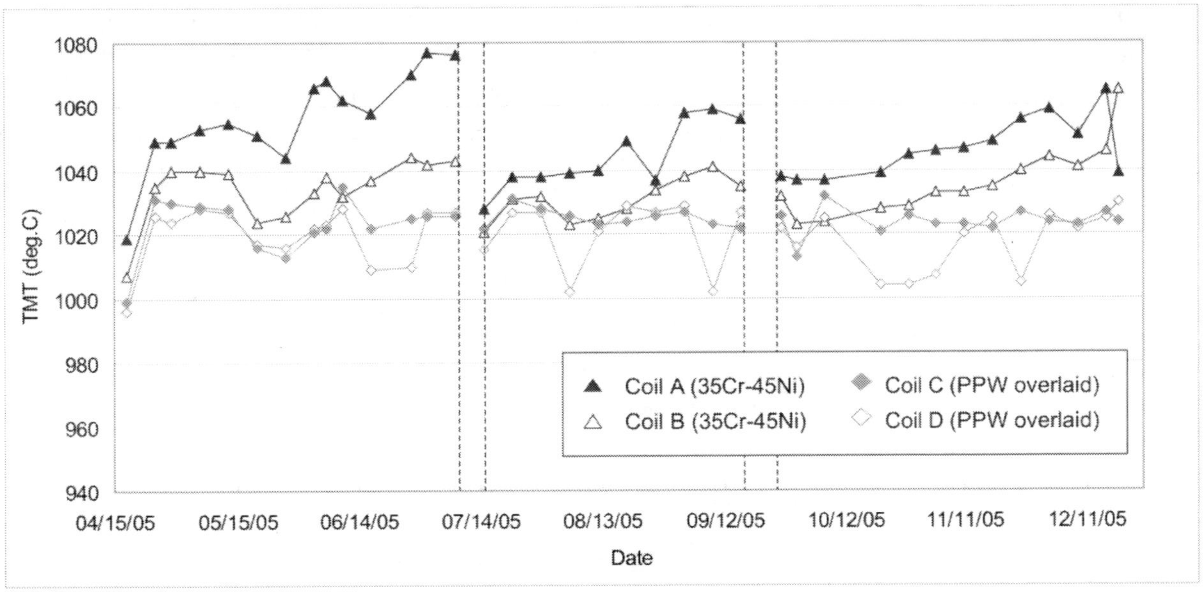

Fig. 4 TMT trend of 4 coils in Furnace A

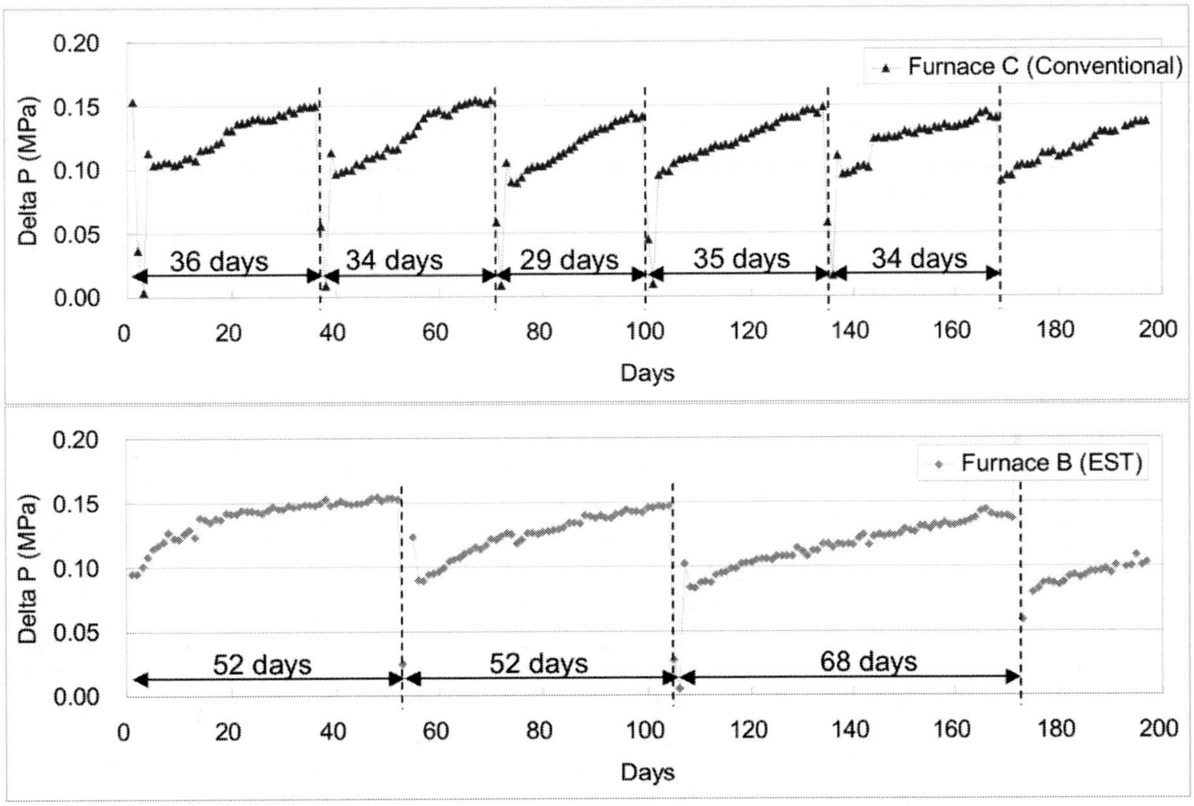

Fig. 5 Trend of pressure drop across the radiant coils in Furnaces B and C

In the case of naphtha feed, the rate of TMT increase was also lower and approximately 30% increase of run length can be expected.

Conversion and Ethylene Yield

In these furnaces, severity of cracking was mainly controlled by keeping the COT (coil outlet temperature) constant and conversion and ethylene yield were monitored. The relative trend of conversion and ethylene yield is shown in **Fig. 6**. While conversion and yield decrease over a period of a run in the coils with conventional tubes, they stay almost constant in the coils with PPW overlaid tubes. When averaged over the run, ethylene yield is 1 - 3% higher in the PPW overlaid tubes.

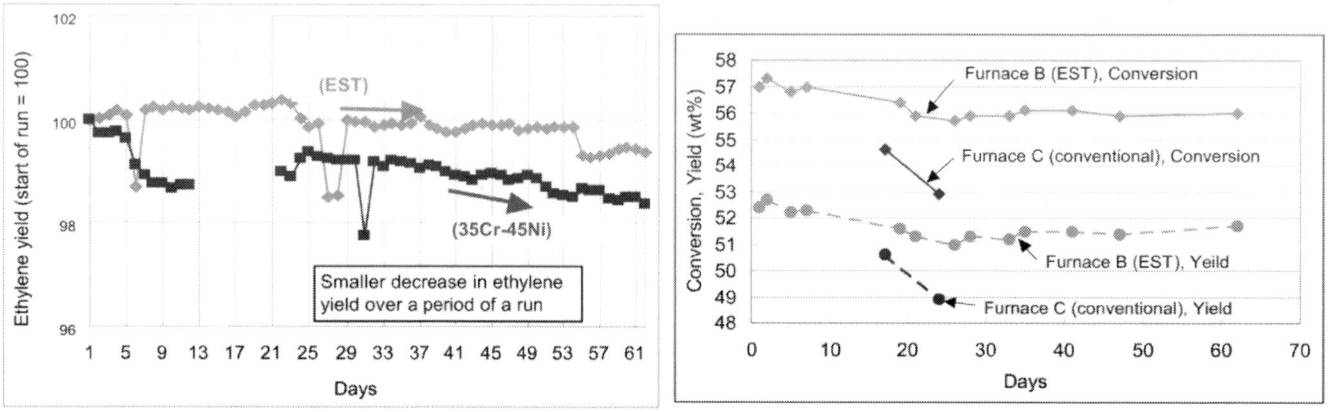

(a) (b)

Fig. 6 Ethane conversion and ethylene yield: (a) Furnace A, (b) Furnaces B and C

Additional Benefits

We have also looked at the pressure TLE (transfer line exchangers) in Furnace A (**Fig. 7**) and found the similar effects as in the case of radiant coils: the temperature (QOT: quench outlet temperature) rise was less and the pressure (COP: coil outlet pressure) rise was also less, indicating that the cooling efficiency of TLE was maintained for a longer period of time. After one year of operation, the number of tubes clogged up in the TLE#2 which was connected to the coils with PPW overlaid tubes was about half of what TLE #1 had.

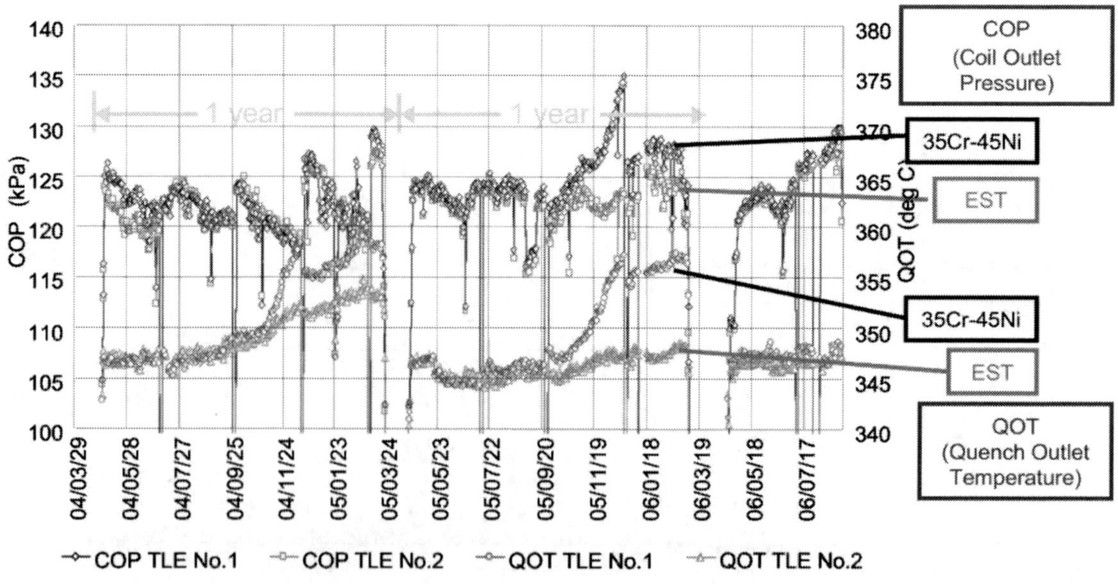

Fig. 7 Trend of TLE temperature and pressure in Furnace A

Analysis of Tube Inner Surfaces

In order to understand the reason for the performance improvements described above, we obtained the tube samples after up to 6 years of operation (naphtha feed) and conducted cross sectional analysis using EPMA (Electron Probe MicroAnalysis). **Fig. 8** shows the compositional mapping of elements for conventional 35%Cr-45%Ni tube (3.3 years) and PPW overlay (6 years) near the inner surface. Both materials have a chromium and oxygen enriched layer very near the surface, indicating the existence of surface chromium oxides. However, if you look at them more closely, you can see that the conventional material's surface oxides contain some iron which is a catalyst for coking, while the PPW overlaid tube has no iron or nickel in the surface oxides. The PPW overlay surface is completely covered by pure chromium oxides, free from catalysts for coking, and this explains the lower TMT and lower pressure drop described above. In addition, the PPW overlay has a much higher chromium content and this enables the self-healing of chromium oxides during the decoking cycle. This corresponds to the pressure drop data shown above and it can be supported by the fact that there is no appreciable difference in the surface structure between after 2 years and after 6 years of operation.

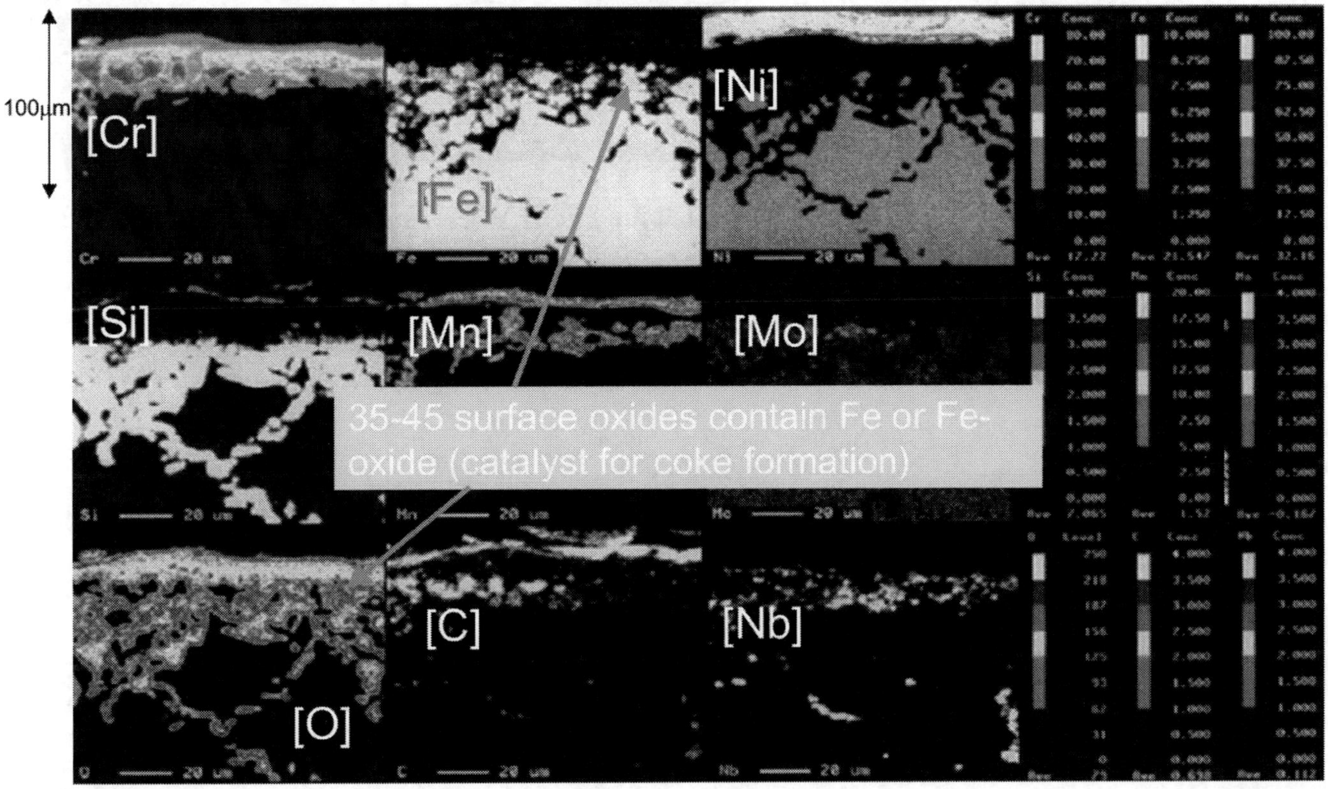

Fig. 8(a) Cross section compositional mapping of 35%Cr-45%Ni tube near the inner surface, after 3.3 years of operation in a naphtha cracker.

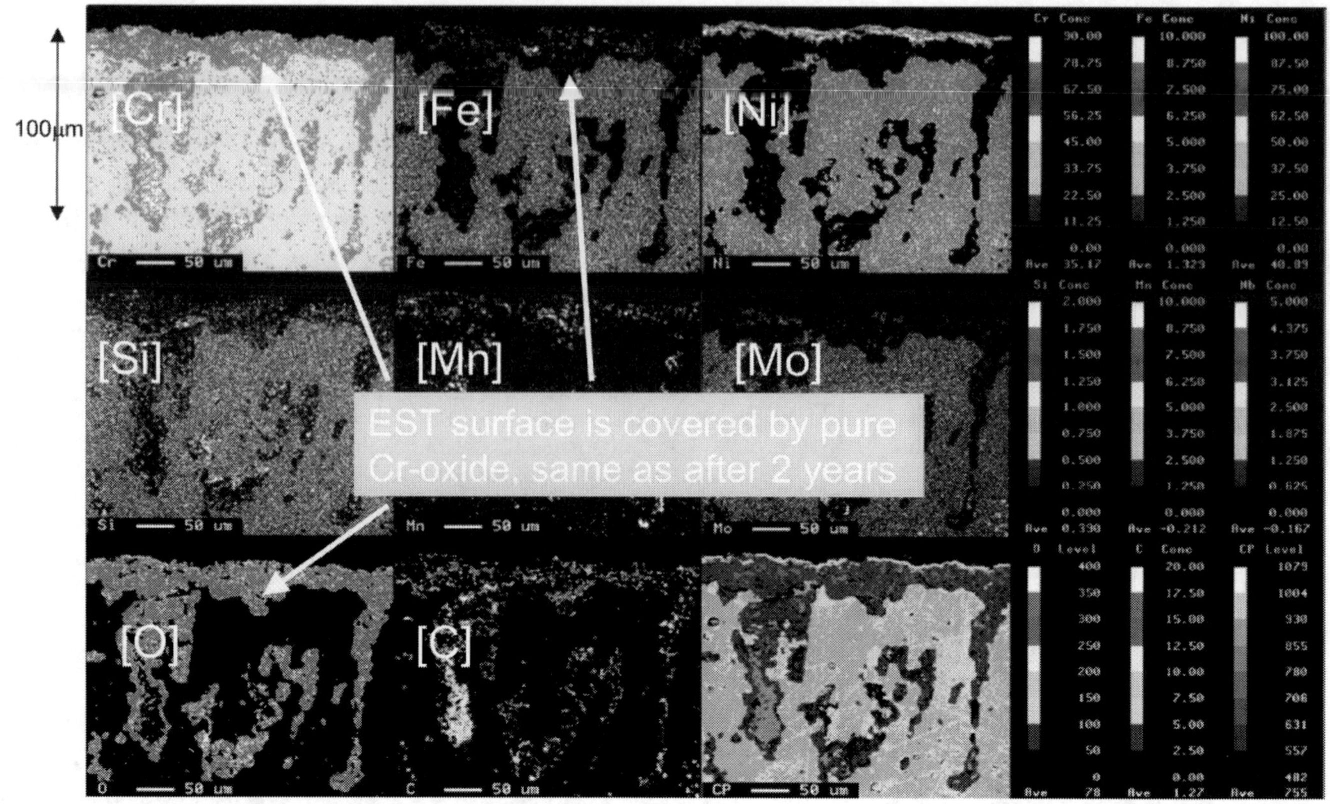

Fig. 8(b) Cross section compositional mapping of PPW overlaid tube near the inner surface, after 6 years of operation in a naphtha cracker.

Discussion of Benefits

The only apparent disadvantage of PPW overlaid tubes is higher cost. Even with an extremely longer tube life (our estimate can be 10 - 20 years), it is not easy to justify the upfront cost penalty against the benefit that can only be reaped after the conventional tube life. However, the following simple but quantitative analysis based on the performance improvements may be able to help demonstrate the short term economical benefits of PPW overlaid tubes.

Assumptions

In this analysis, we take an example ethane furnace with the annual output of 40,000 ton ethylene using 65,000 tons of ethane feed. The furnace margin is $200/ton of ethylene. The radiant coil tube ID is 4" and 9 m in length. The furnace has 4 coils with 6 passes each. The initial run length is 30 days with 2 days for decoking.

As we have seen above, run length extension and yield increase were achieved with PPW overlaid tubes applied to only 20 - 50% of the furnace. In order to reduce the initial cost penalty, it may be a good idea to go with partial application although application to a full furnace may have more pronounced benefits. Here we assume 50% application (3 out of 6 passes). From the data above, we can safely assume the following performance improvements: 2x run length and 1% higher ethylene yield. In addition, due to the lack of carburization to the base tube, we can expect higher severity operation to be possible, in which case additional 2% yield increase with higher COT with the same run length can be achieved.

The additional investment necessary to partially upgrade the furnace tubes from 35%Ce-45%Ni to PPW overlaid tubes (cost penalty) would be:

($1,750 - $600)/m x 9 m x 12 pcs + freight = approx. $135,000. (1)

Case 1: Furnace Limited Plant

When the cracking furnace is the bottleneck of the plant, it is best to increase the on-stream time by increasing the run length. In our case, run length can be 2x longer. The benefit can be calculated as;

Output increase:
R/L (365 days/30 days – 365 days/60 days) x 2 days decoke x 40,000 ton/365 days = 1,316 ton/yr
Yield 40,000 x 1% = 400 ton/yr
Increase in income (return):
(1,316 + 400) ton/yr x $200/ton = $343,200/yr (2)
Time needed for payoff (1)/(2): 0.4 year
ROI (Return On Investment): 154% (first year)

Case 2: Downstream Limited Plant

If the bottleneck is not the cracking furnaces but somewhere in the downstream processes, extending run length or increasing the output from cracking furnaces does not make sense. Instead, what we can do is to take advantage of 1% higher yield and reduce the feed rate while keeping the output the same. We can even lower the feed rate by increasing the severity (additional 2% yield increase), taking advantage of the lack of carburization. The benefit can be calculated as;

Feed rate decrease with 3% yield improvement:
3% x 65,000 ton/40,000 ton = 4.9%
Increase in income (return):
40,000 ton/yr x $200/ton x 4.9%= $392,000/yr (3)
Time needed for payoff (1)/(3): 0.34 year
ROI (Return On Investment): 190% (first year)

ROA (Return On Asset)

It is clear that the additional investment associated with the partial application of PPW overlaid tubes can be paid off in less than 6 months from the performance data alone. In addition, there are added benefits from the lack of carburization to the base tube, such as longer tube life, fewer problems of tube bow and distortion, higher robustness against unplanned shutdowns. These benefits increase furnace availability and potential cost savings in maintenance. Although we are not able to calculate ROA here, it is easily understandable that ROA of the ethylene plant can be improved by the application of PPW overlaid tubes and this should far outweigh the initial cost penalty.

Conclusions

- Partial application (20 - 50%) of high-Cr, high-Ni PPW overlaid tubes to ethylene pyrolysis furnaces can double the run length and improve the ethylene yield by at least 1%.
- The above performance improvements can be attributed to the pure chromium oxides maintained on the inner surfaces that reduces the catalytic coke formation.
- The initial cost penalty of PPW overlaid tubes can be paid off within 6 months and with additional benefits, the plant ROA is expected to be improved.

Acknowledgments

The authors wish to express their thanks to the following people and companies for giving us opportunities to demonstrate our products in their facilities and for giving us permission to publish their operational data: Mr. Takushi Araki (Process Engineering Staff, Ethylene & Energy Production Department, Yokkaichi Manufacturing Complex, Tosoh Corporation), Mr. Hiroki Nabeshima (Senior Staff, No. 2 Maintenance Section, Engineering Department, Kawasaki Operation Center, Nippon Petrochemical Co., Ltd.). We also appreciate the support from the Shell Global Solutions team for their development and marketing efforts.

References

1. Parks, S. B. and Schillmoller, C. M. (1995), "Update in Alloy Selection for Ethylene Furnaces", AIChE Spring Meeting, Paper 24b.
2. Kirchheiner, R., Young, D. J., Becker, P, and Durham, R. N. (2005), "Improved Oxidation and Coking Resistance of a New Alumina Forming Alloy 60 HT for the Petrochemical Industry", NACE/CORROSION 2005, Paper No. 05428.
3. Mullenix, D. and Kurlekar, A (1999), "Results of a Furnace Tube Surface Treatment in a Full Furnace Trial", AIChE Spring National Meeting, Paper No. 63c.
4. Chauhan, A, Moran, W., Casey, E., Si, W., and White, H. (2004), "Limiting Catalytic Coke Formation by the Application of Adherent SiC Coatings via Pulsed Laser Deposition to the Inner Diameter of Tube Material Traditionally Used for Ethylene Pyrolysis Service", Materials Research Society 2004 Fall Symposium, Paper # T3.4
5. Maeda, T. and Terwijn, F. X. (2005), "Carburization Resistance of High-Cr, High-Ni Weld Overlayed Furnace Tubes for Ethylene Pyrolysis", AIChE Spring National Meeting, Paper No. 47.
6. Maeda, T. and Terwijn, F. X. (2007), "The Effect of Plasma Powder Welded Overlay on Carburization and Coke Formation in Ethylene Pyrolysis Furnace Tubes", NACE/CORROSION 2007, Paper No.07419.
7. Kim, D. S., Asrar, N., Maeda, T., Tack, A., and Wolfert, T. (2005), "Performance of Plasma Powder Welded (PPW) Layer on Ethylene Cracking Furnace Tubes – A Technical Evaluation", NACE/ CORROSION 2005, Paper No. 05424.

Multistage Cryogenic Expander Applications

Dalbir Utal, GE Oil and Gas, 3300 Medalist Drive, Oshkosh, WI 54902 and **Kazim Ahktar**, ABB Lummus Global, 3010 Briarpark Drive, Houston, TX 77042.

The cryogenic turboexpander was introduced into the ethylene process design in the 1970's. The expander is generally utilized in ethylene plants to expand hydrogen offgas from the demethanizer feed separator in the LP demethanizer system or to expand methane offgas from the HP demethanizer overhead. Since the offgas is high on hydrogen content and hence has low molecular weight, expansion is normally achieved in multiple stages, mostly in two stages and in some process applications up to four stages.

This paper will discuss the advantages of four stages vs. tradition two stages off expansion.

MPC Fundamentals

Jim Riggs, Texas Tech University, Dept of Chemcial Engineering, Lubbock, TX 79410

A SISO MPC controller will be developed from a step response model by deriving the Dynamic matrix and the prediction vector. Simulation results will be used to demonstrate the effects of design and tuning parameters. The extension of the SISO case to the MIMO case will be also overviewed.

Admc: beyond Classical MPC

Scott Boyden, Cutler Technologies Inc., 70 NE Loop 410, Suite 315, San Antonio, TX 78216

: This tutorial describe Adaptive Dynamic Matrix Control fundamentals. It will provide overviews of ADMC and focus on step by step implementation procedures in industrial settings. Real case studies will be used to provide user's perspective regarding ADMC applications.

The Fundamentals of Oil Fractionator Controls – Advanced or Otherwise

Pete Sharpe, Emerson Process Management, 413 Stuart Circle, #4B, Richmond, VA 23220

Oil fractionation is a very common separation process in all refineries around the world. From atmospheric and vacuum crude towers at the front end to the FCC, Coker and Hyrdocracker main fractionators at the back, these processes often share similar equipment, operating objectives and control strategies. This session will review common fractionation processes, their similarities and differences. Typical control objectives, regulatory and advanced control strategies will be discussed as well as how to find good opportunities for improvement.

Operator Training System for Hydrocracking unit: Real world questions and Answers.

Aleksandr Muravyev and Martin Berutti. MYNAH Technologies, Chesterfield, MO

Abstract

The paper presents a case study on development and implementation of the operator training system for a green-field Hydrocracker unit at a large North American refinery. The users wanted a realistic simulation for operator training and automation design validation. However, the project schedule was compressed and the refinery staff had no experience with developing or maintaining simulators. The challenge of the project team was to find an appropriate balance between a simple simulation model and sophisticated unit operations modeling that mirrored the real process. The solution for this challenge was in the use of a commercially available simulation package designed for operator training, with flexible high-fidelity modeling objects. For instance, the pressure-flow functions for the closed gas track of the unit, implemented using basic simulator, were augmented with models of the compressors and a simple reactor kinetics, taking into account the reactor topology and holdups. The separation block of the unit was simulated using a high fidelity distillation objects, integrated with the process simulator. The distillation package provided simulation of such features, as a real world pressure control for various configurations of the column overhead, weeping at the trays, and "heat shock" in the reboiler. The simulator was successfully used for the plant operator training before the unit start up, as ongoing tool for training in the critical situations (reactor temperature runaways, equipment and control loop malfunctions, e.t.c.), and for testing, tuning and fixing the errors of the DCS control strategies. Customer has used the simulator to find hidden issues in the control strategies that would cause unscheduled downtime as well as certify the operators on the control of the unit.

1. Introduction

Benefits of using Operator Training Systems (OTS) and simulation are proven and compelling. The benefits include faster project startups and time-to-market, improved product quality, reduced plant incidents, and improved operator proficiency In most cases the cost of the simulation system is recovered well before the startup of the process unit, and investment in the simulation software has the greatest return of any capital investment.

This paper presents a case study on development and implementation of the operator training system for a green-field Hydrocracker unit at a large North American refinery. The users wanted a realistic simulation for operator training and automation design validation. The OTS should provide tools [1] that

- Assist in safe, environmentally sound and profitable startups, shutdowns and emergency response of the new Hydrocracking unit
- Reduce the number and severity of unplanned events
- Help with unit control and optimization

The OTS should allow the Hydrocracker process leads to train on the unit before it started up, and later be used to implement ongoing training and qualification of new Hydrocracker process leads.

The challenge before the project team was to develop a simulation model that would be adequate to the purposes of the operator training and provide additional functionality for process control and optimization.

2. Challenges of the Hydrocracking Process simulation

The dynamic model of the process should meet the goals of the operator training. In this section we will give a brief description of the simulated process, then consider the operator training situations (scenarios) user would like to implement, and, finally, evaluate the model functionality and complexity needed.

2.1 Process description

A Hydrocracking Unit (figure 1) catalytically cracks heavier petroleum fractions in the presence of hydrogen to make more valuable products. In the Reactor Section the hydrotreating reactions occur first and remove Sulfur, Nitrogen and the other undesirable components. The hydrocracking reactions convert some of the feed molecules into the smaller molecules that can be separated into valuable naphtha product streams. These components are separated then in the Fractionation Section of the unit into streams that will go into making finished gasoline and distillate products. A Recycle Gas Compressor returns non-reacted hydrogen to the reactors. A Make up Gas Compressor (reciprocal) supplies make-up hydrogen to replace the gas, consumed in the reactions.

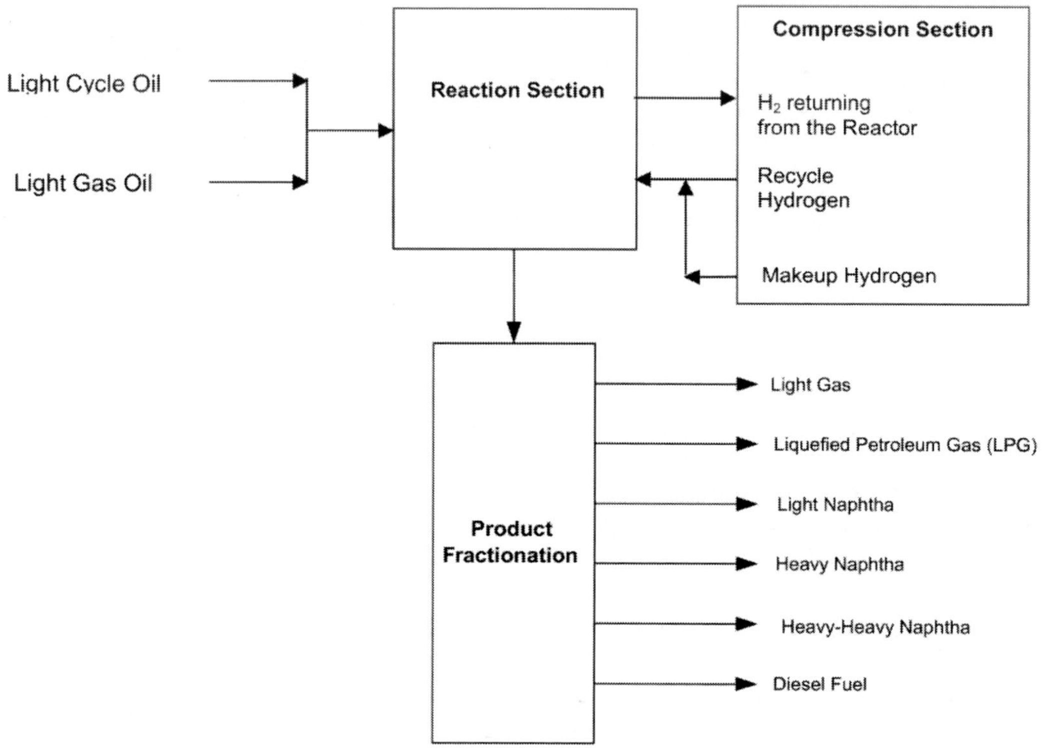

Figure 1. A Hydrocracking Unit schematic

The unit liquid feed can be represented by two generalized components, Light Cycle Oil (LCO) and Light Gas Oil (LGO) having different reactivity and heat of reaction. The unit gas feed, hydrogen, is used as both, a reagent and a cooling (quenching) agent.

2.2 Operator training scenarios

The user was interested in the development of realistic and effective scenarios for basic operator training and response to critical situations of the process operation (associated with unit upsets and malfunctions of equipment or instruments). The list of the upset situations includes:

- Reaction runaway due to the feed disruption or malfunction of the reactor quenching valves
- Recovery of the unit after runaway and depressurizing of the system
- Process upset due to sudden change of the feed composition

In these scenarios the operator should recognize the situation and take the actions to bring the process to normal operation or to safe shutdown conditions. The simulator can be utilized to train the operations engineers on basics of operation and control strategies for distillation, heaters, and/or reactors.

2.3 Model functionality and complexity needed

The operator training scenarios, suggested above, define the requirements for process model functionality and complexity. Particularly, in order to simulate the unit recovery after runaway and depressurizing the model should contain an adequate description of the hydrogen recycle and make up tracks, including the compressor models. The simulation of the reaction runaway due to malfunction of the reactor quenches assumes precise enough modeling of the individual reactor beds, including the heat of reaction and kinetics. The models of the feed composition propagation through the unit and kinetics/heat of reaction for the individual components are needed to simulate the process upsets due to sudden change of the feed composition or feed flow disruption. In order to provide basic operator training in distillation operation and control (product fractionation area) the multivariable column models for the main interacting process parameters (pressures, temperatures, and flow rates and compositions of the inlet/outlet streams) are needed.

A development of these models can require extensive time and resources. However, on this project the schedule was compressed and the workforce limited. Therefore, the challenge was to find an appropriate balance between a simple simulation model and sophisticated unit operations modeling that mirrored the real process.

3. Solutions: Flexible combining of the Process simulation complexity

The solution for this challenge was in the use of a commercially available simulation package designed for operator training, with flexible high-fidelity modeling objects. A balance of high and medium complexity models was selectively applied to the unit operations of the process. MiMiC Simulation Software [2], was used as the simulation solution for the project. MiMiC has utilities for auto generation of the model skeleton and a variety of embedded functions for simulation of common process unit operations (heat exchangers, vessels, mixers, flow rates of the vapor and liquid streams, e.t.c). It provides the tools for fast and easy development of medium complexity process model. The medium complexity models were augmented with high fidelity models for three main process areas: hydrogen recycle and make up tracks, reactors, and product separation.

3.1 Hydrogen recycle and make up tracks

The pressure-flow functions for the closed gas track of the unit, implemented in the simulation, were supplemented with models of the compressors. The Recycle Gas Compressor was simulated by the special models, employed the centrifugal compressor model using approximations of its design pressure- flow curves. The models were implemented in the simulation software User Calculation blocks.

The simulation of the Make up Gas Compressor, which contains several reciprocal compression stages, consists of a series of User Calculation blocks. Each block includes a model of the polythropical compression along with the interstage gas conditioning. The model also simulates the stage loading and unloading functions with realistic compressor performance simulation.

3.2 Reactor section

A number of a high fidelity features, taking into account a simplified reaction kinetics, reactor topology and holdups, were incorporated in the reactor section model.

The overall reactor model consists of the interacting models for the reactor beds. The model for each bed contains material and heat balances, which include transformation balances and heat of reactions for two generalized liquid components, LCO and LGO, and for the hydrogen gas component. The simplified reaction kinetics was modeled using the piece-linear approximation of the Arrchenius type equation for rate of reactions in the working interval of temperatures with correction coefficients for the reactor pressure and liquid holdup. Since the beds different in the roles, hydrotreating or hydrocracking, the model contains two sets of the reaction equations, respectively.

These equations of the reactor model were implemented using a set of interacting User Calculation blocks in the simulation software.

3.3 Distillation Package for the Product Fractionation

The Product Fractionation block of the unit was simulated using a high fidelity distillation objects, integrated with the basic process simulator. These objects were created, initialized, and integrated with the rest of the process simulation using the MiMiC Distillation Package [3].

The Distillation Package allows the user to build dynamic models for a full-scale distillation column simulation, including reboiler, condenser, and reflux receiver (figure2). The column can have up to five feed and side withdraw streams, three tray sections with different diameters, and up to ten components, separated in the column. The features of the column object include two liquid phase separation in the reflux receiver, constant or variable options for column pressure and tray pressure drop, tray weeping.

A distinctive feature of the model is the overhead section layout that makes possible simulation of a variety of the real world pressure control loops (described, for instance, in [4]) used in the distillation. Particularly, the column pressure can be controlled (see the basic column configuration, figure 2):

- By manipulation of cooling agent flow rate to the condenser
- Using control valve on the vapor bypass around condenser to the reflux drum
- Using directly control valve on the vapor stream to the condenser
- By manipulation of flow rate of the outside vapor stream from/to reflux drum (split range control commonly used with the partial condenser)

The Distillation Column Package offers to user a convenient interface to each of the objects, providing:

- Creation and configuration of the object
- Initialization and running of the object
- Visualization tools to monitor the object performance

An example of the run time user interface to the Debutanizer object is shown in figure 3. It contains a real time profile of the column liquid composition, offering a valuable insight into the object.

The distillation column model built with the Distillation Packages has the ability to be initialized and started from a Cold or Warm (near steady-state) status. User controls within the Distillation Modeling Package allow the user to perform initialization off-line and then launch the object to run on-line with a basic simulator.

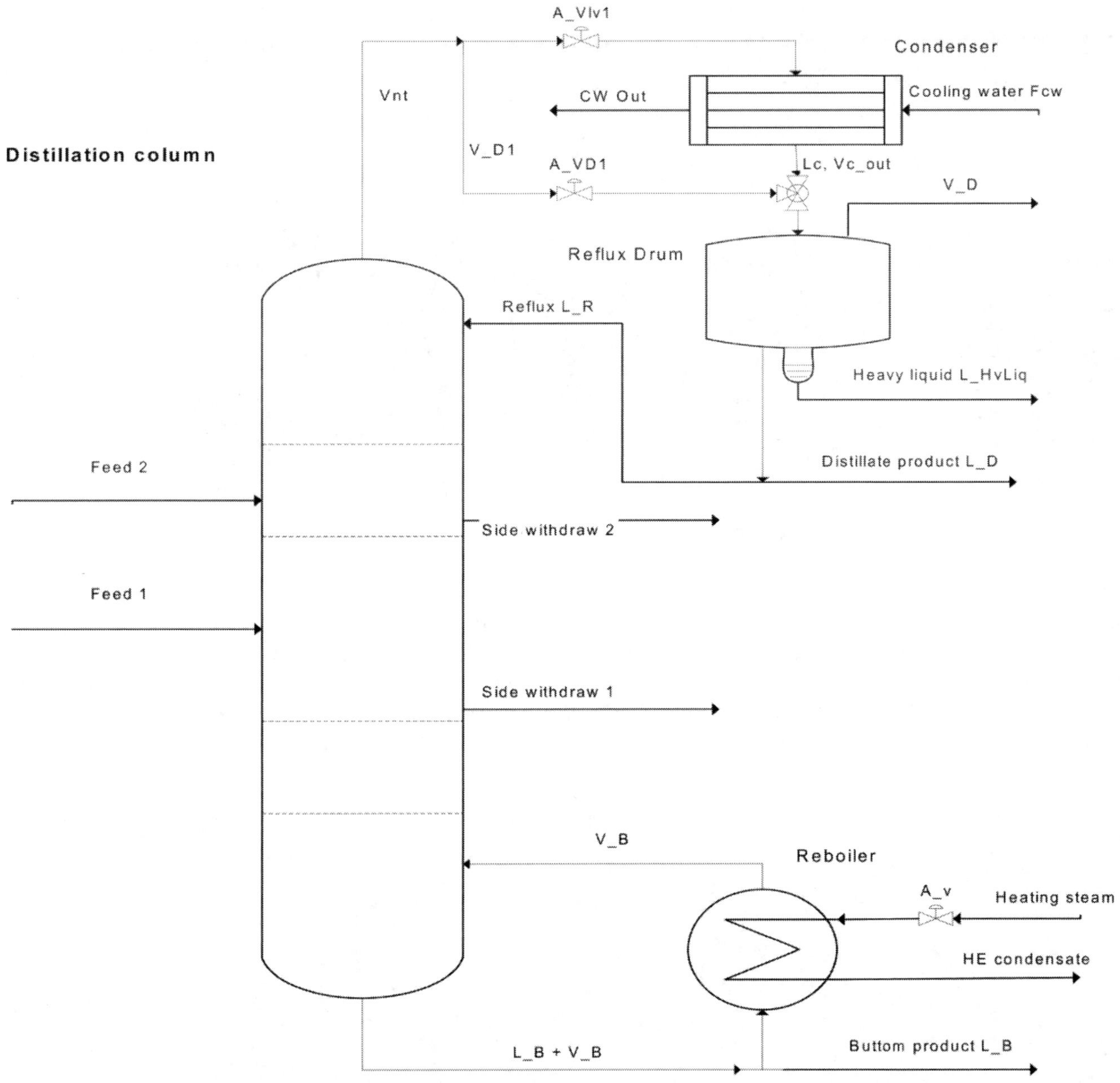

Figure 2. Basic Configuration of the Distillation Column with two liquid phase separation in reflux drum

The Distillation Column models for the Product Fractionation section were configured, initialized and launched on-line using the Distillation Package features, described above. Note, that the package allows creation and initialization of new models while the others are already running on-line, providing flexibility in the process of building and tuning of the entire plant model.

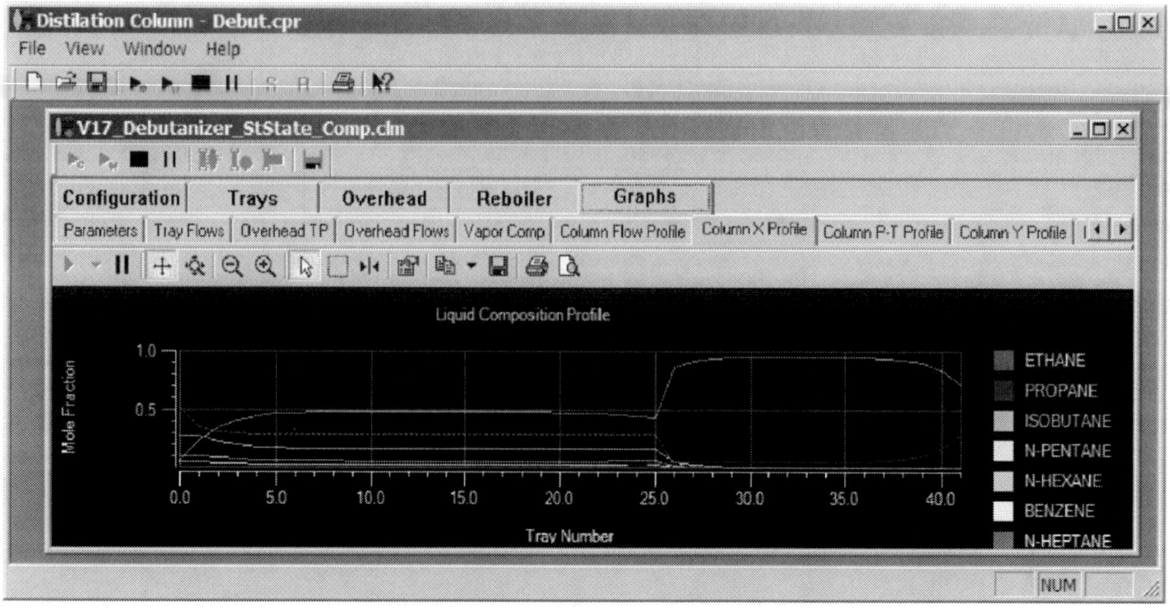

Figure 3. Object user interface with graph of the column liquid composition profile

The distillation package provided simulation of such features, as a real world pressure control for various configurations of the column overhead, weeping at the trays, and "heat shock" in the reboiler. An example of a complex pressure control scheme of the real process is shown in figure 4. It includes the column pressure control loop using the valve at the vapor inlet to the condenser and the pressure difference loop, controlled by the valve on the "hot vapor bypass" around condenser. Due to the flexibility provided by the Distillation Package, this scheme was implemented in the model without compromise.

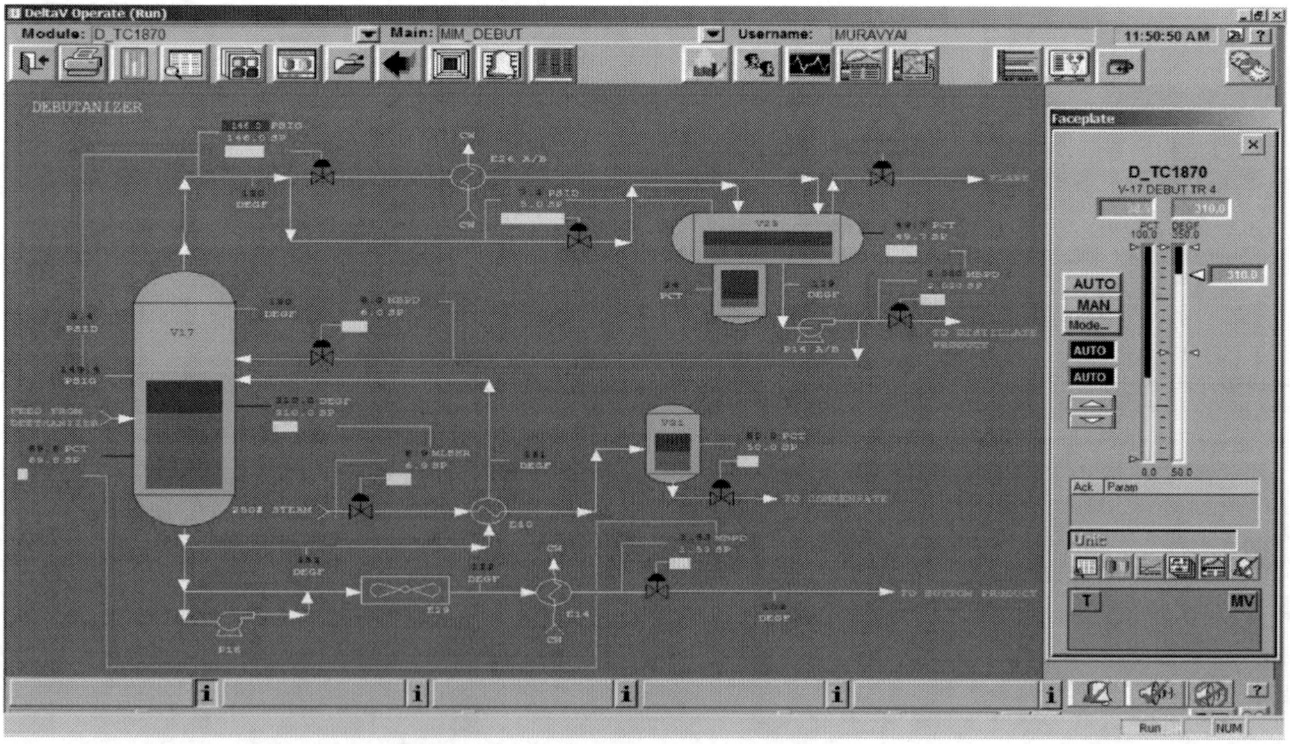

Figure 4. DCS screen of the column with the process controls

4 Conclusions

The Operator Training Systems for the new Hydrocracking unit was developed by means of combining of a commercially available simulation package designed for operator training, with flexible high-fidelity modeling objects. The selective application of high and medium complexity models to the unit operations of the process, provided simulator functionality that completely met or exceeded the customer expectations.

The simulator was successfully used for the plant operator training before the unit start up, as ongoing tool for training in the critical situations (reactor temperature runaways, equipment and control loop malfunctions, e.t.c.), and for testing, tuning and fixing the errors of the DCS control strategies. The customer has used the simulator to find hidden issues in the control strategies that would cause unscheduled downtime as well as certify the operators on the control of the unit.

References

1. Anne Walztoni and Todd Anstine, "The Refinery Behind the Monitor: DeltaV training System for Greenfield Hydrocracker", in *Proceedings of Technical Conference*, 2006 Emerson User Global Exchange, October 2006, Nashville.

2. MiMiC Product Bulletin, http://www.mynah.com/software_documentation.asp, 2006

3. MiMiC Distillation Package Product Bulletin, http://www.mynah.com/index.asp, 2006

4. Béla G. Lipták, "Instrument Engineers' Handbook, Fourth Edition, Volume Two: Process Control and Optimization", ISA, 2005

Integration of Complex Sequences Using Delta V

Dean Taggart[1], **John Kingston**[2], and Stephen Krause[1]. (1) Emerson Process Management, 7175 12 St SE, Calgary, AB T2H 2S6, Canada, (2) Spartan Controls Ltd, 305 27 St SE, Calgary, AB T2A 7V2, Canada

Companies have often struggled with the implementation of complex sequences involving tight integration of the SIS and DCS. Implementation often proved difficult due to the independent nature of SIS, which is problematic when communication between the SIS and DCS is important.

With the release of DeltaV SIS, those issues are no longer a concern. DeltaV SIS along with DeltaV presents options not available before.

For a very complex application like startup and shutdown of an oilsands gasification process, DeltaV DCS and DeltaV SIS are an excellent fit. With the State Transition Diagram function blocks now available with DeltaV SIS, and with the proper implementation methodology, implementation can be done in a logical, efficient manner. With PlantWeb integration, the creation of HMI displays as well as interaction with the highly complex DCS control are greatly simplified.

This paper describes the design issues for integration of the safety, control and operator functions in this new architecture. Technical issues such as the use of equipment (position transmitters, DVC's) and DCS/SIS interaction are discussed, as well as the importance of the implementation framework, which is the heart of the design.

Building a Grassroot Refinery in the 21st Century: Which Way to Go?

Ahmed Khogeer, New Business Development- New Yanbu Export Refinery Project, Saudi Aramco, Houston, TX 77002 and Thomas Phillips, New Yanbu Export Refinery Project, Saudi Aramco, 500 Jefferson Street, Room 1783A, Houston, TX 77002.

Petroleum refining is a capital-intensive business. With the stringent environmental regulation on the process industry, political instability and declining refining margins, higher pressures are exerted on the refining industry. This led to misbalance in the refining capacity growth during the last twenty years. The last grass root refinery was built in the USA was about thirty years ago. In addition, the number of USA refineries shrunk by 40 refineries in 8 years. Moreover, with the increased risk of war and terrorist attacks, refineries and fuel transportations grids are targeted since they supply fuels to essential nodes such as power plants and war efforts which is another constraint on the refining business requiring increased expenditure on safety and security. With the substantial increase in world demand for oil and refined products, the need for refining capacity had changed much during the last a few years. In addition, many of the world refineries will not be able to meet the new stringent product specs in Europe and USA. This work covers the important aspects to take into consideration when deciding to build a new refinery or expanding an existing one including the markets, the crudes used products slate, complexity desired, availability factor, and other economical and financial factors. Case studies were done on several refineries models serving different markets with different products needs. Simulation runs were done using ASPEN PIMS utilizing nonlinear techniques to solve the optimization problem. The results addressed the benefits of processing heavy crudes to utilize the spread between the light and n heavy crudes. It also addressed how exporting high spec products can be more profitable than tailoring the refinery for domestic use even if the products were to be sold domestically at international prices and what needed to be done in order to accomplish that. In addition, the results showed tendency towards more dieselization in expense of gasoline production. The model also covers the problem of controlling benzene content in the gasoline product.

Overhead Corrosion Problems Caused by Opportunity Crude

Paul Eaton[1], Tuyet Le[2], Murray R Gray[2]

1. Champion Technologies, Fresno TX
2. Department of Chemical and Materials Engineering, University of Alberta

Abstract

Processing opportunity crude can play an important role in refinery profitability, but the risks are high because these crude oils usually come laden with contaminants which cause high maintenance costs and equipment losses due to excessive corrosion. In particular, overhead corrosion is caused by the mineral salts, magnesium, calcium and sodium chloride which are hydrolyzed to produce volatile hydrochloric acid, causing a highly corrosive condition in the overhead exchangers. Therefore these salts present a significant contamination in opportunity crude oils.

Other contaminants in opportunity crude oils which are shown to accelerate the hydrolysis reactions are inorganic clays and organic acids. The impact of each of these contaminants on hydrolysis is shown in a simulated crude oil and two crude oils from Canada, classified as "opportunity". Hydrolysis intensity was determined as a function of chloride production under distillation conditions in the laboratory. Chemical additives are shown to be effective in reducing the hydrolysis reaction for various opportunity crude oils, currently on the market.

Taking Some of the Challenge Out of Challenge Crudes

Sam Lordo, Nalco Company
Richard Mathers, Nalco Company

Introduction

Volatility of today's crude prices present opportunities for refineries to significantly increase operating margins for those refiners that are willing to expand the typical crude slates they process. Processing "Opportunity Crudes" or "Challenge Crudes" has become a strategy that many refiners around the world are using to increase their profitability and competitiveness. Opportunity crudes or challenge crudes are considered to be crudes that have discounted prices either due to known processing issues or new crudes that may have the potential for processing challenges. For crudes that have been out on the market for a period of time these issue can be fairly well known. However, for the newer crudes how they behave in refineries is not completely known.

Successful strategies for processing opportunity crudes include processes that are used to predict what type of issues that may arise as well as identify mitigating strategies. These processes allow for a refiner to make a quick assessment on the risks of running opportunity crudes so they can anticipate the processing problems and implement cost effective mitigation measures.

This paper will address several aspects of a successful strategy in predicting the impacts of opportunity crudes on crude operations. This strategy provides reduction in the perceived risk associated with the processing of a new crude allowing a faster uptake of crude for the suppliers and a possible reduction in the cost differential usually associated with the pricing of a new crude. The scope of this paper will not include the effects that can be readily predicted from information commonly found on crude assays, such as metals, product quality, etc.

The term "opportunity crude" can be described as any crude that is traded at a discount price:

- New to market crudes
- Crudes known to cause processing problems
- Distressed cargos

The process used to identify, assess, and mitigate potential impacts encompasses laboratory analysis and interpretation through a combination of standard and proprietary techniques in the 3 key areas affecting refiners:

- Desalting

- Fouling
- Corrosion

A combination of industry experience and proprietary test methods allows the anticipation and understanding of impacts on refineries processing opportunity crude oils.

By developing this information one can potentially lower the perceived risk of processing, leading to a faster uptake of crude for the supplier and more attractive margins for both the supplier and refiner. For the refiner, this information gives a potential to identify alternative crudes that could be viewed as alternatives to their existing crude slate by providing benchmark comparisons.

Below is a summary of some of the testing that can be done to identify the potential risks and identify mitigating steps in processing of crude oils new to a refiner or to the industry.

Desalting

Emulsification Characteristics

Desalting is the process of removing salt and water wettable solids from crude by emulsification with water and then breaking the emulsion and separating the phases. The desalting operation performance is influenced by gravity, viscosity, temperature, conductivity and throughput. Changes in the specific gravity, asphaltene content, sulfur level and crude contaminants such as solids and surfactants will all affect desalter performance but the key characteristic of the crude is its emulsification characteristics.

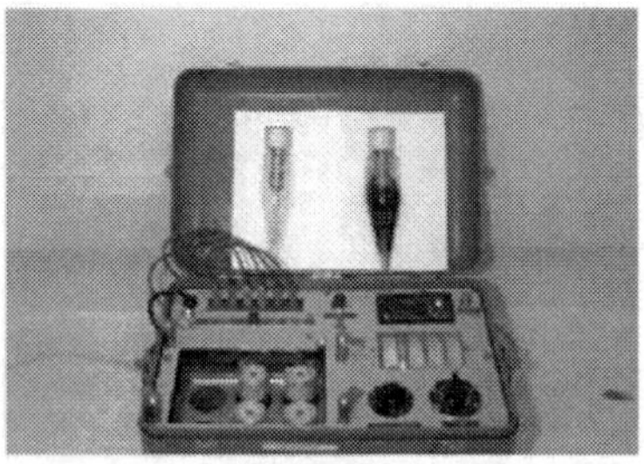

Figure 1: PED equipment

The Portable Electric Desalter (PED) test is a static (i.e. no flow) representation of the desalting process and is designed to compare the relative ease of demulsification of various water-in-oil emulsions and also to compare relative demulsifier chemical performance; see figure 2. It does not reproduce refinery-desalting conditions of mixing energy, temperature, pressure, dosage level and electrical field; however results from this method have been shown to be relevant to actual refinery performance.

Test Description

In the test, a water-in-oil emulsion is prepared under controlled conditions in a homogeniser. The emulsion is poured into glass tubes and the tubes placed into the PED unit. The emulsion is then resolved using a combination of heat, electric field, and chemical treatment, with the resolution of the emulsion being noted by the volume of water separated during the test period. The volume of water and the rate at which it is separated are used as the basis for comparison of the efficiency of the process.

This test can be carried out on any crude or blend of crudes to allow appropriate representation of a refinery crude slate and a direct comparison of individual crudes. It also allows technical comparison of emulsion breaking chemistries to identify the best solution prior to implementation in the refinery.

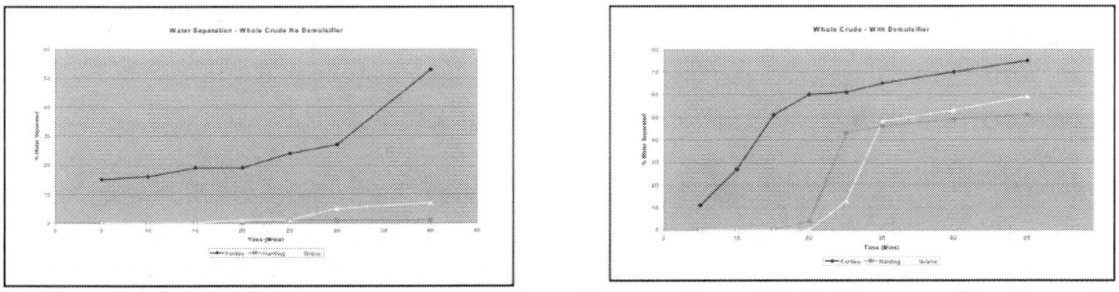

Figure 2: Comparison Of Degree Of Water Separation Of 3 Known Crudes; With And Without A Demulsifier.

As part of the desaltability testing, crude conductivity should also be measured. This helps ascertain whether desalter electric fields will be impacted by processing this crude either at 100% or as part of a pre-determined blend. Measuring the conductivity of a crude blend is important, as conductivity appears not to be a parameter that can easily predicted by just knowing the conductivity of individual crudes.

Figure 3 shows the conductivity of two different crude oils and their resulting blends. As you can see in this case the conductivity of the blend is higher than either of the individual components.

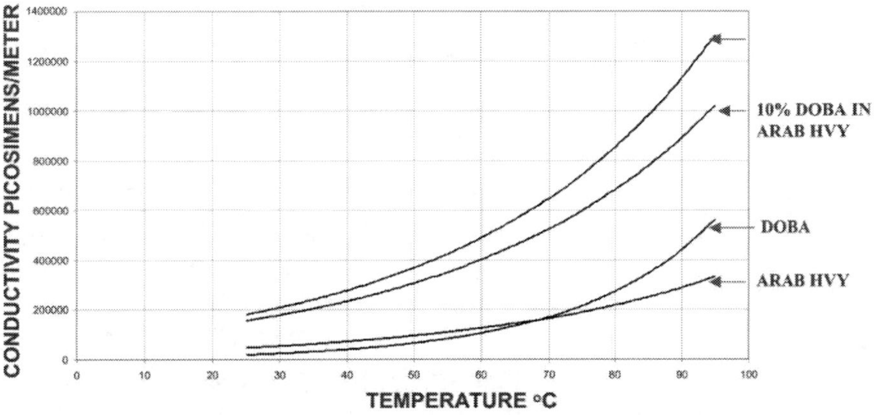

Figure 3: Crude blending impacts on Conductivity

Fouling

Fouling in crude unit preheat exchangers is often, but not exclusively, caused by destabilization of asphaltenes contained within the crude.

Fouling can occur via a number of different mechanisms and at different points throughout a crude unit. Fouling may be caused by organic or inorganic mechanisms involving salts, sediment, iron sulfide, polymers or asphaltenes. The crude contaminants that cause fouling are frequently not identified in conventional crude assays. Some refiners observe that when two crudes are blended together they will cause fouling and yet will not cause fouling when each crude is processed individually or blended with other crudes. This type of fouling is typically caused by the destabilization of asphaltenes. The fouling can broadly be categorized into two types:

- **Crude compatibility**

  Mixing a crude, or multiple crudes, containing asphaltenes with another crude can cause the asphaltenes to be destabilized and agglomerate to form solid foulant particles. This is caused by the second crude solvating the resins that were present to disperse the asphaltenes, leading to asphaltene agglomeration. Typically, this is observed when a heavy asphaltic crude is mixed with a more paraffinic crude. The proportions of each crude type and the order in which they are mixed also strongly determine the potential for asphaltene destabilization.

- **Thermal destabilization**

  Resins in crude oil are bound to the large asphaltene structures, and serve to keep them suspended and dispersed in the crude. The strength of the resin-asphaltene interaction decreases upon heating and the resins are removed from the asphaltenes. This allows the asphaltenes to agglomerate and form particles of foulant. The extent to which the asphaltenes are stabilized at higher temperatures depends upon the strength of the asphaltene-resin interaction. The more strongly the resins are bound to the asphaltenes, the less prone the asphaltenes are to agglomeration and fouling

Fouling Due to Crude Incompatibility

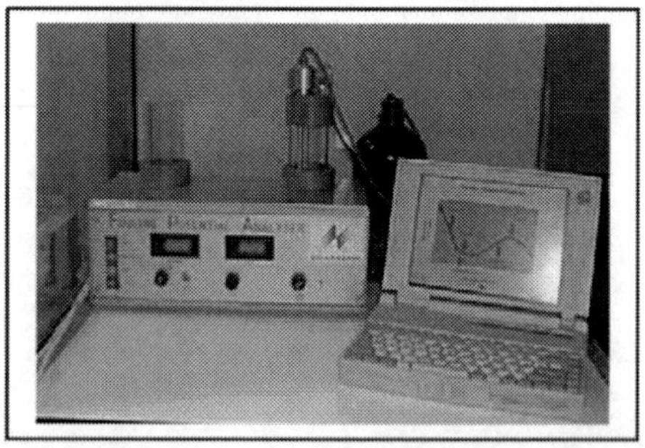

There are a number of methods Industry that are used to predict the potential for destabilizing asphaltenes in a crude matrix. Nalco Ltd. Energy Services Division developed an unique instrument, Fouling Potential Analyzer (FPA), that is used to determine the stability of fouling precursors in the crude (predominantly the asphaltenes) by titration with an aliphatic solvent. The 'peak' in the titration curve (the so-called 'flocculation point') is indicative of their stability. The amount of solvent added at the peak minimum is noted and converted to a 'Value'. This allows a direct comparison of crudes and refinery slates and indicates the fouling tendency due to asphaltene destabilization of the crude prior to refinery processing. The test is done can be done on desalted crude samples and can determine relative crude fouling potential in a crude unit preheat train due to asphaltene instability.

This test method is also used to determine if desalter operations maybe compromised by destabilized asphaltenes.

Test Description

The crude sample is titrated with a paraffinic solvent, which will precipitate asphaltenes (e.g. N-pentane) and the change in absorbance of the crude is monitored in the I.R. wavelength range. As the solvent is added, the absorbance decreases, until the point at which enough solvent has been added to sufficiently remove the resins to allow the asphaltenes to agglomerate. The agglomerated asphaltene solids absorb in the I.R. range and are detected by the FPA. The amount of solvent needed to cause the onset of asphaltene precipitation is dependant on the strength of the resin-asphaltene interaction. It is found that high fouling crudes require small volumes of solvent to cause precipitation, and lower fouling crudes require much more solvent to overcome the resin-asphaltene interaction.

The FPA number, which is a function of the volume of solvent added, is an index of the position of the "inflection point" for each crude blend, as shown in Figure 4. This can be used to compare different crudes with respect to their fouling potential, as shown in Figure 5.

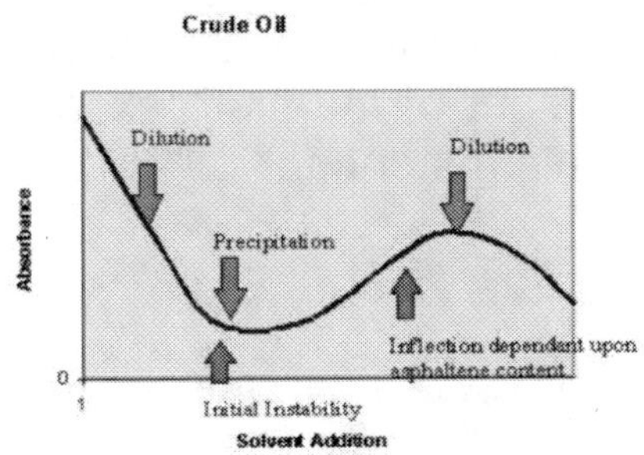

Figure 4: FPA Output

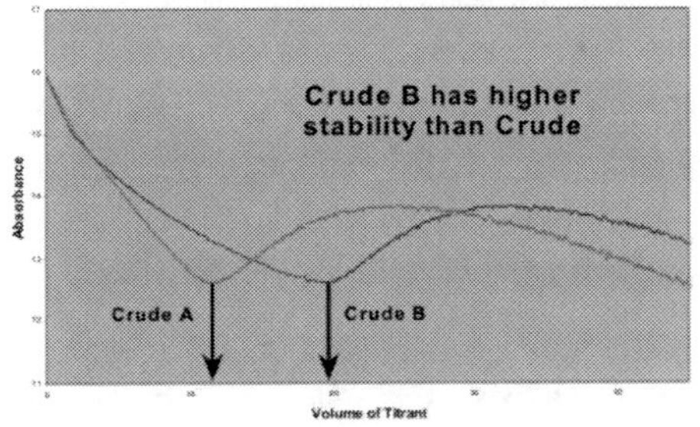

Figure 5: Comparison of high and low fouling crudes

Figure 6 shows several crudes and their corresponding FPA values and the scale that has been developed to interpret the reading from the FPA analyzer. The scale was developed using the ALCOR (discussed further below) and field observations.

Crude	FPA Value
100% Grane	77
30% Grane, 70% Ekofisk	76
REB	63
Kirkuk	45
Urals	56
Maya	<40

FPA Value	Predicted Fouling
<30	Severe
30-40	High
40-60	Medium
60-70	Low
70-100	Very low

Figure 6: Predicted fouling due to asphaltene destabilization of several well-known crudes.

An alternative to the FPA is a crude blending index. These have been developed by several companies and are used to predict asphaltene instability. For the most part these are good to use to get a general guideline for blending. However, these indexes are based on samples of crudes that no longer may be valid as crude source has changed. These indexes also are limited to whole crudes and can not predict how the addition of slop oils may impact asphaltene stability.

Total Quantity of Asphaltenes

In addition to the FPA value, the total asphaltene content, obtained from the hexane insolubles test, is an important parameter. Lower stability crude is much more of a concern when the absolute quantity of asphaltenes in the crude is high. This test is used in conjunction with the FPA test to provide a better understanding of the magnitude that a potentially fouling crude or crude slate might have on desalting or preheat exchanger fouling.

Test Description

In order to determine the total quantity of asphaltenes in a crude blend, a sample of the crude is vigorously mixed with hexane. The hexane is added in excess, to precipitate all of the asphaltenes in the crude samples. This solution is then filtered, dried and weighed. After weighing the filter cake is washed with hot toluene to remove the asphaltenes, dried and weighed again. This gives the filterable solids. The weight of the first hexane is subtracted from the second hot toluene wash to give the amount of asphaltenes.

Asphaltene Dispersion

As previously mentioned asphaltene deposition is a common fouling mechanism observed in refinery heat exchanger networks. Thermal rating calculations are performed to track the effects of exchanger fouling over a period of time. Fouling occurs over a time period that may vary from days to months to years depending on the unit being considered. In the asphaltene dispersancy test (ADT) we can simulate fouling in minutes instead of months.

The ADT provides a rapid method to discern dispersant performance on a specific crude oil. The basis of the ADT is to determine relative effectiveness of dispersants in keeping asphaltenes dispersed in a non-solvent medium. Like several procedures (ASTM D-4124; API, SARA, D-2006 & D-2007), the ADT takes advantage of the insolubility of asphaltenes in alkane diluent (like *n*-hexane, *n*pentane, *i*-octane, and *n*-heptane).

Test Description

The hexane insolubles test is performed as above, but repeated a number of times with varying quantities of the antifoulant dispersant chemical pre-dosed into the crude. Dilution of a fixed volume of crude oil in a fixed volume of alkane diluent results in precipitation of asphaltenes. Dilution of the same fixed volume of crude oil properly dosed with dispersant in the same fixed volume of alkane diluent results in minimal precipitation of asphaltenes during a controlled testing period.

An effective dispersant will prevent the agglomeration and eventual precipitation of asphaltenes when the crude oil sample is diluted in alkane diluent. The effect of a dispersant is measured by comparing the amount of deposit after a half hour period relative to the untreated test. This field procedure does not allow direct comparison with other crudes but does establish the most appropriate chemistry to use to disperse any asphaltenic material.

Fouling Due To Thermal Destabilization

ALCOR Hot Process Simulator testing is used to determine two parameters:

- Fouling potential of crude and atmospheric residue streams
- Chemical program recommendation and effect of antifoulant in mitigating fouling

Crude samples can be evaluated on the ALCOR Hot Process Simulator; Nalco technology center has several of these units. The ALCOR results can be compared with a database of crude samples to rank the fouling potential – although operational factors of this test mean this is only an approximate comparison.

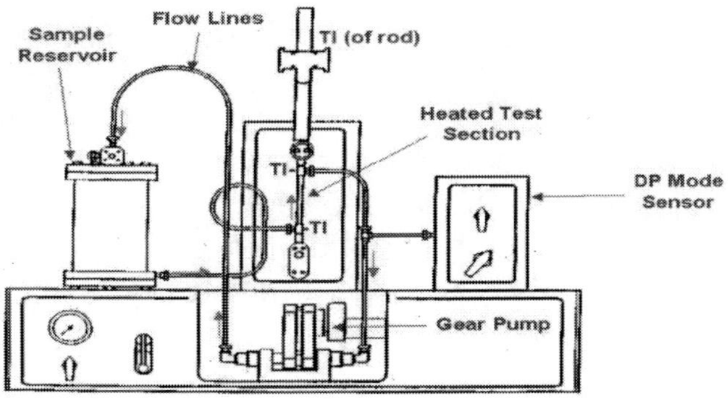

Figure 7: ALCOR Hot Liquid Process

Test Description

The ALCOR apparatus comprises a steel reservoir and a hot pressurized rod over which the oil is pumped. The hot rod simulates the heating effect of the heat exchangers. The rod can be programmed to heat to a predetermined temperature. The oil inlet and outlet temperatures are measured throughout the test. As the rod fouls, the outlet temperature decreases as a result of the decreased heat transfer efficiency of the rod. The speed of the heat transfer decline indicates the fouling potential of the oil in the test. Antifoulant selection can be made using this technique.

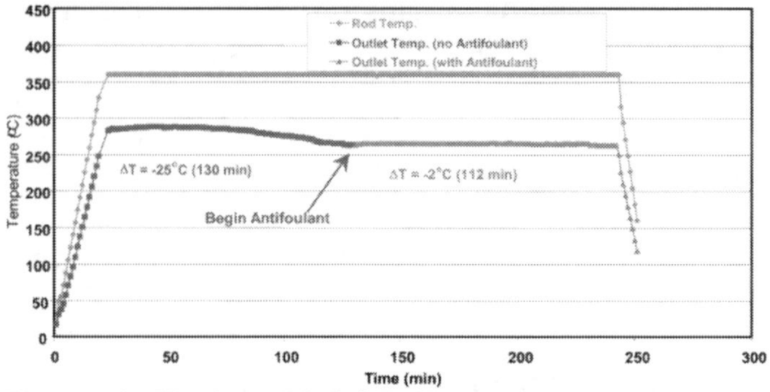

Figure 8: Typical ALCOR Example

Corrosion

For the majority of current opportunity crudes available in the market place the key issue is acid content and the effect on corrosion due to the crude during processing. There are three key analytical tests usually performed on crudes containing naphthenic acids: Total Acid Number (TAN), Naphthenic Acid Titration (NAT), and Simulated Distillation (SimDist).

TAN is a titration of all of the acidic species present in a sample of crude oil except for H2S. The acidic species include all organic acids (primarily naphthenic acids), thiols, phenols, thiophenes, and any other acids. The titration is done with a pH electrode, special solvent and KOH solution as the titrant; the results are reported in mg KOH/g sample.

Separating and isolating the naphthenic acids from the sample and titrating them measures NAT. The separation procedure is a proprietary method that Nalco Energy Services Analytical labs have developed and the titration is the same as the TAN procedure. The simulated distillation takes a sample of the naphthenic acids obtained from the NAT procedure and determines their boiling range by gas chromatography. This procedure (GC column, temperatures, flow rates, etc.) is also proprietary.

Test Description

TAN is the easiest procedure to do and is done routinely in many refinery laboratories. It is a measure of all of the acidic species in the crude, not just the naphthenic acids that cause problems. Since this titration is not specific for naphthenic acids, the results can only be used as a general indicator of problems: If the TAN is high (>1), you should be watching for potential naphthenic acid corrosion problems; if the TAN is low, there is a smaller chance of problems - but there is still a chance of having problems. Some crudes with TANs > 1 t have not had naphthenic acid corrosion problems as they did not contain any appreciable amounts of naphthenic acids. Some crudes with a TAN of 0.5 have exhibited naphthenic acid corrosion because, apparently, almost all of the TAN was due to naphthenic acids.

The NAT is a better measure of corrosion potential of the crude because it measures the naphthenic acids in the crude. If the NAT is greater than about 0.5 mg KOH/gram, you should be watching for naphthenic acid related problems. Although the NAT is a much better

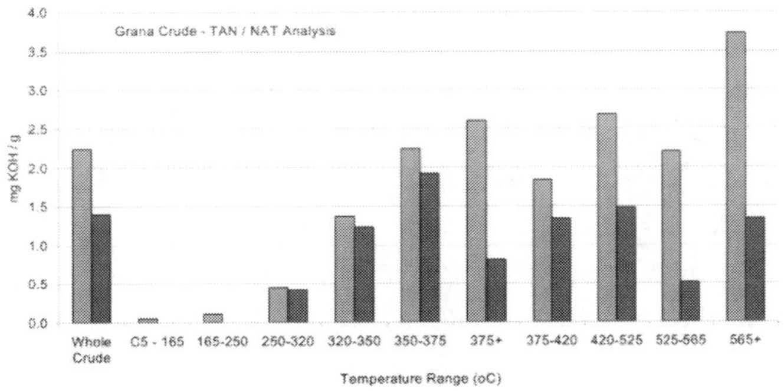

Figure 9: TAN/NAT Comparison

indicator of potential problems, it is not a perfect indicator. Some naphthenic acids appear to be more corrosive than others and so will cause more problems at a lower NAT than other naphthenic acids.

The simulated distillation gives an indication of where the naphthenic acids might concentrate in a distillation tower (see Figure 10). If a plot of percent acid distilled vs. temperature has a relatively small slope (is relatively flat), this indicates that the acids will tend to concentrate in a relatively narrow temperature range. If this were to occur, the NAT of a sidestream in this temperature range could be several times higher than the NAT of the whole crude and so be much more corrosive than even the NAT would indicate.

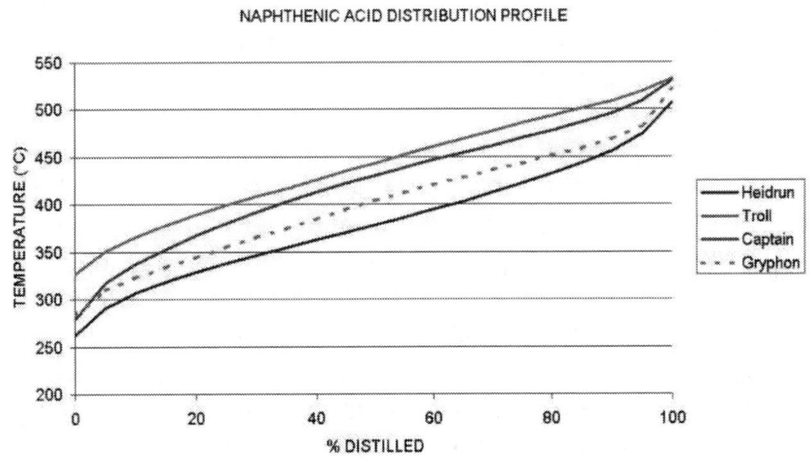

Figure 10: NAT Distribution of Several Crudes

If a sidestream sample is sent for the simulated distillation, the distillation temperature range reported will usually be broad, especially for the higher boiling range sidestreams. This is due to the fact that, in a distillation column, the separation of fractions is far from perfect, and the fraction contains a wide range of materials and that have a wide range of boiling points.

TAN and NAT's can be run on the whole crude and distillation cuts from the individual fields as well as the blend.

Corrosivity Testing

Corrosivity testing can be carried out on various crudes or side cuts and metallurgies, for example a recent corrosivity testing program tested the LVGO, MVGO and HVGO cuts on Carbon Steel, and alloys 5 Chrome, 9 Chrome, and 410 SS (see Figure 11). These cuts were selected based on the typical corrosivity distribution pattern that Nalco has seen with other high acid crudes. The testing included both with and without high temperature corrosion inhibitor and followed ASTM G and G111 procedures for testing corrosivity at high temperatures. Essentially the test uses an autoclave with a series of steel discs of defined geometry and shear stress exposed to the test oil at a constant temperature on a "once through" basis for a defined period of time (usually 24 or 72hrs). The test conditions used on the recent program were:

- Temperature: 315 °C
- Duration: 72 hours
- Spinning Rate: 1000 rpm

Test Description

A pressurized refreshed rotating cylinder rig is used that allows multiple measurements of corrosion rates to be made under controlled fluid flow conditions in hydrocarbon fluids such as crude oils or synthetic mixtures up to 400°C (750°F) and 70 bar (1000 psi). The use of the rotating cylinder allows corrosion data to be obtained using volumes of crude oils or crude fractions of the order of 10 litres per 24 hours without the expense of a rig simulating full pipe flow and the hazards associated with the need to pump large volumes of hydrocarbons at high temperature.

Good reproducibility has been demonstrated in naphthenic acid corrosion studies. Applicability is to materials selection, inhibitor evaluation and crude blending to reduce corrosion from acid crudes. Additionally the use of crude fractions that represent those that are formed at different stages in the distillation process allows the generation of data that can be used to predict corrosion rates at different locations in the refinery.

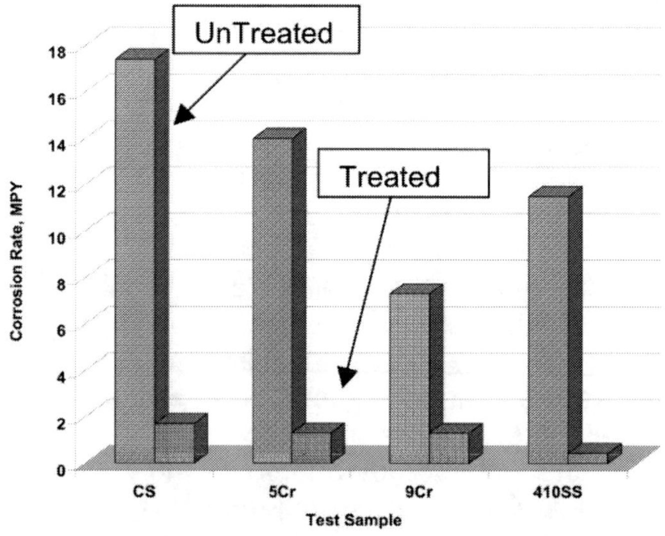

Figure 11: Corrosivity Testing Example

Conclusions

Processing opportunity crudes can significantly improve refinery profitability. Although crude assays do not include all of the information that is needed to predict the impacts on crude unit operation, using the techniques described above the refiner can reduce some of the risk that is associated with processing many of the opportunity crudes in the three key areas of Desalting, Fouling and Corrosion.

Communication between crude buyers, refiners, and companies like Nalco is essential in identifying, assessing and mitigating the potential risks in processing opportunity crudes.

Bibliography

1. Heimbaugh, W.A., Jones, H.B., "Dehydration: The Key to Desalting Crudes", Hydrocarbon Engineering, Nov. 1998
2. "Engineering Economics of Electrical Desalting", Howe-Baker Engineers, Inc.
3. Chaffee, B.A., Comer, D.G., de Jong, A., Sigmon, J.L., "Reducing Energy Costs", Hydrocarbon Engineering, Dec. /Jan. 1998/1999.
4. Johnson, D., McAteer, G., Zuk, H., "Mitigating Corrosion from Naphthenic Acid Streams", Petroleum and Technology Quarterly Winter 2003
5. Mathers, R., David, P., Kettle, A., Jones, P., "Management and mitigation of the issues associated with processing High Acid Crudes using a chemistry based approach with a focus on Doba processing at ChevronTexaco Pembroke", High TAN Crude Conference, Singapore, May 2005.
6. Babaian-Kibala, E., Craig Jr., H. L., Rusk, G. L., Blanchard, K. V., Rose, T. J., Uehlein, B. L., Quinter, R. C., Summers, M. A., "Naphthenic Acid Corrosion in a Refinery Setting", CORROSION/93, Paper No. 631.
7. Hydrocarbon Processing, Sept 2001, "Improve Desalter operations" by P. M. Lindemuth, R. B. Lessard, and M. Lozynski, (Nalco Energy Services) and M. Kremer, (Husky Oil Operations Ltd)
8. Arthur de Jong, Samuel A. Lordo, John L. Sigmon, and Philip R. Petersen, PhD, "Improved Refinery Profitability Through Opportunity Crudes", Today's Refinery

AIChE Spring meeting April 22-24

Effect of Toluene, Resins to Asphaltene Ratio and Surfactant on the Stability of Water-in-Crude Oil Emulsions:

Adel M. Elsharkawy*[1], Taher A. Al-Sahaf[2], Mohammed A. Fahim[2],
[1]Department of Petroleum engineering, and [2]Department of Chemical Engineering
College of Engineering and Petroleum, Kuwait University
P.O. Box 5969, Safat 13060, Kuwait

Abstract

Factors controlling the formation and stabilization of water-in-crude oil (w/o) emulsions in oil fields are of great concern to the petroleum industry for the economic development of underground oil reservoirs. Controlling and minimizing the formation of w/o emulsions and demulsification of water from emulsions are also important for environmental development. Because of its importance, the mechanisms, formation and stability of w/o emulsions have received considerable attention.

This paper deals with some of the factors responsible for the formation and stabilization of w/o emulsions formed in Burgan oil field in Kuwait. Some of the factors investigated in this study are the naturally occurred surface active components of crude oils such as asphaltenes and resins. The effect of additives; toluene and dodecyle benzene sulfonic acid (DBSA) on the stability of various emulsion samples collected from oil field are also reported.

Key words: Asphaltenes; Resins; Surfactant; Water-Oil Emulsions; Kuwaiti oil fields

* Corresponding author, visiting professor, Department of Chemical & Petroleum Engineering, University of Calgary, E-mail address: adel.elsharkawy@ucalgary.ca
Tel.: 1-(403)-220-5742, fax: 1-(403)-284-4852
E-mail address: asharkawy@kuc01.kuniv.edu.kw (A. M. Elsharkawy)
Tel.: +965-483-6059; fax: +965-484-9558.

1. Introduction

Through numerous studies on the stability of water-in-oil emulsion, more light has been shed on the mechanism of emulsion stabilization in the petroleum field. Many researchers now attribute the stability of emulsion to the viscoelastic interfacial film made up of surface active molecules such as asphaltenes, resins, and other organic and inorganic particles (Sullivan and Kilpatrick, 2002).

The degree to which solids increases emulsion stability depends on several factors such as particle size, shape and morphology, density, concentration and surface coverage, and wettability (Gelot, et al., 1984; Yan et al., 1999, 2001; Tambe and Sharma,1993, 1994; Aveyard, et al., 2003). Sztukowski and Yarranton (2005) investigated the role of solids in the stability of oil filed emulsions. They found that emulsion stabilized by fine solids and asphaltenes were most stable at a 2:1 fraction area ratio of asphaltene to solids. There is a strong correlation between asphaltene content and emulsion tightness (Kokal, 2002). The presence of wax in addition to asphaltene and resins are known to promote the stabilization of water-in-oil emulsion (Yarranton et al., 2000, Khristove et al., 2000).

Asphaltenes are flat sheets of condensed polyaromatcic hydrocarbons linked together by sulfide, ether, and aliphatic chain groups. The edges of the sheets are alkyl chains. The polar parts of the asphaltene molecules interact with each others forming aggregates or micelles. As such, these micelles are very much polar. Resins are less polar and made up of smaller molecules with one end being hydrophilic made up of functional groups and the other end hydrophobic made up of alkyl chains. In crude oils, the resins are attached to the asphaltene micelles at the polar end and the non-polar end of the resin interact with crude oil. The resins solvate the asphaltene aggregates and keep them in colloidal suspension in the oil. The aromaticity of the oil also increases the solvency of the asphaltene aggregates. The colloidal aggregates can be interfacially active if the interaction of the asphaltene with the water interface is favored over interaction with resins and aromatics in the crude oil phase. It is proposed that asphaltenes stabilize water-in-oil emulsion if they are near or above the asphaltene precipitation onset point (Gafonova and Yarranton, 2001). This is the point at which the asphaltene start to precipitate, caused by addition of an alkane such as n-heptane to the crude oil.

Formation of mechanically rigid, viscoelastic film around emulsion water droplet in the crude oil prevents droplet-droplet coalescence. It has been found experimentally that strong films made mainly of asphaltene with some resins were responsible for stabilizing water-in-oil emulsions. Resins although they are surface active, they can not stabilize emulsion alone. However, resins can adsorb quickly to water droplet interface, slowing coalescence and providing enough time for asphaltene to migrate to the interface and adsorb on the surface forming the rigid film which prevents coalescence. Therefore, any factor which result in the disturbance of the asphaltene-asphaltene interaction will make the interfacial film weaker and make the droplet coming together easier.

The addition of a highly surface active component such as emulsifier results in replacing the asphaltene aggregates on the water interface with more easily ruptured monolayer of these surface active agents. The amount of emulsifier required for emulsion breakdown is far less than that what is needed to replace the asphaltene structure completely, suggesting that coalescence begins as weak spots in the asphaltene film caused by demulsifier disruption.

The presence of inorganic particles in the emulsion results in increasing emulsion stabilization. Asphaltene and resins are adsorbed on the solid particles making them surface active. The role of solid in emulsion stabilization depends on the resins to asphaltene (R/A) ratio, Sullivan and Kilpatrick, 2002. At low R/A ratio, large asphaltene aggregates form thick interfacial film, bridging particles and preventing adsorption to water-oil interfaces because of bulky shape of the solid-asphaltene aggregates. At high R/A ratio, smaller asphaltene aggregates adsorb to the solid particles and migrate to the water-oil interface enhancing the stability of the emulsion. As the R/A increases further, the adsorption of asphaltene on the solid particles decreases. The amount of surface active aggregates decreases affecting emulsion stability.

Stability of emulsions is assessed in many ways. Gafonova and Yarranton (2001) and Sullivan and Kilpatrick (2002) assessed emulsion stability by measuring the amount of water separated from emulsion as a function of time via heating or centrifuge. Kokal and Wingrove (2000) measure the stability of emulsion by adding a specific concentration of demulsifier which will destabilize the emulsion and measure the amount of water separated from emulsion as a function of time. Another approach to measure the stability of emulsion is the determination of the critical electric field (CEF) for emulsion breakdown (Midttun et al, 2000; Sullivan and Kilpatrick, 2002; Aske et al., 2002).

Because of the complex nature of crude oils, it is difficult to relate the dispersion behavior to the physicochemical properties at the water-crude oil interface. Moreover, each oil well has its own characteristic and offers individual problems (Ali and Aqlam, 2000). Therefore, this paper is aimed at studying several factors which affect water-in-oil emulsion formed in Burgan oil field in Kuwait. Among these factors are the effect of toluene, resin to asphaltene ratio, and addition of dodecyle benzene sulfonic acid (DBSA), an anionic surfactant.

2. Experimental

2.1 Materials

Crude oil used in the preparation of water-in-oil emulsion samples is obtained from Kuwait Oil Company. Properties, characteristics and composition of crude oils are described elsewhere (Elsharkawy et al., 2005). Crude oil samples from Burgan oil field-the second largest oil field in the world, BG 321, BG366, BG75 and RA4 in addition to model oils are used in preparation of emulsion samples. Organic solvents used are: toluene (99.5% purity; Scharlau), n-heptane (99.9% purity; Baker analyzed), benzene (100% purity; baker analyzed), and methanol (99.8% purity; Univar). All chemical were used without further purification. Distilled water provided by Kuwait University facility having a pH 6 was used in emulsion preparation for all samples. Silica gel 60 provided by Fluka was used to absorb the resin fraction.

2.2 Equipments

Shaking bath model SB-16 provided by Techne was used to homogenize crude oil samples before oil is used in preparation of the emulsion samples. Rotary evaporator model BUCHI R-124 was used to remove extraction solvents. PH meter model HNNA HI9017 was used to adjust the pH of the aqueous phase. Silverson laboratory mixer emulsifier model SL2 was used to prepare all emulsion samples and centrifuge model

Labofuge provided by Seraeus Sepatech was used to separate water from emulsion samples.

2.3 General procedure

The oil phase (crude oil) was mixed with water using Silverson Lab mixer emulsifier for 1 minute at 9000 rpm. Equal volume of distilled water of pH 6 was added slowly to oil samples for 5 minutes during mixing. Emulsion samples were transferred immediately to 10 ml centrifuge tubes and left for 1.5 hr at room temperature to settle down. The stability of settled emulsion was studied by measuring the amount of water separated from emulsion as a function of time for 12 hours. Centrifuge tubes were capped to prevent solvent evaporation and centrifuged at 4000 rpm for 5 minutes then transferred to 60 °C water bath for 2 hours, centrifuge again for 5 minutes and heated for 2 hours. Successive heating and centrifuge was repeated for 12 hours or until no appreciable change in the amount of separated water is observed. The amount of water separated from emulsion was reported as the volume percent of initial water content of the emulsion sample.

2.4 Preparation of water-in- crude oil emulsions

2.4.1 Effect of toluene, resins to asphaltene ratio, and surfactant

Various emulsion samples were prepared to study the effect of toluene, resins to asphaltene ratio, and surfactant (dodecyl benzene sulfonic acid, DBSA) on the stability of emulsion samples. To study the effect of toluene, emulsions of different w/w percent of toluene (0, 10, 20, 30, and 50%) in crude oil were prepared by mixing (0, 2, 4, 6 and 8 g) of toluene with (20, 18, 16, 14 and 10 g) of crude oil, respectively.

Emulsions of different weight ratios of resins to asphaltene (3:1, 6:1 ,9:1, 12:1, and 20:1) were prepared by dissolving (0.39, 0.78, 1.17, and 2.47g) of resins respectively in 10 ml toluene followed by 10 g of crude oil. Trials to dissolve resins into crude oil directly were unsuccessful. Therefore, the resins or the surfactants were dissolved first in toluene before they were introduced to the crude oil. The crude oil itself used in this study contains (sample BG321) 3:1 resins to asphaltene ratio and asphaltene content of 1.3%. Molecular weight, density and elemental analysis of asphaltene and resin fractions used in the preparation of emulsion samples are reported in Table 1. These fractions were separated from various oil samples collected from Burgan oil field. To study the effect of surfactant on emulsion stability, emulsion samples of different DBSA weight percent (0, 2, 6, 8, and 10) were prepared by dissolving (0, 0.4, 1.2, 1.6,and 2g) of DBSA in 10 g of toluene followed by (10, 9.6, 8.8, 8.4, and 8g) of crude oil.

3. Results and Discussion

3.1 Effect of toluene, resins to asphaltene ratio, and surfactant

It has been known for a long time that blending of crude oils with aromatic solvents has a dramatic effect on resolving emulsified water. In this section, we study the effect of toluene, resins to asphaltene ratio and surfactant on the stability of water-in-oil emulsions. Kuwaiti crude oils, asphaltenes and resins from Burgan field were used in preparation of all emulsion samples used in this study.

3.1.1Effect of toluene blending

In this study, we report the effect of blending various emulsion samples with toluene in increasing proportions. Figure 1 through 4 shows the degree of destabilization of different emulsions as a function of time for various toluene contents. The various emulsion samples (BG321T, RA4, BG 75, and BG 366T) were mixed with increasing proportions of toluene over the course of 8 to 10 hours. These figures show that amount of separated water increases as a function of destabilization time. Figure 1 through 4 also shows that in order of emulsion tightness; they are BG321, BG366, RA4, and BG75. Emulsion tightness is correlated with the resins to asphaltene ratio content of each sample (3, 4.4, 9.0, and 12.4, respectively). Figure 5 was prepared, as an example, to illustrate the relation between the toluene content and destabilization of emulsion for one of the emulsion sample, BG321T. Similar trend could be drawn for various samples. This figure shows the increasing amount of separated water from emulsion after 6 hours as a function of increasing the amount of toluene blended with crude oil. These results show that increasing the aromaticity of the oil medium will lead to increase solubility of asphaltenes. This in turn indicates that asphaltene molecules will be free to leave its sites at the oil-water interfacial film, rupturing the film, which allows small water droplet to coalescence, and grow overtime, eventually separate dispersed water droplets as a continuous layer at the bottom of the centrifuge tube. The results reported in this section are in good agreement with others (Yarranton, et al., 2000; FØrdedal et al., 1996, and Mclean and Kilpatrick 1997).

3.1.2 Effect of resin to asphaltene ratio

The role of resins to asphaltenes ratio in stabilizing emulsion is investigated via preparation of various emulsion samples using different oil samples (BG321T, RA4, BG 75, BG366T and BG321T) containing varying amount of resins to asphaltene ratio from 3:1 to 20:1. Figures 6 through 9 show the effect of adding resins extracted from Kuwaiti crude oils on the water-in-oil emulsion stability. These figures indicate that at high resin to asphaltene ratios (above 10:1 in many cases) the emulsions formed are mainly loose emulsion since more than 50% of the emulsified water is being resolved. Figure 6 also shows that at R/A ratio of 20:1 no emulsion was formed at all. This phenomenon is realized as it is known that resins are surface active (natural surfactants), hence they could at certain concentration displace the asphaltene molecules at the water-oil interfacial film, rupturing the film surrounding the water droplet and separating the aqueous phase. Figure 7 shows that crude oil RA4 formed the most stable emulsion, and only minimal amount of water resolved upon centrifugation. This could be related to the high percentage of asphaltenes in this sample (2.81% w/w) which is the highest percentage among all the tested samples. Comparing figures 6 through 9, it can be concluded that emulsion tightness is strongly correlated not only to resin to asphaltene ratio but also to origin of asphaltene and resins. Figure 10 shows the effect of increasing R/A ratio for emulsion sample prepared from oil BG321T. This figure shows that increasing the resin to asphaltene (R/A) ratio from 3 to 20 resulted in decreasing emulsion stability, increasing amount of resolve water after 10 hours. This is attributed to increasing solubility of asphaltene in resin-containing oil, there by minimizing the accumulation of asphaltene at the water/oil interfaces. This is in agreement with the results reported by others (Spiecker et al., 2003; Midttun et al., 2000; Schorling et al., 1999). Spiecker et al. (2003) found that addition of resins to asphaltene in model oil of heptane-toluene mixtures reduced the aggregate size by disrupting the π- π and polar bonding interactions between asphaltene monomers. The interaction of resins with asphaltenic aggregates rendered the aggregates less interfacially active and thus reduced emulsion stability. Referring to the results reported in figures 6 through 9, one can conclude that not only the R/A ratio affect emulsion stability but also the type of resins. For example, at the same R/A ratio different resins produce various degree of emulsion stability. Midttun et al (2000) reported that depending on the type of the resins, adding of resins to asphaltene-stabilized w/o emulsion may reduce emulsion stability, increase emulsion stability, or have no significant effect of emulsion stability. They reported that high polarity resins tend to reduce emulsion stability. Schorling et al (1999), however, found that the influence of resins in the presence of asphaltenes on the properties of emulsions is independent of the origin of resins.

3.1.3 Effect of surfactant

Chemical demulsifier are one of the most important tools used to break down tight emulsion in petroleum and refinery industries. The role of surfactant on emulsion stability is investigated by studying the effect of adding dodecyle benzene sulfonic acid (DBSA), an anionic surfactant, to various oil filed emulsions. The optimum amount of DBSA which is required to break down the emulsion completely can be found. In our investigation of tool of breaking emulsions, the most rewarding part of this study was that we manage to resolve all the emulsified water at emulsifier dosage level of only 2% (w/w) in crude oil BG321T. In previous work Al-Sahaf et al. (2002) studied the effect of DBSA on the onset point of asphaltene. They found that the onset point was shifted to very high values at very minute concentration of DBSA. This agrees with the results reported in this study. The tremendous effect of DBSA on destabilizing emulsion could be explained in the view of the acid-base interactions between the surfactant (sulfonic acid moiety) and the polar functional sites on the asphaltene by the replacement of the asphaltene aggregates at the oil-water interface by the surfactant molecules. This causes the rupture of the film surrounding the water droplet and separating the aqueous phase.

4. Summary and conclusions

Four emulsions samples collected from Burgan oil field in Kuwait have been used to study the stability of water-in oil-samples under various conditions. Asphaltenes, resins, and oils separated from emulsion samples (BG321, BG366, BG75, and RA4) were used to study the effect of adding toluene on emulsion stability. The results indicate that increasing aromaticity of the emulsion leads to increase asphaltene solubility which allows asphaltene molecules to be free to leave its sites at the oil-water interfacial film, rupturing the film and small water droplet to coalescence and grow, eventually separate into aqueous phase.

The role of resin to asphaltene (R/A) ratio in stabilizing water-in-oil emulsions has also been investigated. It was found that increasing R/A, generally, decreases the stability of emulsions. Increasing R/A ratio above certain threshold results in asphaltenes at the water-oil interface are replaced by resins, rupturing the film surrounding the water droplet and hence separating the aqueous phase.

The role of dodecyle benzene sulfonic acid (DBSA), an anionic surfactant, in destabilizing emulsion samples has been studied. It was found that 2% (w/w) of this surfactant can completely resolve all water contained in the emulsion sample. The effect of DBSA on destabilization of emulsion sample can be explained by in the view of the acid-base interaction between the surfactant and polar functional sites on the asphaltene by the replacement of the asphaltene aggregates at the w-o interface by surfactant molecules which causes rupturing of the film surrounding the water droplet and hence separation of aqueous phase.

Acknowledgements

This project has been financially supported by Kuwait Foundation for Advancement of Sciences, Grant No. KFAS 2000-09-02, "Role of asphaltene, reins, and wax in the stability of water/oil emulsions in Kuwaiti Crudes". Special thanks go to the Research Administration at Kuwait University for administrating the project, providing assistance in manpower and equipments. The authors also thanks Kuwait Oil Company (KOC) for providing emulsion samples used in this research project. Elsharkawy thanks the Department of Chemical and Petroleum Engineering, University of Calgary for providing computing and other research facilities during his sabbatical leave.

References

Ali, M.F., Aqlam, M.H., 2000. The role of asphaltenes, resins and other solids in the stabilization of water in oil emulsions and its effects on oil production in Saudi oil fields. Fuel, 79, 1309-1316.

Al-Sahaf, T.A, Fahim, M.A., Elkilani, A.S. 2002. Retardation of asphaltene precipitation by addition of resins, deasphalted oil and surfactants. Fluid Phase Equilibria, 194-197, 1045-1057.

Aske, N., Kallevik, H., Sjöblom, J. Water-in-crude oil emulsion stability by electrical field measurements. Correlation to physico-chemical parameters and near-infrared spectroscopy. J. Petrol. Sci Eng. 36, 1-17.

Aveyard, R., Binks, B.P., and John H. Clin, J.H., 2003.Emulsions stabilized solely by colloidal particles. Adv. Colloid Interface Sci. 100-102, 503-546.

Elsharkawy, A.M., Al-Sahaf, T. A., Fahim, M.A., 2005. Characterization of Asphaltenes and Resins Separated from Water-in-Crude Oil Emulsions formed in Kuwaiti Oil Fields. Submitted to J. Petrol Sci. Eng.

FØrdedal, H., , Midttun, Ø., Sjöblom, J., Kvalheim, O.M., Schildberg, Y., Volle, J-L. 1996. A Multivariate Screening Analysis of W/O Emulsions in High External Electric Fields as Studied by Means of Dielectric Time Domain Spectroscopy, II: Model Emulsions Stabilized by Interfacially Active Fractions from Crude Oils. *J. Colloid Interface Sci, 182, , 117-125.*

Gafonova, O.V. and Yarranton, H.W., 2001. The stabilization of water-in-hydrocarbon emulsions by asphaltenes and resins. J. colloidal interface Sci. 241,469-478.

Gelot, A., W. Friesen, W., Hamza, H. A., 1984. Emulsification of oil and water in the presence of finely divided solids and surface-active agents. Colloids Surf. 12, 271-303.

Khristov, Khr., Taylor, S.D., Masliyah, J.H., 2000. Thin liquid film technique: application to water-in-oil bitumen emulsion films. Colloid Surf., A Physicochem. Eng. Asp. 174,183-196.

Kokal, S. and Wingrove, M. 2000. Emulsion separation index: from laboratory to field cases. Paper SPE 63165. Proceedings of the SPE Annual Technical Conference and Exhibition, Dallas, TX, USA, October, 2000.

Mc Lean, J.D., Kilpatrick, P.K., 1997a. Effects of asphaltene solvency on the stability of water-in-oil emulsion, J. Colloid Interface Sci.189, 242-253.
Mc Lean, J.D., Kilpatrick, P.K., 1997b. Effects of asphaltene aggregation in the model heptane-toluene mixtures on the stability of water-in-oil emulsion. J. Colloid Interface Sci.196, 23-34.
Midttun, Ö, Kallevik, H., Sjöblom, J, Kvalheim, O. 2000. Multivariate screening analysis of water-in-oil emulsions in high external electric field as studied by means of dielectric time domain spectroscopy. J. Colloid Interface Sci, 227, 262-271.
Schorling, P.-C., Kessel, D.G., Rahimian, I., 1999. Influence of the crude oil resin/asphaltene ratio on the stability of oil/water emulsions. Colloid Surf., A Physicochem. Eng. Asp.152, 95-102.
Spiecker, P.M., Gawrys, K.L., Trail, C. B., Kilpatrick, P.K. 2003. Effects of petroleum resins on asphaltene aggregation and water-in-oil emulsion formation. Colloids Surf., 220, 9-27.
Sullivan, A.P., Kilpatrick, P.K., 2002. The effect of inorganic solid particles on water and crude oil emulsion stability. Ind. Eng. Chem. Res. 41, 2289-3404.
Sztukowski, D.M., Yarranton, H.W., 2005. Oilfield solids and water in oil emulsion stability. J. Colloid Interface Sci. 285, 821-833
Tambe, D. , Sharma, M.M. 1994. Factors Controlling the Stability of Colloid-Stabilized Emulsions: II. A Model for the Rheological Properties of Colloid-Laden Interfaces. J. Colloid Interface Sci. 162, 1-10.
Tambe, D. E., Sharma, M.M., 1993. Factors Controlling the Stability of Colloid-Stabilized Emulsions: I. An Experimental Investigation. J. Colloid Interface Sci. 157, 244-25.
Yan, N., Gray, M. R., Masliyah, J. H., 2001. On water-in-oil emulsions stabilized by fine solids. Colloids Surfaces A: Physicochemical Eng. Aspects, 193, 97-107.
Yan, Z., Elliott, J.A. W., Masliyah, J.H., 1999. Roles of Various bitumen components in the stability of water-in-diluted-bitumen emulsion. J. Colloid Interface Sci., 220,329-337.
Yarranton, H.W., Hussein, H., Masliyah, J.H, 2000. Water-in-hydrocarbon emulsions stabilized by asphaltenes at low concentrations. J. Colloid Interface Sci. 228, 52-63.

Table 1-Characterization of Asphaltenes (A) and resins (R) separated from emulsions formed in BG Oil fields

					Elemental analysis				
sample	Molar mass	Density @20°C	H/S ratio	H/C ratio	C%	H%	N%	S%	O%
A-BG47T	4246	1.2374	1.1630	1.1623	76.50	7.463	1.159	6.41	8.454
A-BG185T	4052	1.2415	1.1498	1.1487	76.66	7.391	1.221	6.42	8.295
A-BG321T	4432	1.2256	1.1197	1.1706	78.67	7.729	1.233	6.91	5.460
A-BG366T	3997	1.2430	1.1324	1.1541	76.85	7.443	1.255	6.57	7.883
R-BG47T	1150	1.1047	1.7208	1.3684	79.22	9.098	1.143	5.29	5.251
R-BG185T	984	1.0859	1.6387	1.3629	79.36	9.077	1.167	5.54	4.856
R-BG321T	1121	1.1099	1.5628	1.3561	90.98	9.216	1.156	5.89	2.753
R-BG366T	992	1.0860	1.5407	1.3595	81.09	9.252	1.181	6.05	2.468

Fig 1

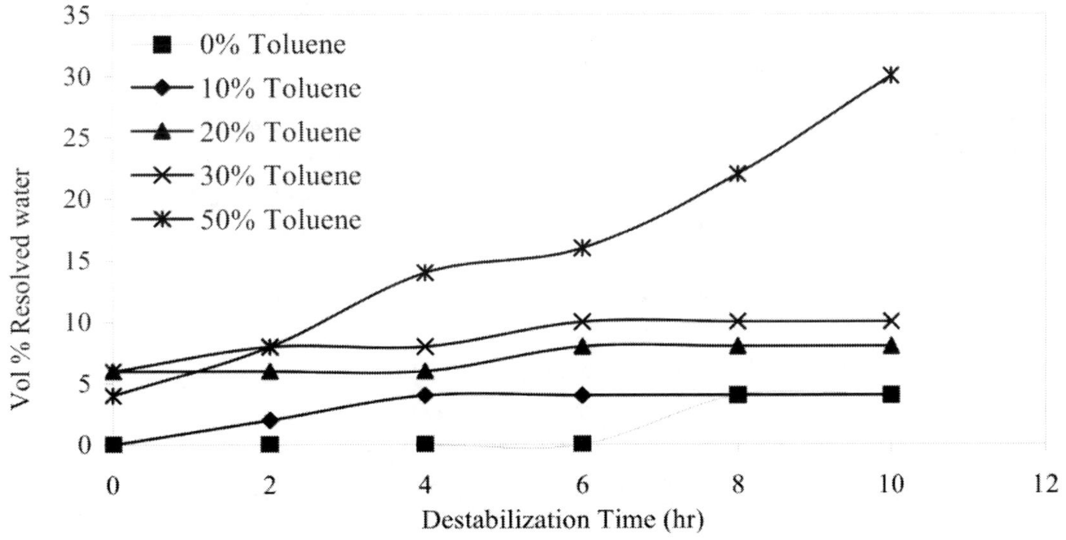

Figure 1. Effect of Toluene Blending on Emulsion Stability Of Kuwaiti Crude Oil (BG 321T).

Fig 2

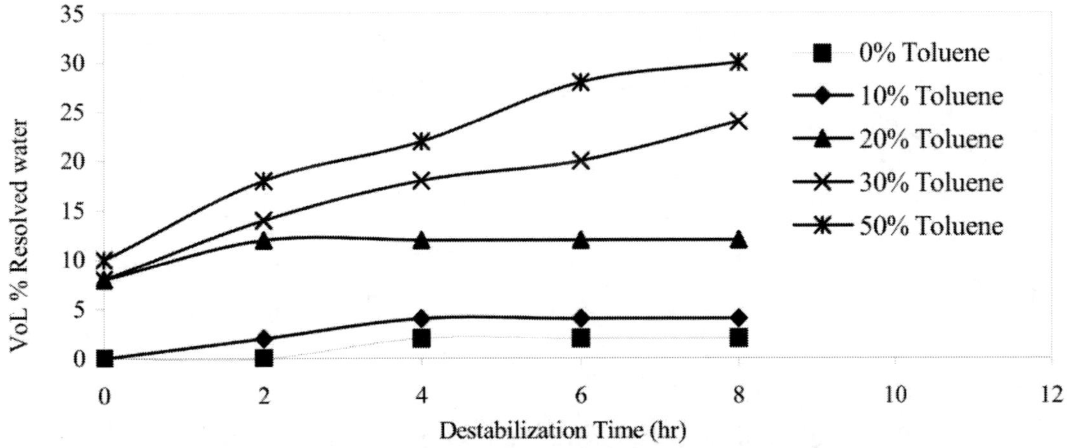

Figure 2. Effect of Toluene Blending on Emulsion Stability Of Kuwaiti Crude Oil (RA 4).

Fig 3

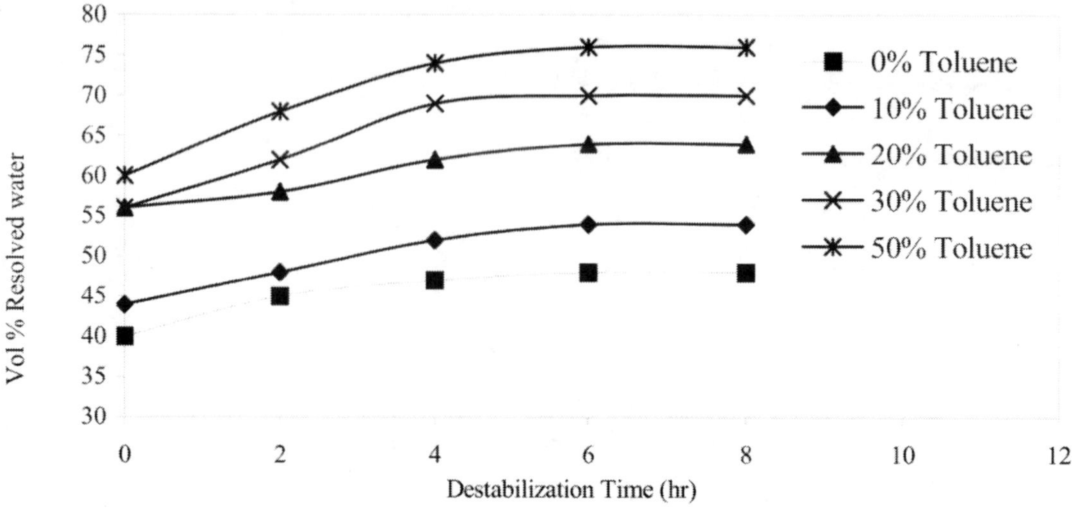

Figure 3. Effect of Toluene Blending on Emulsion Stability Of Kuwaiti Crude Oil (BG 75).

Fig 4

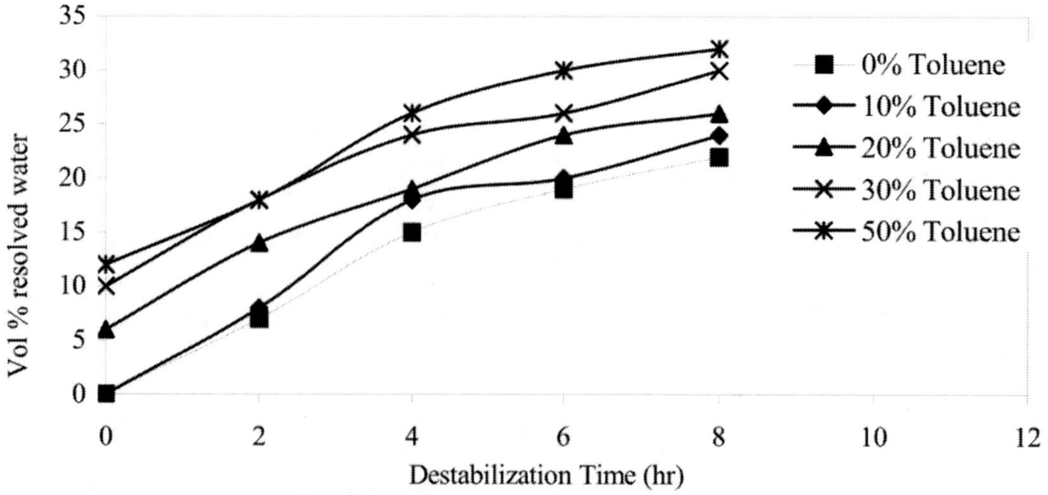

Figure 4. Effect of Toluene Blending on Emulsion Stability Of Kuwaiti Crude Oil (BG 366T).

Fig 5

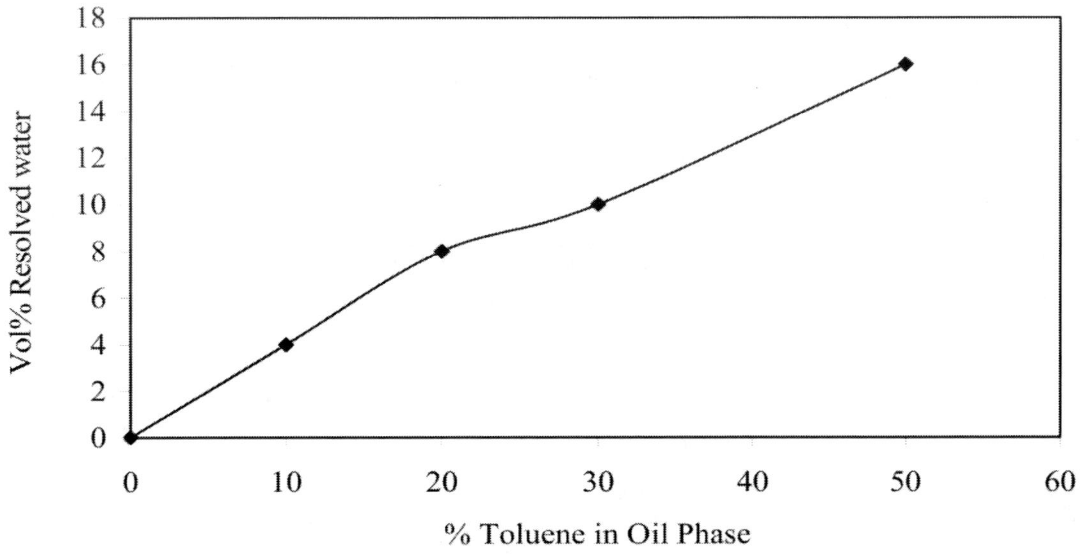

Figure 5. Effect of Toluene Blending on Stability of Water-in-Crude Oil Emulsion formed by Kuwaiti Crude Oil after 6 hours (BG 321 T).

Fig 6

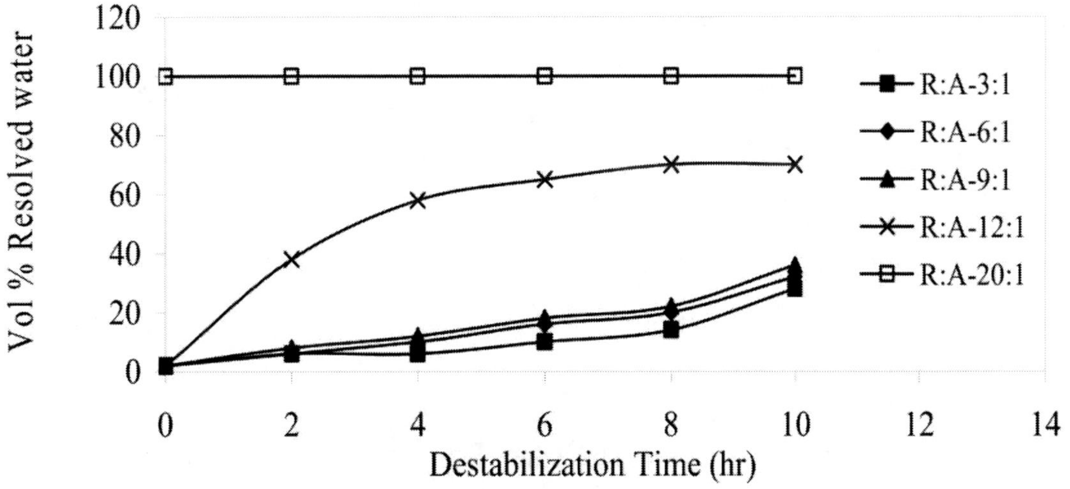

Figure 6. Effect of Increasing R/A Ratio on Water-in-Crude Oil Emulsion Stability Over Time (BG 321T).

Fig 7

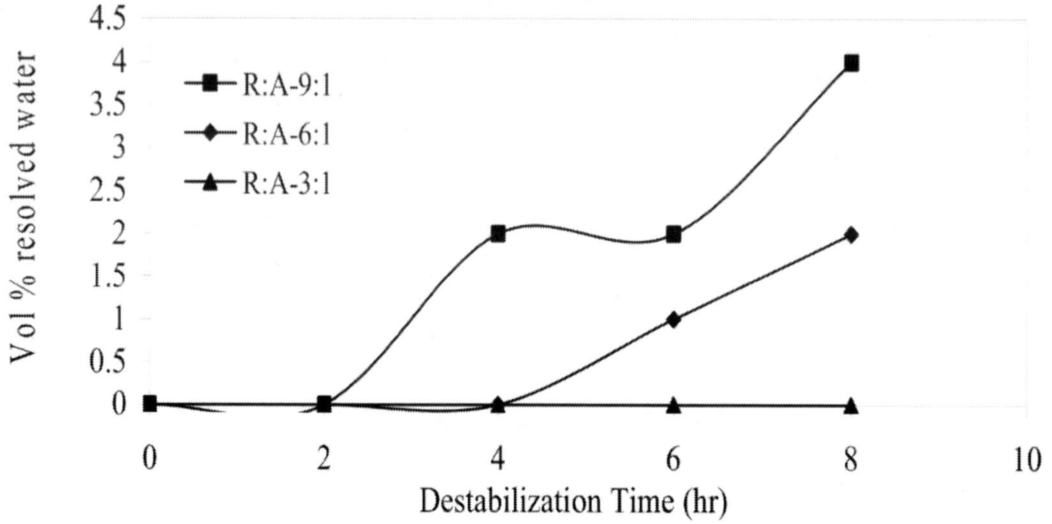

Figure 7. Effect of Increasing R/A Ratio on Water-in-Crude Oil Emulsion Stability Over Time (RA 4).

Fig8

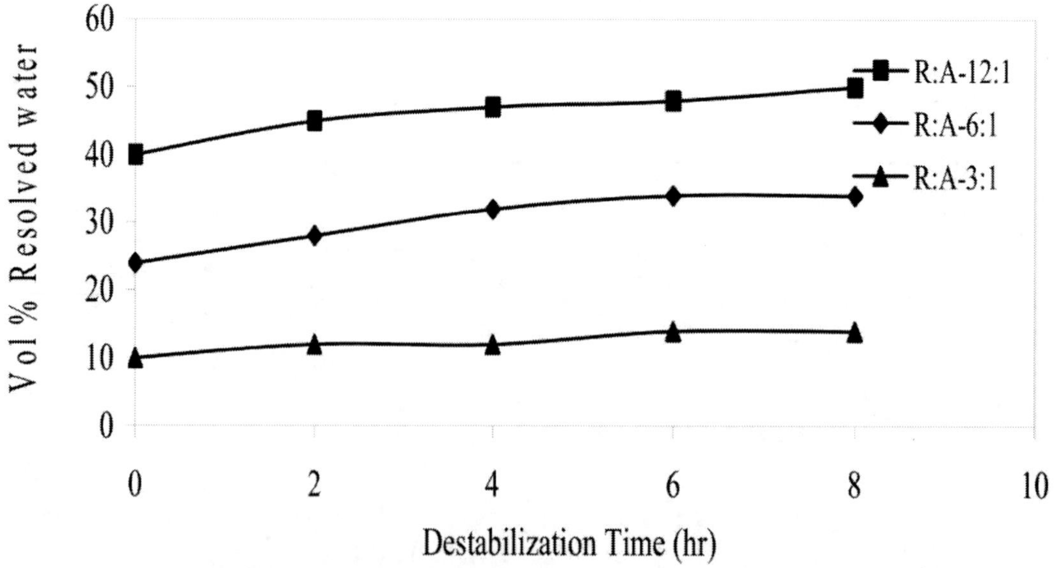

Figure 8. Effect of Increasing R/A Ratio on Water-in-Crude Oil Emulsion Stability Over Time (BG 75).

Fig9

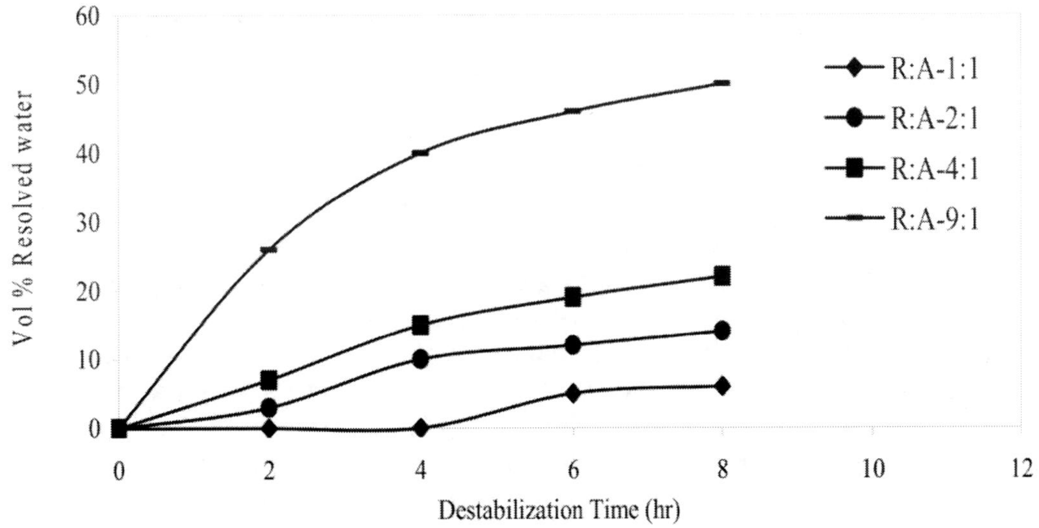

Figure 9. Effect of Increasing R/A Ratio on Water-in-Crude Oil Emulsion Stability Over Time (BG 366 T).

Fig 10

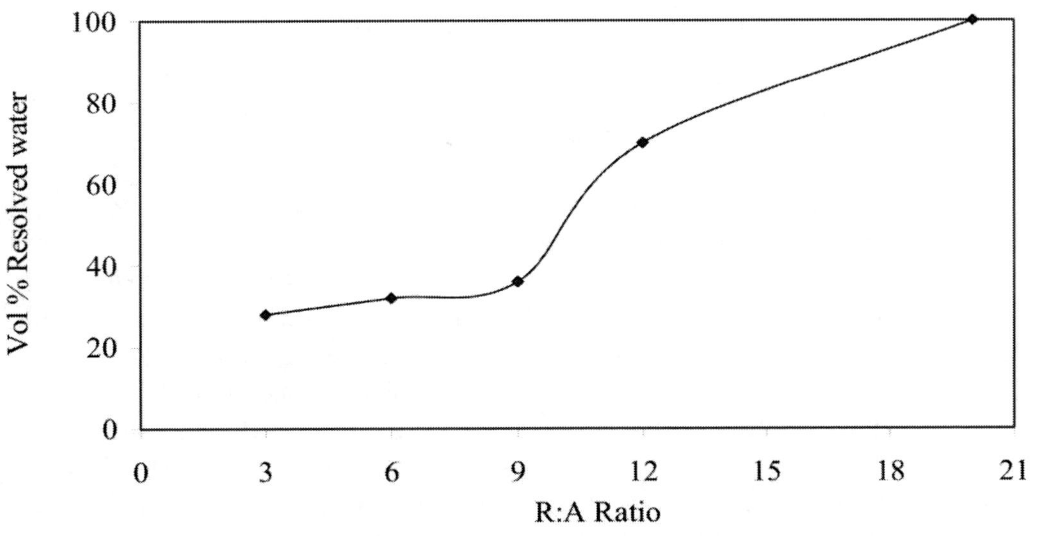

Figure10. Destabilization Caused by Increasing R:A Ratio After 10 Hours (BG 321 T)

FLOCCULATION TEST FOR OILS WITHOUT ASPHALTENES

Irwin Wiehe[1], Parviz Rahimi[2], Youngjun David Oh[2], Remy McNamara[2], and Teclemariam Alem[2]

[1]Soluble Solutions, 3 Louise Lane, Gladstone, NJ 07934
[2]National Centre for Upgrading Technology, 1 Oil Patch Drive, Devon Alberta T9G 1AA8, Canada

Introduction

The use of Oil Compatibility Model and tests have been quite successful[1-4] for predicting and mitigating the asphaltene fouling of heat exchangers and the coking of furnace tubes[5]. This includes self-incompatible oils[4], blends of oils that are incompatible[1,2], and oils that are nearly incompatible[3]. Typically, the insolubility number and the solubility number are determined for each oil that might be blended. Flocculation tests are used that involve the blending of each oil with different proportions of the model asphaltene solvent, toluene, and the model asphaltene nonsolvent, n-heptane, and determining points of incipient asphaltene precipitation. Incompatibility is predicted when the volume average solubility blending number is equal or less than the maximum insolubility number of the oils in the blend[1].

If an oil does not contain asphaltenes, its insolubility number is zero but flocculation titrations on the oil cannot be performed to determine the solubility blending number of the oil. The Oil Compatibility Model may be the only solubility parameter model for petroleum with a procedure that has been developed to evaluate the solubility parameter of an oil without asphaltenes by using a reference oil containing asphaltenes[2]. However, for Athabasca vacuum gas oil with Athabasca bitumen as the reference oil, as will be described, this procedure did not appear to work. Since resins act as asphaltene dispersants and vacuum gas oils contain resins, were these skewing the results for a purely solubility model?

Experimental

The determination of the insolubility number and solubility blending number of an oil containing asphaltenes requires a minimum of two tests: the heptane dilution and the toluene equivalence tests[1]. The heptane dilution test requires determining the maximum number of ml of n-heptane that can be blended with 5 ml of oil without precipitating asphaltenes. To avoid problems with slow kinetics of precipitation and the presence of insoluble wax, a 5 minute wait period in a 60°C oven is allowed after each addition of n-heptane and before determining if asphaltenes precipitated by observing a drop between a glass slide and a cover slip under an optical microscope. The toluene equivalence is on 2 grams of oil and 10 ml of test liquid, composed of mixtures of toluene and n-heptane. The minimum percent toluene in the test liquid required to keep asphaltenes soluble is determined. Again, 5 minute wait periods in a 60°C oven are used before determining asphaltene solubility.

If an oil does not precipitate asphaltenes when 5 ml of oil is blended with 25 ml of n-heptane, the insolubility number is assigned a value of zero and a different procedure is used to evaluate its solubility blending number. This alternate procedure may also be used if the oil contains very soluble or a very low concentration of asphaltenes. A reference oil is selected that has an insolubility number greater than 20 and 5 ml of the reference oil is blended with 25 ml of the sample oil and held at 60°C for 10 minutes. If no asphaltenes precipitate, the solvent oil equivalence test is run. If asphaltenes precipitate, the nonsolvent oil equivalence test is run. The solvent oil equivalence test is run just like the toluene equivalence test on the reference oil except toluene is replaced by the sample oil. The nonsolvent oil equivalence test is run just like the toluene equivalence test on the reference oil except n-heptane is replaced by the sample oil.

Results and Discussion

Compatibility Numbers of Athabasca Bitumen

The toluene equivalence (TE) test was run on Athabasca bitumen, obtaining the value of 19, and the heptane dilution (V_H) test was also run, obtaining the value of 10.25 ml. The insolubility number, I_N, was calculated from[1]:

$$I_N = TE/[1 - V_H/(25\ d)] = 19/[1 - 10.25/\{25\ (1.0107)\}] \quad (1)$$
$$I_N = 32.0$$

where d = density = 1.0107 g/ml.
The solubility blending number, S_{BN}, was calculated from[1]:

$$S_{BN} = I_N\ [1 + V_H/5] = 32.0\ [1 + 10.25/5] \quad (2)$$
$$S_{BN} = 97.6$$

Solvent Oil Equivalence Tests on Vacuum Gas Oil

The example oil that the solubility blending number was desired was a virgin vacuum gas oil (VGO) from Athabasca bitumen. Dilution of VGO with n-heptane showed that the VGO contains no asphaltenes. Athabsca bitumen was selected as the reference oil but it was not an ideal choice because it is very viscous and it contains clay particles. Nevertheless, dilution of the Athabasca bitumen with VGO precipitated no asphaltenes. Although the toluene equivalence test gave consistent results, replacing toluene with the VGO in the solvent oil equivalence test produced more viscous mixtures, smaller insoluble asphaltene agglomerates, and made it more difficult to differentiate insoluble asphaltene particles from clay particles, even with an optical microscope. The results in Table 1 show that repeated tests in two different labs for VGO gave solubility blending numbers that are too variable using[2]:

$$S_{BN} = 100\ TE/SOE \quad (3)$$

Therefore, other methods for evaluating the solubility blending numbers for these gas oils were sought.

Table 1. Solvent Oil Equivalence Test Results on Vacuum Gas Oil Using Athabasca Bitumen as the Reference Oil

Gas Oil	TE Ref.	SOE	S_{BN}
VGO	19	19	100
VGO	19	26	73
VGO	23	29	79

Compatibility Numbers of Blends

Another method to evaluate the solubility bending number is to run the toluene equivalence and heptane dilution tests on a blend of gas oil and Athabasca bitumen and calculate the solubility blending number of the gas oil from the solubility blending number of the blend. The result for thee blends of VGO and Athabasca bitumen are shown in Table 2. The solubility blending number of the VGO, S_{VGO}, was calculated from the solubility blending number of Athabasca bitumen, S_{AB}, and the solubility blending number of the blend, S_{Blend}, using the volumetric blending rule[1]:

$$S_{Blend} = \phi_{VGO} S_{VGO} + (1 - \phi_{VGO}) S_{AB} \quad (4)$$

Where ϕ_{VGO} is the volume fraction of VGO.

Table 2. Solubility Blending Number from Toluene Equivalence and Heptane Dilution Tests on Blends of VGO and Athabasca Bitumen

Vol% VGO	TE	V_H, ml	I_{Blend}	S_{Blend}	S_{VGO}
30	21	10.3	36	110	139
50	20	10.25	34	104	110
80	18.5	10.25	32	97.6	97.6

As can be seen in Table 2 the solubility blending numbers of the VGO determined from blends depend on the percent VGO in the blend. It is also likely that the solubility blending number of the VGO is less than that for the full bitumen because Athabasca Bitumen contains little amount with boiling points below that of the vacuum gas oil boiling range. Thus, the VGO solubility blending numbers in Table 2 seem too high. At this point there was concern that a purely solubility model, like the OIL Compatibility Model, might not describe the action of VGO that contains resins, reported to be asphaltene dispersants.

Heptane Dilution Test on Blends

It was decided to run just heptane dilution tests on blends of VGO and Athabasca bitumen. By replacing toluene with VGO, one could test independently if VGO and Athabasca bitumen obey the Oil Compatibility Model. This model predicts that a plot of volume percent VGO in the mixture of VGO and n-heptane versus 100 times the volume of bitumen divided by volume of VGO plus n-heptane would be a straight line as it was when toluene was used instead of VGO[1]. Considerable heptane dilution data was obtained on VGO-Athabasca bitumen blends in two different laboratories, as given in Table 3. Surprisingly, the heptane dilution values varied little with percent VGO in the blend. Nevertheless, Figure 1 shows that VGO and Athabasca bitumen obey the Oil Compatibility Model as the predicted straight line was obtained. The y-intercept of 33.9 is an insolubility number, $I_{H\text{-}VGO}$, on a heptane-VGO scale. The solubility blending number, $S_{H\text{-}VGO}$, on a heptane-VGO scale is given by Equation 5:

$$S_{H\text{-}VGO} = I_{H\text{-}VGO} (1 + 100/\text{ x-intercept}) \quad (5)$$

$$S_{H\text{-}VGO} = 33.9 (1 + 100/48.8) = 103.4$$

Table 3. Heptane Dilution Tests on Blends of VGO and Athabasca Bitumen

Vol% VGO	V_H (ml)	$\frac{100V_{AB}}{(V_{VGO} + V_H)}$	$\frac{100\,V_{VGO}}{(V_{VGO} + V_H)}$
0	10.25	49.5	0
0	10.1	48.8	0
10	10.1	42.4	4.72
20	10.1	36.0	9.01
30	10.3	29.7	12.7
30	10.25	29.8	12.8
50	10.25	19.6	19.6
80	9.1	7.63	30.5
80	10.25	7.02	28.1

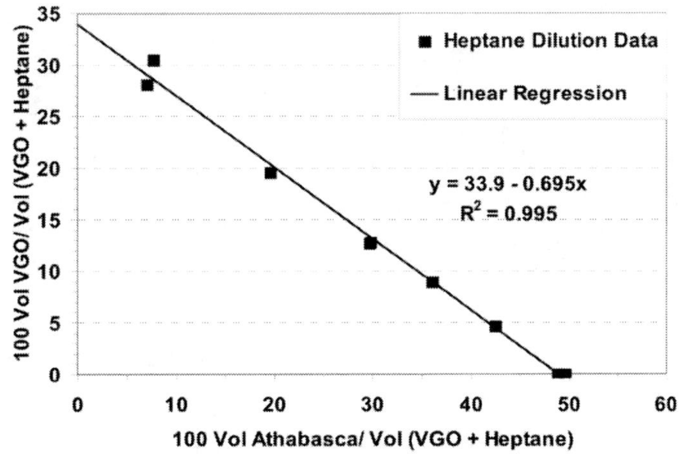

Figure 1. The heptane dilution data on VGO-Athabasca bitumen blends are linear, as predicted by the Oil Compatibility Model.

The insolubility number and the solubility blending number on the heptane-VGO scale need to be converted to the heptane-toluene scale using their definitions[1] in terms of solubility parameters:

$$S_{H-VGO} = 100\frac{\delta_{Ath} - \delta_H}{\delta_{VGO} - \delta_H}; \qquad S_{BN} = 100\frac{\delta_{Ath} - \delta_H}{\delta_T - \delta_H} \qquad (6)$$

$$I_{H-VGO} = 100\frac{\delta_f - \delta_H}{\delta_{VGO} - \delta_H}; \qquad I_N = 100\frac{\delta_f - \delta_H}{\delta_T - \delta_H} \qquad (7)$$

$$S_{VGO} = 100\frac{\delta_{VGO} - \delta_H}{\delta_T - \delta_H} = 100\frac{S_{BN}}{S_{H-VGO}} = 100\frac{I_N}{I_{H-VGO}} \qquad (8)$$

where δ_{Ath} = solubility parameter of Athabasca bitumen
δ_H = solubility parameter of n-heptane

δ_{VGO} = solubility parameter of VGO
δ_T = solubility parameter of toluene
δ_f = flocculation solubility parameter of Athabasca bitumen

Substituting the values into Equations 8, one obtains that both equations give the solubility blending number of VGO, S_{VGO}, equal to 94.4. While, as expected, this is less than the solubility blending number of the full Athabasca bitumen (97.6), it much closer than anticipated. However, since the two solubility blending numbers are so close, it explains why the heptane dilution values did not vary much for the VGO – Athabasca bitumen blends.

Change of Reference Oil

As a check on the procedure of measuring the heptane dilution of blends, this test was repeated for Athabasca VGO but using Arabian Heavy crude oil as the reference oil. On Arabian Heavy crude oil alone, the toluene equivalence was measured to be 21 and the heptane dilution was 6.25 ml. Using Equations 1 and 2, the insolubility number was calculated to be 29.2 and the solubility blending number to be 65.7. However, to parallel the multipoint determination used for VGO blends, two additional toluene-heptane flocculation tests were run on Arabian Heavy crude oil, resulting in Figure 2. The y-intercept of 29.5 is equal to the insolubility number[1]. The solubility blending number was calculated from the x-intercept of the linear regression line using Equation 5 but in the heptane-toluene scale, obtaining 65.8. Therefore, the multipoint determination gives essentially the same compatibility numbers as the two point determination for a more typical crude oil, like Arabian Heavy, without the high viscosity and clay particles of Athabasca bitumen.

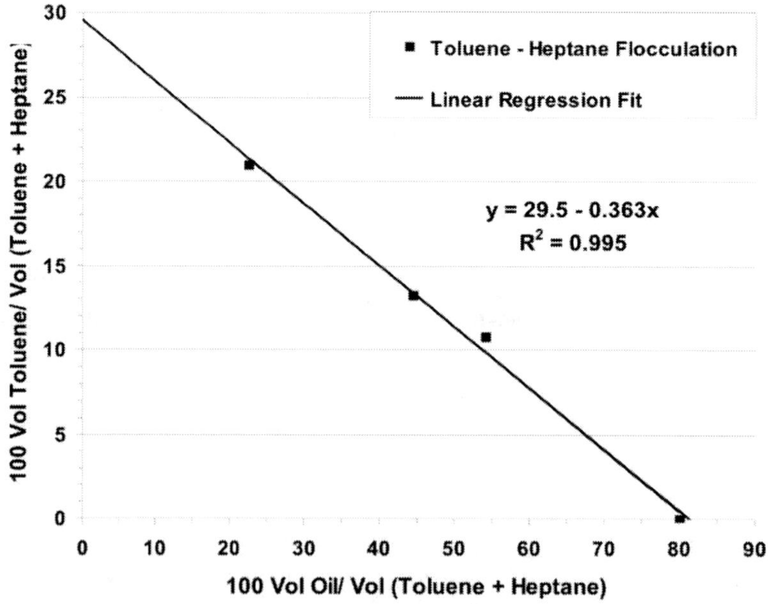

Figure 2. Toluene - Heptane Flocculation Points for Arabian Heavy Crude Oil

The solvent oil equivalence of VGO with Arab Heavy crude oil as the reference oil was measured to be 23.5. Using Equation 3, the solubility blending number of VGO was calculated to be 89.4. The heptane dilutions of blends of VGO and Arabian Heavy crude oil were measured and plotted in Figure 3. While the linear regression was done only on the heptane dilution data, the solvent oil equivalence point is also shown in Figure 3. The intercept of 30.6 is equal to the insolubility number of Arabian Heavy crude oil on the heptane-VGO scale and from Equation 5 the solubility blending number was calculated from the x-intercept to be 68.2 on the same scale. Using Equations 7 and 8, these were used to calculate the solubility blending number of VGO equal to 96.4. Therefore, the solubility blending number of VGO of 95.4 +/- 1.0 for the two reference oils are within experimental error. Therefore, the heptane dilution of blends with reference oils has proven to be a much more accurate method than the solvent oil equivalence test for measuring the solubility blending number of oils without asphaltenes.

Conclusions

It is recommended that the heptane dilution test on blends of an oil without asphaltenes and a reference oil with asphaltenes be used to measure the solubility blending number of an oil without asphaltenes. Even with a difficult reference oil, Athabasca bitumen, this method was found to be consistent with multiple data, unlike methods based upon the solvent oil equivalence and on the compatibility numbers of blends. In addition, the heptane dilution test on blends was found to be more accurate than the solvent oil equivalence test on a more ideal reference oil, Arabian Heavy crude oil. The requirement that the data be linear, like that in Figures 1 and 3, provides a check with the heptane dilution test data on blends that is not available with other methods. In addition, the heptane dilution test on blends can be run with automatic titration methods, unlike the solvent oil equivalence method.

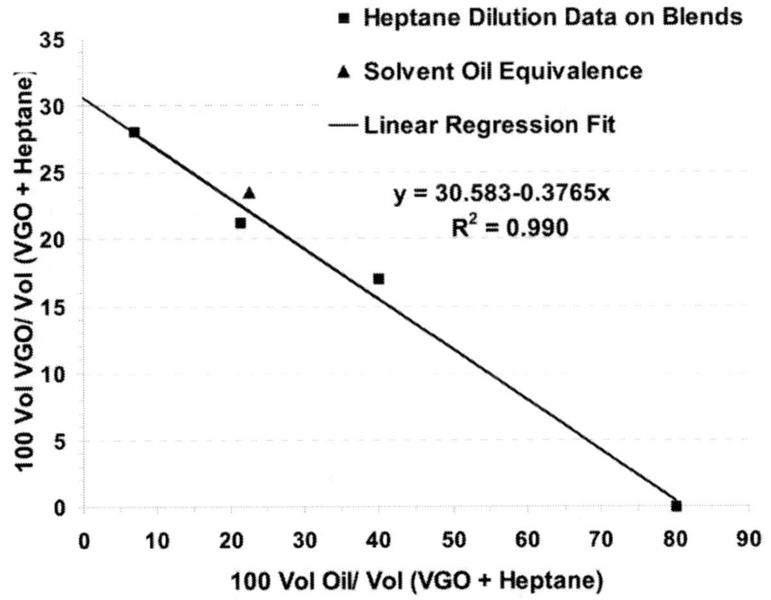

Figure 3. The Heptane Dilution and Solvent Oil Equivalence Data on Blends of Athabasca VGO and Arabian Heavy Crude Oil

Acknowledgement

Partial funding for NCUT has been provided by the Canadian Program for Energy Research and Development (PERD), the Alberta Research Council (ARC) and the Alberta Energy Research Institute (AERI).

References

1. Wiehe, I.A. and R.J.Kennedy (2000), "The Oil Compatibility Model and Crude Oil Incompatibility" *Energy & Fuels*, 14, 56-59.
2. Wiehe, I.A. and R.J.Kennedy (2000), "Application of the Oil Compatibility Model to Refinery Streams", *Energy and Fuels*, 14, 60-63.
3. Wiehe, I.A. and R.J.Kennedy and G. Dickakian (2001), "Fouling of Nearly Incompatible Oils", *Energy and Fuels*, 15, 1057-1058.
4. Wiehe, I.A. (Oct. 2001), "Are You Processing Incompatible Crude Oils?", *World Refining*, 24-28.
5. Wiehe, I.A. (2006), "Petroleum Fouling: Causes, Tools, and Mitigation Methods", *Proc. 2006 Spring AIChE Meeting*, Paper 220a.

Chemical Structure of Petroleum Asphaltenes and Implications for Processing

Murray Gray, Chemical & Materials, University of Alberta, T6G 2G6, Edmonton, AB T6g 2G6, Canada

The properties and chemistry of the vacuum residue fraction are dominated by the behavior of the asphaltenes. This fraction contains components that contribute strongly to coke formation, to the formation of sediment in hydroconversion, and to incompatibility in feed blending. This paper will review recent results on asphaltene behavior, and emphasize the insights available from model compounds for asphaltene behavior. This chemical information leads to suggestions for process opportunities for conversion of the bottom of the barrel.

THE PROCESSING OF RESIDS AND HEAVY OILS

Irwin A. Wiehe, Soluble Solutions, 3 Louise Lane, Gladstone, NJ 07934

Introduction

The golden era of heavy oil processing in the petroleum industry that many predicted in the 1970's has finally arrived. The supply of light crude oils has not been able to keep up with demand. Meanwhile, the quantity of extra heavy oil in Venezuela and in Alberta, Canada is vast at known locations. However, nothing is easy about processing heavy oils: production, transportation, and upgrading/refining are all difficult to those accustom to light crude oils. Nevertheless, with the world's strong appetite for petroleum products, those that are prepared to produce, to upgrade, and to refine extra heavy oil are amply rewarded. We do not need new refineries but instead we need more resid conversion capacity. Unfortunately, despite the early prediction of this heavy oil era, most have waited until now to construct upgraders and/or to add resid conversion capacity to refineries. While Venezuela had its recent heavy oil boom period, an unfriendly government is discouraging further foreign investment. As a result, nearly every petroleum company is rushing to the tar sands of Alberta, Canada to construct mining and upgrading capacity. The hamlet of Ft. McMurray resembles a gold rush city that cannot attract enough construction workers because they cannot build houses, roads, and schools fast enough. The high cost of housing has the locals calling it "Fort McMoney". Meanwhile, little research was done on resid processing since the 1970's. As a result, the petroleum industry is still using resid processes from the 1970's or earlier. Thus, the point of this tutorial is to show that there is plenty room for improvement of resid conversion processes. Even reserves of extra heavy oil will be depleted some day. By moving to more efficient resid and heavy oil processes, that day will be pushed much further in the future.

Petroleum resids are the heavy fraction remaining after distilling petroleum crudes at atmospheric pressure (atmospheric resid) or at reduced pressure of 25 to 100 mm Hg (vacuum resid). Even most light crude oils contain some atmospheric and vacuum resid. Thus, heavy and extra heavy oils just contain a greater fraction of atmospheric and vacuum resid than light crude oils and upgrading and refining of extra heavy oils are essentially resid conversion processes. Most extra heavy oil was a lighter oil during its history but it migrated close enough to the surface to be biodegraded. Microorganisms consumed the lighter, more paraffinic fractions that normally can be distilled directly into gasoline and diesel fuel. The exposure to oxygen and microorganisms usually formed naphthenic acids in the extra heavy oil that can cause corrosion and stable water emulsion problems when the oil is processed. As a result, not much extra heavy oil was processed as long as light crude oil was available. However, with the present shortage of light crude oil and the greatly increased price of all crude oils, the upgrading and refining of extra heavy oil makes good business sense.

Not only do resids have high molecular weight ($\approx$1000 amu number average for vacuum resids) but they contain polynuclear aromatics (PNA's). The PNA's greater than 3-4 rings provide the greatest limitation to the conversion of resids. This is because of their high thermal stability and because desirable aromatics in transportation fuels are limited to alkylated single rings. In addition, the high concentrations of heteroatoms (sulfur, nitrogen, vanadium, and nickel) in petroleum resids act along with PNA's to poison catalysts. Therefore, care must be taken to protect catalysts, to replace and regenerate catalysts, to use throwaway catalysts, or to use purely thermal processes. No matter which type of process is used, a substantial fraction of resid molecules can be cracked off as fragments to become liquids in the transportation fuel and vacuum gas oil boiling ranges. However, one should not try to overly convert resids or the PNA limit will force the selectivity to go to the thermodynamically favored[1] but lower valued products: coke and/or hydrocarbon gases. As a result, all economical resid conversion processes have an even heavier carbonaceous byproduct. It is critical to have an economical and environmentally acceptable use for this byproduct for the resid conversion process to be viable.

Characterization of Resids

The understanding of the chemical transformations of the macromolecules during resid conversion is greatly limited by the wide diversity (over a million chemical structures) of these complex macromolecules. One way to sample this molecular diversity is to separate the resid and its conversion products into diverse fractions using solubility/insolubility in volatile liquids as well as adsorption/desorption on solids. In this way a number of resids and resid conversion products were separated into coke (toluene insoluble), asphaltenes (toluene soluble/n-heptane insoluble), resins (n-heptane soluble/adsorbs on Attapulgus clay), small ring aromatics (n-heptane and cold methylethyl ketone soluble/does not adsorb on Attapulgus clay), and saturates (n-heptane soluble/ cold methylethyl ketone insoluble/does not adsorb on Attapulgus clay). A plot of molecular weight versus hydrogen content, the solvent-resid phase diagram in Figure 1, shows that each of these fractions occupies an unique area that better defines the fraction than the separation method[2]. Thus, to a first approximation resids differ in the relative amounts of similar fractions: saturates, small ring aromatics, resins, and asphaltenes (unconverted resid contains no coke).

While saturates and small ring aromatics contain sulfur, they contain little or no nitrogen, oxygen, vanadium, nickel, or PNA's of 4 rings or larger[2]. Thus, they are relatively catalyst friendly. During conversion, the saturates can be completely converted to lower molecular products without forming a carbonaceous byproduct. However, the small ring aromatics combine during thermal cracking to form resins. The resins crack off side chains from PNA's to form lower molecular weight, lower hydrogen content byproducts. When these PNA containing molecules combine, they form higher molecular weight asphaltenes. The asphaltenes crack off side chains from PNA's to form lower molecular weight, lower hydrogen content byproducts until they combine to form higher molecular weight coke[2]. The addition of hydrogen and the presence of catalysts can alter the rates of these aromatic combination reactions and thus the selectivity to products. However, at high enough conversion, coke and gas are the principal products, independent of conversion chemistry as the remaining molecules approach partially saturated PNA's with very short side chains.

Hydrocarbon Solubility

Above the wax melting point (40 to 60°C) where resids are processed, all fractions with lower hydrogen content and higher molecular weight[3] become less soluble in hydrocarbons. Figure 1 shows this with the solubility parameter increasing with lower hydrogen content (higher aromaticity) because the attractive force between PNA's is the strongest intermolecular force in resids and resid conversion products. This is very important in the mechanism of coke and sediment formation that often plagues resid conversion processes.

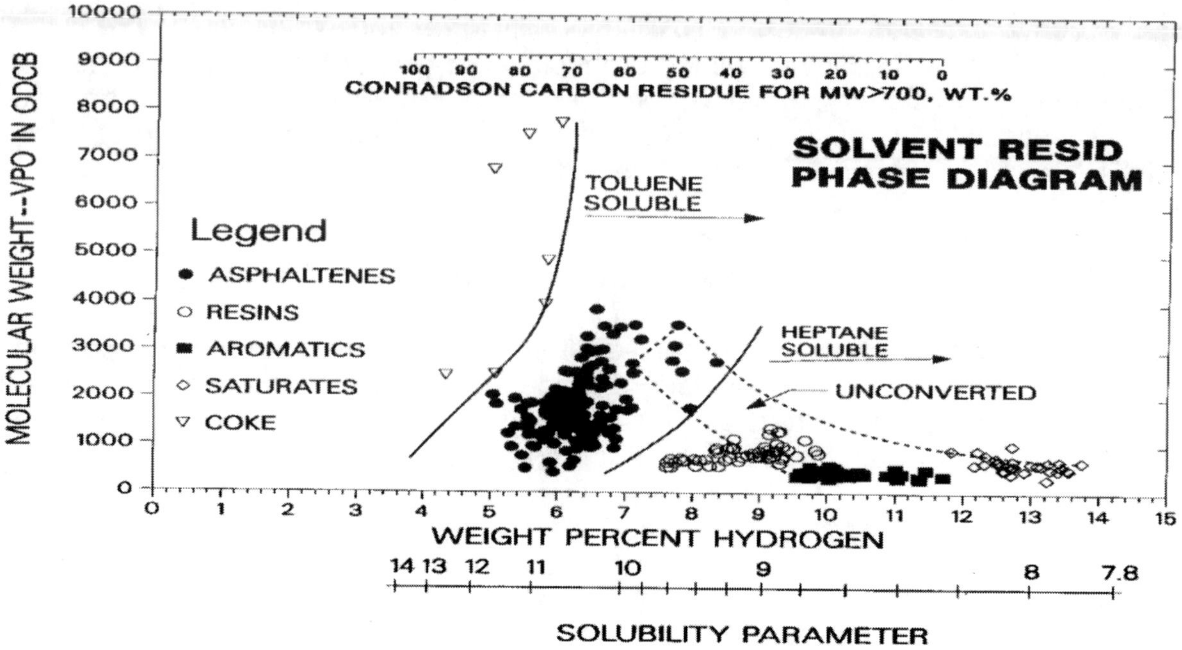

Figure 1. The Solvent - Resid Phase Diagram

Pendant-Core Building Block Model and Conradson Carbon Residue

Models of petroleum with detailed chemical structures can be misleading with the present characterization uncertainty. However, one does not approximate reality if the wide diversity of molecular types is not included. A compromise that satisfies both of these requirements is to represent molecules as combinations of pseudo building blocks. The simplest model has two building blocks[4]. Since thermally stable polynuclear aromatics are such an important part of the mechanism of coke formation, they are represented by the aromatic core building block. The other building block represents the fragments that become volatile liquids when cracked off the core, called pendants. Then a vacuum resid can be represented by a wide distribution of species made up of various combinations of these two building blocks (Figure 2). If each building block is assigned a hydrogen content and a molecular weight, the diversity of the solvent-resid phase diagram (Figure 1) will be represented.

Conradson carbon residue (CCR) is a standard petroleum industrial coking test for characterizing the coke forming tendency of petroleum liquids that today is measured by the Microcarbon method[5] (ASTM D4530). Since this test is widely used, it is extremely important for those working in resid conversion to understand what this test does or does not reveal. In the terms of the Pendant-core Building Block Model, CCR measures the percent fraction of the resid that is cores. A hydrogen balance around the Microcarbon test (Figure 3), shows that the CCR is a linear function of hydrogen content. This enables determining that the hydrogen content of the core is 3.8 wt.% and the hydrogen content of the pendant is 11.6

Saturates	**Aromatics**	**Resins**	**Asphaltenes**
P-P	**P-•-P**	**P-•-P-•**	**P-•-•-P-•-P-P**

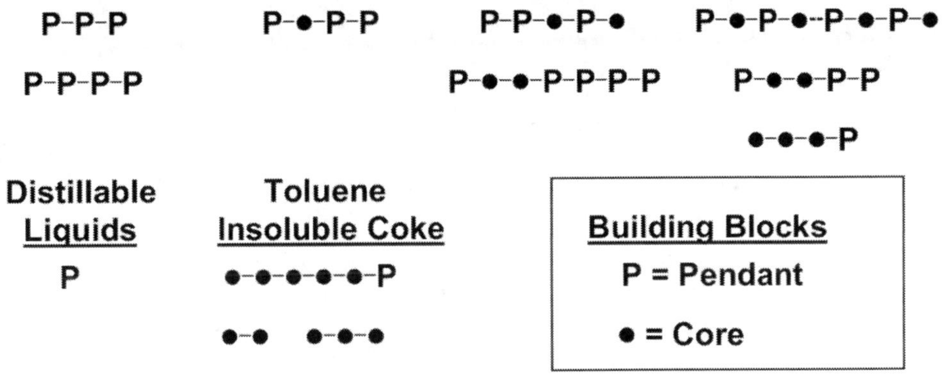

Figure 2. The Pendant - Core Building Block Model Allows for a Wide Distribution of Species.

wt.% (Figure 4). These values were independently verified by directly measuring the hydrogen content of the coke in the microcarbon test and the conversion independent hydrogen content of distillable liquid product of vacuum resid thermolysis[4]. In addition, the pendant-core model explains why CCR is conserved for physical separations (hydrogen is conserved), increases for thermal conversion (hydrogen escapes as part of hydrocarbon gas byproduct), and decreases for hydroconversion (hydrogen is added).

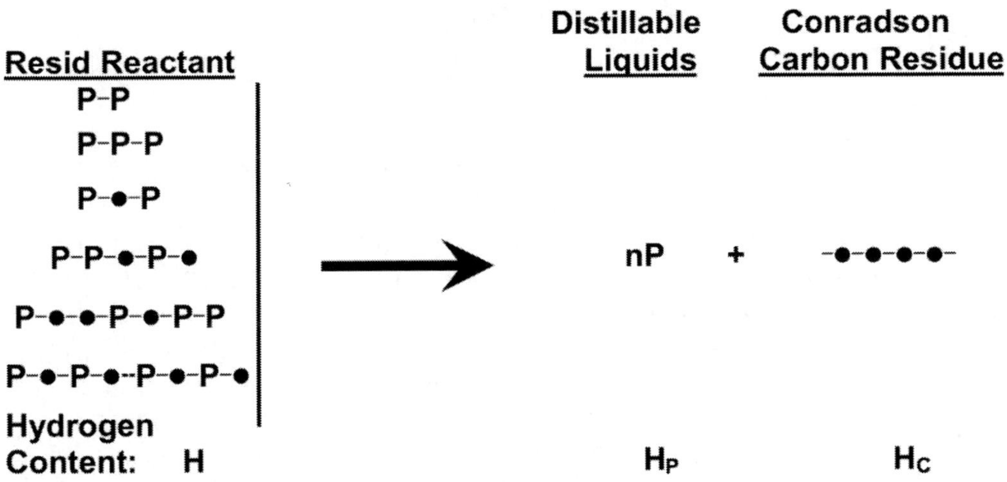

H Balance: $100\ H = (100\text{-}CCR)\ H_P + CCR\ H_C$

$$CCR = 100\ (H_P - H)\ /\ (H_P - H_C)$$

Figure 3. Model predicts that Conradson carbon residue (CCR) is a linear function of hydrogen content.

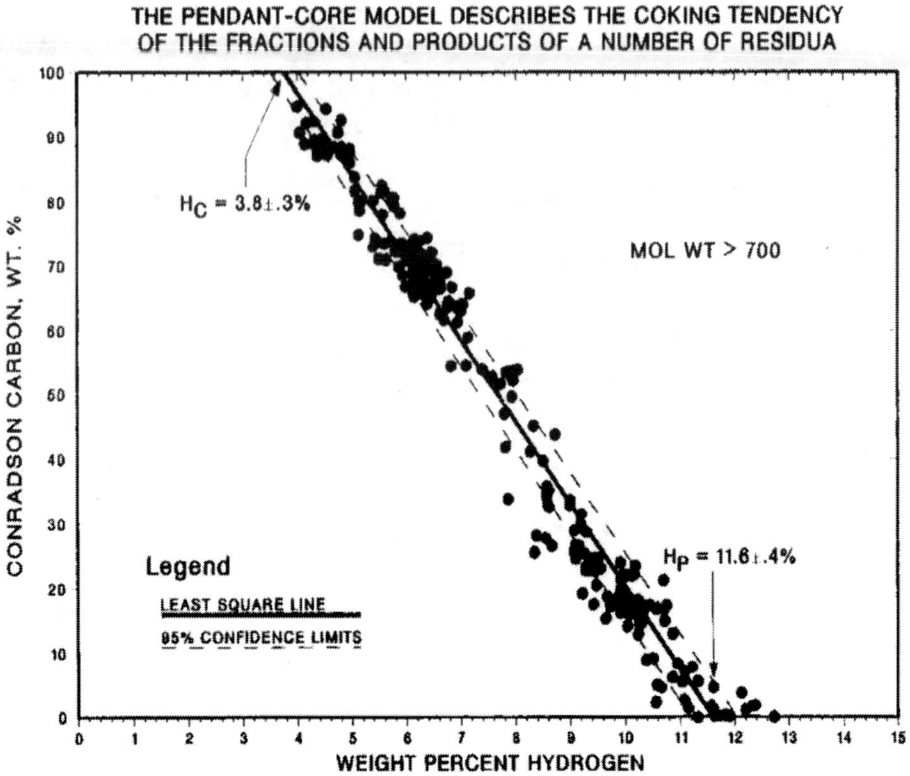

Figure 4. Conradson Carbon is a Linear Function of Hydrogen Content.

Selection of Resid Conversion Processes

Most petroleum resid processes can be made up of combinations of separations, thermal conversion, hydroconversion or hydrotreating, and fluid catalytic cracking. Distillation is the most common separation choice because of its low cost. However, fuels deasphalting is becoming more popular because it can separate saturates and small ring aromatics out of resids. Nevertheless, the selection of the conversion process depends on the quality of the resid. Resids high in saturates and small ring aromatics and low in resins and asphaltenes are more appropriate for catalytic processing. On the other hand, resids high in resins and asphaltenes, as found in extra heavy oils, are more suited for coking.

Resid Catalytic Cracking[6]

Resid catalytic cracking has much better selectivity to desired products (high gasoline and low gas yields) than for coking or hydroconversion. In fluid catalytic cracking feed is sprayed on zeolite catalyst within a porous silica alumina support in a short contact time, riser reactor. The vaporized product flows to a fractionator while the catalyst with coke and adsorbed hydrocarbons flow to a fluidized bed regenerator where the coke and hydrocarbons are burned off the catalyst. Fluid catalytic cracking requires much higher quality feeds than coking or hydroconversion. This is because of expensive zeolite catalysts, intolerance to sodium, nickel, vanadium, and basic nitrogen as well as limitations on the amount of coke that can be burned in the regeneration step by cooling capacity. As a result, the feed in resid catalytic cracking is at worst an excellent quality atmospheric resid but mixtures of vacuum gas oil and atmospheric resids are more common. The total feed is limited to Conradson Carbon Residues of 3 to 8 wt.% , depending on the cooling capacity. This and other typical feed limitations are found in Table 1.

Table 1. Resid Fluid Catalytic Cracking Feed Guidelines[6]

Feed Property	Limit
Conradson Carbon Residue	3 - 8 wt%
Nickel Content	20 - 25 ppm
Vanadium Content	10 - 15 ppm
Sodium Content	5 - 10 ppm
Basic Nitrogen	800 ppm
Sulfur Content	3 wt%
Maximum Density	0.940 g/cc
Minimum Hydrogen	11.5 wt%

Deasphalting[7,8] and Separation Processes

Deasphalting can be used to separate the higher quality oil from vacuum resids for making a resid feed for fluid catalytic cracking (FCC). The deasphalted oil is typically blended with the vacuum gas oil and hydrotreated before catalytic cracking. This combination of deasphalting and hydrotreating can enable producing a good FCC feed from a low cost, poor quality atmospheric resid. Disposing of the asphalt can be problem because asphalt products and heavy fuel oil usually cannot absorb such high volumes. The most common choice is to add the fuels asphalt to coker feed. However, since the resulting coke is becoming difficult to sell, gasification of asphalt and/or coke is probably the better long term choice in many locations.

Deasphalting is done in a countercurrent treater (or separator) using n-propane, n-butane, isobutane, or n-pentane as the precipitating liquid. For fuels deasphalting the uniform quality is not as important as for lubes. What is more important for fuels is processing large volumes at low cost, both capital and expense. Hydrotreating further cleans up the deasphalted oil and FCC can tolerate some impurities. Therefore, fuels deasphalter treating towers are shorter (order of 40 ft. high by 13 ft. diameter) and contain packing instead of trays or a combination of packing and trays. Fuels deasphalters typically use butane or pentane at a low solvent to oil ratio (4-8) and do not heat the top of the tower.

<u>Separation of Vacuum Resid</u> Curves representing the separation of several vacuum resids are shown in terms of Conradson carbon residue (Figure 5) and vanadium (Figure 6) in the deasphalted oil as a function of yield of deasphalted oil[7]. The lowest concentration of Conradson carbon and metals are in the propane deasphalted oil but the maximum yield is about 50 wt%. As the carbon number of the alkane solvent is increased, the possible yield of deasphalted oil is increased but so are the Conradson carbon and the metals in the deasphalted oil. The overlapping curves for the lowest concentration of each contaminant as a function of yield of deasphalted oil approximates the separability of vacuum resid. There are limits imposed by the molecules in the resid of how well we can separate according to Conradson carbon residue and this overlapping curve approaches this limit.

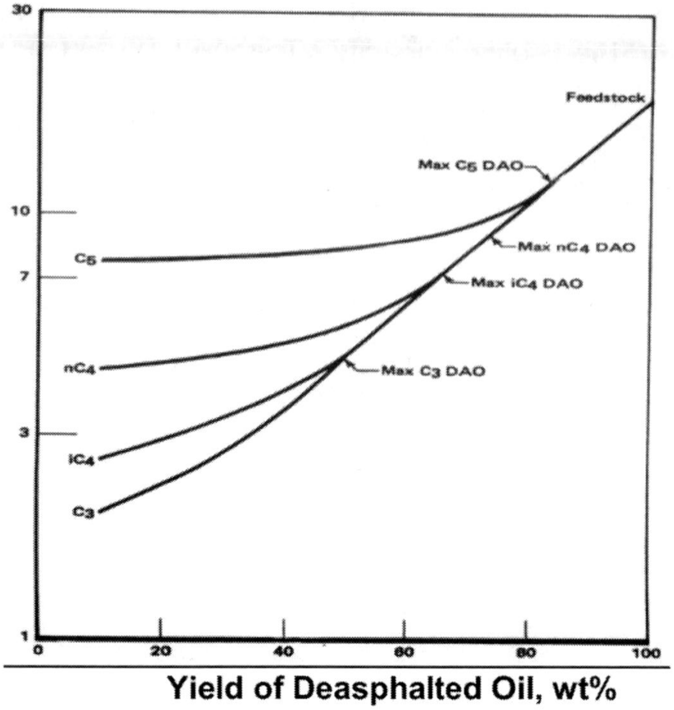

Figure 5. Conradson Carbon Residue (wt%) vs. Yield of DAO for Varying Solvent[7]

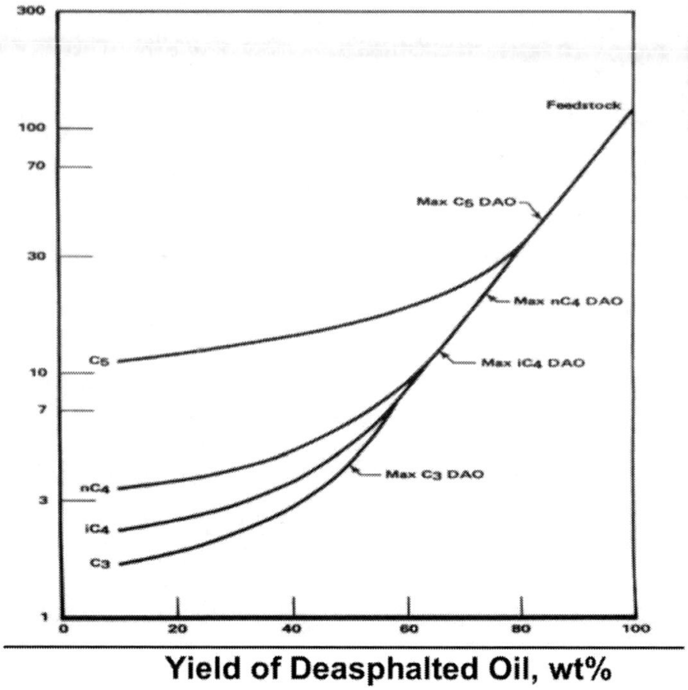

Figure 6. Vanadium Content (ppm) vs.Yield of DAO for Varying Solvent

<u>Molecular Limitation to Separation[9]</u> The limit for separation on the basis of Conradson carbon is a molecular limit as demonstrated by the pendant-core building block model. While there is a large fraction of vacuum resids with no cores (zero Conradson carbon), there is little or no fraction with Conradson carbon above 50 wt%. Since there is a large fraction with pendants and cores covalently bonded in the same molecules, there is a molecular limit of how well heavy oils can be separated on the basis of Conradson carbon. Thus, it makes no sense to research other ways to obtain a better separation than by deasphalting. Instead, one needs to crack bonds between pendants and cores by visbreaking or by hydroconversion prior to deasphalting. Alternatively, one could separate the low Conradson carbon fraction by deasphalting and rely on coking of the asphalt to reactively separate the pendants and cores. This alternative was already discussed.

Although the metals also appear to have a separation limit, an analysis of the concentrations of metals and the molecular weights present shows there cannot be a molecular limitation. Typically, all the vanadium and nickel in a vacuum resid can be contained in less than 1 wt% of the resid. However, the molecules containing the metals are in large polynuclear aromatics (porphrins) that separate much like other large polynuclear aromatics. It is this similarity that has limited the selective separation of metals.

Coking

In coking processes the resid is heated to high temperatures (450-500°C) with the light boiling products distilling overhead and leaving the solid coke behind. Coking is the most popular conversion choice for resids high in cores (low hydrogen, high heteroatoms, and high Conradson carbon) because it reactively separates the cores, the catalyst poisons, into a low value byproduct (coke) leaving the pendants for catalytic upgrading. Therefore, coking should be called a "catalyst poison rejection" process because a principal reason for coking is to improve the liquids for further upgrading by catalytic processing[10]. Unfortunately, coking has been mistakenly called a "carbon rejection" process, a term that continues to be repeated without critical examination. For example, Table 2 shows that the coke product in a microcarbon tester has a much higher concentration of sulfur, nitrogen, and metals than the vacuum resid feed and by atomic balance, the volatile liquid products. However, the concentration of carbon in the vacuum resid, in the coke, and in the liquid product are nearly the same. The rejection of heteroatoms is counter balanced by an increase of hydrogen content in the liquid product. This highly desirable result is because most of the polynuclear aromatic building blocks in the petroleum macromolecules contain heteroatoms that are reactively separated with the polynuclear aromatics into the coke.

Kinetics of Coke Formation and Mesophase The thermal cracking of resid is a first order reaction with a constant activation energy over a large range of temperatures[11]. There is an induction period before coke begins to form (Figure 7). It is now understood[12] that coke formation is triggered by the phase separation of a second liquid phase formed from partially converted asphaltene cores. Evidence of the liquid - liquid phase separation of asphaltene cores, promoted by the stacking of polynuclear aromatic structures, is the presence of spherical liquid crystalline coke[12], called the carbonaceous mesophase. Mesophase in a coke sample can be used as a diagnostic tool to show that it was caused by thermal cracking and insolubility at high temperatures[13].

Table 2. Elemental Balance Around Microcarbon Test

Element	Arabian Heavy Vac. Resid	22.3 wt% Microcarbon Coke	77.7 wt% Liquid Balance	Resid to Liquid % Change
Carbon	83.5%	85.8%	82.9%	-0.8
Hydrogen	9.9%	3.7%	11.7%	+18
Sulfur	5.8%	8.6%	5.0%	-14
Nitrogen	0.45%	1.3%	0.21%	-53
Vanadium	165 ppm	808 ppm	0 ppm	-100
Nickel	40 ppm	165 ppm	4 ppm	-90

Delayed Coking[14,15] The delayed coking process is the oldest and most popular choice. The resid is heated by flowing through a long tube in a furnace and then reacted by flowing into the bottom of a high, cylindrical, insulated drum. The volatile products go overhead into a fractionator and coke accumulates in the drum. The heavy liquid product from the fractionator is recycled back to the furnace. Meanwhile, when the drum fills up with coke, the feed is switched to a second drum. The coke is mined out of the drum by drilling a hole down the center with high pressure water and cutting out the remainder also with high-pressure water to get the drum ready for the next 16 hour reaction cycle. Anode grade coke for making electrodes for manufacturing aluminum commands a higher price than fuels grade coke. However, the low specification on sulfur, vanadium, and nickel content for the coke requires a higher quality (high in saturates and small ring aromatics) and more expensive resid feed. Needle coke for making electrodes for steel manufacturing is priced even higher

($500/ton) than anode grade coke but it requires low sulfur, fluid catalytic cracker bottoms, not resid, as a feed.

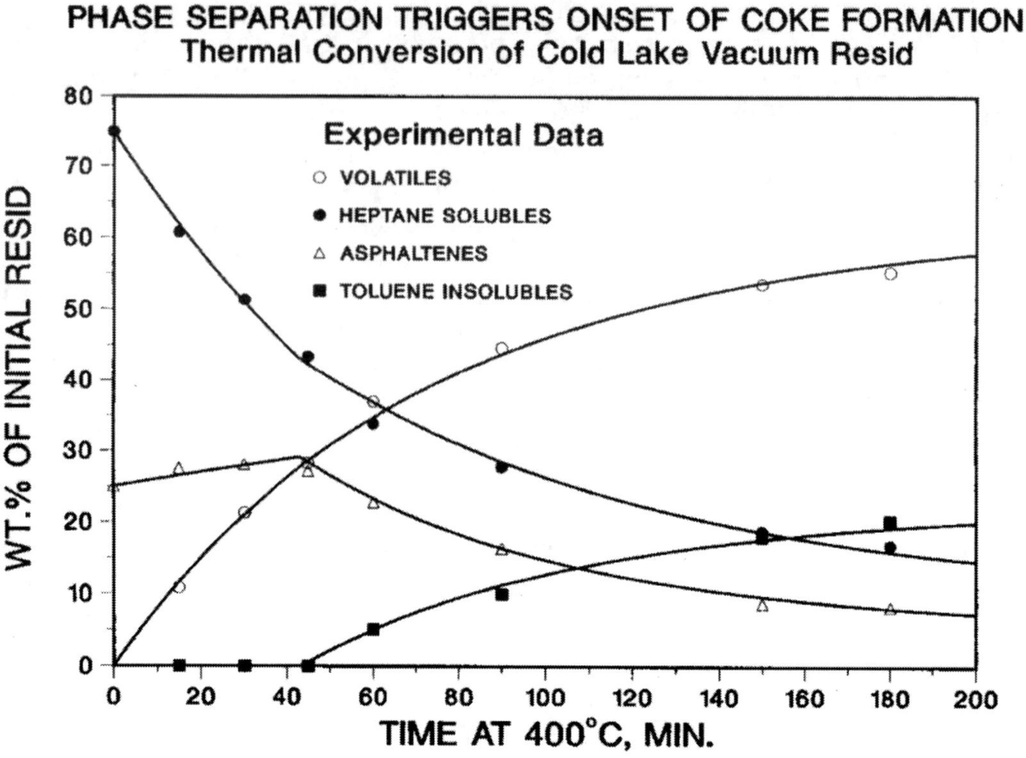

Figure 7. Kinetic Data Has a Coke Induction Period when Asphaltenes are Soluble.

Fluid Coking and FLEXICOKING[16] These processes are licensed by ExxonMobil. In Fluid Coking the resid is sprayed on a hot, fluidized bed of coke particles in a vessel, the reactor. The volatile products go overhead to a fractionator while the coke particle are removed out of the bottom and transferred to another vessel, the burner, where the coke is partially burned with air to provide the heat for the process. This coke then is recirculated back to the reactor. Since this process produces much more coke than is required for heat, fluid coke is withdrawn at the bottom of the reactor. This is a high sulfur, fuels grade coke of low value. A heavy liquid product is recycled back to the reactor from the fractionator. This is used to scrub coke fines from the reactor vapors and to improve the quality of the liquid product. In FLEXICOKING a third vessel, the gasifier, is added to the Fluid Coker. In the gasifier coke is gasified with steam and air in net reducing conditions to produce a low BTU gas containing hydrogen, carbon monoxide, nitrogen, and hydrogen sulfide. After adsorbing out the hydrogen sulfide, this low BTU gas is burned as a clean fuel within the refinery and/or in a nearby power plant. Only small amount of coke (about 3%) needs to be withdrawn in FLEXICOKING to keep vanadium and nickel from accumulating in the gasifier. With the trend of decreasing high sulfur coke prices, gasification of coke to form fuel and syngas to be used within the refinery may be the only economical and environmental option for refineries in the future.

Reactive Separation by Coking Coking has no limitation on separation due to the bonds between pendants and cores because these bonds are thermally broken during the coking process to reject the cores at 100% CCR. However, previously[17] we discussed three limitations to coking chemistry:

1. Secondary cracking of volatile liquid products
2. Formation of 5+ ring polynuclear aromatics by combining of reaction fragments containing smaller ring aromatics
3. Aromization of hydroaromatics to form 5+ ring polynuclear aromatics

Any significant improvement in the yield of high quality liquids depends on one devising ways to overcome these limitations. Secondary cracking degrades the quality of the most desirable product while forming hydrocarbon gases, an undesirable byproduct. Previously[18], we showed that the secondary cracking of volatile liquid products can be minimized by using a coking reactor with a very short vapor residence time but with a long resid residence time. This latter feature is necessary to go to complete conversion to coke of 100% CCR. Figure 9 shows the result using four different ideal laboratory-coking reactors with these two features and a feed of Arabian Heavy vacuum resid[18]. Hydrogen content of the volatile liquids is used as the measure of quality and the resulting yield-quality from the different reactors is compared with a coking model prediction for a commercial coking process. Either the quality can be improved at the same yield or the yield can be increased at the same quality. While this 7.9 wt% increase in yield (69.3 wt% total yield) provides incentives to improve existing processes, it falls short of the requirement for a new coking process. Of course, one must realize that no commercial coker is going to match the yields obtained from a small, ideal laboratory reactor. Therefore, to justify developing a new coking process, one needs to both reduce secondary cracking and minimize the remaining limiting chemistries.

Coking-Separation Synergy The chemical combination of aromatics and aromization of hydroaromatics are two ways that coking degrades the resid feed. Actually, both of these chemistries occur during the microcarbon test for measuring Conradson carbon residue (CCR). Therefore, part of the CCR (5+ ring polynuclear aromatics) is formed during the measurement. As a result, if these chemistries are minimized, one should form less coke than CCR while commercial cokers form 1.2-1.6 times as much coke as CCR[14,15,16]. Since coking degrades the feed, it makes sense to separate the portion of the resid that already meets the quality requirement before coking. At least this portion will be spared from being degraded. In the laboratory we decided to use high cut-point vacuum (Distact) distillation as the only separation process. Deasphalting could also be used on the feed. Thus, the overhead of high cut-point distillation to an atmospheric equivalent boiling point of 1300°F is combined with the overhead product of coking the 1300°F+ resid in an enhanced microcarbon tester[18], that is capable of short vapor and long resid residence times while collecting the volatile liquid product. Since the microcarbon tester measures CCR, the yield of coke is the CCR of the 1300°F+ resid or on a resid basis it is this value times the fraction of the resid that is 1300°F+. A schematic of the process and the calculation of the coke yield for three different vacuum resids are in Figure 8. For each resid the coke yield is less than the CCR of the resid feed (average of 0.8 times CCR) but since it also depends on the fraction of 1300°F+, the coke yield is not proportional to resid CCR. Therefore, Syrian Light resid, that has a higher CCR than Arabian Heavy resid, has a lower coke yield.

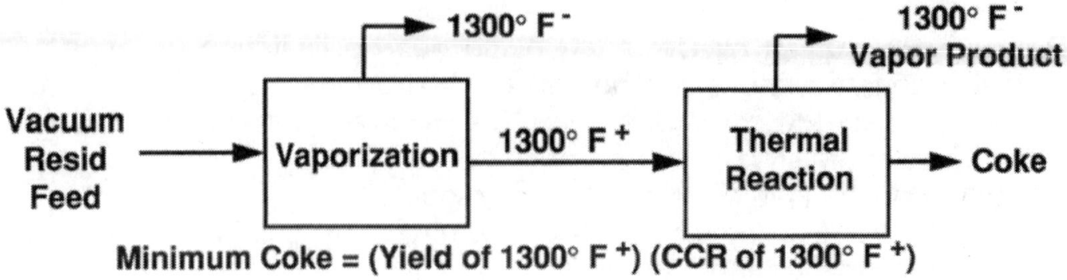

Berri Vacuum Resid (CCR = 18%): Coke = (.382)(35.1) = 13.4 wt%

Arabian Heavy Vacuum Resid (CCR = 22%): Coke = (.546)(35.9) = 19.6 wt%

Syrian Light Vacuum Resid (CCR = 23%): Coke = (.541)(31.9) = 17.3 wt%

Figure 8. Separation before low vapor residence time coking reduces coke yield[18]

In order to measure the yield and quality of the liquid product, the 1300°F⁻ fraction is added to the liquids collected in the enhanced microcarbon tester as shown in Table 3. The result is that the liquid yield is increased by 14.8 wt% over the commercial coker at only a slightly lower quality (11.20 wt% hydrogen as compared with 11.28 wt% hydrogen predicted for a commercial coker). Even while separating part of the resid without coking, the process removes 21% of the sulfur, 44% of the nitrogen, 86% of the vanadium, 93% of the nickel, and 80% of the Conradson carbon residue. Since the gas yield is 3.6 wt.%, there is still room for further improvement towards the upper bound of a liquid yield of 85.3 wt.% for no gas using hydrogen balance (Figure 9). Therefore, this result shows that the combination of the separation of the higher quality fraction of resid without coking and coking of the lower quality fraction of the resid at long residence time, but with short vapor residence time, can achieve a substantially higher liquid yield over commercial cokers at nearly the same liquid quality. Thus, despite the current perception, the coking of petroleum resids has the potential for significant improvement.

Table 3. Yields and analyses of products of separation plus low vapor residence time coking on Arabian Heavy vacuum resid

	Yield (wt%)	C (wt%)	H (wt%)	S (wt%)	N (wt%)	V (ppm)	Ni (ppm	CCR (wt%)
1300°F- Feed	45.4	84.33	11.11	4.12	0.23	45	7	7.4
Coking Liquids	31.4	84.33	11.32	3.99	0.24	0	0	0
Total Liquids	76.8	84.33	11.20	4.07	0.24	27	4	4.4
Coke	19.6	84.99	3.85	8.85	1.5	835	260	100
Gas (By Diff.)	3.6							
Arab Heavy Resid	100	83.67	10.18	5.13	0.42	190	55	22.4
% Reduction				20.7	43.8	85.9	92.6	80.5

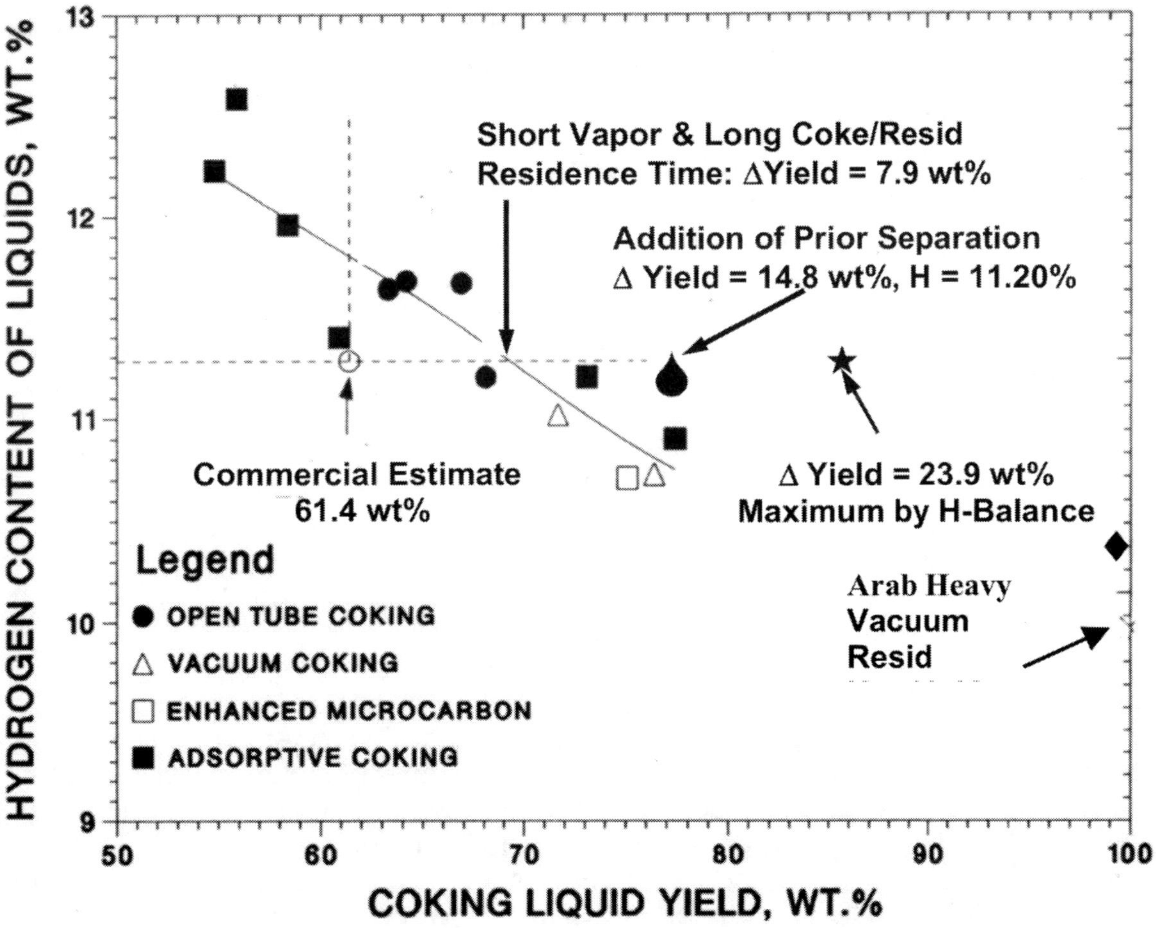

Visbreaking

This is a low conversion thermal process used originally to reduce the resid viscosity for heavy fuel oil applications. However, today it uses a resid that exceeds minimum heavy fuel oil specifications and converts just enough to obtain 15-30% transportation boiling range liquids and still have the heavy product meet heavy fuel oil specifications. Since this process cannot tolerate coke formation, it is required to be within the coke induction period that may limit conversion. Increasing the solubility of converted asphaltenes in the reacting oil can increase the induction period and thus, the conversion. An example is to strip out the volatile liquid product, a nonsolvent for converted asphaltenes[13]. A visbreaker reactor may be a furnace tube followed by a soaker drum or a long tube coiled within a furnace. Upsets cause coke to form and accumulate on visbreaker walls so that they need to be decoked periodically.

Even a low conversion process, like visbreaking, has potential for much greater resid conversion when coupled with a separation process, like deasphalting. Figure 10 shows a process scheme where the deasphalter separates the heavy visbreaker product into three streams: a high quality deasphalted oil, a moderate quality resin, and a low quality asphalt. Since the resin stream is not high enough quality to be part of the deasphalted oil but too good of quality to reject into the asphalt product, it is recycled back to the visbreaker to be thermally cracked into high and low quality products. In the laboratory we took the overhead from vacuum distillation as the only high quality product and used cyclohexane on the vacuum resid to reject cyclohexane insolubles (CCR = 70) and recycle the cyclohexane solubles.

The result was that the overall conversion of Arabian Heavy vacuum resid was 67 wt%, 6 wt% higher than Fluid Coking.

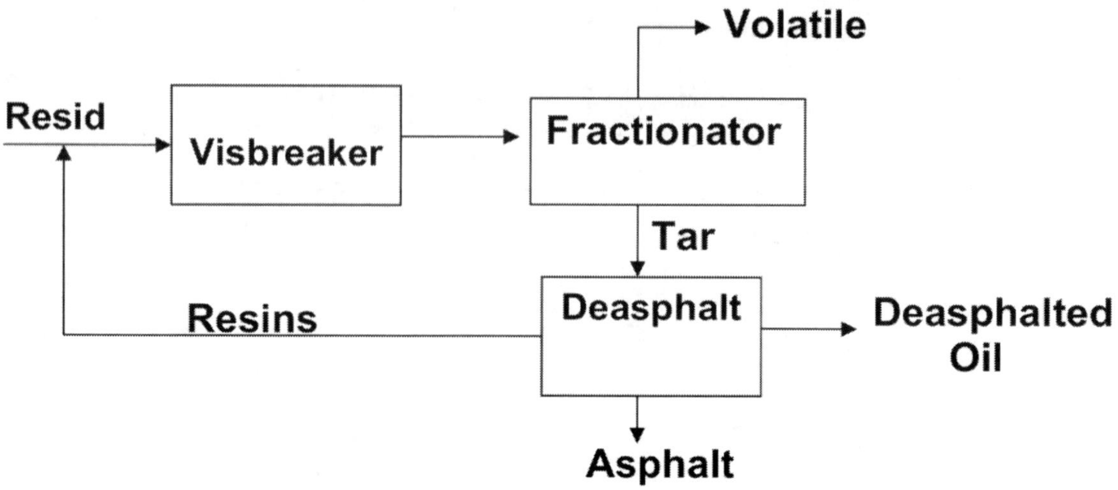

Figure 10. Three way separation of tar and recycle can greatly increase visbreaker conversion.

Hydroconversion

Hydroconversion combines thermal cracking with hydrogenation. The active catalyst needs to be a transition metal sulfide because the high level of sulfur in resid feeds will poison other hydrogenation catalysts. The addition of hydrogen increases the coke induction period, primarily by terminating free radicals and reducing the frequency of aromatics from combining to form larger PNA's. Thus, hydroconversion of vacuum resids to volatile liquids can be over 85% as opposed to 50-60% for commercial coking. However, one has to deal with the cost of hydrogen and catalyst, high-pressure vessels, poisoning of catalysts, the difficulty of asphaltenes to diffuse through small pores, and the intolerance to coke and sediment formation. Since metals removal is one of the fastest reactions and since the metals accumulate in the pores of supported catalysts, it is common to have a guard bed in front of the fixed bed. In order to hydrogenate the largest macromolecules in the resid, the asphaltenes, some or all of the catalysts need to have pores 100 microns or greater in diameter. Even with these precautions, it is difficult to get longer than one year run lengths on fixed bed hydrocoversion units with vacuum resid feeds and conversions to volatile liquids of 50% or more. This is because of catalyst deactivation with coke or by coke and sediment formation downstream of the reactor. The LC Finer and H-Oil units mechanically ebullate the catalyst so that it can be mixed and replaced onstream. The conversion is greatly dependent on the feed but conversions of vacuum resids to volatile liquids of the order of 70% are possible. Often these units are limited by the deposition of coke and sediment downstream of the reactor in hot and cold separators[19].

One way to avoid pore plugging of catalysts is to use finely dispersed catalysts[20]. Dispersed catalysts also can be continuously added in low enough amounts (i.e. 100 ppm) to consider them throwaway catalysts with the carbonaceous by-product. However, economics usually dictate some form of catalyst recycle to minimize catalyst cost. Nevertheless, by designing the reactor to maximize the solubility of the converted asphaltenes[13], the conversion of vacuum resids to gas and volatile liquids can be above 95% with greater than 85% volatile liquids. However, the last 5 to 10% conversion may not be worth the cost of hydrogen and reactor volume to produce hydrocarbon gases and very aromatic liquids from this incremental conversion. The answer depends on the value and use of the unconverted carbonaceous liquid byproduct (essentially all PNA cores). Recently, ENI[21] is developing a dispersed catalyst, hydroconversion process using deasphalting to remove the deasphalted oil product and recycling a major part of the asphalt with the dispersed catalyst. This satisfies the need to recycle catalyst and to maintain solubility of converted asphaltenes. By not requiring hydrogenated rings to be cracked, the reactor can be run at lower temperatures and still obtain reasonable kinetic rates.

Summary

Table 4 compares the conversion of Arabian Heavy vacuum resid using the many alternatives discussed. However, laboratory conversions are mixed with commercially available conversions, the conversions are in weight% rather than the more practical volume%, and the quality of the desirable products are variable and unknown. Nevertheless, it is clear that by combining conversion with selective separation, such as deasphalting, conversions can be greatly increased over the most popular resid conversion process, delayed coking.

Table 4
Liquid Yields by Resid Conversion Processes on Arabian Heavy Vacuum Resid

Process	Liquid Yield (wt%)
Conventional Visbreaking	30
Visbreaking + Cyclohexane Sol. Recycle	67
Conventional Delayed Coking	59
Once-Thru Delayed Coking	61
Conventional Fluid Coking	61
Once-Thru Fluid Coking	66
Short Vapor Res. Time Coking	69
Sep.+ Short Vapor Res.Time Coking	76
Ebullated Bed Hydroconversion	70-80
ENI Slurry Hydroconversion	89

Future Improvements of Resid Conversion Processes

We are still on the steep part of the learning curve as to the characterization, phase behavior, and conversion chemistry of petroleum resids. As a result, much room remains to improve resid conversion processes. As fuel grade coke continues to reduce in value, gasification of coke, deasphalted bottoms, or even poor quality vacuum resid to produce electrical power, refinery fuel, and syngas for chemicals will become a more popular alternative[22].

References

1. Fitzer, E., Mueller, K., Schaefer, W. (1971), in "*Chemistry and Physics of Carbon*", P. L. Walker, ed., Marcel Dekker, 7, 237.
2. Wiehe, I.A. (1992),Ind. Eng. Chem. Res., 31, 530.
3. Wiehe, I.A., Liang, K.S., *Fluid Phase Equilibria*, 117, 201 (1996) and *Fuel Sci. & Tech. Int'l*, 14, 289 (1996).
4. Wiehe, I.A. (1994), *Energy and Fuels*, 8, 536
5. Noel, F. (1984), *Fuel*, 63, 931.
6. Barnes, P.H. (1998), *Proceedings of 1st International Conference on Refinery Processes*, AIChE, 227.
7. Sprague, S.B. (1986), *NPRA Annual Meeting*, Paper # AM-86-36.
8. Low, J.Y., Hood, R.L. and Lynch, K.Z. (1995), *PREPRINTS ACS Div. Pet. Chem.*, 40, 780-784.
9. Wiehe, I.A. (2000), *Proceedings of Conference on Energy and the Environment*, AIChE, 34.
10. Wiehe, I. A. (1995), *ACS Div. Pet. Chem. Meeting*, Chicago.
11. Olmstead, W. N., Freund, H. 1998, *AIChE Spring National Meeting*, New Orleans.
12. Wiehe, I.A. (1993), Ind. Eng. Chem. Res., 32, 2447.
13. Wiehe, I. A., ACS Div. Pet. Chem. Meeting, San Francisco (1997).
14. Ellis, P. J. and Paul, C.A. (1998), *Proceedings of 1st International Conference on Refinery Processes*, AIChE, 151.
15. G. C. Hughes, B. J. Doerksen, A. M. Englehorn and S. Romero (2000), *Proceedings of the 3rd International Conference on Refinery Processing*, AIChE, 66.
16. Hammond, D.G. (1998), *Proceedings of 1st International Conference on Refinery Processes*, AIChE, 170.
17. I. A. Wiehe, "The Enhanced Microcarbon Tester as a Laboratory Coker", Presentation at AIChE Spring Meeting, Atlanta (2000).
18. Wiehe, I.A., Proceedings of 4th International Conference on Refinery Processes, AIChE, 245 (2001).
19. Nowlan, V.J. and Srinivasan, N.S. (1996), *Fuel Sci. & Tech. Int'l*, 14, 41.
20. Bearden, R. and Aldridge, C.L., US Patent 4,134,825 (1979) and *ACS Div. Pet. Chem. Meeting, ACS Award in Petroleum Chemistry Address*, San Francisco (1997).
21. Delbianco, A., Panariti, N., et al. (2002), *Proceedings of 3rd International Conference on Petroleum Phase Behavior and Fouling*, AIChE, 520.
22. Penrose, C.F., Rodarte, A., Wallace, P.S., and Preston, W.E. (2000), *Proceedings of 3rd International Conference on Refining Processing*, AIChE, 385.
23. Corma, A. and Wojciechowski, B.W. (1985) "The Chemistry of Catalytic Cracking", *Catal. Rev.- Sci. Eng.*, 27(1), 29-150.
24. Dale, G.H. (1983), "Heavy Oil Cracking", *World Pet. Congress*, PD8(3).
25. Ekholm, E. and Self, F. (1999), "*Refining Fundamentals*", South Texas Section AIChE.
26. Venuto, P.B. and Habib, E.T. (1978), "Catalyst-Feedstock-Engineering Interactions in Fluid Catalytic Cracking", *Catal. Rev.-Sci. Eng.*, 18(1), 1-150.

Residue Upgrading by Solvent Deasphalting and Pitch Gasification-a Review

James F. McGehee, Research & Development, UOP, LLC, 25 E. Algonquin Rd., Des Plaines, IL 60056

Upgrading of vacuum residue, regardless of origin, has complex economics related to the fact that it is a highly contaminated, hydrogen-deficient fraction. To varying degree, solvent deasphalting, coking, ebbulated bed hydrocracking and more recently, slurry hydrocracking can practically achieve 50-95 % vacuum bottoms conversion. In many recent projects, the goal has been to achieve a zero-residue position by gasification of the remaining residue to produce hydrogen and power, leading to a very interesting technoeconomic problem.

Solvent Deasphalting (SDA) typically uses C3-C5 hydrocarbon under pressure at near-critical conditions to break the colloidal structure of heavy oil, causing a phase separation. Recovered deasphalted oil (DAO) is substantially reduced in metals, polar aromatics and asphaltenes. SDA alone or in combinations can be a low capital cost and very energy-efficient upgrading system.

This tutorial will discuss several recent project examples of solvent deasphalting in conversion refineries and discuss potential hydrogen production from pitch gasification as well as control schemes and the balancing of SDA lift against hydrogen demand, cost of gasification and net value for DAO upgrading. Future trends, such as very deep lift SDA, DAO upgrading and pitch utilization in nonconventional ways will also be briefly discussed.

Joint Ebullated Bed Catalyst Development Program Results in over 30% Reduction in Catalyst Usage

Craig McKnight & Wayne Schiewe (Syncrude Canada, Ltd.), Josiane M. Ginestra, Ph.D., Wessel Ijlstra, Dorian Rauschning & David E. Sherwood, Jr., Ph.D. (CRI/Criterion US LP)*
** Presenter*

1 Introduction

1.1 Syncrude's Operation and Drivers

Alberta's recoverable oil sand bitumen reserves are equivalent to those of Saudi Arabia. The oil sand deposits are located in the northern part of Alberta, and are concentrated in three major regions: Cold Lake, Athabasca, and Peace River. Of course, extraction and processing of the bitumen presents unique technical challenges. Over the past few years, many companies have invested in proven and developing technologies to rapidly ramp up production of this valuable resource for the North American market.

Syncrude Canada Ltd. is the operator of the Syncrude Project, a joint venture owned by eight participants, operating near Fort McMurray, Alberta, in the Athabasca region. It has produced Syncrude Sweet Blend (SSB) since 1978. Oil sand is excavated in open pit mines, using some of the biggest equipment in the world, then crushed and slurried in hot water. Each grain of sand is covered with a thin film of water and then by a layer of bitumen, which makes it possible to separate the organic and inorganic phases. The clean sand, after the layer of oil has been removed, is eventually sent to a settling basin and the extraction water reused.

Whole bitumen is a very heavy, sour, high metals crude that is almost solid at normal room temperature. Most North American refineries are not equipped to process bitumen, so Syncrude upgrades the material in its own large refinery Upgrading complex to produce high quality sweet crude.

Bitumen is recovered as an aerated froth, still containing significant amounts of water and fine clay solids. The extracted bitumen froth is diluted with naphtha and almost all the remaining solids and water are removed by either centrifugation or gravity settling. This clean diluted bitumen is processed in a large refinery Upgrading complex to produce high quality SSB. The diluent is recovered in an atmospheric distillation tower, where light components are also extracted from the bitumen. The heavy portion is further treated in a vacuum distillation unit. A portion of the vacuum topped bitumen (VTB), mixed with some atmospheric topped bitumen (ATB), is processed in an LC-Finer^{SM} unit. The bottoms from the LC-Finer^{SM} hydrotreater are processed in the fluid coking units along with the balance of the bitumen. The various product streams from primary upgrading are then directed to the appropriate hydrotreater: naphtha, light gas oil or heavy gas oil. All products are combined to produce the Syncrude Sweet Blend crude, which is an attractive sweet, bottomless feedstock for conventional refineries. The product is distributed in the North American pipeline networks to refineries in Canada and the USA.

The ebullated bed hydrocracker (LC-Finer[SM]) increases the overall up-grading yield and product quality. This unit with two parallel reactors operates at moderate conversion (60%), moderate pressure (75 bar H_2 Partial Pressure), and high temperature (440°C). It was originally designed to process 40,000 barrels per day of ATB, but is presently processing 50,000 barrels per day of a 60/40 blend of ATB and VTB.

Economic drivers of the LC-Finer[SM] unit are complex. In the past, the Upgrader was usually not constrained by the amount of bitumen available to be processed, so LC-Finer[SM] unit throughput was constrained by hydraulics (i.e. unreacted bottoms pumping capacity). As a result, pitch conversion was important for relieving this bottleneck. Occasionally, there were periods of time when bitumen available to Upgrading was limited because of restrictions in the upstream mining and extraction operation. In the case of limited bitumen supply, yield considerations dominated. For this case, higher MCR (Micro Carbon Residue) conversion would lead to higher coker yields and hence higher overall yield of valuable hydrocarbon liquids. However, with the new expansion project in operation, primary upgrading is no longer the bottleneck. Higher activity catalyst would be attractive for the yield benefits, but higher conversion would be accompanied by higher heat release. As there is no ability to quench the ebullated bed reactors, higher heat release cannot be fully accommodated. In fact, the Syncrude[SM] reactors operate near maximum heat release, because the mechanical design of the equipment limits the maximum temperature difference between the reactor inlet piping and the reactor (i.e. the "exotherm"). As a result, achieving higher conversions by increasing catalyst addition rate or using higher activity catalyst is not possible. The economics for the Syncrude LC-Finer[SM] unit make it desirable to operate at the exotherm limit, and then utilize higher activity catalyst to reduce the catalyst consumption. This then translates to a significant operating cost savings.

1.2 Syncrude/Criterion Alliance

The Syncrude and Criterion Catalyst Alliance, formed in 1992, has two major focus areas: logistics (catalyst supply and handling) and integrated research. The objective of the joint research program is to design catalysts specifically for each of Syncrude's hydroprocessing units.

This paper describes the latest success of the Alliance: the new ebullated bed catalyst has reduced catalyst consumption in the LC-Finer[SM] unit by 33%. This catalyst resulted from many years of dedicated research, including the development and commercialization of a new powder technology, optimization of the support pore structure, and optimization of the active metals formulation. The catalyst development program was supported by pilot plant testing at both the Syncrude Research Centre in Edmonton and the Shell Technology Center in Houston.

2 Catalyst Development Strategy

In the late 90's the joint Criterion-Syncrude development team decided that the most valuable catalyst improvement would be higher pitch conversion performance. The R&D objective thus became to produce a catalyst with superior pitch conversion activity and stability, without necessarily looking for improved performance in MCR, sulfur or nitrogen conversion.

The research effort focused on the hypothesis that the alumina support plays a vital role for pitch conversion. The ideal support pore structure should have a high pore volume for maximum vanadium capacity and an optimized pore size distribution for better utilization of the alumina surface. It was also postulated that if the alumina surface played a direct role in pitch conversion, its morphology could be improved to better promote interaction with asphaltene molecules. The alumina support development was done without modifying the active metals formulation.

The second part of the strategy was to then optimize the active metals for the new surface chemistry. Total metal loading and nickel to molybdenum ratio were thus optimized to carefully balance initial activity and stability. The catalyst preparation technique was also used as a tool to boost metal utilization for this severe application.

3 Pilot Testing Conclusions

Pilot plant testing was conducted with ATB feed, for ease of operation, even though the current LC-Finer^{SM} unit feed contains forty percent VTB. Typical feed properties of both feeds can be seen in Table 1.

Table 1

Typical Feed Properties

	ATB (Design Feed)	ATB/VTB (Current Feed)
Density °API / sg	6.5 /1.025	4.5 / 1.040
Sulphur, wt%	5.0	5.3
Nitrogen, wppm	5000	5300
MCR, wt%	14.9	18.0
V, wppm	200	270
N, wppm	75	100
524C+, LV%	56	70

A continuous stirred tank reactor (CSTR) equipped with Mahoney-Robinson internals was used in Shell Westhollow Research Center, while a mini ebullated bed reactor (MEBR) was used in Syncrude Research Laboratory.

A new alumina process was developed and commercialized for this application. Optimization of the resulting alumina extrudate pore structure identified a fairly narrow region with improved pitch conversion performance. Pilot plant testing showed that the commercially produced catalyst (RN-610) performed as well as the lab prototype. Plant evaluation confirmed better pitch conversion and the new Alliance catalyst replaced the incumbent catalyst (RN-510).

Research efforts immediately turned to the second part of the program: optimization of the metal formulation for the new extrudate. A series of three catalysts were prepared and tested, each probing a specific variable. From these results, a catalyst formulation was proposed as the prototype for the improved catalyst. MEBR results indicated that the latest prototype appeared to be significantly more stable than RN-610.

A detailed analysis of the pilot plant results showed that RN-610 initially yielded a significantly higher pitch conversion than RN-510, and a somewhat higher MCR conversion, with no real gain on HDS. Stability for pitch conversion was enhanced, thus leading to an even larger pitch conversion advantage under end of run conditions. The small advantage for MCR reduction however disappeared. RN-610 thus appeared to be a better pitch conversion catalyst, as intended, which validates the assumption that the alumina support plays a key role for pitch conversion.

RN-615 demonstrates an initial activity very similar to that of RN-610. However, its stability is significantly better for all functions. This assumption, that optimization of the metal formulation could improve catalyst stability, thus appeared to be validated.

4 Commercial Performance

Even though the new prototype (RN-615) had been tested only once, the team decided to aggressively pursue commercial evaluation of this catalyst, without confirming R&D findings on the pilot scale. Results from the commercial test looked good and the plant quickly transitioned from RN-610 to RN-615.

Looking back at the last four to five years of plant data, it is obvious that this catalyst development translated into a real commercial success.

Quality of the feed to the LC-Finer^{SM}, measured by any of the normal parameters (density, boiling ranges, pitch content, MCR content, sulfur content, etc..) remained unchanged over this period. Production rate stayed high, especially in the last couple of years, and pitch conversion has not declined. Reactor temperatures have been unchanged, but the catalyst addition rate (CAR) has been reduced by a third, as illustrated in Figure 1.

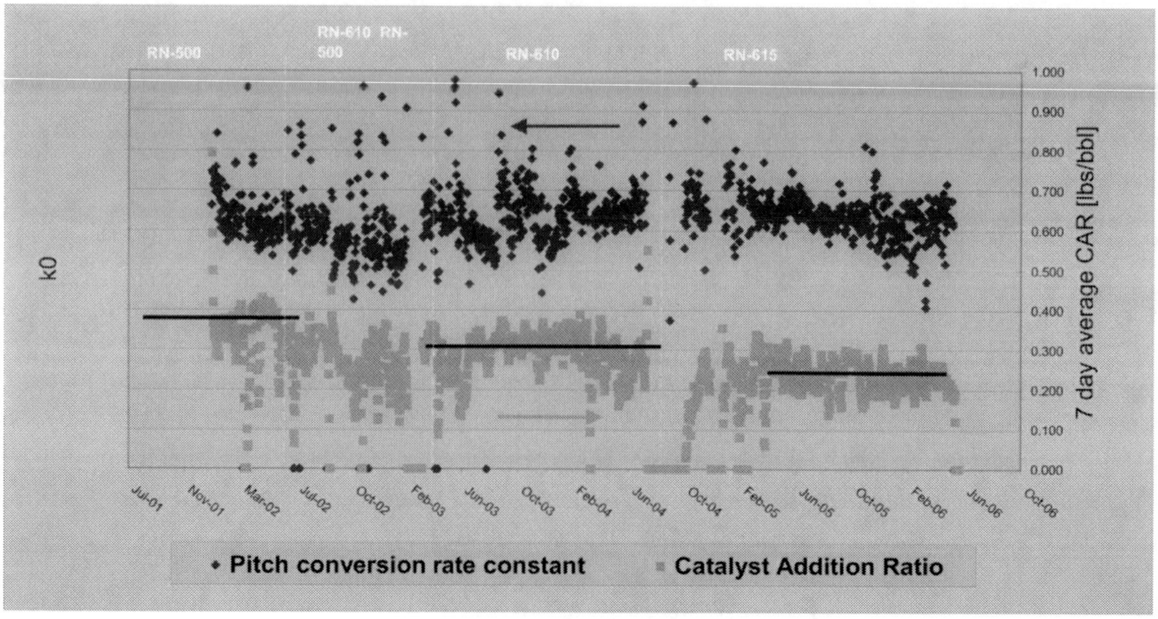

Figure 1
Commercial catalyst addition rate (CAR)

Analysis of the spent catalyst shows that coke formation (expressed on a fresh catalyst basis) constantly remains in the lower portion of the historical range, even though the average residence time is significantly longer (Figure 2).

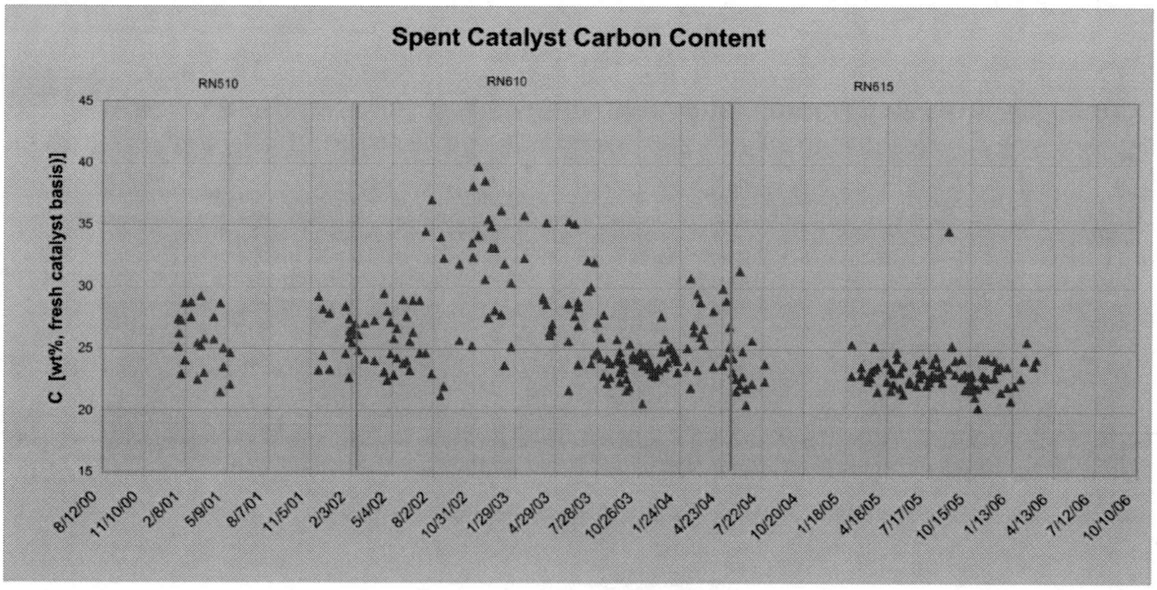

Figure 2
Carbon on spent catalyst

The amount of deposited metals is higher, as should be expected from the longer residence time (Figure 3).

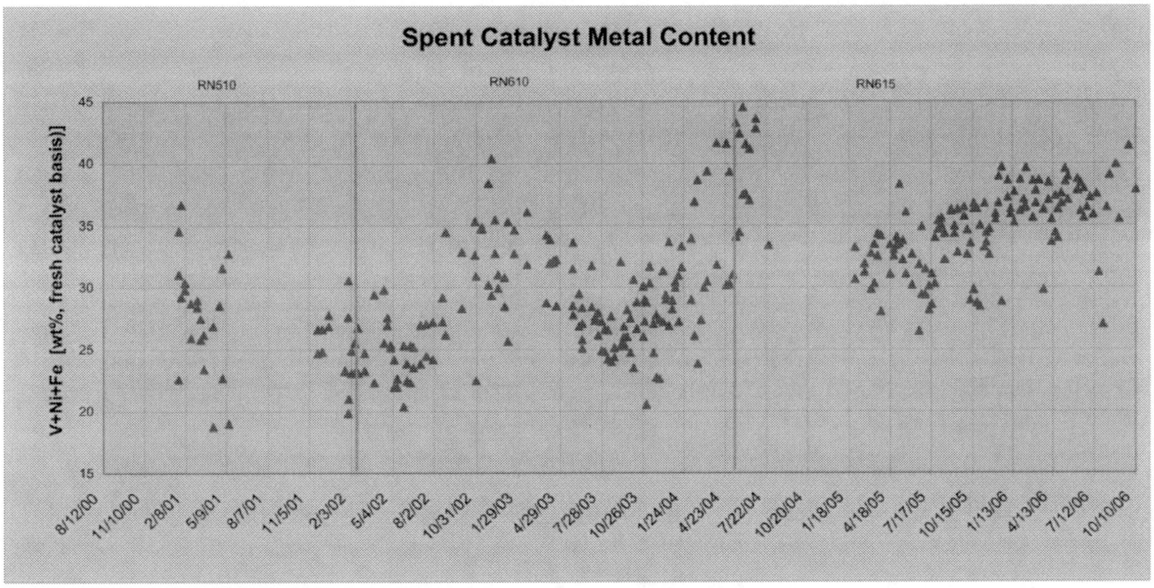

Figure 3
Metals on spent catalyst

This indicates that the catalyst indeed performs as expected, capitalizing on the improved support properties and the optimized metal formulation to significantly increase pitch conversion and catalyst stability.

5 Summary

This very significant improvement in catalyst performance allows the unit to operate at about the same catalyst addition rate used in the early 1990's when conditions were significantly less severe, with a feed rate 20% lower than the current feed rate and a diet of ATB only (the unit feed composition is currently 40% VTB & 60% ATB). The decrease in catalyst consumption, which amounts to more than a thousand tonnes per year, translates to multi million dollar savings annually.

Of course a reduction in fresh catalyst consumption also implies a similar reduction in spent catalyst being sent for reclamation, with a proportional reduction in transportation needs, thus adding environmental benefits to this attractive financial picture.

Encouraged by successes such as RN-615, Criterion and Syncrude continue to jointly develop customized catalysts for both ebullated and fixed bed applications in the Syncrude refinery.

Catalytic Aqua and hydro processing of Athabasca bitumen heaviest fractions

Pedro Pereira-Almao, Gustavo Trujillo, Carlos Scott, Redescal Gomez, Alejandro Vasquez, John Thompson.
Department of Chemical and petroleum, Schulich School of Engineering, University of Calgary, Calgary, AB. Canada

INTRODUCTION

The use of well dispersed catalytic particles for hydro-upgrading of heavy oils has been proposed since two decades already. Varied technical propositions during the period have progressively reduced the particle size of the catalysts employed.

Here are presented some of the most significant variables that ensure particle size control in the preparation of ultra dispersed (UD) catalysts for use in both hydro processing and aqua-processing of bitumen heaviest fractions. The setup below was used for this study. Emulsions and micro emulsions containing or not precursor salts of the different UD catalysts for Aqua and hydro processing o Athabasca heavy fractions were prepared, introduced in pumps A (blanc emulsions) and B (catalytic emulsions). Space velocity, decomposition temperature, aqueous dispersion (drop size), salt concentration and system pressure were all variables studied with this system.

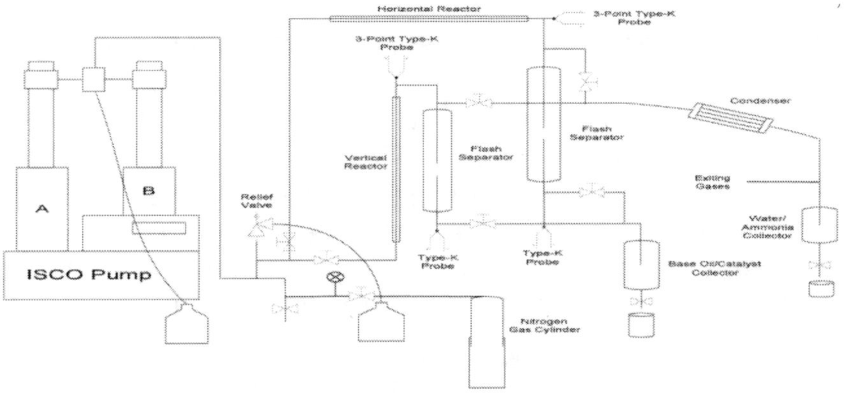

Figure 1: Ultradispersed catalysts synthesis setup

PARTICLE SIZE DETERMINATION: *Dynamic Light Scattering*

Figure 2 illustrates typical DLS output in measuring both the average polydispersivities (PDI) and average particles sizes of a single experimental run at two distinct dispersant ratios, each measured 4 times in succession with intervals of 3 minutes between readings.

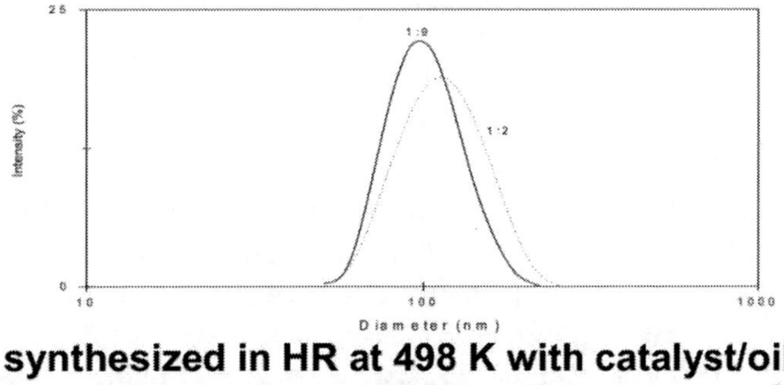

Figure 2. Particles synthesized in HR at 498 K with catalyst/oil product-to-base oil dispersant ratios of 1:9 and 1:2, corresponding to Z-Averages of 98.5 nm and 110.4 nm, respectively and PDIs of 0.136 and 0.175, respectively.

CATALYST CHARACTERIZATION

All XRD spectra of many different catalysts formulations exhibit amorphous patters consistent with those obtained from previous work [13,14]. As an example, Figure 4 is a typical representation of the XRD spectra and indicates that the particles are both amorphous and partially composed of MoO_2 and MoO_3 species. The sharp peak observed at 12 degrees observed for the catalyst prepared at 473 K (denoted by 'x') demonstrates that the precursor salt did not fully decompose at this low temperature as the peak represents a compound containing molybdenum and ammonium [16].

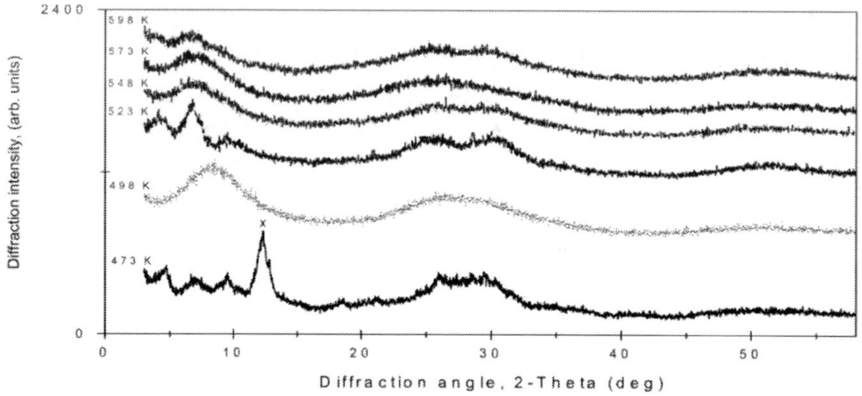

Figure 3. XRD analysis of Mo-particles produced from HR

EFFECT OF DECOMPOSITION TEMPERATURE ON PARTICLE SIZE

As an example, in the synthesis of Mo oxide particles followed by Dynamic Light Scattering (DLS) the Z-average diameter of the particles was measured as a function of the decomposition temperature, the results of which are presented in Table 1. The configuration of the decomposition vessel has a relatively minor effect on the particle size . In general, catalyst particle size increases as precursor salts decomposition temperature increases for both vertical and horizontal reactor.

Table 1: Particle diameter (Z-ave) for horizontal and vertical synthesis vessels as a function of precursor emulsion flashing and decomposition temperature for the synthesis of Mo oxide particles

Decomposition Temperature (K)	HR Z-Ave (nm)	VR Z-Ave (nm)
498	98.5	n/a
523	115.3	398.5
548	462.8	498.1
573	472.8	610
598	658.9	n/a

A series of bimetallic UD catalysts were prepared following the controlled procedure indicated, and tested in a micro reactor unit, using between 500 to 2000 mg. of catalyst. Model reactions such as the hydrogenation and steam dealkylation of toluene and the HDS of thiophene, allowed to find the optimal ratios of active phases for UD catalysts. For example an optimal ratio Co/(Co+Mo) of 0.3 was found corroborating the optimal proportion published in the literature for supported hydrodesulphurization catalysts [1-5]; whereas a Ni/(Ni+W) optimal ratio of 0.5 was found, slightly diverging from what others have found in the literature for that combination of metal sulphide supported catalysts[6].

Reactivity tests using full range Athabasca Vacuum Gas Oil (VGO) as feedstock were performed in a mini pilot plant (0.05 bpd) designed and built at our facilities, under both steam processing and hydro processing conditions. In the case of catalytic steam processing the relevance of particle size, quality of active phases and specificities related to the composition of the feedstock are found of to be of impact on the global reactivity performance. Table 2 presents a comparison of thermal steam cracking with catalytic steam cracking and with an equivalent fixed bed version of the same catalytic formulation. In this particular case the fixed bed version of the process is superior in the sense that less coke and better quality of the unconverted VGO was found.

Table 2 Catalytic steam processing of Athabasca VGO

Process	Steam	Steam+UD cat	Steam + solid cat	Pat US 6,030,522
Feed:	Athabasca VGO	Athabasca VGO	Athabasca VGO	Bachaquero VGO
Temperature	430C	430C	430C	430C
Pressure	260 psi	260 psi	260 psi	260 psi
WHSV hr -1	1	1	1	1
Yields				
Gas + coke	23	13	4.9	5.4
Yield 360- liq.	27.3	37.8	30.2	55.9
Conversion 360+C	31.9	42.4	33.3	60.9
CoC 360+C remanent	7.3	4.4	2.9	
gas ratio CH4 / CO2	21.9	7.7	5.8	

The relevance of catalyst particle size for the performance of both hydro and steam processes was confirmed while testing this specific feedstock. The results in general enlightened the significant room for improvement of ultra dispersed formulations that is still expectable via increasing its span of applications in the near future. Progresses on the science and technology of UD catalysts will definitely impact heavy oil upgrading options and economics.

References

1. Vrinat, M., M. Breysse, and R. Frety, *Unsupported cobalt molybdenum sulfide catalysts Part I: Catalytic activities for hydrodesulphurization of dibenzothiophene.* Applied Catalysis, 1984. **12**(2): p. 151-163.
2. Candia, R., B.S. Clausen, and H. Topsoe, *The origin of catalytic synergy in unsupported Co---Mo HDS catalysts.* Journal of Catalysis, 1982. **77**(2): p. 564-566.
3. Chianelli, R.R. and G. Berhault, *Symmetrical synergism and the role of carbon in transition metal sulfide catalytic materials.* Catalysis Today, 1999. **53**(3): p. 357-366.
4. Hagenbach, G., P. Courty, and B. Delmon, *Catalytic activity of cobalt and molybdenum sulfides in the hydrogenolysis of thiophene, hydrogenation of cyclohexene, and isomerization of cyclohexane.* Journal of Catalysis, 1971. **23**(2): p. 295-298.
5. Inamura K. and Prins R., *The Role of Co in Unsupported Co-Mo Sulfides in the Hydrodesulfurization of Thiophene.* Journal of Catalysis, 1994. **147**(2): p. 515-524.
6. Lacroix, M., M. Vrinat, and M. Breysse, *Unsupported nickel tungsten sulfide catalysts : Part 1: Catalytic behaviour in hydrogenation and hydrodesulfurization reactions.* Applied Catalysis, 1986. **21**(1): p. 73-83.

NATURAL HYDROGEN DONORS IN PETROLEUM RESIDS

Kenneth A. Gould[1] and Irwin A. Wiehe[2]
[1]92 Briarwood Dr. West, Berkley Heights, NJ 07922
[2]Soluble Solutions, 3 Louise Lane, Gladstone, NJ 07934

Abstract

The phase-separation kinetic model for coke formation during resid thermal conversion assumes the presence of natural hydrogen donors in the heptane soluble portion of the resid but not in the heptane insoluble asphaltenes. This explains why coke does not form while the converted asphaltenes remain soluble in the rest of the resid but rapidly forms when asphaltenes become insoluble. To test this assumption, a new method of measuring the donor hydrogen concentration and reactivity relative to the model compound hydrogen donor, tetralin, was devised using dehydrogenation by 2-3-dichloro-5,6-dicyano-p-benzoquinone. Indeed, petroleum resids were confirmed to contain high concentrations of natural hydrogen donors but many surprises were revealed. First, the asphaltene fraction was found to contain approximately the same concentration of donor hydrogen as the full resid but the most reactive. This requires modifying the phase-separation mechanism for coke formation and the concept of the chemical structure of asphaltenes. Second, even more hydrogen donors are generated during thermolysis by substituted naphthenes in the resid. Third, reactive hydrogen donors can reduce the thermal reaction rate and increase the selectivity to distillable liquid products. Based on these and previous results, methods to increase visbreaker conversion and to inhibit thermal cracking are described and the concept of an economic optimum conversion in resid hydroconversion is proposed.

Introduction

In 1993 the phase-separation mechanism of coke formation during the thermal conversion of converted asphaltenes was published[1]. A summary of this mechanism is as shown in Figure 1. Asphaltenes in the oil contain thermally stable, polynuclear aromatic cores with pendant groups connected to the cores by thermally unstable bonds. When exposed to thermal cracking temperatures, these bonds break to form free radicals. As long as the asphaltenes are dispersed in the rest of the oil, these free radicals abstract hydrogen from hydroaromatics in the resid and terminate the free radicals. As a result, no coke forms. However, the loss of pendant groups make the asphaltenes less soluble. Eventually, the converted asphaltenes become insoluble in the reacting resid and undergo a liquid-liquid phase separation. This asphaltene rich phase has little or no donor hydrogen to be abstracted by free radicals. As a result, the asphaltene free radicals combine to form high molecular weight and insoluble coke. However, before coke is formed, the polynuclear aromatics in the converted asphaltenes tend to orient parallel to each other. This orientation can be detected by observing the coke with an optical microscope under crossed polarized light where ordered structures show up as bright (usually orange for coke) while unordered (amorphous) structures are dark. Thus, direct proof that coke was initiated by a liquid-liquid phase separation is the presence of spherical, ordered particles (carbonaceous mesophase) in the coke[1]. The kinetic model based upon this mechanism has since been modified from reactions in parallel to reactions in series and hydrogen balance constraints have been used to evaluate all but one stoichiometric coefficient in the model[2]. However, the hypothesis that the heptane soluble fraction of resids contain high concentrations of hydrogen donors while the asphaltene fraction contains little or no donor hydrogen has not been validated. Since many naturally occurring substances, such as cholesterol, are hydrogen donors[3], the hypothesis of natural hydrogen donors in petroleum is reasonable. Nevertheless, the process of checking the natural hydrogen donor hypothesis led the authors to new concepts of resid conversion and of asphaltene chemical structure as is reported here.

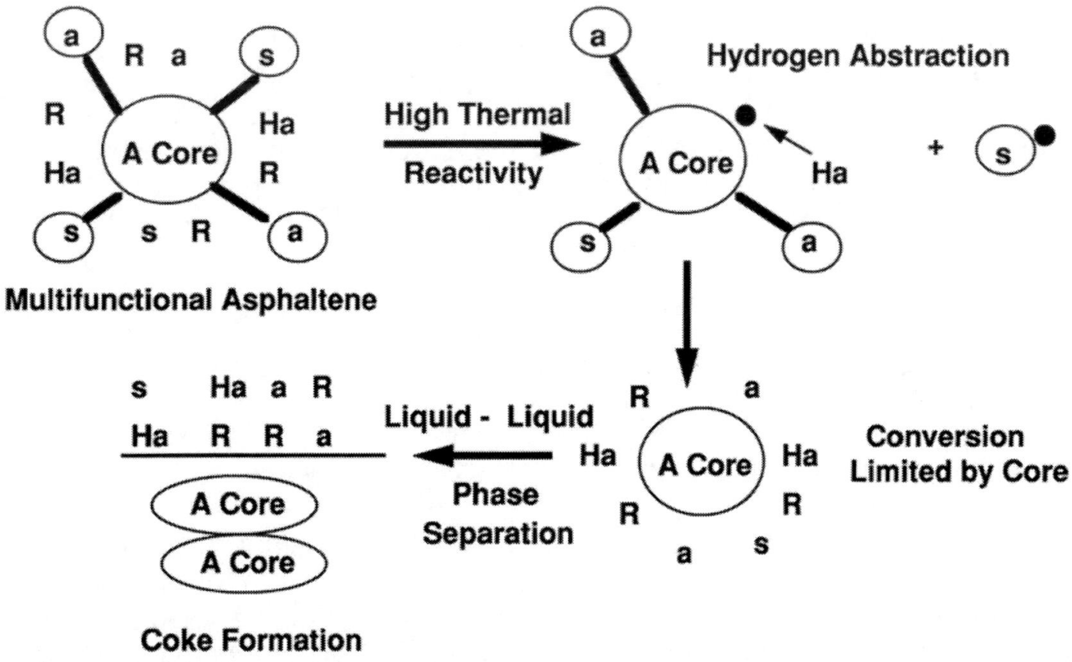

Figure 1. A phase-separation mechanism for coke formation during resid thermolysis.

The ability of hydrogen donor diluents to enable higher thermal conversion of resids to distillable liquids without forming coke has long been recognized[4,5,6]. Typically, hydrogen donors are compounds with a hydrogenated ring fused to an aromatic ring with tetralin as an ideal model hydrogen donor[4]. Thus, partially hydrogenated, aromatic process streams from refineries are considered as practical alternatives to tetralin[5,6]. However, it is not generally recognized that natural hydrogen donors in resids enable resids to be thermally converted to a limited extent without forming coke (visbreaking) as implied by the mechanism in Figure 1.

Measurement of Donor Hydrogen Concentration

The compound, 2-3-dichloro-5,6-dicyano-p-benzoquinone (DDQ), is commonly used in organic synthesis[7] to dehydrogenate hydroaromatic compounds. As is shown in Figure 2, when DDQ reacts with a hydroaromatic compound like tetralin, DDQ forms the corresponding hydroquinone while abstracting two hydrogens per molecule. Thus, this reaction was used to measure the quantity of hydrogen donors in petroleum resids and their fractions. However, the procedure was made more complex because DDQ formed strong charge transfer complexes with polynuclear aromatics in the petroleum resids at room temperature. Therefore, 1-2 grams of sample was reacted with 3-4 grams (an excess) of DDQ in 200 ml of dry toluene at reflux temperature for two hours. To this mixture 2.78 grams (an excess) of 9,10-dihydroanthracene, a model compound hydrogen donor, was added and reacted at reflux temperature for an additional 2 hours. Thus, the amount of unreacted DDQ could be measured by how much 9,10-dihydroanthracene was dehydrogenated. The product mixture was centrifuged at room temperature to remove solids that might be formed because of the dehydrogenation of petroleum resids. The solids free solution was analyzed by gas chromatography for 9,10-dihydroanthracene and the dehydrogenation product, anthracene. Since the moles of unreacted DDQ equals the moles of anthracene formed and the moles of donatable hydrogen in the feed equals two times the moles of DDQ reacted, the grams of donatable hydrogen per 100 grams of sample and the percent hydrogen that is donatable could be calculated. This procedure was successfully checked with the model compound hydrogen donors: tetralin, 1,2,3,4-tetrahydroquinoline, 9,10-dihydroanthracene (using tetralin in the procedure), and 9,10-dihydrophenanthrene (chemical structures in Figure 3). Then the procedure was applied to other model compounds, various resids, and the saturate, aromatic, resin, and asphaltene fractions of Arabian Heavy vacuum resid, using the separation procedure given elsewhere[8]. The results are shown in Table 1.

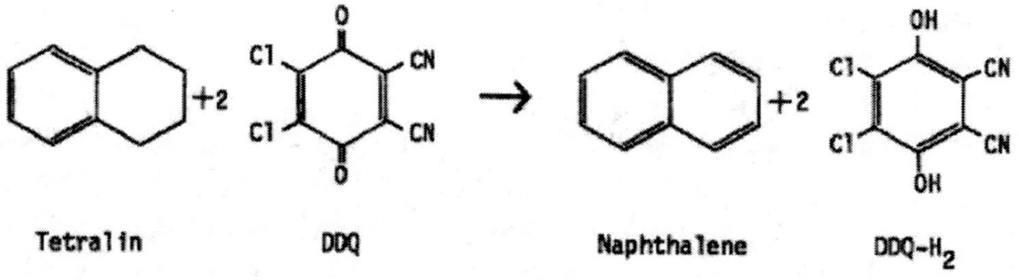

Figure 2. DDQ abstracts two donor hydrogens per molecule.

Clearly, the natural hydrogen donor hypothesis is validated as the 6 resids in Table 1 have higher concentrations of donor hydrogen than model compound hydrogen donors: dihydroanthracene and dihydrophenanthrene. Hondo vacuum resid even has almost as high a concentration of donor hydrogen as tetralin and tetrahydroquinoline, much higher than expected. The greatest surprise is that Arabian Heavy asphaltenes have as high a concentration of donor hydrogen as the total Arabian Heavy resid and Hondo has a concentration of donor hydrogen in its asphaltene fraction only slightly less than its full resid. This contradicts the phase-separation mechanism of coke formation that assumed that the asphaltene fraction has little or no donor hydrogen. With 25% of the hydrogens in Arabian

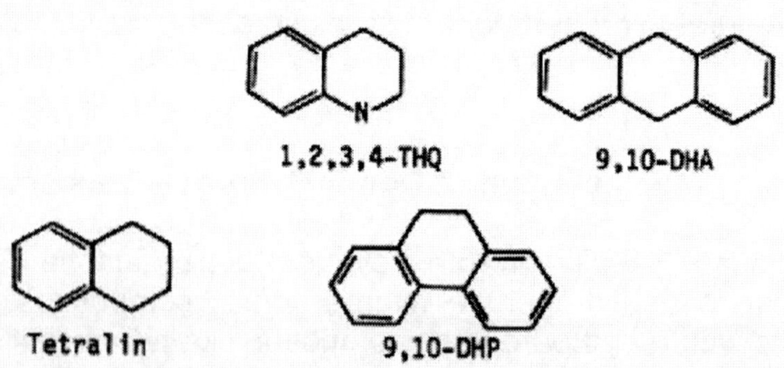

Figure 3. Chemical structures of model compound hydrogen donors: tetralin, 1,2,3,4-tetrahydroquinoline, 9,10-dihydroanthracene, and 9,10-dihydrophenanthrene.

Table 1. Concentration of Donor Hydrogen in Various Model Compounds and Resids

Sample	DDQ Donor H (g/100 g Sample)	DDQ Donor H (% of Total H)
Tetralin	3.0	33
1,2,3,4-Tetrahydroquinoline	3.0	33
9,10-Dihydroanthracene	1.1	17
9,10-Dihydrophenanthrene	1.1	17
Cholesterol	1.3	11
Cholestane	0.0	0.0
Deacalin	0.0	0.0
Brent Atmospheric Resid	1.2	11
Dan-Gorm Atmospheric Resid	1.6	14
Hondo Vacuum Resid	2.9	30
Hondo Asphaltenes	2.7	32
Arabian Heavy Vacuum Resid	1.9	19
AHVR Saturates	1.1	9.6
AHVR Aromatics	2.1	19
AHVR Resins	2.0	21
AHVR Asphaltenes	1.9	25
Arabian Light Vacuum Resid	1.6	16
Brent Vacuum Resid	1.4	13

Heavy asphaltenes and 32% of the hydrogens in Hondo asphaltenes being on saturated rings fused to aromatic rings, most will have to modify their concept of an average asphaltene chemical structure. In comparison, by proton NMR only 10% of the hydrogens in Arabian Heavy asphaltenes are bonded to the 54% aromatic carbons (^{13}C NMR). Completely saturated compounds cannot donate hydrogen to free radicals in agreement with DDQ dehydrogenation values of zero for decalin and cholestane (fully hydrogenation product of cholesterol). However, the saturate fraction of Arabian Heavy contains donor hydrogen, albeit a lower concentration than the other fractions. This is a result of the separation procedure producing a fraction more saturated than the other fractions but still containing small ring aromatics[8].

Measurement of Hydrogen Donor Relative Reactivity

The reactivity of hydrogen donors were measured relative to the hydrogen donor reactivity of tetralin by setting up a direct competitive reaction with DDQ. Thus, the sample (1-2 grams), an equivalent amount of donor hydrogen of tetralin, and a large excess of dry toluene for complete solubility (50-200 ml) were mixed. Sufficient DDQ to abstract 25% of the total donor hydrogen was added and this mixture was reacted at reflux temperature for 2 hours. This product mixture was centrifuged at room temperature and the solids free solution was analyzed by gas chromatography for tetralin and naphthalene. The hydrogen donor reactivity relative to tetralin, R_{HD}, was calculated from:

$$R_{HD} = [50 - \%\ \text{Conversion of Tetralin}] / [\%\ \text{Conversion of Tetralin}]$$

The resulting hydrogen donor reactivities relative to tetralin are shown in Table 2 for many of the model compounds, resids, and resid fractions in Table 1. One immediately finds that tetralin is not an ideal hydrogen donor in terms of reactivity. All the resids and resid fractions and two of the other three model hydrogen donors are more reactive hydrogen donors than tetralin. Only dihydrophenanthrene is a less reactive hydrogen donor than tetralin of the samples measured. Tetrahydroquinoline (THQ) is a super reactive hydrogen donor being more than 6 times as reactive as each of the samples in Table 2 except for dihydroanthracene and 50 times as reactive as tetralin. Not only does Hondo vacuum resid have the highest concentration of donor hydrogen of the resids tested but also the highest hydrogen donor reactivity. The hydrogen donor reactivity of the fractions of Arabian Heavy increase in going from saturates to aromatics to resins to asphaltenes. This is in the same direction as increasing aromaticity and probably as increasing ring size of the aromatic fused to the saturated ring with the donor hydrogen. Since the hydrogen donors in asphaltenes are more reactive than those in the other resid fractions, the fast depletion of donor hydrogen in asphaltenes is likely the cause of asphaltenes forming higher molecular weight coke when in a separate phase than the other resid fractions. Not only does hydrogen donation terminate free radicals but it can increase the ring size of the aromatic core in the asphaltene. Since the asphaltene core cannot be converted in thermal conversion, one would have preferred asphaltenes to have the least donor hydrogen of all the fractions.

Table 2
Hydrogen Donor Relative Reactivities of Various Model Compounds and Resids

Sample	Hydrogen Donor Relative Reactivity
Tetralin	1.0
1,2,3,4-Tetrahydroquinoline	50
9,10-Dihydroanthracene	28
9,10-Dihydrophenanthrene	0.63
Dan-Gorm Atmospheric Resid	6.0
Hondo Vacuum Resid	7.9
Arabian Heavy Vacuum Resid	2.7
AHVR Saturates	1.2
AHVR Aromatics	2.0
AHVR Resins	2.9
AHVR Asphaltenes	4.2
Arabian Light Vacuum Resid	5.4
Brent Vacuum Resid	5.7

Effect of Resid Thermolysis on Hydrogen Donors

Arabian Heavy vacuum resid was reacted at various times at 400°C in tubing bomb reactors and the n-heptane insoluble asphaltenes were separated. The donor hydrogen concentration was measured on the asphaltene fraction as well as the percent aromatic carbon by ^{13}C NMR. The result is in Table 3. Just heating the resid to 400°C and immediately quenching causes the asphaltenes to decrease in hydrogen donor concentration and to increase in aromaticity.

Table 3. Thermolysis of AHVR Rapidly Decreases H-Donor Concentration of Asphaltenes

Time at 400° (min)	DDQ Donor H (g/100 g Sample)	% Aromatic Carbon
Unreacted	1.89	54
0[a]	1.78	63
55	1.64	67
90	1.51	71

[a] Taken to temperature and quenched.

Further reaction time at 400°C causes the asphaltenes to lose hydrogen donor concentration rapidly while increasing aromaticity. In contrast Table 4 shows that the full Arabian Heavy vacuum resid only slightly decreases the concentration of hydrogen donors after reacting for 55 minutes at 400°C even though the aromaticity is increased. Heating for longer times causes the full resid to decrease hydrogen donor concentration more rapidly but not as rapid as the asphaltene fraction. The decrease in hydrogen donor concentration of the full resid after 180 minutes is about equivalent to the decrease of the hydrogen donor concentration of the asphaltene fraction after 90 minutes. If one fraction of the resid is decreasing hydrogen donor concentration rapidly and the full resid is barely changing hydrogen donor concentration for the first 55 minutes, at least one fraction of the resid must be generating hydrogen donors.

Table 4. Prolonged Thermolysis of Arabian Heavy Vacuum Resid Required to Decrease Donor Hydrogen Concentration

Time at 400° (min)	DDQ Donor H (g/100 g Sample)	% Aromatic Carbon
Unreacted	1.94	31
55	1.90	38
120	1.70	----
180	1.55	----

Table 5 shows that the saturate fraction does indeed generate hydrogen donors as the hydrogen donor concentration increases when Arabian Heavy saturates are reacted for 100 minutes at 400°C while the aromaticity increases. A hypothesis of how this might occur is shown in Figure 4. Fully saturated ring structures are not hydrogen donors. If a two ring saturated structure (2-decalyl phenyl sulfide) is attached to a pendant group by a sulfur bridge, this sulfur bond might break thermally and form a free radical at the saturated ring. Donating a hydrogen could form an olefin instead of the saturated ring. Donating four more hydrogens would form tetralin that can donate four more hydrogens. Thus, the process of converting a substituted saturated ring structure into an aromatic can generate considerable donor hydrogen. This mechanism was verified by thermally reacting 2-decalyl phenyl sulfide in a tubing bomb reactor for 60 minutes at 400°C and forming 36 wt% tetralin and 5 wt% naphthalene, as analyzed by gas chromatography. Therefore, the generation of donor hydrogen by saturates and the high hydrogen donor reactivity of asphaltenes causes the large difference between the hydrogen donor concentration in the reacting resid and the insoluble asphaltene phase that causes the phase separation of converted asphaltenes to trigger the onset of coke formation.

Table 5. H-Donors Are Generated In Situ During Thermolysis of AHVR Saturates

Time at 400° (min)	DDQ Donor H (g/100 g Sample)	% Aromatic Carbon
Unreacted	1.10	13
100	1.29	18

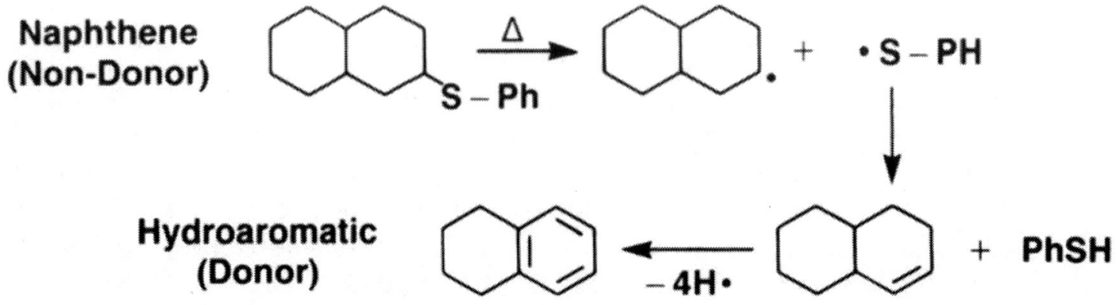

Figure 4. Mechanism for generating hydrogen donors during thermolysis demonstrated by thermally reacting 2-decalyl phenyl sulfide.

Effect of Very Reactive Hydrogen Donors on Resid Thermolysis

Tetrahydroquinoline (THQ) is such a reactive hydrogen donor that when it is with resid during thermolysis, very little molecular growth reactions or dehydrogenation reactions of the resid is expected. Therefore, equal weight mixtures of the asphaltenes and heptane solubles of Arabian Heavy vacuum resid and THQ were reacted in closed tubing bomb reactors at 400°C by immersing the reactors in a large sand bath at various times, followed by quenching the reactors in water at room temperature. THQ, quinoline, and distillable liquid product were distilled directly out of the tubing bomb. This distillation was done at 315°C and 1.4 mm Hg using a small laboratory sand bath and a vacuum pump to simulate a vacuum resid cut-point. The remaining converted resid was separated into toluene insoluble coke, toluene soluble – n-heptane insoluble asphaltenes, and n-heptane solubles as described previously[1]. The amount of distillable liquids were determined by subtracting the starting quantity of THQ from the total quantity of liquids distilled out of the tubing bomb.

Previously[1], the kinetic model for asphaltene conversion without THQ was more straight forward because asphaltenes started in a separate phase and therefore, all reacted asphaltenes formed coke or lighter products. The kinetic model for asphaltenes dissolved in THQ is similar as the asphaltenes in the full resid in that unreacted asphaltenes can form asphaltene cores without forming coke. The difference is that THQ does not form asphaltenes like the heptane soluble fraction of the resid. However, like the full resid, toluene insoluble coke forms after an induction period, indicating that asphaltene cores are not soluble beyond a certain concentration in the mixture of THQ, quinoline, heptane solubles, and distillable liquids in the closed reactor. The data was not sufficient to determine a solubility limit so the change in the sum of asphaltenes and toluene insolubles was modeled. Likewise, the formation of distillable liquids were odd when asphaltenes were reacted in 50% THQ in that no distillable liquids formed at 60 minutes, although considerable nonvolatile heptane solubles were formed. In contrast, without THQ, asphaltenes from Arabian Heavy form nearly no nonvolatile heptane solubles at 400°C. As a result, the sum of distillable liquids and heptane solubles were modeled. A comparison of the kinetic models for the thermolysis of Arabian Heavy asphaltenes with and without THQ are shown below:

No THQ: $A^+ \longrightarrow 0.611\ TI + 0.389\ (H + D)$; rate constant = 0.0204 min^{-1}

50% THQ: $A^+ \longrightarrow 0.442\ A^* + 0.558\ (H + D)$; rate constant = 0.0103 min^{-1}

Where A^+ represents unreacted asphaltenes, A^* represents asphaltene cores, TI represents toluene insoluble coke, H represents nonvolatile heptane solubles, and D represents distillable liquids. The stoichiometric coefficients determine the weight fraction of converted asphaltenes that form each of the products. Based upon this reaction in 50% THQ, the equations for the sum of heptane solubles and distillable liquids and the sum of asphaltenes and coke are given below:

$$H + D = 100\ [0.558]\ [1 - \exp(-\ 0.0103\ t)]$$

$$A + TI = A^+ + A^* = 100\ \exp\ (-0.0103\ t) + 100\ [0.442]\ [1 - \exp(-\ 0.0103\ t)]$$

where t is the reaction time in minutes. The stoichiometric coefficient and rate constant were determined by fitting the data in Table 6 for the thermolysis of asphaltenes with 50% THQ in closed reactors at 400°C. The stoichiometric coefficient and rate constant for Arabian Heavy asphaltenes without THQ are part of the series reaction kinetic model for Arabian Heavy vacuum resid given previously[2].

Table 6. Comparison of Experimental Data and Calculated Yields (wt%) for Arabian Heavy Asphaltenes Reacted in 50 wt% THQ in Closed Reactors at 400°C

Product	60 min Data	60 min Calc.	120 min Data	120 min Calc.	240 min Data	240 min Calc.
Distillable Liquids	0.0		16.1		23.0	
Heptane Solubles	25.8		23.5		20.5	
Dist Liq + Hept Sol	25.8	25.7	39.6	39.6	49.2	51.1
Asphaltenes	74.6		53.9		37.8	
Toluene Insolubles	0.0		3.7		13.0	
Asph + Tol Insol	74.6	74.3	57.6	60.4	50.8	48.9

It is clear that the super reactive hydrogen donor, THQ, can reduce the fraction of asphaltenes that form asphaltene cores and increase the fraction that forms lower molecular weight products. This is probably a result of more effectively decreasing or eliminating molecular growth of aromatic fragments of cracking. Although, 50% THQ delays the onset of coke formation from asphaltene thermolysis, coke eventually forms when asphaltene cores become insoluble in the reaction medium. In addition, THQ greatly reduces the thermal reaction rate constant. This is the disadvantage of greatly reducing the free radical concentration when one wants conversion and not suppression of thermal reactions.

The heptane soluble fraction of Arabian Heavy vacuum resid was reacted separately alone and in 50 wt% THQ for 120 minutes at 400°C in closed tubing bomb reactors. The products were separated into distillable liquids, nonvolatile heptane solubles, asphaltenes, and toluene insoluble coke with yields as is shown in Table 7. It is clear that THQ greatly reduces the rate of conversion of heptane solubles as after the same reaction conditions, 61.5% of the heptane solubles converted without THQ and 22.3% of the heptane solubles converted in the presence of 50% THQ. Of the converted heptane solubles, 25.0% formed asphaltenes and coke without THQ and 13.9% formed asphaltenes with 50% THQ. Thus, while the super reactive hydrogen donor did not eliminate the formation of asphaltenes from heptane solubles, it did improve the selectivity to lighter products. This indicates that the heptane solubles contain polynuclear aromatic cores that become n-heptane insoluble when the pendant groups are cracked off without requiring molecular weight growth reactions to form asphaltenes.

Table 7. Comparison of Experimental Yields for Arabian Heavy Heptane Solubles Reacted in Closed Reactors at 400°C, Neat and with 50 wt% THQ

	Neat Heptane Sol	50% THQ
Product	Yield (wt%)	Yield (wt%)
Distillable Liquids	46.4	18.4
Heptane Solubles	37.2	77.7
Asphaltenes	14.9	3.1
Toluene Insolubles	0.5	0.0

Implications for the Conversion of Resids

The important role of natural hydrogen donors in resid conversion is now much clearer. In order to extend the period of conversion without forming coke in visbreaking, one needs to increase the solubility of converted asphaltenes without diluting the natural hydrogen donors in the resid. Thus, adding an aromatic oil with less donor hydrogen concentration than the resid could decrease the coke induction period, rather than increase it. Since the hydrogen donor concentration of resids are already quite high, finding a practical co-processing oil to increase visbreaker conversion without forming coke is not easy unless the oil has been hydrogenated. A more practical solution may be to more effectively strip out the distillable liquid product during conversion. The distillable liquid product of resid thermal conversion is usually a nonsolvent for converted asphaltenes and is low in concentration of donor hydrogen.

Another problem with donor diluent cracking is that swamping the resid with reactive hydrogen donors greatly slows the cracking rate because most thermal cracking is caused by radical attack, not by radical initiation. As we saw with THQ, reactive hydrogen donors reduce the thermal cracking rate by reducing the free radical concentration. However, also we learned from THQ that reactive hydrogen donors do not just delay the onset of coke formation but reduce the formation of asphaltenes from heptane solubles and reduce the formation of asphaltene cores from asphaltenes in favor of a greater selectivity to distillable liquids. Thus, one would like to add the reactive hydrogen donor slowly during the conversion to maintain a minimum concentration for better selectivity without greatly slowing the reaction rate. The saturate fraction provides this for a limited extent by generating donor hydrogen during the conversion. An alternative is to use hydrogen pressure and hydrogenation catalysts as is practiced by hydroconversion processes. This is why hydroconversion can attain much higher conversions of resids than visbreaking (≈30%) without forming coke. However, hydroconversion processes usually run up against a compatibility limit near 50-60% for packed beds and 70-80% conversion with catalyst replacement processes. As for visbreaking, more effectively stripping out the distillable liquid product away from the asphaltene cores is usually the best way to remove the compatibility limit in hydroconversion[9,10]. As a result, it is possible to attain greater than 95% hydroconversion[10]. In spite of this, it is very questionable that conversion of the polynuclear aromatics greater than four rings is worth the price. The problem is illustrated in Figure 5. If a ring of a polynuclear aromatic greater than four rings is hydrogenated, it can be thermally cracked only slowly because it is competing with the much faster hydrogen donation reaction. If the hydrogenated ring does thermally crack, it only forms hydrocarbon gases and a polynuclear aromatic with methyl groups. When this is cracked down to a four ring aromatic, it becomes part of the vacuum gas oil product. However, four ring aromatics are the least valuable part of vacuum gas oil because they need to be hydrogenated and cracked with acid catalysts to be converted into transportation fuels with hydrocarbon gas byproducts. Therefore, in the end the conversion of large polynuclear aromatics consumes an enormous amount of hydrogen, can double the size of the hydroconversion reactor because of the slow kinetics, and produces a low quality liquid product with a large quantity of low valued hydrocarbon gases. Thus, there is an economic optimum hydroconversion that is in the range of 85-90% for Arabian Heavy vacuum resid because of this polyaromatic core limit. Even the super reactive hydrogen donor, THQ, could not convert the polyaromatic cores in either the asphaltenes or the heptane soluble fraction of Arabian Heavy vacuum resid but it is questionable if such conversion is wise even for hydroconversion.

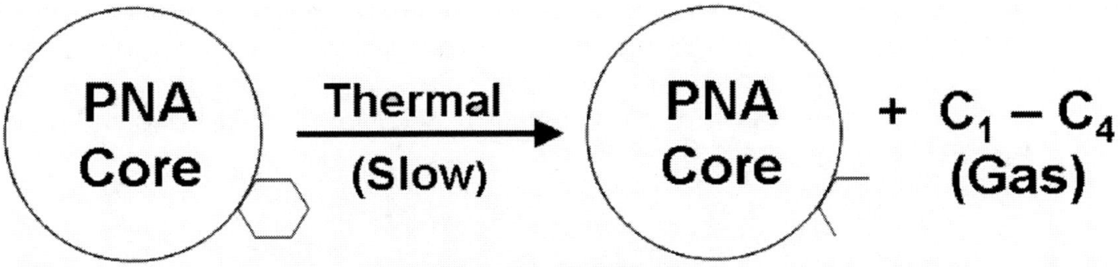

Figure 5. Illustration of the problem of converting large polynuclear aromatics in hydroconversion.

The ability of reactive hydrogen donors to inhibit thermal cracking reactions can be an advantage as a coprocessing oil when thermal cracking is undesirable. Examples are the furnace coils during vacuum distillation of reactive resids and the heaters of delayed cokers. However, one needs to assess the thermal reactivity of the coprocessing oil. For instance, although Hondo contains the most reactive hydrogen donors of the resids reported here, it also has the highest thermal cracking rate constant of the resids reported here. This is the result of a very high concentration of aliphatic sulfur in Hondo vacuum resid[11]. Therefore, Hondo vacuum resid would likely increase the thermal cracking rate of another resid if the two were co-processed.

References

1. Wiehe, I. A. (1993), "A Phase-Separation Kinetic Model for Coke Formation", I&EC Research, 32, pp. 2447-2454.

2. Wiehe, I. A. (2000), "Hydrogen Balance Constraints on Pseudocomponent Kinetic Models", *Proc. 3rd Int'l Conf. on Refinery Processing*, AIChE, New York, pp 607-612.

3. Morrison, R.T. and R.N. Boyd (1959), *Organic Chemistry*, Allyn and Bacon, Boston, p. 811.

4. Carlson, C.S., A.W. Langer, J. Steward, and R.M. Hill, (1958), *Ind Eng Chem*, 50, pp. 1067-1070.

5. Langer, A.W., J. Steward, C.E. Thompson, H.T. White, and R.M. Hill, (1961), *Ind Eng Chem,* 53, pp. 27-30.

6. Langer, A.W., J. Steward, C.E. Thompson, H.T. White, and R.M. Hill, (1962), *I&EC Proc Des Dev*, 1, pp. 309-312.

7. Fu, P.P. and R.G. Harvey, (1978), "Dehydrogenation of Polycyclic Hydroaromatic Compounds", *Chemical Reviews*, 78, pp. 317-361.

8. Wiehe, I.A. (1992), "A Solvent-Resid Phase Diagram for Tracking Resid Conversion", *I&EC Research*, 31, pp. 530-536.

9. Wiehe, I.A. (1997), *ACS Div. Pet. Chem. Meeting*, San Francisco.

10. Bearden, R. (1997), *ACS Div. Pet. Chem. Meeting*, ACS Award in Petroleum Chemistry Address, San Francisco.

11. Wiehe, I.A. (1994), "The Pendant-Core Building Block Model of Petroleum Residua", *Energy & Fuels*, 8, pp. 536-544.

Adsorption, Diffusion and Reaction in FCC Catalyst for Enhanced Selectivity Towards Environmentally Friendly Gasoline

Mustafa Al-Sabawi, Chemical Enginnering, University of Wetern Ontario, London, AB N6A5B8, Canada

The demand for advancement in zeolite science and technology has been persistent over the past 40 years, especially today since refineries need to improve the processing of heavy feedstocks into valuable gasoline and other lighter products while reducing costs. The existence of bulky molecules in heavy feedstocks presents a major challenge to fluid catalytic cracking (FCC) due to possible hindered transport of hydrocarbon species inside the catalyst micropore network. To meet processing demands and needs of manufacturing environmentally friendly gasoline (low in aromatics), catalysts used in FCC operations will have to be more active, and in particular, more selective for desired products. This study examines the roles of intracrystallite diffusion, adsorption, and reaction phenomena in the catalytic cracking of vacuum gas oil (VGO). Catalytic cracking experiments of VGO on FCC-type catalysts are carried out in a fluidized bench-scale CREC riser simulator reactor. These experiments are conducted under FCC operating conditions in terms of temperature (510oC to 570oC), reaction time (3 to 7 seconds), partial pressure of reactant and products, and catalyst-to-oil ratio. The crystallite size of the supported zeolite is varied between 0.4 and 0.9 microns, with both activity and selectivity being monitored. A five-lump kinetic model describing the catalytic cracking of VGO to light cycle oil, gasoline, light gases, and coke is considered. The model accounts for diffusional constraints experienced by hydrocarbons while evolving in the zeolite pore network. Results show that the catalyst with the smaller crystallites provided higher activity and selectivity towards desirable intermediate products (gasoline with low aromatics) and lower selectivity for terminal products (coke), indicating that diffusion plays a significant role in catalytic cracking. Diffusivity and kinetic parameters including modified Thiele modulus and effectiveness factor are calculated to determine the effect of crystallite size and temperature on the operating regime of the catalyst. It was found that, in 510-530°C range, the overall cracking rate is controlled by the highly temperature-sensitive intracrystalline gas oil transport, while in 550-570°C range, the overall cracking rate is dominated by a mildly temperature-sensitive intrinsic cracking rate.

Key words: Catalytic Cracking, Y zeolites, Kinetic Modeling, Diffusion, Environmentally Friendly Gasoline

Hydrocracking experience in a Gulf Refinery

Author: **Kamal Rashid, P.Eng., Senior Process Engineer**
Former Employee of Saudi Aramco Central and Western Region Projects Department (SA C&WRPD)
Diesel Hydrotreater Project Division (DHTPD)
Riyadh, Saudi Arabia

Abstract: The paper deals with the challenging but rewarding refining process of hydrocracking. With demand for white oil products (Gasoline & Diesel) increasing all the time and the corresponding decrease in demand of black oil product (Fuel oil), the Refiners are more and more taking the help of processes to convert black oil into white oil products. Hydrocracking comes in the forefront of such processes. This conversion process helps Refiners in improving significantly refining margin since it converts low value products into very high value products. In the process, Refiners are able to keep a favourable balance between white oil products and black oil products to maintain an optimum refinery throughput. The author had the opportunity to be intimately involved in the operation of an unique Hydrocracker in a Gulf Refinery. The paper will deal with the theory, process details and the challenges this high temp. and pressure process pose in handling very difficult to hydrocrack heavy feed. The author's extensive experience in this unique Hydrocracker will highlight the importance of proper catalyst selection. A case study will be presented to show the disastrous effect of wrong catalyst selection.

INTRODUCTION: Hydrocracking is a key refining process available to Refiners to increase yield of the valuable white oil products to improve significantly their refining margin. To put this benefit in proper perspective, consider this: hydrocracking results in about 20% volume expansion and over 40% increase in product value compared to feed value. In the process, it helps Refiners to reduce Low Value Black-oil yield and maintain an optimum Refinery throughput. This very rewarding process is at the same time very challenging, especially, for hard to crack heavy feed. Such was the case with an unique Hydrocracker in a Refinery in the Gulf region, where the author had the opportunity to be intimately involved for a number of years. This Paper will focus on the author's extensive experience in this unique Hydrocracker in an "Operation Technical Support" role. This particular refinery is a 120,000 BPSD Refinery, designed to process Arabian Light (AL) or Khurais crude oil, and is a deep hydrocracking refinery that produces Liquefied Petroleum Gas (LPG), Gasoline, Jet Fuel, Kerosene, Diesel, and Asphalt (Refer to Fig. 1 in the Appendix for the Refinery Block Flow Diagram).

This Paper will deal in detail with the theory of Hydrocracking, the salient feature of the Hydrocracker in question, and a Case Study, providing the catalyst selection process, the failure of the selected catalyst batch, the consequences thereof, and an analysis of reasons behind the catalyst failure.

HYDROCRACKING THEORY: Hydrocracking is a versatile process for converting heavy petroleum fractions into lighter, more valuable products. Typical feedstocks include heavy atmospheric and vacuum gas oils, and catalytically or thermally cracked gas oils. These products are converted to lower molecular weight products, primarily naphtha or distillates, by catalytic cracking in the presence of hydrogen at elevated pressures and temperatures. Sulfur, nitrogen, and oxygen removal and olefin saturation occur simultaneously with the hydrocracking reaction, thereby yielding a product mixture of essentially pure paraffins, naphtenes, and aromatics. The benefits of hydrocracked products are: these are saturated, isomerized, and free of contaminants.

Typical reactor operating conditions for a hydrocracker require moderately high temperatures of 600-840°F (315-450°C) and very high reactor circuit pressures ranging from 1500-2650 psig (105-186 kg/cm^2 g). The reactions that occur in the process consume hydrogen and are highly exothermic.

Hydrocracking is characterized by two types of reactions: Treating – removal of contaminants (S, N2, O2, halides, metals, saturation of olefins), and cracking – create usable products. The key to these hydrotreating and hydrocracking reactions is by bi-functional catalysts. These catalysts contain an acid function and a metal function. A hydrocracking catalyst has four key performance criteria: (a) initial activity, which is measured by the temperature required to obtain the desired product at the start of run. (b) stability, which is measured by the rate of increase of temperature required to maintain conversion. (c) product selectivity, which is a measure of the ability of a catalyst to produce the desired product slate. (d) product quality, which is a measure of the ability of a catalyst to make products with the desired specifications, such as pour point, smoke point, or octane number.

Process Chemistry: The Hydrocracking process that is discussed here is an UOP process, commercially known as HC Unibon. This process makes use of two distinct types of reactions: hydrotreating and hydrocracking, to make clean, saturated high value products. The hydrotreating reactions remove contaminants from the feed, while the hydrocracking reactions create usable lighter molecular weight products. These reactions are carried out over catalyst at elevated temperatures and pressures, mentioned above. The HC Unibon catalysts are typically composed of metals on an amorphous or zeolitic silica-alumina base.

Hydrotreating Chemistry: Some of the first reactions to occur in the HC Unibon unit are the hydrotreating reactions. These hydrotreating reactions are catalyzed by the metal sites on the catalyst and, in general, are more rapid than the cracking reactions. Most zeolitic hydrocracking catalyst systems have a bed or a reactor of hydrotreating catalyst in front of them to perform this function. In an amorphous catalyst system as well as some lower activity zeolite systems, the same catalyst will be performing hydrotreating reactions concurrently with cracking reactions.

The primary hydrotreating reactions are sulfur and nitrogen removal as well as olefin saturation. The products of these reactions are the corresponding contaminant-free hydrocarbon, along with H_2S (Hydrogen Sulfide) and NH_3 (Ammonia). Other treating reactions include oxygen, metals and halide removal, and aromatic saturation. In each of these reactions, hydrogen is consumed and heat is liberated. Some of these reactions are outlined below:

1. Sulfur (S) Removal: Sulfur is typically the largest single contaminant in hydrocracking feedstocks. The sulfur can be found in various forms across the entire feed distillation. The lighter compounds such as mercaptans and disulfides are easily converted to H_2S. The heavier aromatic sulfur compounds that are found in the higher boiling end of the feed can be more difficult to convert.

Mercaptan: C-C-C-C-SH + $H_2 \rightarrow$ C-C-C + H_2S

Disulfide: C-C-S-S-C-C + $3H_2 \rightarrow$ 2 C-C + $2H_2S$

2. Nitrogen (N_2) Removal: Denitrogenation is generally more difficult than desulphurization because the nitrogen in the feed is typically present only in heteroatomic aromatic compounds and side reaction can yield nitrogen compound that are more difficult to hydrogenate than the original reactant. The denitrogenation of pyridine starts with aromatic ring saturation, followed by ring hydrogenolysis, and finally, denitrogenation.

Amine: C-C-C-C-NH_2 + $H_2 \rightarrow$ C-C-C-C + NH_3

3. Olefin Saturation: Olefin saturation reactions proceed very rapidly and have a high heat of reaction.

Linear Olefin: $H_2C{=}CH{-}CH{=}CH_2 + 2H_2 \rightarrow H_3C{-}CH_2{-}CH_2{-}CH_3$

4. Oxygen Removal: Organically combined oxygen is removed by hydrogenation of the carbon-hydroxyl bond forming water and the corresponding hydrocarbon.

Paraffinic Oxygenate: C-C-C-C-C-OH + $H_2 \rightarrow$ C-C-C-C-C + H_2O

5. Metals Removal: Almost all the metals removal occurs at the top bed of the first reactor. The mechanism of the decomposition of organo-metallic compounds is not well understood. However, it is known that metals are retained on the catalyst by a combination of adsorption and chemical reaction.

6. Hallides Removal: Organic halides, such as chlorides and bromides, are decomposed in the reactor. The inorganic ammonium halide salts which are produced when the reactants are cooled are then dissolved by injecting water into the reactor effluent, as shown below. Decomposition of organic halides is considered difficult and the maximum possible removal is about 90%.

$HCl + NH_3 \rightarrow NH_4Cl$

7. Reaction Rates: All of the reactions discussed above are exothermic and result in a temperature rise across the reactor. Olefin saturation is the most rapid reaction, with desulfurization somewhat slower and denitrification significantly slower than either olefin saturation or desulfurization. The approximate relative reaction rates for the three hydrotreating reactions are:

Olefin Saturation	100
Desulfurization	80
Denitrification	20

While olefin saturation and some desulfurization reactions have similar rapid reaction rates, it is the saturation of olefins that is the most exothermic. The relative heats of reaction per unit of hydrogen consumption for these reactions are:

Desulfurization	1
Olefin Saturation	2
Denitrification	1

Hydrocracking Chemistry: Hydrocracking is essentially the acid catalyzed carbonium ion reactions of catalytic cracking coupled with hydrogenation reactions. C4 fractions, rich in isobutene, are formed because of a great tendency to form tert-butyl-carbonium ions. This is illustrated in Fig. 3 in the Appendix. The mechanism for cracking of Naphthenes and multiring aromatic compounds is illustrated in Fig. 4 and 5 respectively in the Appendix. Fig. 5 in the Appendix shows possible pathways for multiring aromatics in the cracking process. These mechanisms for multiring aromatics cracking process is especially important in understanding why certain catalysts cannot cope with very heavy compounds with multiring aromatics and will be the topic of discussion in the case study in this paper. It also shows the mechanism of formation of the problematic heavy polynuclear aromatic (HPNA's) containing more than 11 rings.

Process Variables: The objective of hydrocracking is to convert (crack) big molecules to small molecules. Therefore, conversion is at the root of most process variables. For a hydrocracker, apparent "conversion" is described as:

Conversion, vol% = (FF-Frac. Bottom)*100/FF

Where: FF = Fresh Feed rate, m^3/hr; Frac. Bottom = Product Fractionator Bottom to storage, m^3/hr

The amount of conversion which takes place in the reactors is determined by several variables:

Catalyst: The most important part of the cracking process is the catalyst system. Both amorphous and zeolitic catalysts have been used commercially in hydrocracking processes. The moderate activity of amorphous catalyst offers a significantly higher middle distillate selectivity. The acidic nature of zeolitic catalysts produces high cracking activity, thereby increasing naphtha selectivity. The author's experience was in a two stage UOP licensed hydrocracker with an amorphous catalyst (UOP DHC-8) in the first stage, and a zeolitic catalyst (HC-100) in the second stage.

Temperature: Temperature is the main control variable for hydrocracking units as most of the other variables are fixed by the design. The temperature must be increased throughout the cycle to maintain conversion as the catalyst deactivates due to the unavoidable formation of coke on the catalyst surface.

Fresh Feed Rate (LHSV): The amount of catalyst loaded into the reactors is based upon the quantity and quality of design feedstock, the desired conversion level and the desired catalyst life. Liquid hourly space velocity (LHSV) is the ratio of volumetric feed rate per hour to the total catalyst volume. Increasing the fresh feed rate with a constant catalyst volume increases LHSV and a corresponding increase in catalyst temperature will be required to maintain a constant conversion.

Hydrogen Partial Pressure: The reactor section operating pressure is controlled by the pressure of the high pressure separator. This pressure, multiplied by the recycle gas hydrogen purity fraction, determines the hydrogen partial pressure at the High Pressure (HP) Separator. The hydrogen partial pressure required for the operation of the unit is determined by the type of feedstock, the desired product quality, and the amount of conversion desired for a specific catalyst life.

Recycle Hydrogen/Hydrocarbon (H_2/HC) Ratio: In addition to maintaining the partial pressure of hydrogen in the reactor section, it is equally important to maintain the physical contact of the hydrogen with the catalyst and hydrocarbon. Continuous circulation of the recycle gas through the reactor circuit ensures that hydrogen is available at the catalytic sites where the reactions are taking place. The recycle gas also helps maintain good flow distribution through the catalyst. More importantly, the mass provided by the gas flow is a large contributor to the heat sink for this highly exothermic process. This hydrogen to hydrocarbon ratio is expressed in Nm^3 H_2/m^3 FF. As with hydrogen partial pressure, this ratio is to be maintained at or above the design ratio.

Make-up H_2/HC Ratio: The hydrogen is constantly being depleted in a hydrocracking unit, as it is consumed by the reactions. The constant depletion of hydrogen requires the addition of fresh hydrogen to the system. The amount required, expressed in Nm^3 H_2/m^3 FF, depends on the type of Feed handled.

Hydrogen Purity: The purity of hydrogen in the make-up gas to the hydrocracker will have a major influence on the hydrogen partial pressure and Recycle H_2/HC ratio.

Combined Feed Ratio (CFR): Most hydrocrackers are designed to recycle unconverted feed from the product fractionator bottom back to the reactors. This stream is often a 600-700°F (315-370°C) plus boiling range material. The liquid recycle rate is reported as a ratio with the fresh feed, called the combined feed ratio (CFR), and is defined as: CFR = (FF Rate + Liquid Recycle Rate)/FF Rate. An important function of liquid recycle is to reduce the severity of operation. This severity reduction is a function of conversion per pass, defined as: Conversion per pass = (FF Rate – Frac Bottoms Rate to storage)*100/(FF Rate + Liquid Recycle Rate). If a unit were operating once-through (CFR = 1.0) and 100% of the feed were converted into products boiling below 700°F (370°C), the conversion per pass is 100% since the feed only makes one pass through the catalyst. At the other extreme, if a unit is designed at a CFR of 2.0 and 100% of feed converted into products, the conversion per pass is only 50%. It can be shown that as the CFR increases, the conversion per pass decreases. Therefore, the catalyst temperature requirement is reduced as the CFR is increased at constant conversion. Conversely, reducing the CFR below the design value can lead to higher catalyst temperature and shorter life between regenerations.

Wash Water Rate: The rate of wash water injection to the Reactor Effluent Air Coolers is critical to the operation of a hydrocracker. Organic nitrogen compound in the feed is converted to ammonia in the reactor. Since ammonia is a basic compound, it competes with the normal reactants for the acid sites on the catalyst. This competition acts to inhibit the cracking activity of the catalyst forcing higher temperatures and higher deactivation rates. The ammonia also combines with chlorine and hydrogen sulfide to form salts that will foul the colder, lower pressure section of the unit. The majority of the ammonia and salts are removed from the system by the injection of wash water at the reactor effluent.

A CASE STUDY: In this case study, the catalyst selection process for a hydrocracker will be discussed and the disastrous consequence that resulted from the failure of the wrong Catalyst Package selected. The reasons behind the catalyst failure, in light of the theory discussed above, will also be highlighted, indicating the pointers for Refiners to follow a fail-safe catalyst selection process.

Catalyst selection process: The first thing to do is to prepare a "Vendor Enquiry Package" to get "Catalyst Proposal" from selected vendors. This Enquiry Package should contain the following:

(1) System Description: This particular hydrocracker is an UOP designed unit (Unibon). The unit is designed to process 30,000 BPSD of a combination of Vacuum Gas Oil (VGO), and Demettalized oil (DMO) from an UOP Propane/Butane Deasphalting unit (Vacuum residue raffinate from an UOP Demax unit). The unit comprises of two stage series flow reactor utilizing UOP's DHC-8 (amorphous) catalyst in the first stage & HC-100 catalyst (zeolitic) in the second stage. The unit was revamped to utilize an Indirect recycle mode to alleviate HPNA (Heavy Polynuclear Aromatics) build-up problem. There are two trains of equal capacity, with each train having four down flow reactors (two each in the first and second stage). A schematic Flow Diagram of this hydrocracker is given in the Appendix (Fig. 2).

(2) Feed Stock Definition: VGO: Specific Gravity (S.G) 0.92-0.93;Total N_2 700-900 wppm; Total Sulfur 2-3 wt%; Con.Carbon 0.3-0.5 wt%; ASTM Distillation, D-2887, range 255-590°C; Nickel (Ni) + Vanadium (V) <1 wppm; Vol% in FF, 75-85. DMO: S.G 0.96-0.97;Total N_2 1300-2100 wppm; Total Sulfur 3-3.5 wt%; Con.Carbon 7-12 wt%; ASTMD-2887, range 216-699°C; Ni + V 8-13.5 wppm; Vol% in FF, 15-25.

(3) Unit Operation: FF Rate, avg/max: 195/210 m^3/hr; Reactor (Rx) max temp.: Rx 1/Rx 2/Rx 3/Rx 4 - 465/465/440/440°C; EOR WABT First stage/Second stage: 440/420°C; Rx 1/Rx 2/Rx 3/Rx 4 Pressure: 149-154/146-151/143-148/140-145 cm^2g;

HP Separator Pressure: 135-140kg/cm^2g; Make-up Gas Flow: 62,000-67,000 Nm^3/hr; Make-up Gas purity: 96-97 mol% H_2; Catalyst volume, First Stage/Second Stage: 452/146 m^3; LHSV, First Stage/Second Stage: 0.43/1.34 hr^{-1}; Recycle H_2/HC ratio, min/avg: 1,100/1,200 Nm^3/m^3 FF; Recycle H_2 purity: 78-84 mol%; CFR: 1.2-1.4 vol/vol; Heavy Diesel ASTMD 86 90% point: 370°C max; Conversion, liquid vol%: 96-98.

(4) Hydrocracker Finished Products Specifications at EOR: Light Naphtha: Total S (ASTMD-4045) 5 wppm max; RON (Clear) (ASTMD-269) 80 min; RVP (ASTMD-323) 15 psia max. Heavy Naphtha: Total S (ASTMD-4045) 0.5 wppm max; N_2 (ASTMD-3431) 0.5 wppm max; Napthenes (ASTMD-1319) 40vol% min; Aromatics (ASTMD-1319) 10vol% min; Kerosene: °API (ASTMD-287) 38.8-51.0; Total S (ASTMD-4045) 200 wppm max; Aromatics (ASTMD-1319) 22 vol% max; Olefins (ASTMD-1319) 5 vol% max; Freezing Point (ASTMD-2386) -53°F max; Smoke Point (ASTMD-1322) 25mm min; Diesel: Total S (ASTMD-4045) 200wppm max; Ash (ASTMD-482) 0.02 wt% max; Cloud Point (ASTMD-2500) 35°F max; Pour Point (ASTMD-97) 25°F max; Colour (ASTMD-1500 3.0 max; Cetane Index (ASTMD-4737) 45 min; Viscosity @40°C (ASTMD-445) 1.9-4.1 cst; Carbon Residue, 10% Bottoms (ASTMD-524) 0.20 wt% max; 90% Distillation (ASTMD-86) 370°C max.

(5) Predicted Performance: The catalyst vendors are to predict performance of their Catalyst Package for 15 vol% and 25 vol% DMO in the feed, indicating: CFR, Recycle H_2/HC ratio, Make-up H_2/HC ratio, overall conversion (vol%), and FF Rate. Performance to be shown as wt%/vol% yield on FF for the following products: NH_3, H_2S, C1-C2, C3, C4, C5-85°C cut (Light Naphtha), 85-185°C (Heavy Naphtha), 185-240°C (Kerosene), 240-315°C (Light Gasoil), 315-375°C (Heavy Gasoil), and >375°C (Bottoms). The vendors to also indicate the predicted catalyst life in years.

(6) Desired Criteria For the Catalyst Package: (a) Proven commercial experience on Feed System similar to that indicated in the Enquiry Package. (b) Cycle Length to be of 3 years minimum duration. (c) Regenerability: Ability to retain stability (cycle length) & activity (yield pattern) to acceptable levels after repeated regenerations (Vendors to specify the expected loss of stability and activity after each regeneration). Here the objective is to use the catalyst package for a minimum of 4 cycles (three regenerations), each of three years minimum run length. (d) The product yield of at least 118 vol% of FF (for 96% overall conversion) of which 45-55% is Diesel & 50-60% gasoline component. (e) Maximization of C5+ yield. (f) To meet the indicated product specifications. (g) Minimization of Make-up H_2 consumption. (h) Flexibility to vary yield of Naphtha (C5-85°C & 85-185°C cut) and Middle Distillates (185-375°C cut). Vendors to indicate how to attain the switching between Naphtha and Distillate mode of operation.

Evaluation & Selection of Catalyst Package: For this particular Refiner, the Catalyst Proposal for its hydrocracker was sought to replace the current batch of UOP catalyst after its EOR. The objective was to have two batches of catalyst available all the time such that one batch is operating while the other sent for ex-situ regeneration. This was determined as the most cost effective approach as it saved downtime for the hydrocracker. After a thorough evaluation, based on the criteria set, the Refiner accepted an amorphous/zeolitic catalyst package from a vendor other than UOP (due to sensitivity of the case, the name of the vendor and the catalyst package used cannot be divulged). The selection was based, apart from lowest cost, on the potential yield advantage, superior product quality, and longer cycle length. It is to be pointed out here that in the matter of these advantages, the refiner believed the vendor's claim, supported by supposedly conducted

Pilot Plant data with their catalyst package on the Refiner's feed system.

Catalyst Performance Guarantee Agreement: Before purchasing the catalyst package, a comprehensive Catalyst Performance Guarantee was put in place. This Performance Guarantee set criteria for determining a catastrophic failure of the catalyst package (within 6 months of operation), as follows: "the point in time during the first six months following SOR wherein any of the following criteria cannot be achieved: (1) the required 96% conversion cannot be achieved with maximum WABTs (440°C for 1st stage reactors & 420°C for 2nd stage reactors) and/or maximum allowable reactors' Peak temperatures (465°C for 1st stage reactors, 440°C for 2nd stage reactors (2) C5+ yield falls below 103 vol% (3) any of the stipulated product specifications (e.g. kerosene smoke point) cannot be achieved. The Performance Guarantee also made it clear that, in case of a catastrophic failure of the catalyst package, the vendor is liable to pay full invoice price of the catalyst Package to the refiner. In case of catalyst failure after 6 months, sliding scale was in place to determine the extent of liability for the vendor.

Catalyst Failure and Failure Analysis: Soon after the start-up, it was observed that the expected WABT was higher by 20°C, and products not meeting guaranteed specifications. The vendor's representative mentioned that WABT would decrease after 1000 hours of operation. But it never did.

After about three months of start-up, in a formal meeting with the vendor representative, the refiner pointed out the non-performance of their catalyst package. In this meeting, the vendor representative recommended to cut DMO completely from the feed in order to sustain hydrocracker operation until the refiner's planned plant turnaround after about three months. This recommendation by the vendor representative clearly indicated the vendor's admission that their catalyst package could not handle the heavy feed component of DMO in the feed. This is in line with the theory discussed above with respect to the multi-ring aromatic compounds present in DMO. Thus, catalyst in the first stage failed to crack these multi-ring aromatics to smaller molecules such that the second stage catalyst could complete the cracking process. The failure of the first stage catalyst caused these heavy molecules of muti-ring aromatics to pass to the second stage and plug its pores necessitating higher temperature in a short time spell, heavy coke lay-down on the catalyst, and the ultimate failure of the catalyst package.

Soon after the first formal meeting with the vendor, the Refiner served a formal notice to the vendor stating that it plans to exercise the remedy under the terms of the Performance Guarantee Agreement, which calls for the full reimbursement of the total invoiced amount.

After the shutdown of the unit, approximately six months of operation, during the refiner's planned turn-around, the reactor internal and catalyst inspection revealed, contrary to vendor's expectation, no abnormality. The catalyst was free flowing during unloading, the distribution trays at the top of the reactors were in their places and were free of any deposit.

In their failure analysis report, submitted to the refiner, the vendor claimed that hydrogen starvation as a result of emergency shutdown, approximately one month after the start-up, due to the loss of make-up hydrogen is the main cause of the non-performance of their catalyst package. This claim by the vendor is incorrect since the catalyst non-performance was evident prior to this emergency shutdown of the unit. The activity loss claimed by vendor during the emergency shutdown should result in shorter cycle length, but should not have resulted in the catalyst failure, indicated by products that failed to meet the guaranteed specifications throughout the time the catalyst was operational.

In summary, the vendor's catalyst system failed from day one, evidenced from the SOR WABT higher than expected by 20°C and products not meeting guaranteed specifications. The failure was due to its catalyst package unsuitable to handle the heavy component of DMO in the feed. The catalyst free flow during unloading and the evidence of no coke ball, contrary to vendor's expectation, are indications of very good operation by the Refiner's operating staff throughout the period of operation with the vendor's catalyst package. The low metal concentration in the spent catalyst samples indicated good feed quality.

At the end, five years after the incident, the vendor took full responsibility of the failure of their catalyst package and agreed to pay the refiner the full invoice price of the catalyst package. Although, this is a token gesture by the vendor since the payment was to be made five years after the incident, the author feels that the Refiner is fully vindicated. The author feels proud about it since he played a big part in the settlement exercise with the vendor.

Pointers For Catalyst Selection: When treading in unknown territory, the Refiners are best advised to take the Pilot-testing route. After the failure of the catalyst package, mentioned above, this particular Refiner took this route in their selection of their next batch of catalyst. To this effect, they took the help of Mobil, USA, to conduct pilot test with three short-listed vendor's (Chevron, IFP, and UOP) catalyst package on their feed system. For the purpose, Mobil, USA, has a comprehensive pilot testing facility. As expected, the UOP catalyst package (DHC-8/HC-100) came on top. Again to stress that this UOP catalyst package is the most suitable catalyst package for this Refiner's feed system, the Refiner have had numerous successful cycles with new as well as regenerated catalyst of this UOP catalyst system.

The other important thing to consider is to have a comprehensive Performance Guarantee Agreement in place, clearly indicating vendor's liability in different catalyst failure scenarios, including a catastrophic catalyst failure.

Appendix

Figure	Description	Remarks
1	Refinery Block Flow Diagram	Shows Major Units' Inter-link
2	Hydrocraker Schematic Flow Diagram	Shows Salient Feature of the UOP Unibon Unit
3	Hydrocracking of Paraffins	Shows mechanism of hydrocracking Paraffins
4	Hydrocracking of Naphtenes	Shows mechanism of hydrocracking Naphthenes
5	Hydrocracking of multi-ring aromatics	Shows mechanism of hydrocracking multi-ring aromatics & formation of HPNA's

Nomenclature

BPSD : Barrels per Stream Day
UOP : Universal Oil Product Company, Des Plaines, Illinois, USA
ASTM : Associated Standard Testing Method
ppmw : Parts per million on weight basis
C1 : Methane
C2 : Ethane
C3 : Propane
C4 : Butane
C5 : Pentane
Nm^3 : Gas Volume in m^3 at 1 atm/0°C
WABT : Weighted Average Bed Temperature
SOR : Start of Run
EOR : End of Run

Acknowledgement

The author acknowledges that the material for this paper includes items from an UOP Unicracking Process Technology Course that author attended.

Author Profile: Kamal Rashid is a chemical engineer and holds a BSc degree in chemical engineering from Bangladesh University of Engineering and Technology, Dhaka, Bangladesh, and a post-graduate degree in petroleum refining and chemical engineering (Ing. ENSPM) from Ecole Nationale Superieur du Petrole et des Moteurs, an affiliate of IFP (French Petroleum Institute), Rueil-Malmaison, France. He has more than 40 years of process engineering and operational experience in petroleum refining industries in Bangladesh, Canada and Saudi Arabia. He spent nine years in the tar-sand industry with Suncor in Alberta, Canada; seven years with Newgrade Energy/CCRL, Regina, Saskatchewan, Canada, with the first heavy oil upgrader in Canada; 14 years with Samarec/Saudi Aramco, Riyadh, Saudi Arabia, with participation in the design, construction and operation of the first low-sulfur diesel hydrotreater in Saudi Arabia.

Mr. Rashid is a licensed professional engineer in the Province of Alberta, Canada (APEGGA) and a member of AIChE. He can be reached by e-mail at rashka0d@yahoo.com.

FIGURE 1: REFINERY BLOCK FLOW DIAGRAM

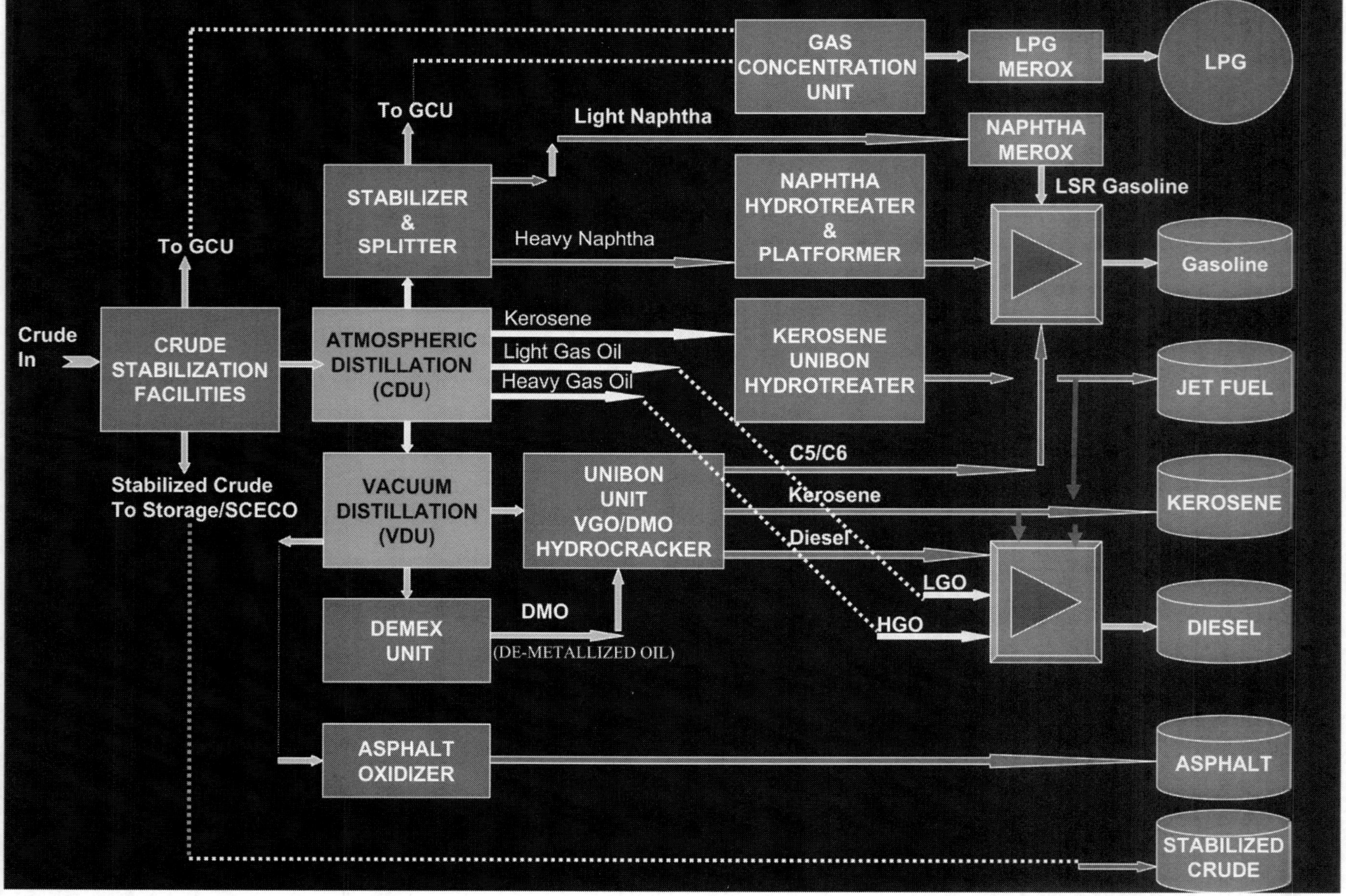

FIGURE 2

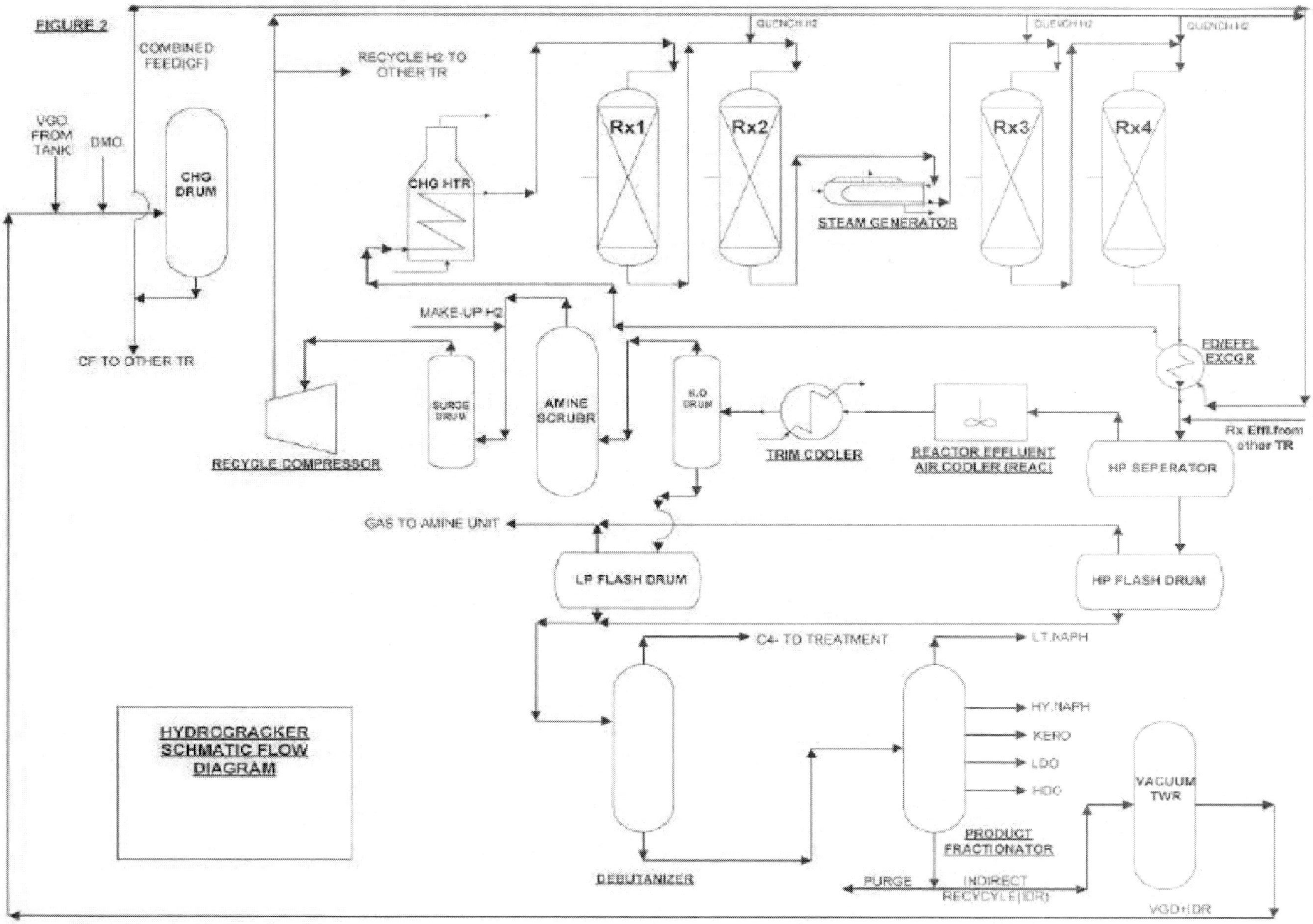

COMBINED FEED(CF)
VGO FROM TANK
DMO
CHG DRUM
CF TO OTHER TR
RECYCLE H2 TO OTHER TR
CHG HTR
Rx1
Rx2
Rx3
Rx4
QUENCH H2
QUENCH H2
QUENCH H2
STEAM GENERATOR
MAKE-UP H2
SURGE DRUM
AMINE SCRUBR
K.O DRUM
TRIM COOLER
REACTOR EFFLUENT AIR COOLER (REAC)
FD/EFFL EXCGR
Rx Effl.from other TR
HP SEPERATOR
RECYCLE COMPRESSOR
GAS TO AMINE UNIT
LP FLASH DRUM
HP FLASH DRUM
C4- TO TREATMENT
LT.NAPH
HY.NAPH
KERO
LDO
HDO
PRODUCT FRACTIONATOR
DEBUTANIZER
VACUUM TWR
PURGE
INDIRECT RECYCLE(IDR)
VGO+IDR
HYDROCRACKER SCHMATIC FLOW DIAGRAM

Figure 3

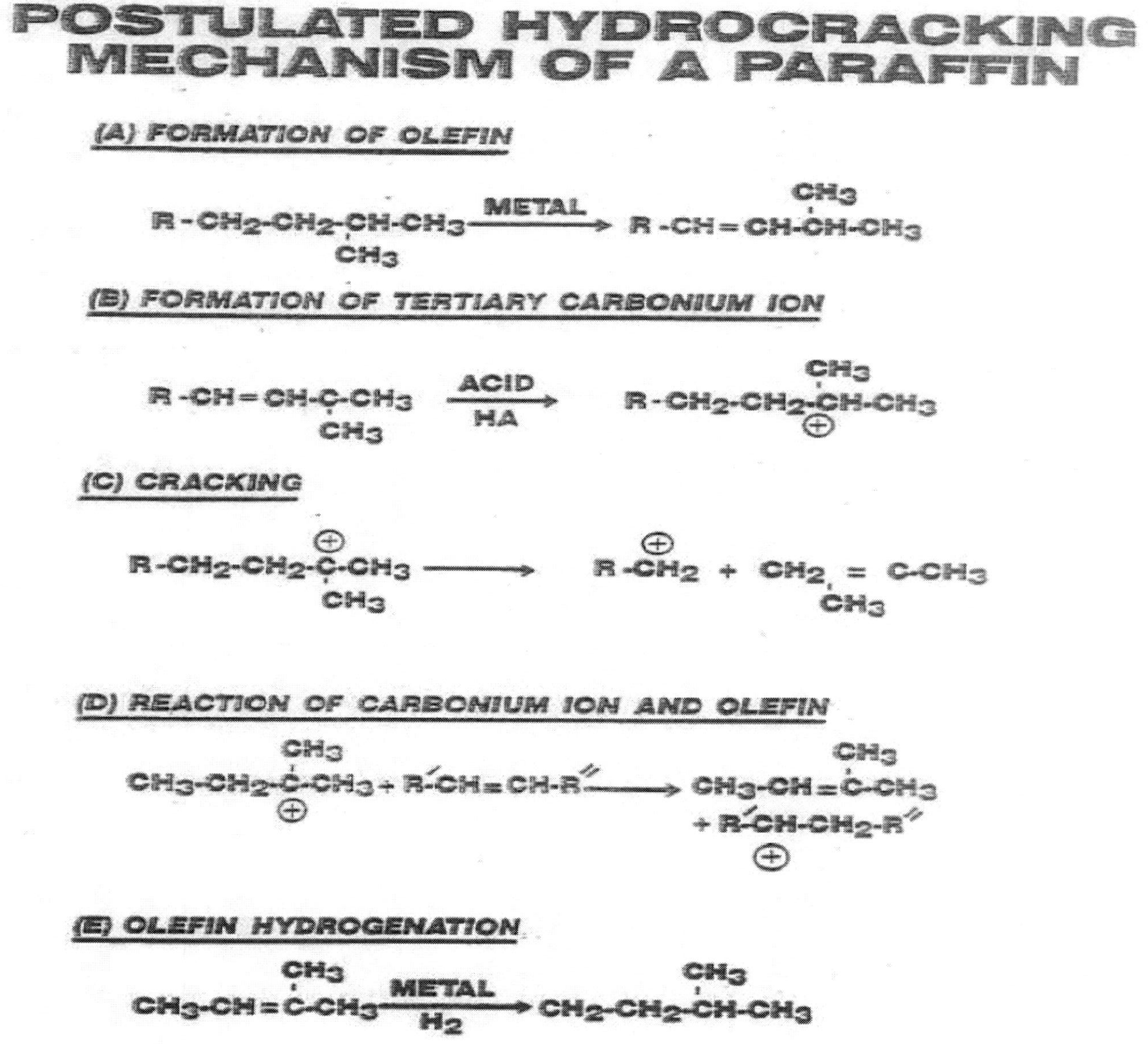
POSTULATED HYDROCRACKING MECHANISM OF A PARAFFIN
(A) FORMATION OF OLEFIN
METAL
(B) FORMATION OF TERTIARY CARBONIUM ION
ACID
HA
(C) CRACKING
(D) REACTION OF CARBONIUM ION AND OLEFIN
(E) OLEFIN HYDROGENATION
METAL
H2

Figure 4

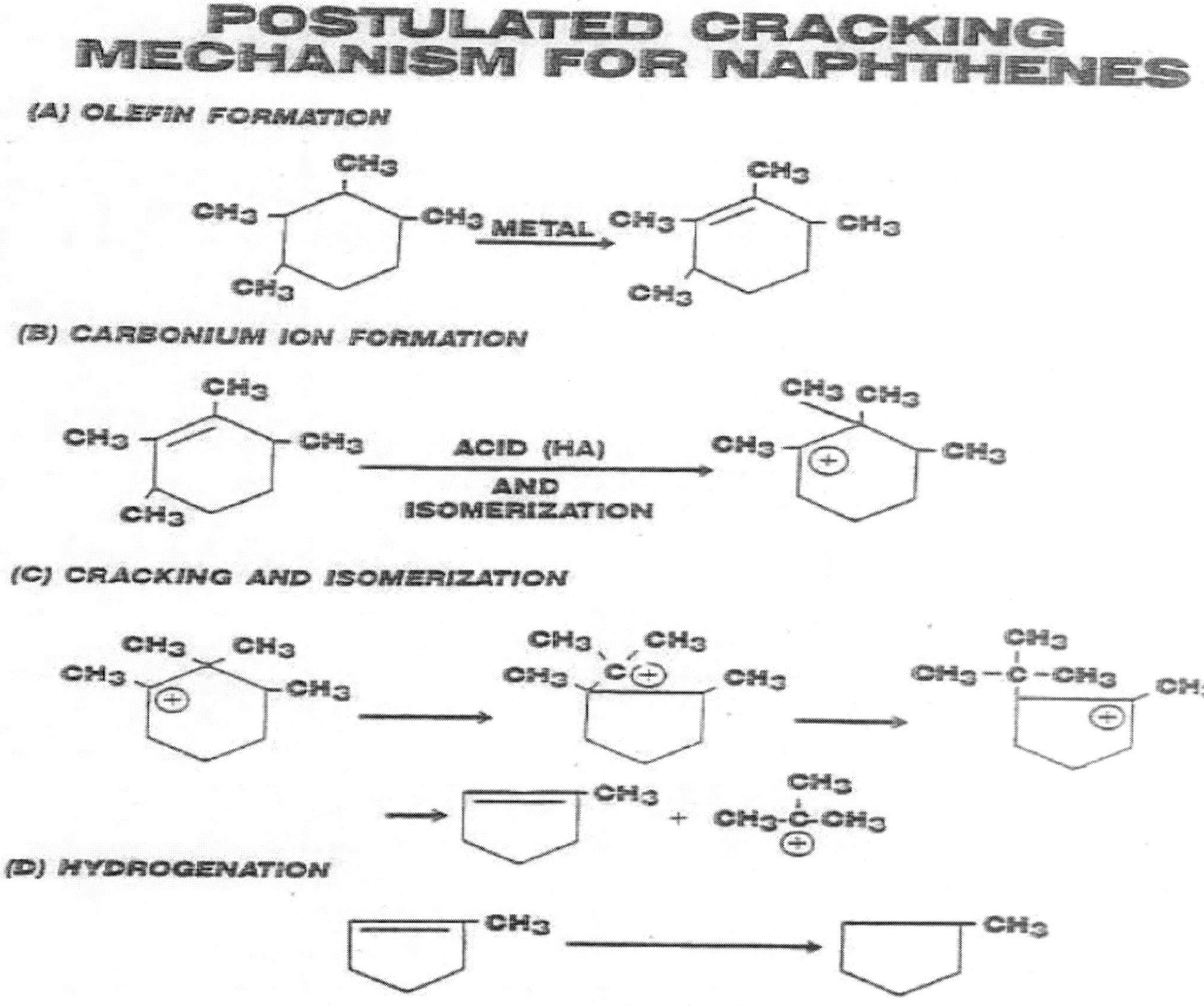

POSTULATED CRACKING MECHANISM FOR NAPHTHENES
(A) OLEFIN FORMATION
CH3
METAL
(B) CARBONIUM ION FORMATION
ACID (HA)
AND
ISOMERIZATION
(C) CRACKING AND ISOMERIZATION
CH3-C-CH3
+
(D) HYDROGENATION

Figure 5

POSSIBLE PATHWAYS FOR MULTIRING AROMATICS

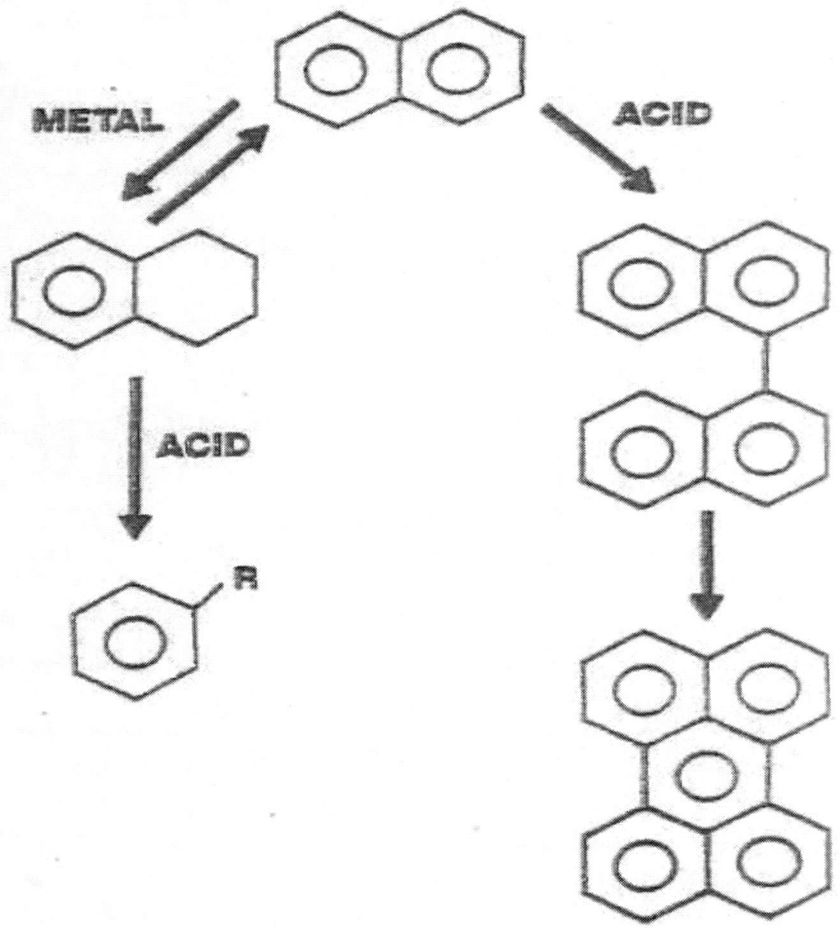

OIL/WATER SEPARATION: A REVIEW OF THE HISTORICAL DEVELOPMENT AND THE EVOLUTION OF NEW TREATMENT TECHNIQUES

ROBERT W. PETERS, Ph.D., P. E.
Professor of Environmental Engineering
Department of Civil, Construction, and Environmental Engineering
University of Alabama at Birmingham
1075 13th Street South
Birmingham, Alabama 35294-4440

M.P. SHARMA, Ph.D.
Professor of Chemical Engineering
Department of Chemical and Petroleum Engineering
University of Wyoming
Laramie, Wyoming

TRACY WILLIAMS
Graduate Student
Department of Civil, Construction, and Environmental Engineering
University of Alabama at Birmingham
1075 13th Street South
Birmingham, Alabama 35294-4440

Abstract

Current environmental regulations require that ships can not discharge bilge water containing >15 parts per million (ppm) oil. Defense missions include operations that generate significant amounts of bilge water; a consistently performing treatment system is paramount to sustain these missions and operations. Bilge water is a highly complex wastewater containing various organic and inorganic contaminants in which the oil may exist in several phases, each behaving differently in conventional oil water separation treatment. Due to the nature of this oily phase, a single treatment method can not efficiently treat bilge water to meet these regulatory limits. A series of process steps should be employed to treat each of the oil phases in order to achieve <15 ppm in the final effluent in accordance with international maritime, federal, and defense discharge requirements. A literature review of current technologies and systems revealed a number of potential candidate methodologies. A qualitative assessment of performance criteria, advantages, and disadvantages of these various technologies and systems was conducted to determine theoretical process trains for simulated bilge water types. The qualitative assessment approach was useful as a preliminary screening tool to condense the compilation of technologies to a manageable number for further study, analysis, evaluation, testing, and demonstration.

Introduction

Current environmental regulations require that ships can not discharge bilge water containing >15 parts per million (ppm) oil. The International Maritime Organization (IMO) originally adopted in 1973, and later modified in 1978, provides standards governing the discharge of oily mixtures from seagoing vessels, referred to as MARPOL 73/78. Under these standards, IMO prohibits oily discharges from such vessels unless such discharges are below 15 ppm; discharges are prohibited altogether when vessels are within 12 nautical miles of land [1]. The U.S Code, Title 33, Sections 1901-1911 implements MARPOL 73/78 under the Clean Water Act. The CWA, amended in 1996, authorized the U.S. Environmental Protection Agency and the Department of Defense to jointly develop Uniform National Discharge Standards to control certain discharges incidental to normal operations of Armed Forces vessels [2].

The objective of this research is to develop a theoretical process or processes for the effective onboard treatment of bilge water for the defense community to comply with current regulatory discharge limits of 15 ppm oil. Such theoretical processes should be considered for future evaluation, testing, and demonstration. Research tasks involved: (1) conducting literature review to characterize the nature, concentration, and generation rates of bilge water from defense operations; (2) conducting literature review to identify the various treatment technologies for onboard applications of bilge water treatment; (3) synthesizing the information compiled through the literature review to evaluate the various treatment technologies identified, (4) determining the most promising and eliminating the least suitable technologies, and; (5) developing preliminary recommendations of theoretical treatment options for onboard bilge water applications. The results presented in this paper are based on a qualitative assessment of the operating parameters, performance criteria, and limitations presented in the literature reviewed for this research.

An extensive literature review was conducted, examining scholarly journals, industry and technical journals, textbooks, government reports (U.S government and foreign), patent reports, research and demonstration proposals and reports, and government, vendors, university and other websites to identify the composition of bilge water and the types of treatment technologies and systems currently available for treating bilge water. Approximately 50-75 resources were reviewed, spanning from 1973 to 2006.

A number of operations onboard Armed Forces vessels generate wastewaters containing oily wastes which must be treated prior to disposal, the most significant being bilge water, ballast water, and tank cleanings. Bilge water results from oils, fuels, and other petroleum products leaked, spilled, or otherwise released from ship machinery where it pools in the bottom of the vessel. Ballast water is used for stabilizing ships when fuel levels are low by filling fuel tanks or oil cargo spaces with seawater. Fuel tanks are periodically cleaned, requiring the disposal of the rinsate [3]. While ballast water and tank cleanings pose problems for ships and vessels, bilge water presents complex operational and environmental issues and is the focus of this manuscript.

While housekeeping and maintenance practices can be changed to reduce the frequency of discharges, these steps alone could not completely eliminate the need for discharging bilge water. This is of particular concern to the defense community as its operations generate significant volumes of bilge water and a consistently performing onboard treatment system is needed to ensure minimal impacts to its missions. Such systems are often operated in confined, congested portions of the vessel and can not always be attended to by ship personnel. Residual processing of wastewater and associated wastestreams must be a minimum since onboard space availability is a premium. Finally, shrinking budgets constrain resources to operate and maintain these systems. Consequently, high-performance, high-yield reliable treatment is needed to ensure environmental compliance and mission sustainment.

BACKGROUND

Bilge water is a highly complex wastewater containing various organic and inorganic contaminants in which the oil may exist in several phases, each behaving differently in conventional oil water separation treatment. This 'oily water cocktail' exists in varying composition and concentrations of a wide variety of contaminants including, but not limited to: fuels, oils, detergents, greases, solvents, corrosion inhibitors, surfactants, rusts, paints, insulation material, and other materials and substances which accumulate in the bottom of a ship or vessel. The presence of water and cleaning fluids is the result of cleaning the engine room, and oil products result from leaks from engine room or breaks in engines or oil lines. If there is a crack in the outer hull, seawater or freshwater will enter the bilge. Rust and other deposits are created by the corrosive seawater. Mixing by pumps, and the addition of surfactants, corrosion inhibitors, detergents, and other surfactants result in mechanical and chemical emulsions.

The USEPA and the Navy are jointly working to characterize each of the vessel discharge subject to the UNDS to develop performance standards for devices to control these discharges. Although composition of bilge water varies considerably among vessels, preliminary results presented in a draft EPA report are based grouping vessels fleet-wide according to similarities in engineering and discharge characteristics. For example, **Tables 1 and 2** below summarizes bilge water from surface vessel oil-water separators and submarines, respectively [4, 5].

Table 1. Summary of Composition of Surface Vessel Bilge Water/OWS Discharges

Constituent	Concentration	Annual Mass Loading, (lbs)
Classicals	**(mg/L)**	
Ammonia as nitrogen	BDL – 0.17	67
Nitrate/Nitrite	0.2 – 0.17	202
Total Nitrogen[A]	1.77*	1,304
Total Phosphorus	1.2 – 2.7	1,353
TPH	6 – 16	7,208
Mercury	**(ng/L)**	
Mercury	32.05 – 79.8	0.04
Metals	**(μg/L)**	
Copper		
Dissolved	116 – 201	122
Total	277.5 – 426	255
Iron		
Total	432 – 531	353
Nickel		
Dissolved	109 – 247	132
Total	97.75 – 245	126
Zinc		
Dissolved	511 – 1,260	640
Total	514 – 1,330	657

* – Log normal mean concentration.
A – Total nitrogen is the sum of nitrate/nitrite and total Kjeldahl nitrogen.
BDL – below detection limits.

Table 2. Summary of Composition of Submarine Bilge Water Discharges

Constituent	Concentration, (mg/L)	Annual Mass Loading, (lbs)
Cadmium	<0.005 – 0.2	0.75
Chlorine	0.0 – 1.6	7.9
Copper	0.065 – 15	53.4
Cyanide	<0.01 – 0.03	0.15
Heptachlor	0.000005	0.00019
Heptachlor epoxide	0.000003	0.00014
Mercury	<0.0002 – 0.0007	0.0026
Nickel	<0.04 – 11	36.9
Oil	<5 – 820,000	1130
Phenol	<0.01 – 5.4	7.15
Silver	<0.01 – 0.035	0.23
Zinc	<0.02 – 11	51.2

The oil content in bilge water is generally estimated to be 20-25%, although lower concentrations (several parts to million to 0.1%) are not uncommon [6]. The new generation of ships with dryer bilges [generating <1,000 gallons per day (gpd) of bilge water] will produce bilge water with higher concentrations of oil and detergents [7]. Oftentimes, oil in bilge water is not 'pure'; it can exist in several phases, each behaving somewhat differently in an oil/water separator. Also, the bilge water is often contaminated by other materials (e.g., detergents, cleaning solvents, particulate matter, etc.) which are not removed by physical separation and/or can impede gravity settling. The oil can exist as free, dispersed, solubilized, dissolved or chemically emulsified [6]. In an oil-water mixture, the oil particle size may vary considerably, depending on the phase:

- Free - >150 μm
- Dispersed – 20-50 μm
- Emulsions - <20 μm [8]

The volumes of bilge water generated are controlled to a certain extent by ensuring hull tightness and minimizing the amount of water used when cleaning engine rooms. Estimates of bilge water generated range from 0.5-50 m^3/day, and bilge water accounts for about 20% of several tons oily wastewater discharged annually from vessels [9].

Results and Discussion

Treating bilge water by conventional gravity settling requires a long residence time, and consequently large treatment tanks or vessels. Designing an OWS to promote particle size growth (i.e., coalescence) reduces the residence time and vessel size. This improves capture of the oily phases dominated by physical separation (i.e., increasing density differences would promote oil particle rise to the water surface), primarily the free oil, dispersed oil, and some of the emulsified oil. Other treatment methods are required to more effectively treat and remove the solubilized and more stable emulsified phases [10]. Because the oil in bilge water can exist in different states, the water can not be effectively treated by a single treatment method, but rather by using a combination of staged treatment methods. Preferably, the free oil phase is removed first, followed by one or more stages to remove the dispersed and chemically interacting phases (solubilized, dissolved and emulsified phases) [6]. Consequently, a series of process steps should be employed to treat each of the oil phases in order to achieve <15 ppm in the final effluent in accordance with international maritime, federal, and defense discharge requirements.

A literature review of current technologies and systems revealed a number of potential candidate methodologies. This inventory was divided into seven broad categories of technologies and systems:

a. Gravitational methods;
b. Centrifugal technology;
c. Flotation;
d. Membrane methods;
e. Electrochemical advanced oxidation processes;
f. Biological treatment, and;
g. Miscellaneous treatment methods [to include combination/hybrid methods and those methods not meeting the other six categories].

The miscellaneous treatment methods reviewed included: thermal treatment (incineration); filtration utilizing organoclay media; air-sparged hydrocyclone; and sonication; and combination/hybrid systems such as electroflocculation/ ultraviolet treatment; ultrafilitration/reverse osmosis (UF/RO), ultrafiltration/nanofiltration composite ceramic membranes; and microfiltration/ultrafiltration systems.

In order to develop a treatment process train, the nature of the wastestream being treated (physical states, volumes, contaminants, concentrations, etc.) and the conditions under which that wastestream is produced, must be determined. In the absence of analytical data, a waste characterization can be developed based on a process owner's knowledge of the operations producing the waste stream. Using the literature values and data presented herein, two bilge water types were hypothesized: (1) a low-volume, highly concentrated bilge ('dry' bilge water) and (2) a high-volume bilge water containing low concentrations of oil ('wet' bilge water). Both cases can be reasonably encountered inasmuch as older vessels tend to generate large volumes of low-concentration bilge water, while newer vessels are being designed and operated to minimize water generation, thus producing a more concentrated wastestream. For the purposes of this research, the presumptive characteristics of these hypothesized "dry" and "wet" bilge waters are as follows and used as the benchmarks for comparing the various treatment technologies and methods:

Hypothesized 'Dry' Bilge Water

- <1000 gallons of bilge water generated per day
- High oil content - approximately 15%
- Chemically produced emulsions (as a result of cleaning with surfactants)

Hypothesized 'Wet' Bilge Water

- Approximately 100,000 gallons of bilge water generated per day
- Minimal oil content - 0.1% (approximately 1,000 ppm) in primarily dispersed form
- Some settable solids present

Each treatment method or technology was evaluated based on its capability of separately treating each of these components of the respective bilge water type.

In addition to wastestream considerations, the conditions (physical, environmental, and operational) under which these wastewaters are generated must be considered as well. A qualitative assessment of performance criteria, advantages, and disadvantages of the major categories of technologies and systems presented in this paper (gravity/coalescing separators; centrifuges/hydrocyclones; flotation; membrane technology; and electrochemical advanced oxidation processes) was conducted to evaluate for their treatment suitability for each of the proposed hypothesized bilge water types and determine theoretical process trains for these hypothesized bilge water types. The physical, environmental and operational conditions considered in this quality assessment were:

1. Final effluent meets dischargeable limits (<15 ppm);
2. Ambient water temperature (20°C or approximately 68°F) [11];

3. Low salinity environment (3.1.-3.8%) [12];
4. Capable of being installed in a small area;
5. Simple to operate;
6. Minimal maintenance requirements;
7. Minimize residuals (sludge, etc.) produced; and
8. Capable of recovering oil of such a quality that can be recycled or reuse as processed or with minimal reprocessing.

The first seven factors above, as well as the individual components of the hypothesized bilge water types, were used as the basis for the primary selection criteria for the theoretical process train for each wastewater. First, the bilge water must be treated to an acceptable level to discharge – in the case of bilge water, that level (<15 ppm) is defined by regulation. The salinity content and water temperature represent the mean conditions found at sea. Space is a premium in the bilge area of a ship or vessel, so the selected process train must be capable of being installed in a confined space. Furthermore, the system should not be complicated to operate or maintain since operators may not always be able to attend to the treatment system and vessels and ships may be away from port for extended periods of time. The limited onboard space and operator attention also dictate that minimal waste is generated by the process. The last factor was used as a balancing criterion to make the final processing determination in the event more than one potential process train of equivalent performance is identified. A treatment system that can produce a reusable product (in this case, oil) can be very beneficial to DoD ship operations. Effective and efficient recovery of oil from bilge water could produce cost savings and operational efficiency by reducing the amount of virgin oil procured and stored onboard.

Because the oil in bilge water can exist in one or more phases, the sequencing of the process steps is critical to optimal treatment efficiency. As such, it is practical to sequence the processes so that the free oil is removed first, then provide secondary processing to treat the dispersed and chemically interacting phases (solubilized, dissolved, and emulsified phases) [6]. This sequence represents the level of increasing difficulty in treating each phase.

A qualitative assessment of the performance criteria, advantages and disadvantages of the major treatment technologies, and methods reviewed was used to determine the suitability of each technology or method for treating each of the hypothesized bilge water type. The results of this assessment are summarized in **Tables 3 and 4** for the hypothesized 'dry' and 'wet' bilge water types, respectively. This determination was based solely on the literature values and data obtained in this research. Independent verification of the literature values and data presented herein was not performed as this was outside the scope of this research.

Treatment suitability was determined based on the capacity of each treatment method or technology to treat the various bilge water components, achieve the discharge limit, and operate under the prescribed physical, environmental and operational conditions. Based on these factors, a total of eleven criteria were developed to evaluate each treatment technology or method. Each technology or method was assigned a 'pass/fail' rating [equally weighted (an assumed value of unity)] for each of the eleven criteria. The preliminary score was based on the ratings of the primary criteria noted in **Tables 3 and 4**. Those treatment technologies and methods which demonstrated the capability of producing a reusable or recyclable oil were rated slightly higher by weighting that factor slighter higher than the other ten criteria. The final rating in each table reflects a presumed 'degree' of suitability of that process for treating the hypothesized bilge water in question; a value less than 100% demonstrates that an individual process in each column would not completely treat the hypothesized bilge water under all of the physical, environmental, and operational conditions presented. That is, the treatment method or technology would not treat one or more of the bilge water components and/or satisfy one or more of the criteria listed, and as such, additional treatment would be warranted. Hence, the three processes with the highest total 'scores' demonstrated the highest degree of suitability, and consequently the most promising technologies, to develop a theoretical treatment process train for treating the respective bilge water type.

Summary and Conclusions

For the hypothesized 'dry' bilge water (concentrated and chemically-emulsified oil under low-flow regimes), the theoretical process train involved:

- Gravity separation in a coalescing unit, followed by an electrochemical advanced oxidation process (AOP), and membrane technology for final polishing of the effluent. This process train could be further improved by:
  - Air-sparged hydrocyclone (ASH) technology as primary treatment in lieu of gravity separation, and;
  - UF/RO membrane technology contingent upon effluents from upstream processes achieving an oil concentration <400 ppm.

Although centrifuges and hydrocyclones shows a tendency to remove the free oil phase more effectively than gravity/coalescing separators, the latter was selected given hydrocyclones' susceptibility to seawater.

For the hypothesized 'wet' bilge water (low oil concentrations mainly in dispersed phase under high flow regimes); the theoretical process train involved:

- Centrifuges or hydrocyclones for primary treatment, followed by biological treatment, and membrane technology for final polishing of the effluent. Membrane technology offers the added benefit of recovering useable product. This process train could further improved or modified by:
  - Gravity separation in a coalescing unit with subsequent filtration using organoclay filter media as primary treatment in lieu of a centrifuge or hydrocyclone;
  - Incorporation of air-sparged hydrocyclone technology as a more efficient hydrocyclone;
  - Hybrid membrane systems utilizing (1) ultrafiltration/nanofiltration composite membranes, (2) microfilration/ultrafiltration membranes, or

ultrafiltration/reverse osmosis membranes (particularly if upstream effluent concentrations are <400 ppm) for final polishing of the effluent.

- Incineration for complete treatment as this eliminates all the wastewater and generates little residual waste.

Table 3. Qualitative Analysis of Treatment Processes for Hypothesized 'Dry' Bilge Water

Treatment Method/Technology		**Gravity/ coalescing OWS**	**Centrifuge/ hydrocyclone**	**Flotation**	**Membrane**	**Biological**	**Electrochemical AOP**
PRIMARY CRITERIA							
Type of Bilge Water	**Bilge Water Characteristics**						
Dry	<1,000 gpd	P	P	F	P	P	P
	High oil content (~15%)	P	P	P	F	P	P
	Oil primarily in free phase (>150 μm droplet size)	P	F	P	F	P	F
	Chemical emulsions present (<20 μm)	F	P	P	P	F	P
Operational and Other Factors							
Final effluent meets dischargeable limits (<15 ppm)		F	F	F	P	F	F
Ambient temperature (20°C or ~68°F)		P	P	P	P	P	P
Low salinity environment		P	F	P	P	P	F
Capable of being installed in a small area		P	P	P	P	P	P
Simple to operate		P	P	P	P	P	F
Minimal maintenance requirements		P	P	F	P	F	P
Minimize residuals (sludge, etc.) produced		F	P	F	F	F	F
PRELIMINARY SCORE		**72.7%**	**72.7%**	**63.6%**	**72.7%**	**63.6%**	**54.5%**
BALANCING CRTERION							
Capable of recovering oil of such a quality that can be recycled or reused as processed or with minimal reprocessing		F	P	F	P	F	F
TOTAL SCORE		**72.7%**	**81.8%**	**63.6%**	**72.7%**	**63.6%**	**54.5%**

Table 4. Qualitative Analysis of Treatment Processes for Hypothesized 'Dry' Bilge Water

Treatment Method/Technology		Gravity/ coalescing OWS	Centrifuge/ hydrocyclone	Flotation	Membrane	Biological	Electrochemical AOP
PRIMARY CRITERIA							
Type of Bilge Water	Bilge Water Characteristics						
Dry	<1,000 gpd	P	P	F	P	P	P
	High oil content (~15%)	P	P	P	F	P	P
	Oil primarily in free phase (>150 μm droplet size)	P	F	P	F	P	F
	Chemical emulsions present (<20 μm)	F	P	P	P	F	P
Operational and Other Factors							
Final effluent meets dischargeable limits (<15 ppm)		F	F	F	P	F	F
Ambient temperature (20°C or ~68°F)		P	P	P	P	P	P
Low salinity environment		P	F	P	P	P	F
Capable of being installed in a small area		P	P	P	P	P	P
Simple to operate		P	P	P	P	P	F
Minimal maintenance requirements		P	P	F	P	F	P
Minimize residuals (sludge, etc.) produced		F	P	F	F	F	F
PRELIMINARY SCORE		**72.7%**	**72.7%**	**63.6%**	**72.7%**	**63.6%**	**54.5%**
BALANCING CRTERION							
Capable of recovering oil of such a quality that can be recycled or reused as processed or with minimal reprocessing		F	P	F	P	F	F
TOTAL SCORE		**72.7%**	**81.8%**	**63.6%**	**72.7%**	**63.6%**	**54.5%**

Capital, and operational and maintenance (O&M) costs were not included as selection criteria due to the lack of and/or incomplete data for identified in the literature review. It should be noted that cost data is essential in evaluating treatment technologies, and as such any assessment of treatment options would be incomplete without such data. Consequently, the preliminary results presented herein are subject to change, given how cost data will be used in any subsequent assessment and evaluation of the treatment trains proposed in this paper.

The qualitative assessment approach was useful as a preliminary screening tool to condense the compilation of technologies to a manageable number for further study, analysis, evaluation, testing, and demonstration.

References

1. U. S. Navy Military Sealift Command, (1995). "Oil Pollution Prevention: Oily Water Discharges Corrections", *Environmental Issues Newsletter (No.16, 14 April)*, Washington Navy Yard, DC.
2. U. S. Environmental Protection Agency (USEPA), (1999). *Uniform National Discharge Standards (UNDS) for Armed Forces Vessels Factsheet/EPA-821-F-99-009, April*, U.S. EPA, Washington, DC.
3. Harvey, A.; Guzdar, A., *et al.*, (1973). *Centrifugal Coalescer for Separating Oil from Water Discharged from Ships, U. S. Coast Guard Office of Research and Development Project No.430*, Waltham, MS: Foster-Miller Associates, Inc.
4. U. S. Environmental Protection Agency (USEPA), (2003). *DRAFT Characterization Analysis Report Surface Vessel Bilgewater/Oil Water Separator/Section 1.0 – Introduction and Table of Contents/August 2003*, Washington, DC.
5. U. S. Environmental Protection Agency (USEPA), (2003). *DRAFT Discharge Assessment Report: Surface Vessel Bilgewater/Oil Water Separator/June 2003*, Washington, DC.
6. Finger, S. M., and Tabakin, R.B., (1973). *Development of Shipboard Oil/Water Separation Systems". Proceedings for the SAE-ASME-AIAA-ASMA-AIChE Intersociety Conference on Environmental Systems (July 16-19)*, London, England, American Society of Mechanical Engineers.
7. Tompkins, K. and Murphy, L., (2002). *Ultrafiltration Membrane Polishing Systems for Shipboard Treatment of Oily Wastewater*, Raleigh, NC, North Carolina Department of Environment and Natural Resources Division of Pollution Prevention and Environmental Assistance.
8. Cheryan, M., and Rajagopalan, N., (1998). "Membrane Processing of Oily Streams: Wastewater Treatment and Waste Reduction", *Journal of Membrane Science, 15*(1): 13-28.
9. Tomaszewska, M.; Orecki, A.; and Karakulski, K., (2005). "Treatment of Bilge Water Using a Combination of Ultrafiltration and Reverse Osmosis", *Desalination, 185*(1-2): 203-212.
10. Plat, R., (1994). *Gravitational and Centrifugal Oil-Water Separators with Plate Pack Internals*, The Netherlands, Delft University Press.
11. Wikimedia Foundation, Inc., (2007a). *Sea Surface Temperature*, St. Petersburg, FL.
12. Wikimedia Foundation, Inc., (2007b). *Seawater*, St. Petersburg, FL.

In Situ Asphaltenic Film Characterization of W/O Emulsions by Sans: the Role of Film Thickness and Composition in Emulsion Stability

Vincent J. Verruto, Chemical and Biomolecular Engineering, North Carolina State University, 911 Partners Way, Engineering Building 1, Rm 2-102, Raleigh, NC 27695-7905 and Peter K. Kilpatrick, Department of Chemical and Biomolecular Engineering, North Carolina State University, College of Engineering 1, 911 Partners Way, Raleigh, NC 27695-7905.

We describe here the use of small angle neutron scattering (SANS) for the characterization of interfacial films directly related to the stabilization of water-in-oil emulsions frequently encountered in the petroleum industry. These emulsions are often detrimental in crude oil transport as they are typically of high viscosity and they tend to transport so-called poisons (usually inorganic salts) to downstream catalytic processes. Asphaltenes, the heaviest component in crude oil, have been established as a likely culprit in the formation of the elastic interfacial films that protect these emulsion systems from coalescence. The physicochemical properties of these films are expected to correlate with mechanical properties and, ultimately, emulsion stability.

Presented here are the results of SANS experiments aimed at ascertaining the thickness and composition of the stabilizing films formed on the surface of aqueous droplets dispersed in asphaltene-rich organic solvents. Three liquids of increasing solvency – 1-methylnaphthylene, toluene, and decalin – were used as the organic phase along with 1 wt% Hondo whole (HOW) asphaltenes. In this work we describe two avenues for handling aggregate scattering that occurs in these systems: 1) subtraction of the bulk phase scattering from the total emulsion scattering and 2) the physical removal of the bulk aggregates by sequential replacement with fresh organic solvent.

Using a polydisperse core/shell model, we have determined the average thickness and scattering length density (SLD) of each emulsion system tested. The film thickness values were 9-11 nm, and did not appear to directly correlate with emulsion stability – determined by centrifugation @ 14000•g for 1 hr – but do trend with solvent strength/aggregate •RG•. However, from the fitted film SLD values, and known SLDs for the solvents and asphaltenes, we have been able to determine the average film composition in each system. In general, the interfacial films tended to be quite solvent-rich, with only 15-45% of the film thickness and SLD accounted for by asphaltene. Surprisingly, the strongest emulsions, those made with decalin, had the thinnest films (9.3 nm), but exhibited the highest surface concentration (4.4 mg/m2), more than twice that of toluene (1.9 mg/m2) or 1-methylnaphthalene (2.0 mg/m2). These results demonstrate the power of SANS as a tool for probing physicochemical properties of asphaltenic W/O emulsions in situ, enhancing our understanding of how asphaltenic films lead to emulsion stability.

The Benefits of Optimizing Your Desalter Operation

Sandra N. Garcia-Swofford, Downstream, Nalco Company, 7705 Hwy 90A, Sugar Land, TX 77478 and Sam A. Lordo, Energy Services - Downstream, Nalco Company, P.O. Box 87, Sugar Land, TX 77487.

Refiners continue to come across new challenges related to crude oil desalting. As new fields are developed and transported all over the globe refiner's are looking at opportunities to process lower cost crudes. In an industry with constant changes it is sometimes difficult to dedicate the appropriate time to ensure that systems are operating at Optimum Conditions. Desalting systems are all different and each one requires a customized program.

This paper will focus on an approach that a refiner can use to optimize their current operation in a continuously changing environment. It will cover practices in the area of tank management, slop oil recovery, desalter operation and chemical programs.

Tutorial on Handling Slop Oil in the Refinery

Lawrence Kremer and Kerlin Byers
Baker Petrolite
12645 West Airport Blvd
Sugar Land, TX 77478

Abstract

Waste oil streams are produced by many processes in a refinery. It is challenging to recover oil from the waste streams or slop oil with minimum disruption to the operation of a refinery. The first step in reprocessing these waste oils streams is to characterize them sufficiently to know how to treat the slop oil. The slop oils can be either oil in water or water in oil emulsions and this nature determines how it can best be handled. Discussions in this paper will be restricted to water in oil emulsions. Various equipment and operational options will be discussed along with examples of chemical treatment programs showing how problem slop oils are handled in refineries.

Waste Oil Handling Process

Waste oil or slop oil has many names and many sources in the refinery (Ref. 1). It can range from off-specification products that contain relatively small amounts of water (<500ppm) to skim oils from amine units or other gravity separators that may contain more water than oil. The first step in processing a waste oil stream is to characterize the material so that its disposition can be determined. There is a tendency to mix all of the slop oil together. However, it is generally a best practice to segregate the waste oil streams as much as possible to minimize upsets and maximize the value to the refinery at minimal cost and effort. Some streams can simply be re-introduced into the appropriate process unit. Other streams should be treated chemically and mechanically before reprocessing. With some streams, the negative impacts on operation outweigh any value that might be extracted from the material, so disposal becomes the best option.

Problems Associated with Slop Oil Reprocessing

Before discussing characterization, it is useful to understand problems associated with reprocessing waste oil. Potential problems suggest relevant characterization testing.

1. *Cost of Reprocessing* – There are various direct costs of reprocessing oil, for example, energy costs, equipment, chemicals, and manpower. In general, there are also opportunity costs, since introduction of slop oil on a unit may back out virgin feed.
2. *Desalter Upsets* – Often slop oil is injected into the crude unit charge either in line or in tankage. Improper additions of slop oil can lead to

high water in the desalted crude oil, stable emulsions, high current draw and grid loss, and oil contamination of the effluent water.

3. *Catalyst Poisons* – Some slops contain heavy metals such as nickel and vanadium as well as alkali metals such as sodium and potassium that are catalyst poisons.
4. *Corrosion* – Many contaminants in slop oil can cause corrosion in the units where they are sent.
5. *Fouling* – There are both inorganic solids and organic materials, such as asphaltenes, that can lead to fouling if not properly handled.
6. *Pressure Surges* – Excess water or light ends can flash off when the slop oil is introduced into a tank or vessel. This can lead to pressure surges that can result in tray damage and back out throughput on process units. Introduction of light ends into hot oil or product storage tanks could result in release of flammable vapors.
7. *Environmental Releases*- Many potential components in slop oil can adversely impact the waste water system if not handled properly. If slop oil is introduced into a system that is vented to the atmosphere, there is a potential for release of hazardous gases, such as hydrogen sulfide, and compounds that produce nuisance odors.

Characterizing the Waste Oil

There are many tests that can be run to characterize slop oil. It is not practical to run all these tests on a regular basis. An understanding of the source of the slop oil will likely suggest what parameters are most useful to test. For example, slop oil recovered from the oil under carry of the desalter will be chemically similar to crude oil being charged to the unit. Once a waste stream is characterized a few key tests can be run to monitor the quality of the stream over time. The list is not exhaustive but is more or less in the order of the frequency used.

1. There are many dark liquids in the refinery setting that can be called slop oil. Both oil continuous (w/o) and water continuous (o/w) emulsions have been called slop oil, but this paper will be restricted to water in oil emulsions. Oil in water emulsions are generally processed in the waste water treatment plant using combinations of gravity separation and air flotation units (Ref 2). These processes are generally improved by the addition of chemicals that produce clean water and slop oil that is an oil continuous emulsion. This oil continuous emulsion may still contain 10 to 50% water suspended as droplets in the oil. The simplest way to determine which phase is continuous is to pour some of the emulsion into a beaker of tap water and some into a beaker of aromatic solvent, like xylene. In whichever beaker the emulsion dissolves tells which phase is the continuous or external phase. If the "emulsion" dissolves in the water then the emulsion is water external and if it dissolves in the solvent it is oil

external. There are cases where combination emulsions occur, i.e. oil in water in oil; the best way to tell in this case is microscopic examination.

2. Most all slop oil has some water in it. Off-specification product and intermediate streams will likely have less than 1% water which can be measured by distillation or Karl Fischer. Skim oil is likely to have considerably more than 1% water, which can best be measured by centrifugation. It is very common for the % water in slop oil streams to change with time. Thus it becomes important to monitor this parameter on a regular basis.
3. Solids in slop oil can cause fouling and desalter upsets. Some solids such as coke fines and catalyst fines are very effective at stabilizing emulsions, so slop oils with these solids should not be injected upstream of the desalter.
4. Metals can poison many catalytic processes. If the refinery is producing metallurgical or anode grade coke, metals in the slop oil can impact product quality. If the slop oil is to be burned, metals such as sodium, potassium, and vanadium can cause fireside fouling and corrosion. Metals are usually measured by inductively coupled plasma emission spectroscopy (ICP) or atomic absorption (AA).
5. Salts are usually measured as the chloride. This is done by titration in the field or by ICP or ion chromatography in a well equipped lab. Chloride salts can hydrolyze in thermal processes, such as the crude unit atmospheric furnace, and lead to corrosive conditions in overhead systems. While chloride is the most common concern, the author is aware of two refineries where entrained fluoride in slop from HF alky units led to severe corrosion. In one case, the fluoride attacked a crude unit overhead, and the other resulted in corrosion of stainless steel trays in a coker fractionator. This indicates the importance of measuring the appropriate species in the slop oil.
6. Distillation is probably not done as frequently as it should be. It is important to anticipate the evolution of light ends.
7. Asphaltene instability can lead to fouling and emulsion stability. This occurs primarily when asphaltenic oil is mixed with paraffinic oil. Numerous tests have been developed to predict when unstable conditions might occur including various spot tests and the Baker Petrolite Asphaltene Stability Index Test (ASIT®) (Ref. 3 and 4). For example, asphaltene instability leading to desalter upsets and fouling can occur when a paraffinic naphtha is added to a crude oil high in asphaltenes.
8. Amines can destabilize desalters and form corrosive salts in the crude unit atmospheric column and overhead condensing system, or in the coker fractionator. Slop oil from the sulfur plant is likely to be high in amines if the slop is not properly skimmed or if stabilizers are emulsifying amine solution into the slop oil.

Managing the Waste Oil Streams

Once the slop oil is properly characterized, a plan can be developed for ultimate disposal. In some cases, the slop oil should be treated chemically and mechanically to improve the quality before it is worked off on a unit. The following are some aspects of a remediation program.

1. *Segregation* – It is better to work off the slop in the appropriate unit than to blend all of the slop together making one large complex problem. Contaminated ultra low sulfur diesel (ULSD) could be processed through the hydro desulfurization unit to make it pass the sulfur spec. Blending it with slop from the desalter effluent will mean that it will probably have to be reprocessed through the crude unit or coker, with a higher total cost of reprocessing.
2. *Repair* – Not all wasted oil needs to be reprocessed. At one refinery location, gasoline that failed the spindle corrosion tests was being injected in the crude charge to be re-refined. We showed that the appropriate corrosion inhibitor could be added so that the gasoline would meet all specifications without rework.
3. *Find the best spot* – Slop that does not contain water or high levels of salt does not need to be rerun though the desalter. If the desalter is a bottle neck, inject the slop after the desalter.
4. *Avoid dumping the problem into the crude storage tank* – Many refineries allow various units to send slop to the crude storage or crude charge tank. This often converts a small problem into a large problem. Dumping turnaround cleaning waste in the crude storage tank may take months to filter out in the heat exchangers and tower trays. During this period, the operators will be fighting desalter problems and waste treatment problems. Baker Petrolite has developed special treatment programs that have enabled some refineries to work off cleaning waste with minimal desalter upsets. In some cases, disposal or processing by an outside contractor is the best option.
5. *Slow and steady* – It is better to inject slop oil at a low continuous rate, than to inject it at high or uncontrolled rates in batches. Operators can make adjustments for a constant addition rate. At uncontrolled rates upsets occur.

Processing the Slop Before Injection

Many waste oil streams are high in water or solids and should be processed to improve the quality before they are charged to a process unit. Two types of equipment are commonly used; gravity separation tanks or vessels and centrifuges. In the case of centrifuges, an artificially high gravitational field is produced by rapidly spinning the emulsion. In both cases, chemical additives can be used to cause the small droplets to flocculate and coalesce into larger drops that settle much faster.

Centrifuges resolve the emulsion much faster than gravity settling in tanks. However, they are complex pieces of equipment that require constant operator attention. For this reason, it is common for refineries to contract specialists to operate the centrifuge equipment. Tank settling is much slower, but requires little operator attention. Refineries frequently conduct slop treats in tankage with their own staff. Design of tank treating equipment has been described in detail elsewhere (Ref 1).

Selection of the proper emulsion breaker chemistry determines at what dosage the treatment will be successful and if the solids will partition to the oil or water phases. The performance of the chemical treatment is often greatly accelerated by heating the slop oil. This reduces the viscosity of the slop oil enabling the droplets to settle faster and also collide with each other more frequently, thus facilitating flocculation and coagulation.

Chemical Treatment

The most common treatment for slop oil is to chemically treat it, preferably in a heated tank. The best location for the chemical injection is to the suction side of the transfer pump going into the tank. It is very difficult to add chemical directly into the slop oil tank and obtain adequate mixing. The tank is generally allowed to settle for a period of a few days up to several weeks. Once the emulsion has been separated, the oil can be skimmed from the top for reprocessing in the crude unit and the water can be drained to the waste water treatment plant. The optimum chemical treatment and dosage can be easily determined using a simple laboratory screening test. However, it is extremely important to note that the effectiveness of any chemical treatment is heavily dependant on the slop sample that is used. Testing on a representative sample of the composite slop helps predict not only the proper demulsifier product, but also the outcome. In order to obtain a representative sample, the vessel containing the slop oil should be circulated prior to sampling or that a top, middle, and bottom sample be taken and blended together. One of the biggest reasons for unsuccessful treatments stem from non-representative samples being used to determine the most effective treatment chemicals and their dosages.

Because there is a large variation in the composition of slop oil, it has been found that a wide variety of chemical additives are required for optimal treatment of the different types of slop oil. Baker Petrolite's emulsion breaking chemical product line encompasses a wide variety of chemistries to treat very complex slop oils. That technology includes:

- Non-ionic emulsion breakers for refinery slop treatments with typical dosages of 500 to 10,000 ppm.

- Anionic emulsion breakers which have been successfully used on many refinery slop oils. The typical dosage range is 500 to 10,000 ppm.
- Inorganic slop oil treatment chemical, which produces clean dry oil (< 1% BS&W) and very dirty water since much of the solids partition to the water phase. The typical dosages are 1 – 3%.
- Organic demulsifiers consisting of various surface active chemicals. They promote rapid coalescence of water and water wets particulate matter to resolve water in oil emulsions. The typical dosages are 100 to 1,000 ppm.

Laboratory Example: Heat vs. Residence Time

A customer needed to treat a slop oil tank at room temperature with no mixing, which is one of the hardest to treat scenarios. The advantage in this case was the long residence time for the treatment due to the tank flexibility. The BS&W of the raw sample was 2% water, 3% solids, and 95% rag layer (unresolved emulsion). Table 1 shows the water, oil, and unresolved emulsion content of the original sample. Baker Petrolite DMO 46X emulsion breaker, known as knock out drops, was added to the sample, as called for in the ASTM D-96 centrifuge test, to resolve the emulsion.

Table 1. Oil, water, and emulsion in untreated slop oil by centrifuge

	NO KNOCK OUT DROPS	**KNOCK OUT DROPS**
BS (unresolved emulsion)	**95%**	**6%**
Water	**2%**	**58%**
Solids	**3%**	**6%**
Oil	**0%**	**30%**

The initial testing was carried out at room temperature (25°C). A preliminary screening test selected Additive A and Additive B as the best candidates. Then, a ratio test demonstrated that Additive A at 4000 ppm was the best option, achieving 30% oil recovered and 12% rag layer, after 3 days of settling time (residence time). Table 2 shows the amount of oil and water recovered at room temperature as a function of time. Treatment of 1000 ppm and 2000 ppm were no better than the blank. At 4000 ppm treatment with Additive A, the emulsion slowly resolved at room temperature. After 3 days, there was still 12% unresolved emulsion.

Table 2. Water separation in room temperature treatment

ADDITIVES	DOSAGE (PPM)	(12 HOURS)	(24 HOURS)	(48 HOURS)	(72 HOURS)
A	2000	0%	0%	0%	0%
A	3000	0%	0%	0%	0%
A+B	3000+1000	0%	5% w / 43% rag layer	10% w / 38% rag layer	30% w / 18% rag layer
A	4000	5%	10% w / 60% rag layer	35% w / 35% rag layer	58% water / 12% rag layer
BLANK	0	0%	0%	0%	0%

After 72 hours, a sample of the recovered oil from the slop oil treated with 4000 ppm of Additive A at room temperature was analyzed using the centrifuge. The quality of this oil is shown in Table 3.

Table 3. Quality of recovered oil treated with 4000 ppm Additive A after 72 hours at room temperature

	NO KNOCK OUT DROPS	**KNOCK OUT DROPS**
BS (unresolved emulsion)	**0.8%**	**0.8%**
Water	**0%**	**0%**
Solids	**0%**	**0%**
Oil	**99.2%**	**99.2%**

The second phase of the evaluation was increasing the temperature of the test to show the advantage of heat in this kind of treatment. A ratio test was run placing the bottles into a water bath at 150°F (66°C). After only 15 minutes, 30% water separation was achieved with 3000 ppm of Additive A and 50% with 4000 ppm. Sixteen hours (16h) later, 42% water separation and 12% rag layer was observed with 3000 ppm of the Additive A, and 50% water separation and 10% rag layer at 4000 ppm. Finally, after 24 hours of settling time, 50% water separation and 10% rag layer was achieved with 3000 ppm of Additive A, and 60% water separation and no rag layer at 4000 ppm. The clarity of the produced water was also much better at elevated temperatures than that produced at room temperature. This represents an outstanding improvement just by adding some heat. Table 4 shows that water is removed much more rapidly when the test is run at 66°C. After 15 minutes the performance is roughly equivalent to the room temperature treatment after 72 hours.

Table 4. Water separation when treated at 66°C

ADDITIVES	DOSAGE (PPM-V)	(15 MINUTES)	(16 HOURS)	(24 HOURS)
A	1000	0%	5%	38% w / 27% rag layer
A	2000	10%	35%	40% w / 25% rag layer
A	3000	30%	42% w / 18% rag layer	50% w / 10% rag layer
A	4000	50%	50% w / 10% rag layer	60% w / 0% rag layer
BLANK	0	0%	0%	0%

A comparison of the results of the room temperature and heat treated slop oil samples after 24 hours is given in Figure 1. The data was taken from Table 2 and Table 4. The first four columns are dosages of 0, 2000 ppm, 3000 ppm, and 4000 ppm at room temperature. Columns 5 through 8 are 0, 2000 ppm, 3000 ppm, and 4000 ppm run hot at 66°C. The dark band of each column represents the amount of unresolved emulsion or rag layer, the hashed area on the bottom represents the % of water that is recovered, and the white area on top represents the amount of recovered bright oil.

For the slop oil treated at room temperature, only the sample with 4000 ppm emulsion breaker additive (#4) showed any signs of oil water resolution, and that sample still had 60% unresolved emulsion. Raising the temperature to 66°C did not cause the emulsion to be resolved in the untreated sample (#5). However, when heated to 66°C the sample treated with 2000 ppm additive (#6) out-performed the room temperature sample treated at 4000 ppm (#4). At 66°C the sample treated at 4000 ppm (#8) was completely resolved into oil and water with no residual rag layer.

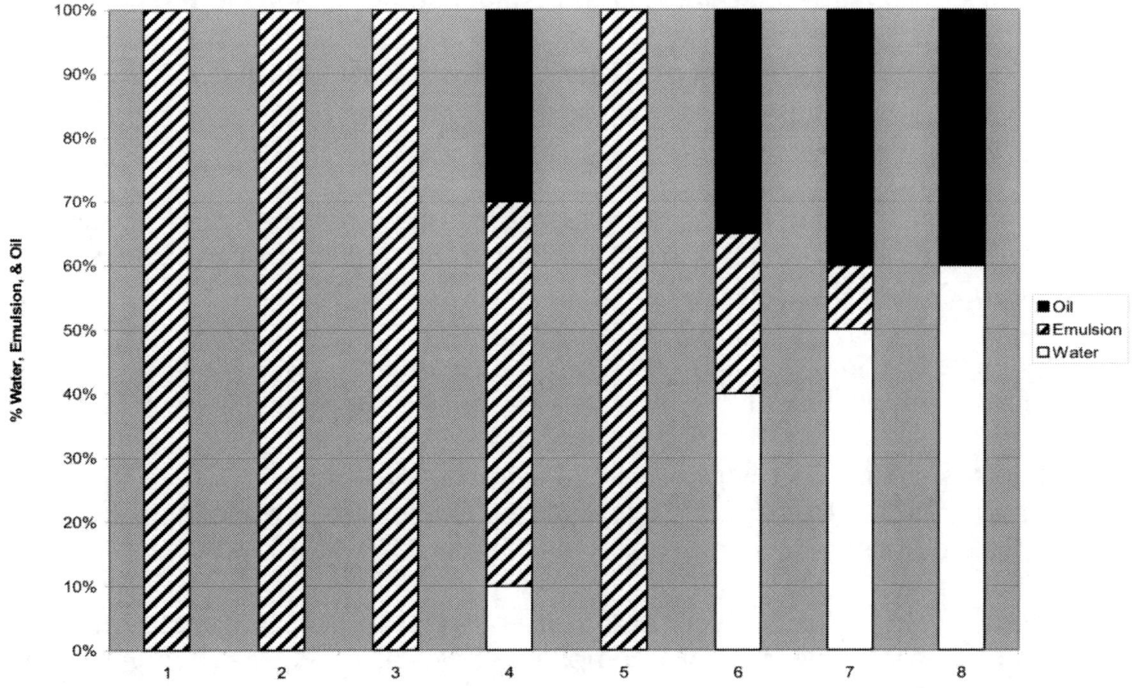

Figure 1 Comparison of heat treated and cold treated slop oil after 24 hours

A sample of the slop oil treated at 4000 ppm of Additive A for 24 hours at 66°C was analyzed by centrifugation. The recovered oil has only 0.4% water as compared with the room temperature treat that had 0.8% water remaining in the oil. The quality of the recovered oil is given in Table 5.

Table 5. Quality of recovered oil after 24 hours treatment with 4000 ppm Additive A at 66°C

	NO KNOCK OUT DROPS	**KNOCK OUT DROPS**
BS (unresolved emulsion)	**0.4%**	**0%**
Water	**0%**	**0%**
Solids	**0%**	**0%**
Oil	**99.6%**	**100%**

Summary

The laboratory example had results that are commonly obtained when treating slop oil chemically. These observations are summarized in the following bullet list.

- Chemical treatment generally produces much better oil water separation than the blank (no treatment).
- Different slop oil treatment additives often show greatly different performance for a given slop oil.
- Heating the slop oil greatly accelerates the rate at which oil and water are separated.
- Higher temperature often results in drier oil and clearer water.

ASIT is a registered trademark of Baker Hughes Incorporated

References:

1. Lawrence N. Kremer, "Using Emulsion Breaking Chemicals to Recycle Waste Oil" presented at 1st International Congress on Liquid Waste Recycling San Francisco CA, May 23-27 1994.

2. T.E. Schultz, "Improve oily wastewater treatment," *Hydrocarbon Processing* November 2006, pp 103-111.

3. Joseph L. Stark, Joe Nguyen and Lawrence N. Kremer, "Crude Stability as Related to Desalter Upsets," 5th International Conference on Refinery Processing AIChE Spring National Meeting 2002, New Orleans, LA March 11-14.

4. Joseph L. Stark, Joe Nguyen, and Lawrence N. Kremer, "New method prevents desalter upsets from blending of incompatible crudes," *Oil & Gas Journal,* March 18, 2002, pp 89-91.

A New Concept for Simultaneous Reduction of Alkalinity, Oil and Undesirable Gases Such as CO2 and H2S

James Chen and Dallas Douglas. Petreco Process Systems, Cameron, 3711 Shodow Cove Dr., Houston, TX 77082

In a SAGD application, produced water is normally recovered and treated to alevel suitable as feed to a "wet stream generator" The produced water typically contains oil, alkalinity,SiO_2 and hardness.

Zeolite softening or hot lime softening is used to reduce the hardness depending on the concentration of TDS and silica.Due to the high concentration of TDS and/or silica, lime softening is often utilized in a SAGD application.

However, it should be noted that for every part of alkalinity reduced, two parts of $CaCO_3$ will be produced as sludge in a lime process. Furthermore, more sludge will be produced if CO_2 exists in the produced water.

Acid may be added to the produced water upstream of the lime softening to convert the alkalinity to water and CO_2 which would be stripped in a mew WEMCO flotation machine. This process will greatly reduce the consumption of lime and production of lime sludge.

This paper will present a "Patent Pending" flotation machine which will achieve several functions: Oil Removal, Alkalinity Reduction and Stripping Gases such as CO_2 ans H_2S.

Stabilizing Hydraulic Fluids with Membrane Drying

Stuart Nemser, Sudipto Majumdar, Peter Dell, Shannon Williams, Dan Avery and Don Stookey, Research, Compact Membrane Systems, Inc., 325 Water Street, Wilmington, DE 19804.

The need to remove small amounts of water from oil is of significance in many stages of oil processing as well as other industrial processes.

This program introduces an in-line membrane/vacuum system to remove water from oil. The chemically inert membrane withstands harsh thermal and chemical BHF environment. Since membranes are in-line they remove water in real time. We demonstrated rapid 85% removal of water from oil and showed this water removal dramatically reduced deleterious acid formation. System showed excellent stability.

Figure 1 shows membranes ability to rapidly remove water from oil in real time. Figure 2 shows that by using membranes in-line that acid formation (e.g. TAN) from biodegradable esters can be kept to minimum.

Membrane technology associated with stabilizing biodegradable ester hydraulic fluid is a platform technology. The membrane technology should be of significant value in broader oil processing situations. Some of these other opportunities will also be discussed.

Figure 1 - Drierite Air Sweep (~1L/min), 50°C, 200 ml/min, vacuum, 8 L reservoir

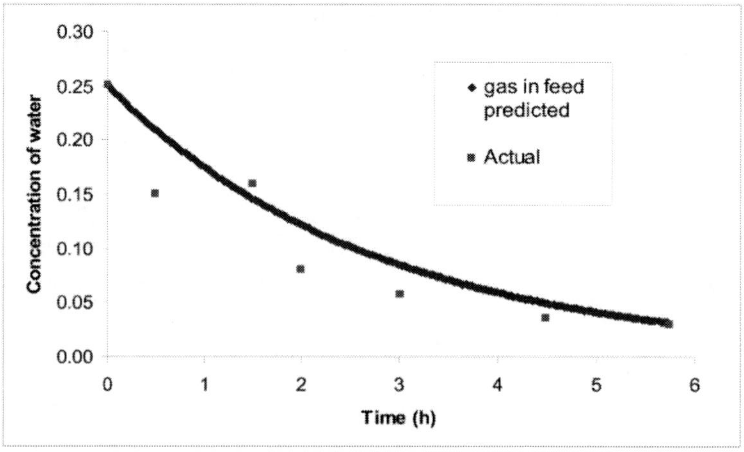

Figure 2
Using Dewatering Membrane
High Temperature Testing - TAN

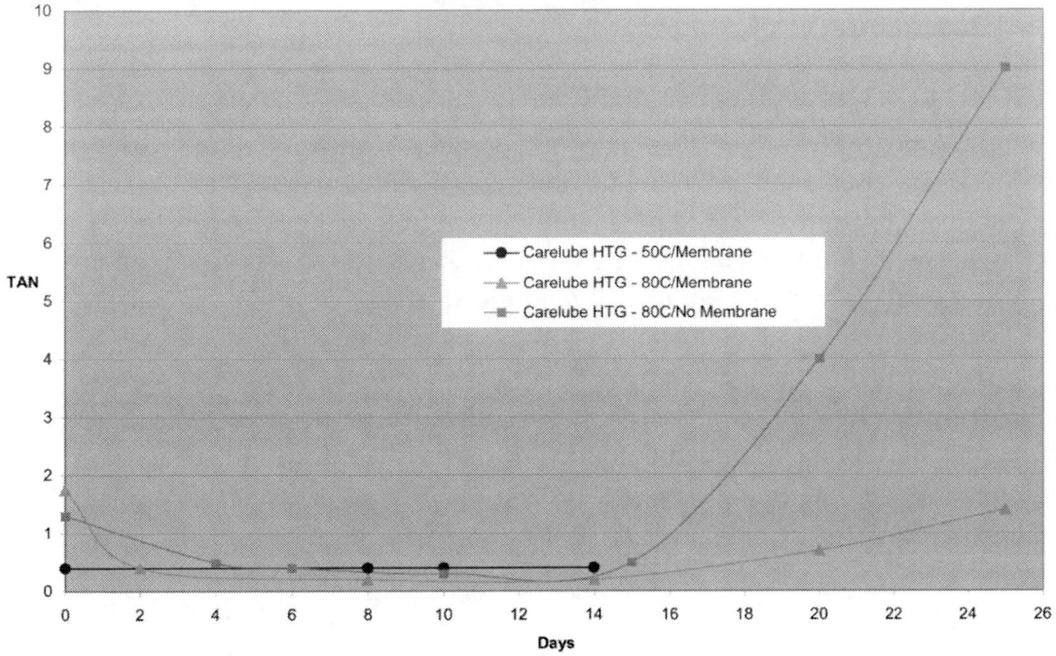

Solidification Mechanism and Solidifier Effectiveness for Fuels and Oils

Thomas R. Hanley[1], Byung-Hwan Um[1], and Dan Parker[2]. (1) Chemical Engineering, Auburn University, 149 Ross Hall, Auburn, AL 36849, (2) Immediate Response Spill Technologies, LLC, 11760 Commonwealth Drive, Louisville, KY 40299

Regulatory agencies have pre-approved a solidifier, CIAgent™, on the NCP products schedule as a chemical countermeasure for use on spills in the United States. Compared to the performance of skimmers and absorbent booms in recovering oil and other hydrocarbon spills from water, this solidifier accomplishes astounding results in record time, cutting both time and labor costs by 50 to 75 percent with a reduced exposure to wildlife and the environment. The solidifier has also been used to meet EPA requirements for mineral oil containment in electrical transformers.

This paper investigates the mechanism for solidification using temperature and volume changes during solidification to characterize solidifier effectiveness, resulting in improved solidifier performance at lower cost. Gasoline, diesel fuel, mineral oil, and crude oil are included in the analysis.

CASE STUDY ON THE IMPLEMENTATION OF ISO 14001 AT THE JORDAN OIL REFINERY CO.

N. P. CHEREMISINOFF, Ph.D.
Princeton Energy Resources International, LLC
1700 Rockville Pike, Suite 550, Rockville, MD, USA 20852
Email: ncheremisi@aol.com

A. HIJAZI, Ph.D.
U.S. Agency for International Development,
U.S. Embassy, P.O. Box 354, Amman 11118, Jordan
Email: ahijazi@usaid.gov

M. ABU-TARBUSH and M. HADDADIN
CDM International Inc.
No. 6 Ahmad Zaki St., Al-Shmaisani, Amman, Jordan
Email: abutarbushm@cdm.jo

Presentation at the Spring 2007 Session of the American Institute of Chemical Engineers

PROJECT BACKGROUND

- REUSE FOR INDUSTRY, AGRICULTURE AND LANDSCAPING (RIAL)
  - FUNDED BY THE U.S. AGENCY FOR INTERNATIONAL DEVELOPMENT
    - TECHNICAL ASSISTANCE PROGRAM TO GOV. OF JORDAN ON WATER CONSERVATION AND EFFICIENCY
  - PRIME CONTRACTOR – CDM INTERNATIONAL, INC.
  - SUBCONTRACTOR – PERI
- TASK 3 – INDUSTRIAL COMPONENT
  - INTRODUCE EMS THROUGH P2

APPROACH

- CREATED PARTNERSHIPS (PARTNER ENTERPRISES)
- WORKED WITH SENIOR MANAGEMENT
  - IN-COUNTRY CASE STUDIES ON P2/EMS
  - FOCUS ON THE RELATIONSHIP BETWEEN ENVIRONMENTAL PERFORMANCE AND THE BOTTOM LINE
  - RAISED SENSITIVITY TO THE CHANGING REGULATORY LANDSCAPE IN JORDAN
- CREATED A MOU
  - DEFINED SCOPE OF WORK
  - COMMITTED RESOURCES
  - WORK PLAN AND SCHEDULE
  - DEFINED THE RESOURCES AND COSTS

MAJOR STEPS

- INITIAL RETOOLING THROUGH TRAINING
- CREATED ENVIRONMENTAL POLICY STATEMENT
- FORMED EMS MANAGEMENT TEAM
- CONDUCTED AN IER/P2 AUDIT
  - CREATED AND REFINED TOOLS
- PERFORMED COST/BENEFIT ANALYSES
- ESTABLISHED ENVIRONMENTAL ACTION PLANS
- CREATED ENVIRONMENTAL MANAGEMENT PLAN
  - DEFINED METRICS
  - ESTABLISHED MONITORING AND REPORTING PROTOCOLS
  - INTRODUCED SIMPLE 'IT' TOOLS FOR IMPLEMENTING CORRECTIVE ACTIONS AND PERFORMANCE TRACKING

MAIN TASK

- INTEGRATED IER/P2 AUDIT
  - ENABLED EARLY PRIORITIZATION OF ENVIRONMENTAL ASPECTS (EAs)
  - LOW TO MODERATE COST P2 OPPORTUNITIES IDENTIFED
    - ACHIEVED GREATER LEVEL OF SUPPORT BY MANAGEMENT

AUDIT PRACTICES TRANSLATED INTO A COMMERCIALLY AVAILABLE TOOL AFTER THE PROJECT

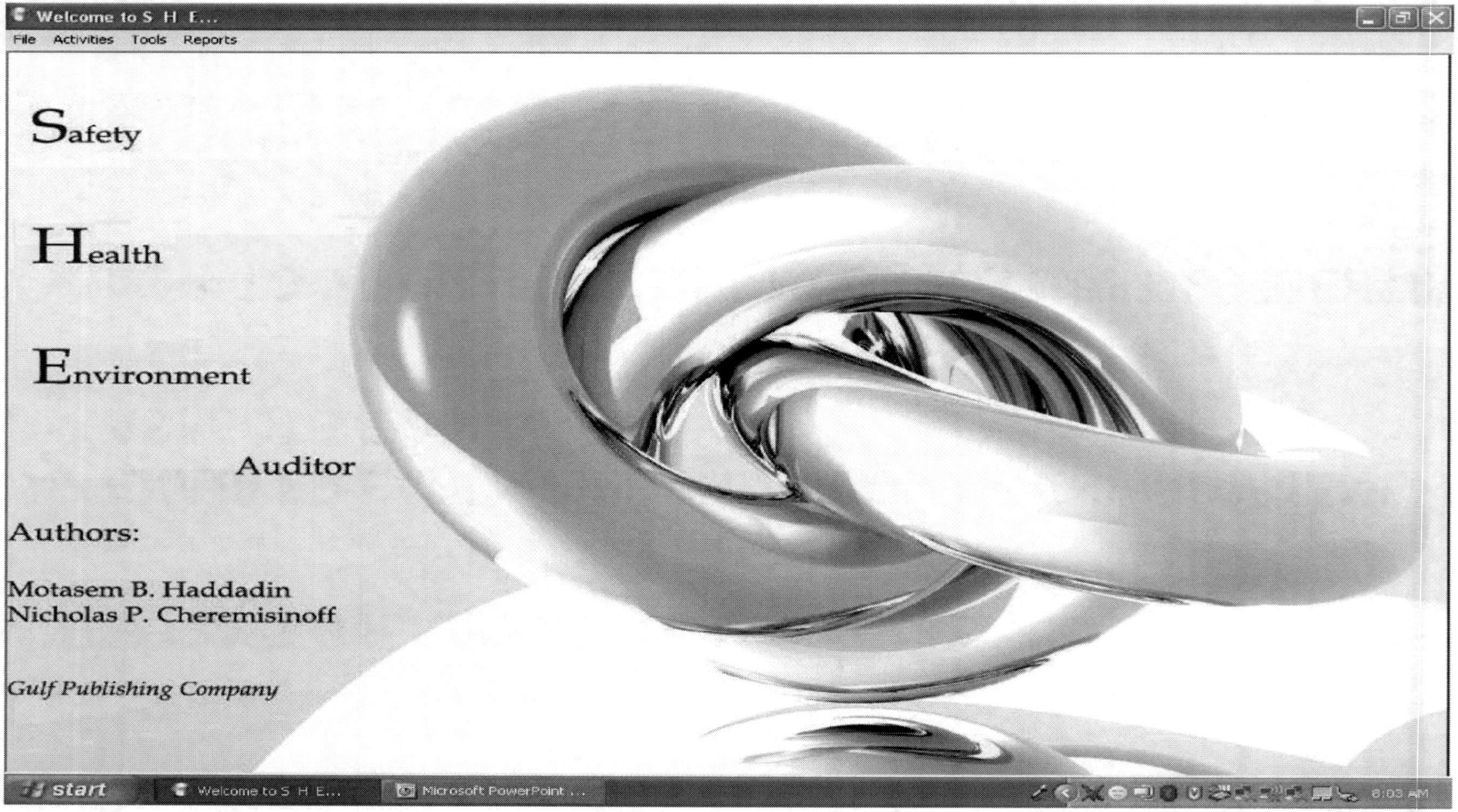

OUR OBJECTIVE WAS TO OBTAIN RESOURCES AND ESTABLISH PROCEDURES TO ENABLE THE EMS/P2 TEAM TO MULTI-TASK

Safety, Health and Environment (SHE) Auditor is a multi-task software tool designed to assist EH&S (Environmental, Health and Safety) personnel to both better manage their facility operations and to track performance of any part of a plant, an entire plant, or multiple facilities.

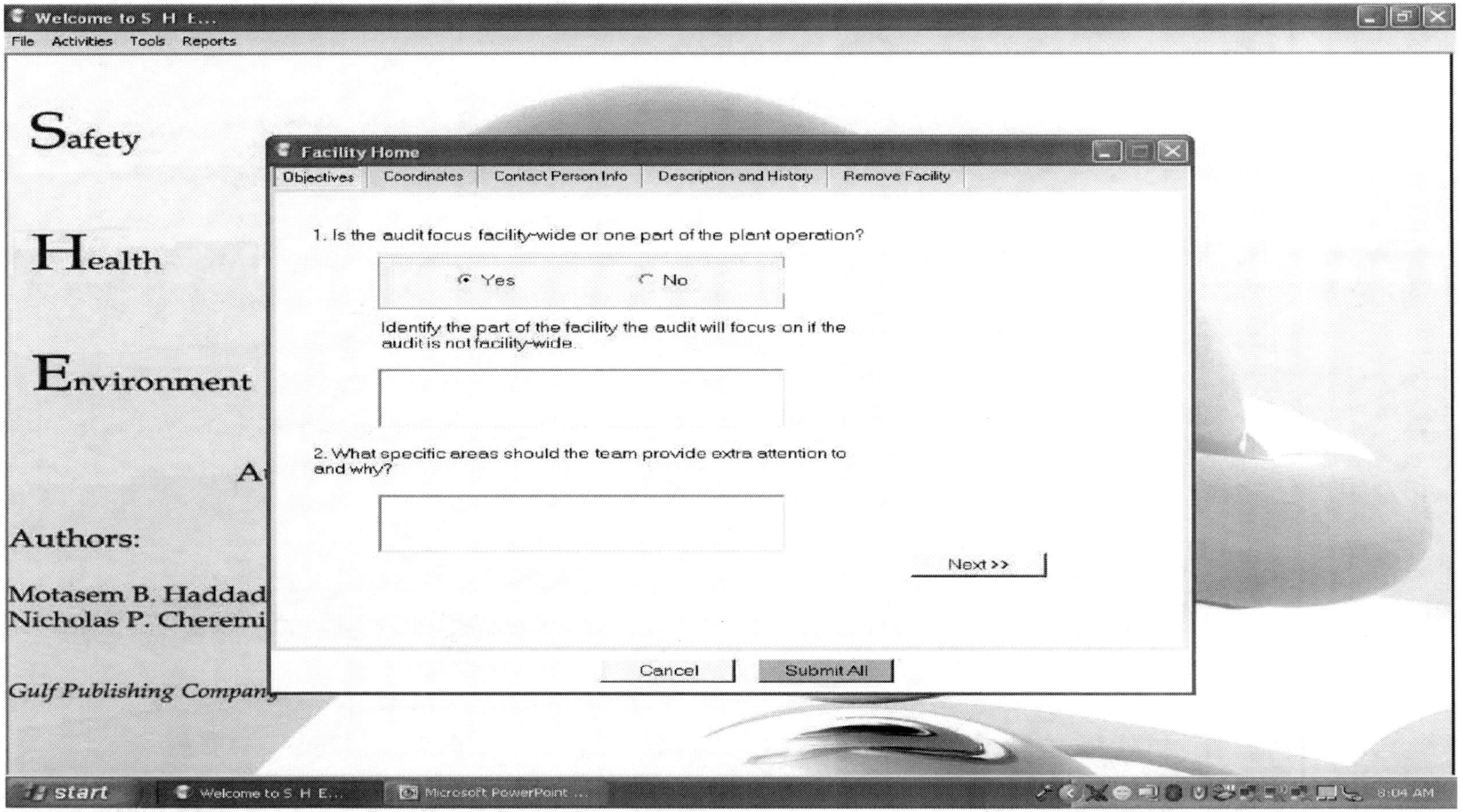

IER FOCUSED ON FIRST IDENTIFYING AND PRIORITIZING THE EAs, AND THEN GETTING THE STAFF TO TRACK THEM

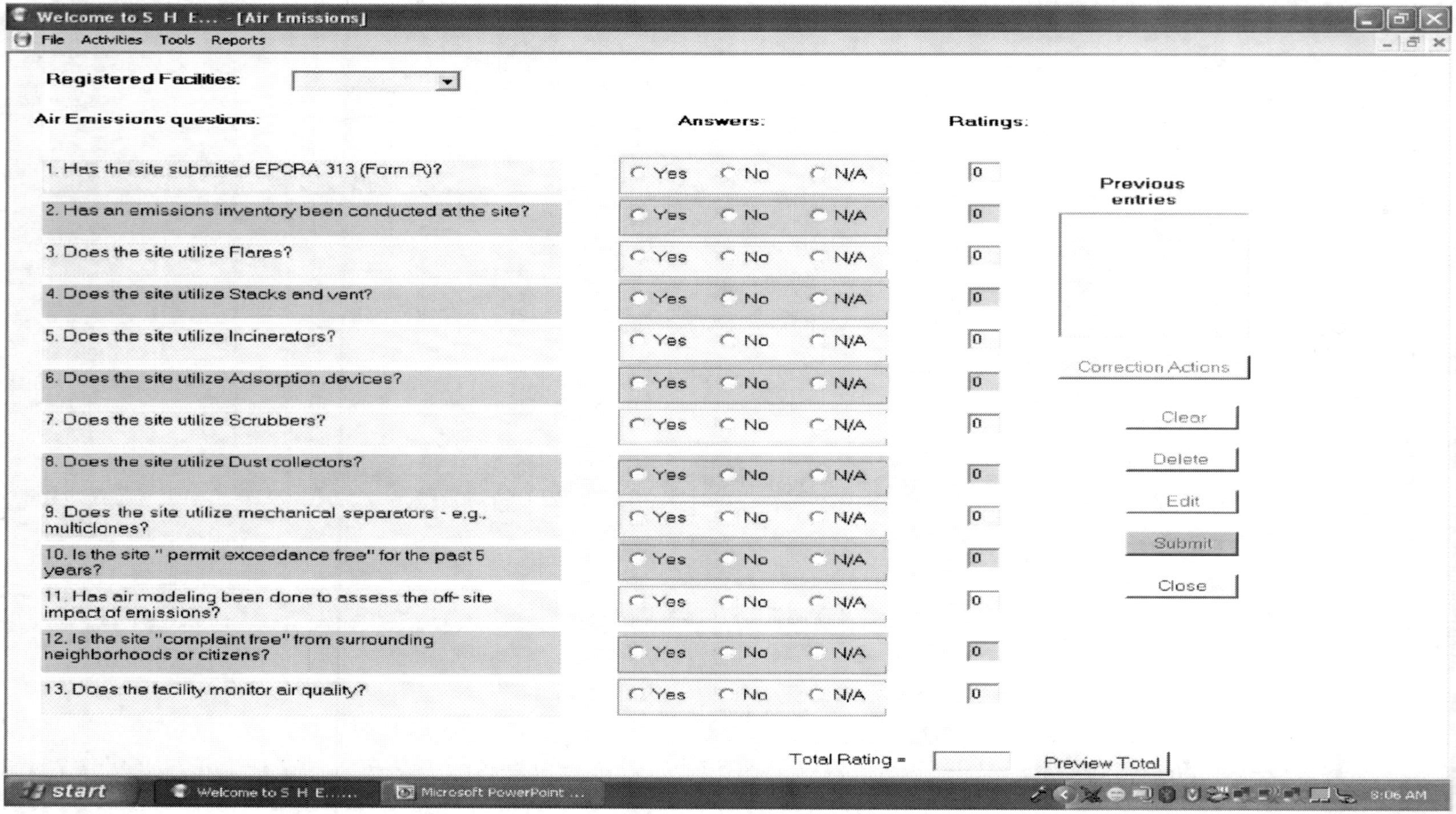

WASTE MANAGEMENT

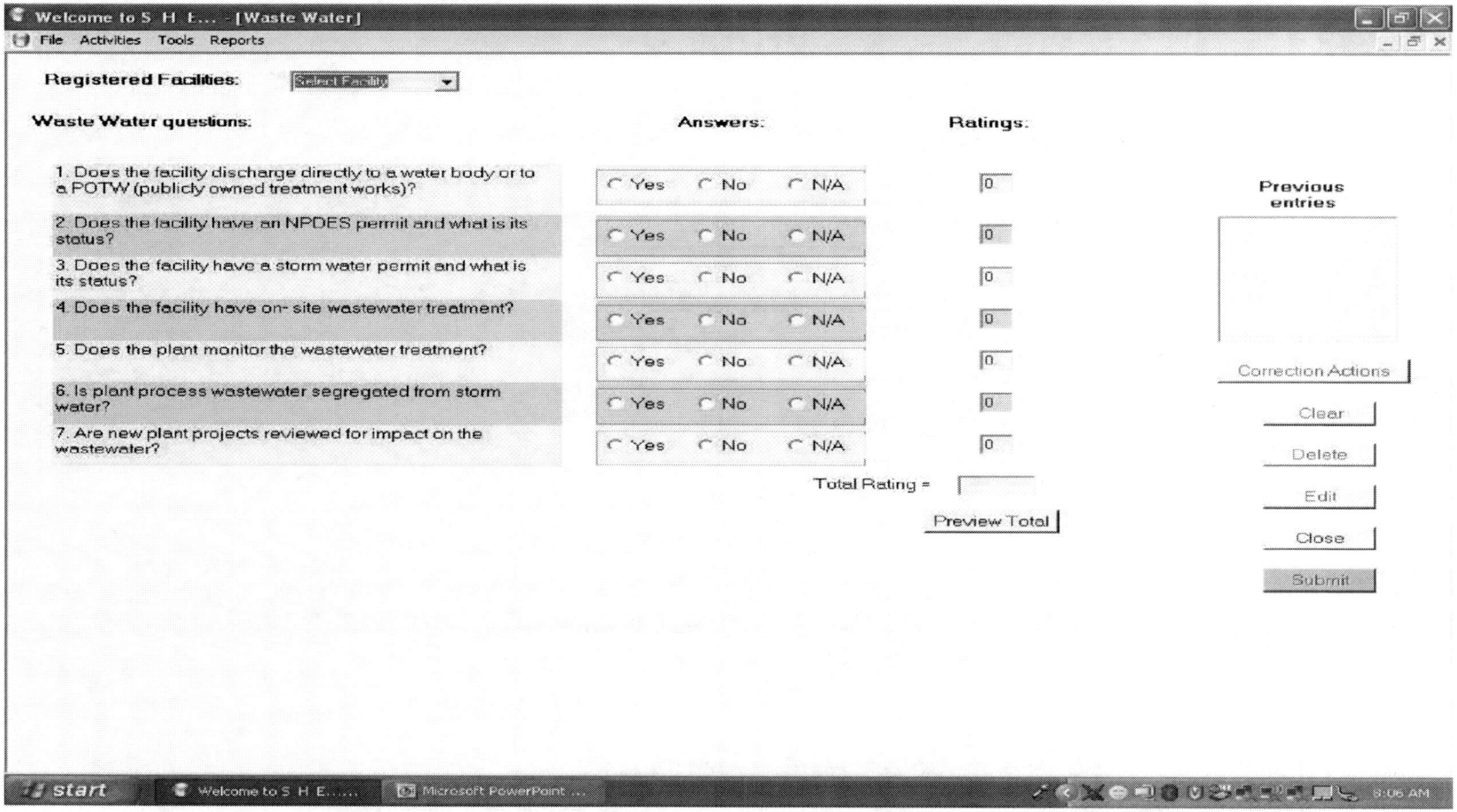

GROUNDWATER

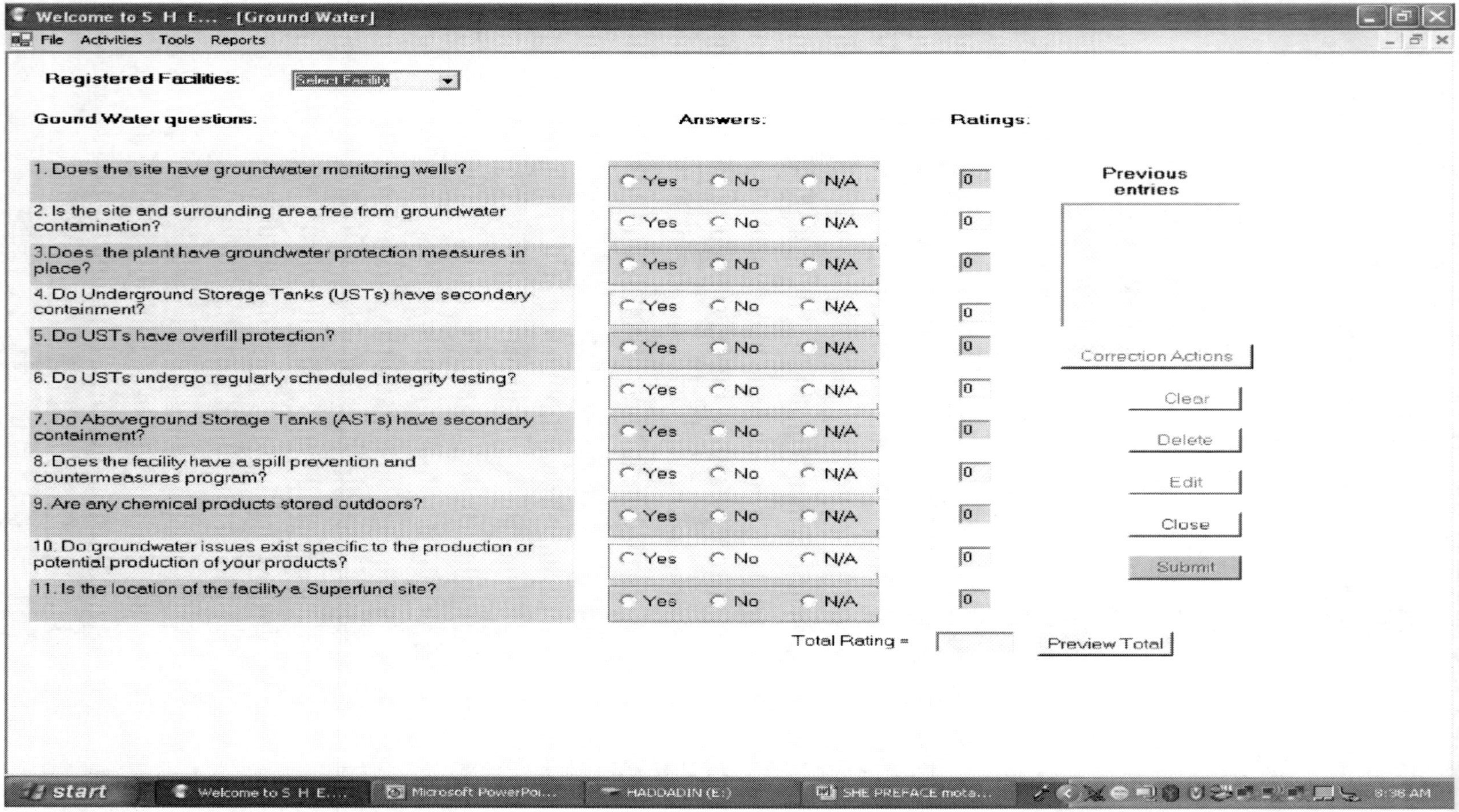

HAZARDOUS WASTES

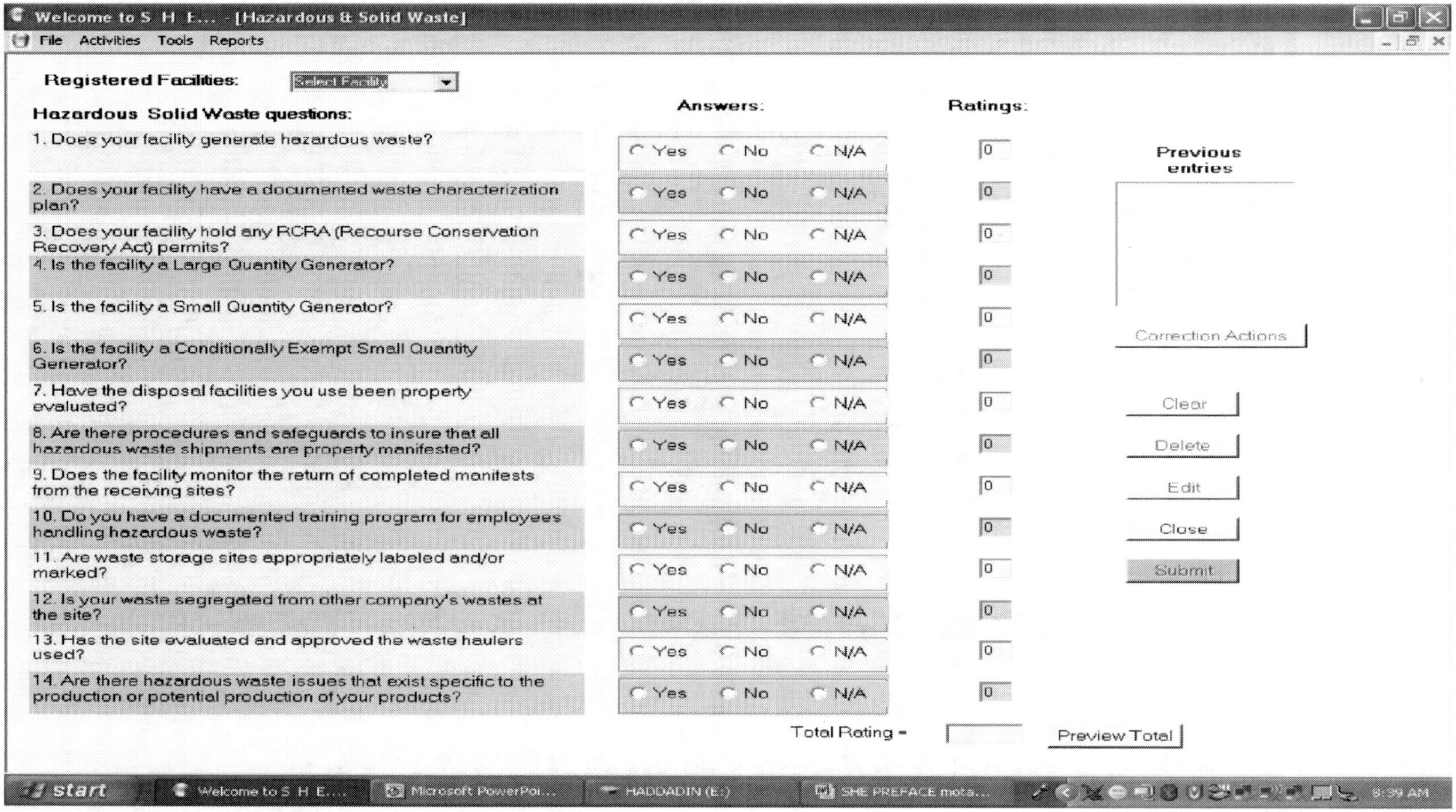

WE FOCUSED THEM ON SPILL PREVENTION

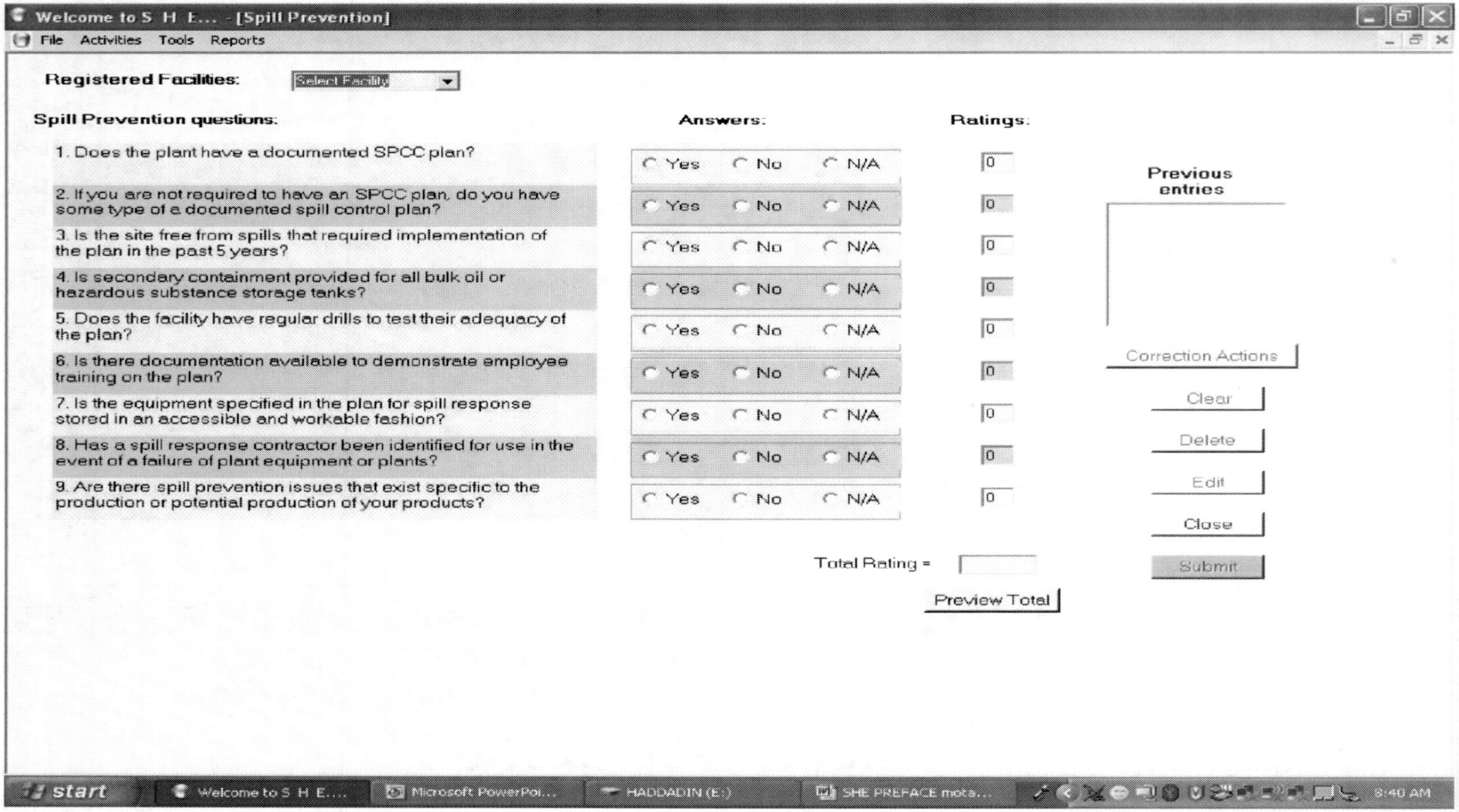

WE GOT THE TEAM LOOKING AT MANY EAs

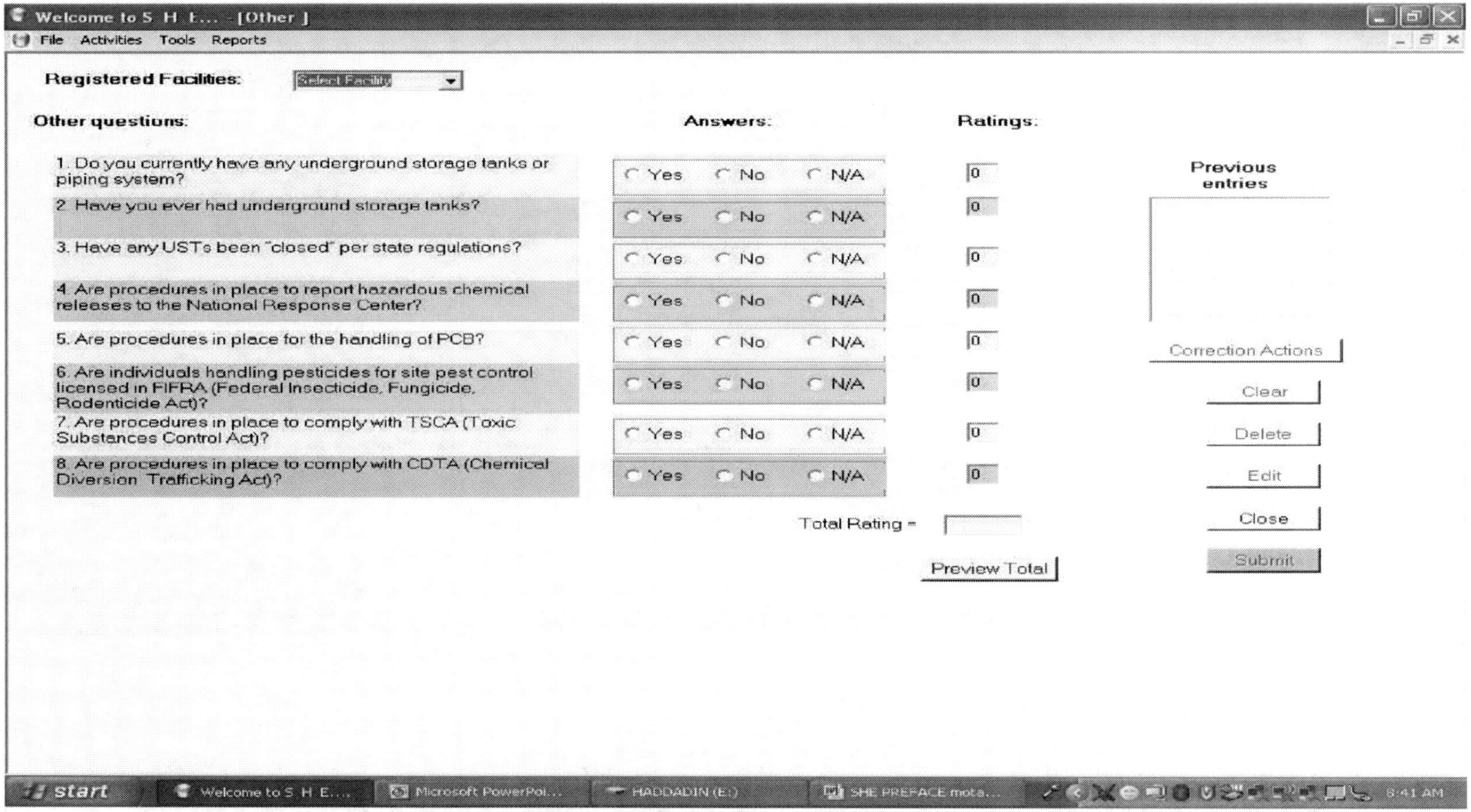

WE INTEGRATED HEALTH AND SAFETY WITH THE ENVIRONMENTAL COMPONENTS

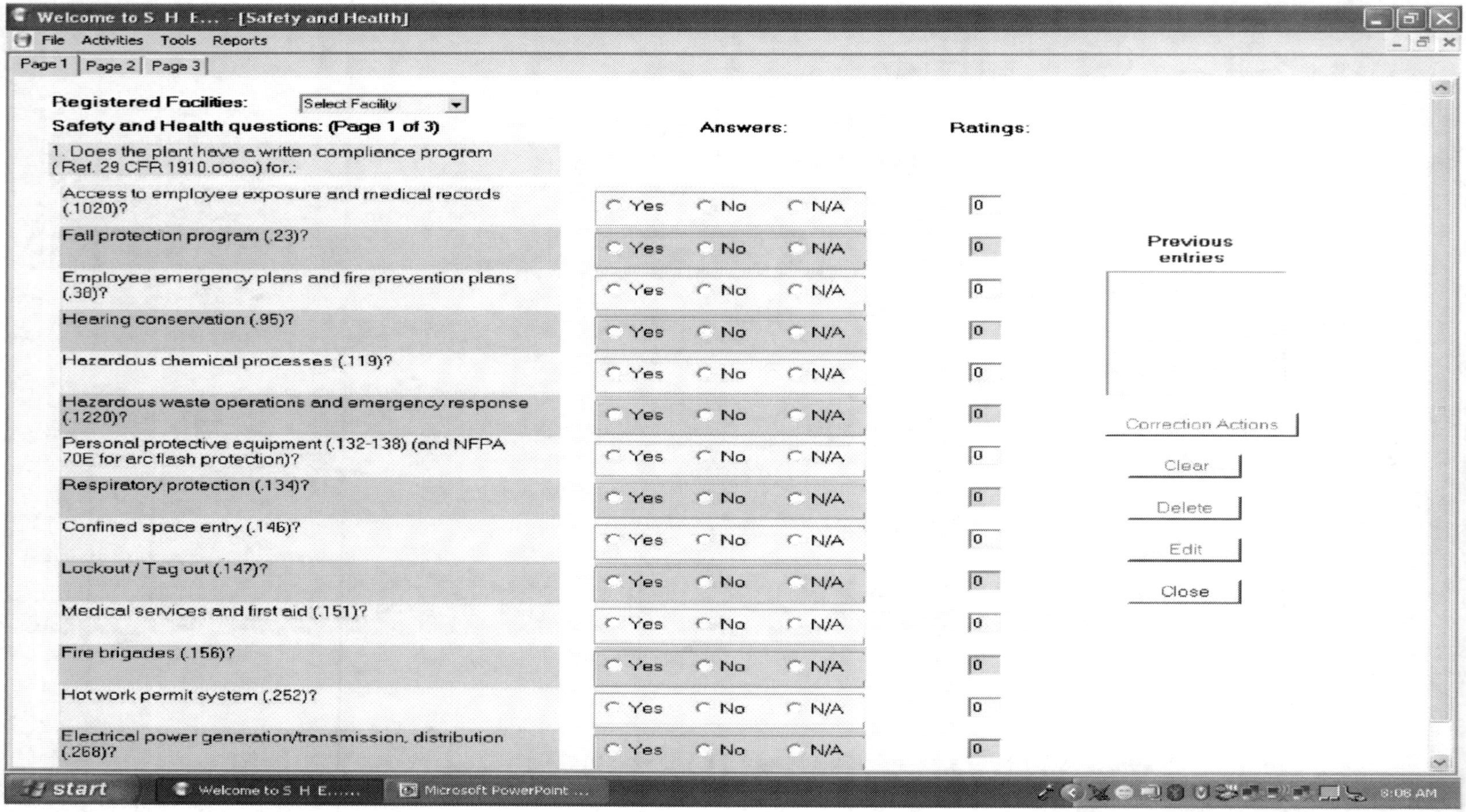

FIRE SAFETY

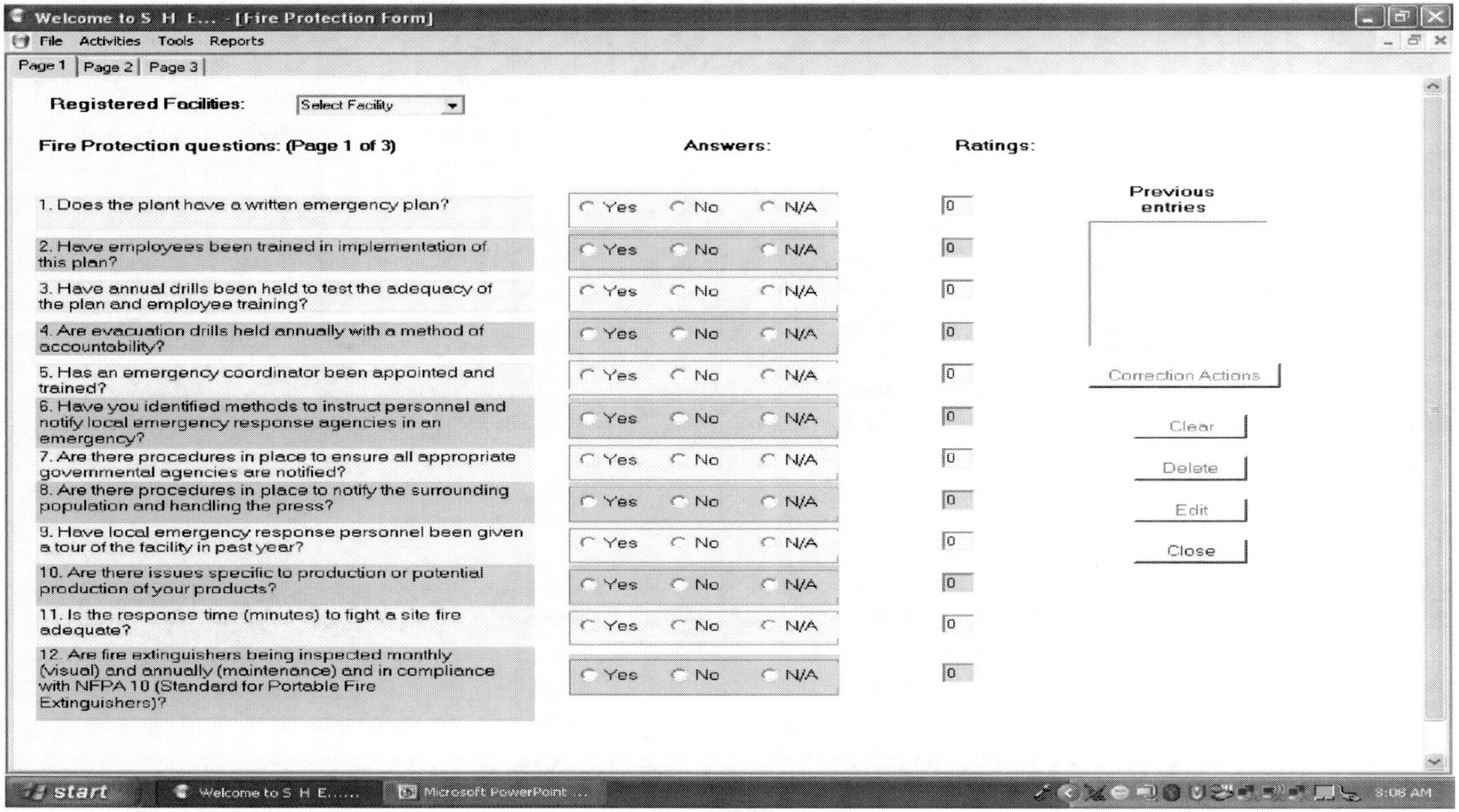

Welcome to S H E... - [Fire Protection Form]
File Activities Tools Reports
Page 1 Page 2 Page 3
Registered Facilities:
Select Facility
Fire Protection questions: (Page 1 of 3)
Answers:
Ratings:
1. Does the plant have a written emergency plan?
2. Have employees been trained in implementation of this plan?
3. Have annual drills been held to test the adequacy of the plan and employee training?
4. Are evacuation drills held annually with a method of accountability?
5. Has an emergency coordinator been appointed and trained?
6. Have you identified methods to instruct personnel and notify local emergency response agencies in an emergency?
7. Are there procedures in place to ensure all appropriate governmental agencies are notified?
8. Are there procedures in place to notify the surrounding population and handling the press?
9. Have local emergency response personnel been given a tour of the facility in past year?
10. Are there issues specific to production or potential production of your products?
11. Is the response time (minutes) to fight a site fire adequate?
12. Are fire extinguishers being inspected monthly (visual) and annually (maintenance) and in compliance with NFPA 10 (Standard for Portable Fire Extinguishers)?
Yes No N/A
0
Previous entries
Correction Actions
Clear
Delete
Edit
Close
start
Welcome to S H E......
Microsoft PowerPoint ...
8:06 AM

PROCESS SAFETY AND LINKS TO EAs

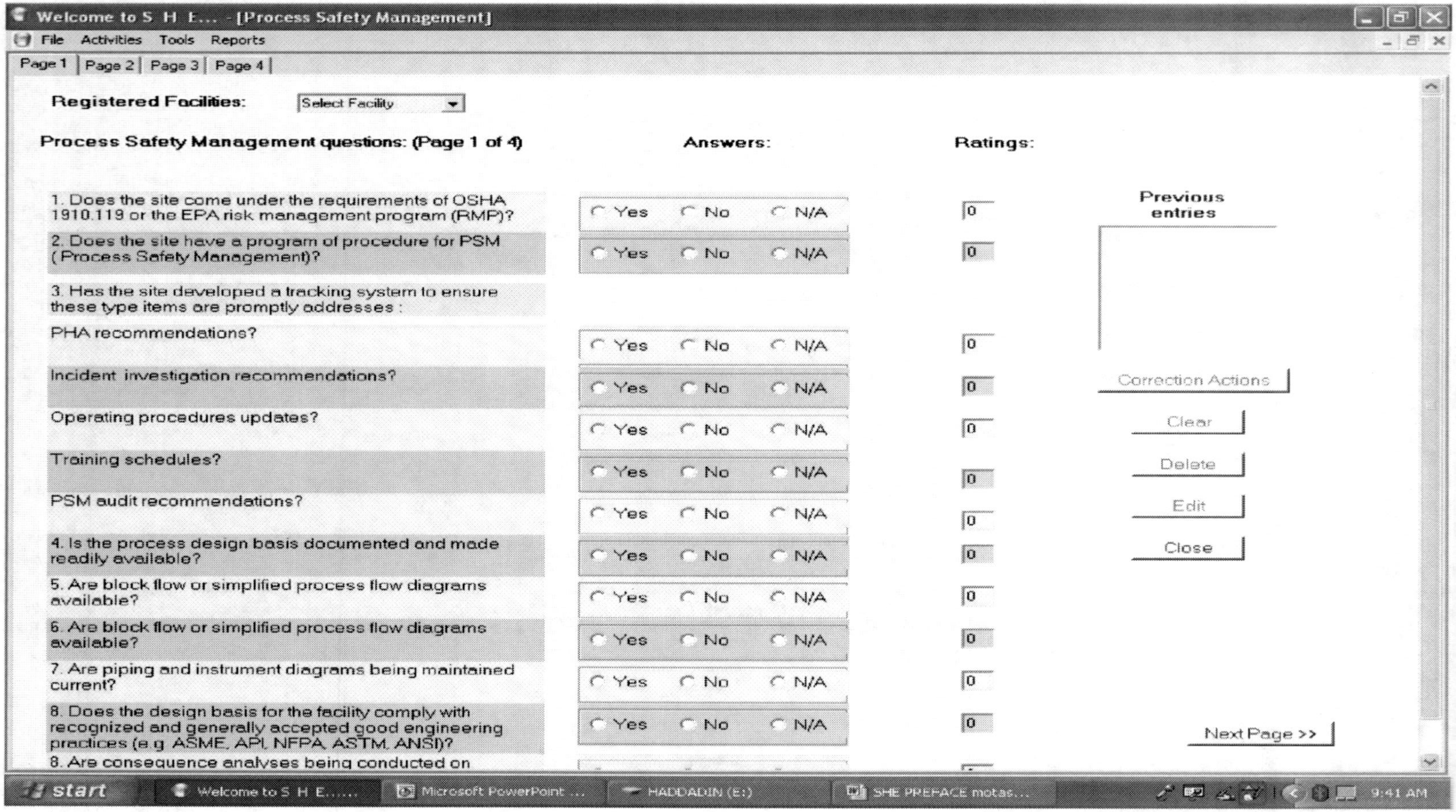

THE AUDIT FORCED THEM TO THINK ABOUT HOW TO MANAGE THE EAs

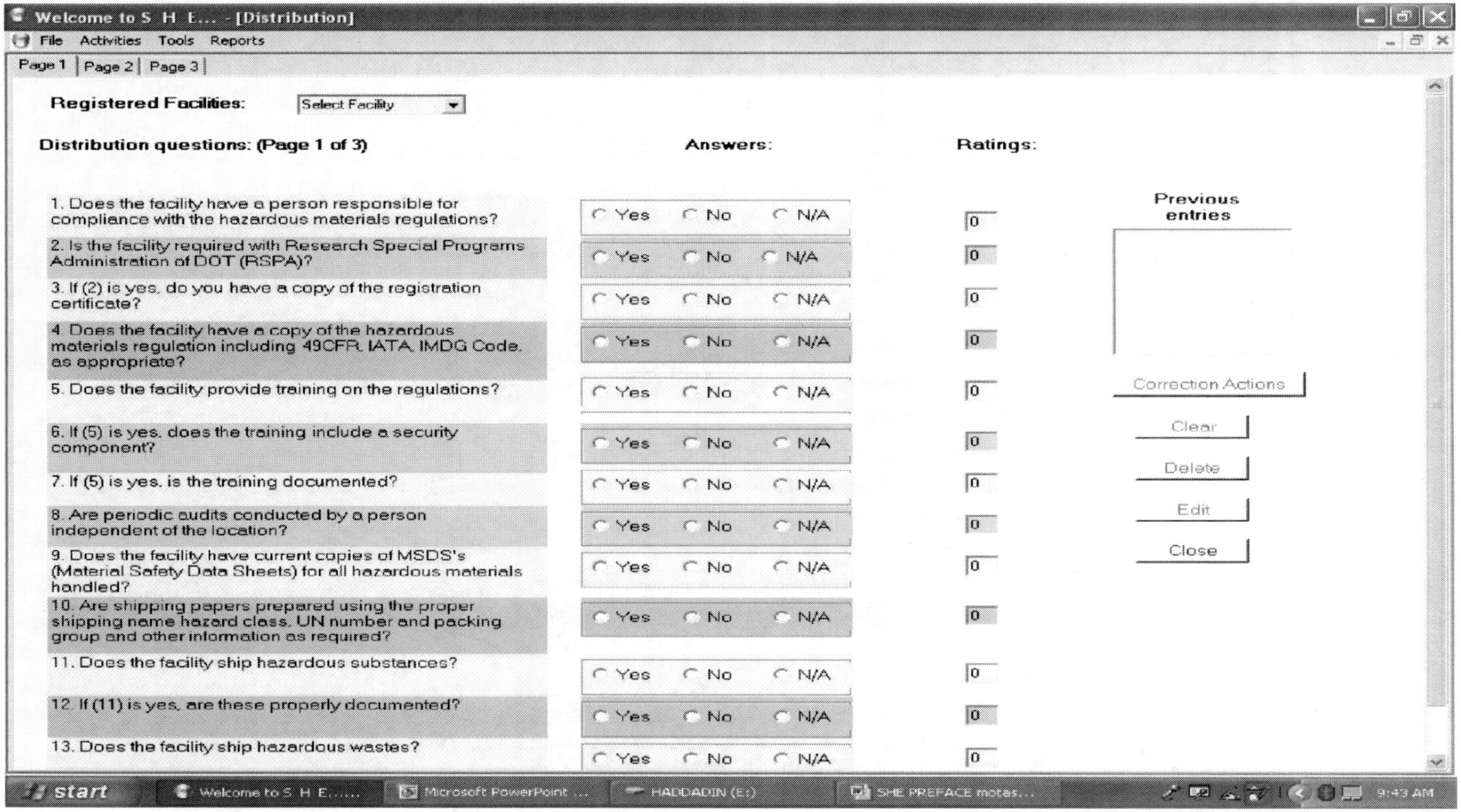

PRODUCT HANDLING OPERATIONS
INITIALLY THEY WERE NOT LOOKING AT THE EAs

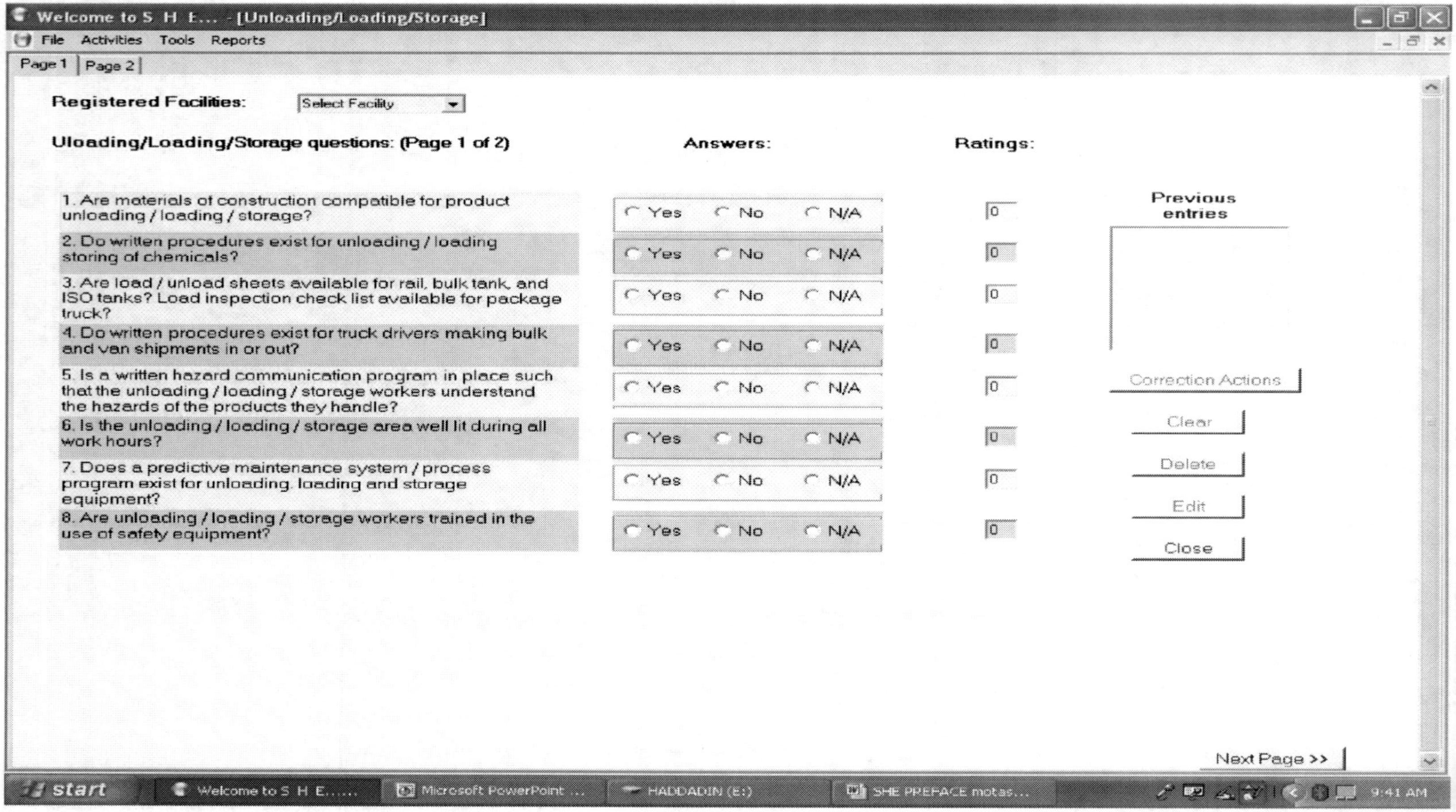

PRODUCT STEWARDSHIP
INITIALLY THEY WERE NOT LOOKING AT THE EAs

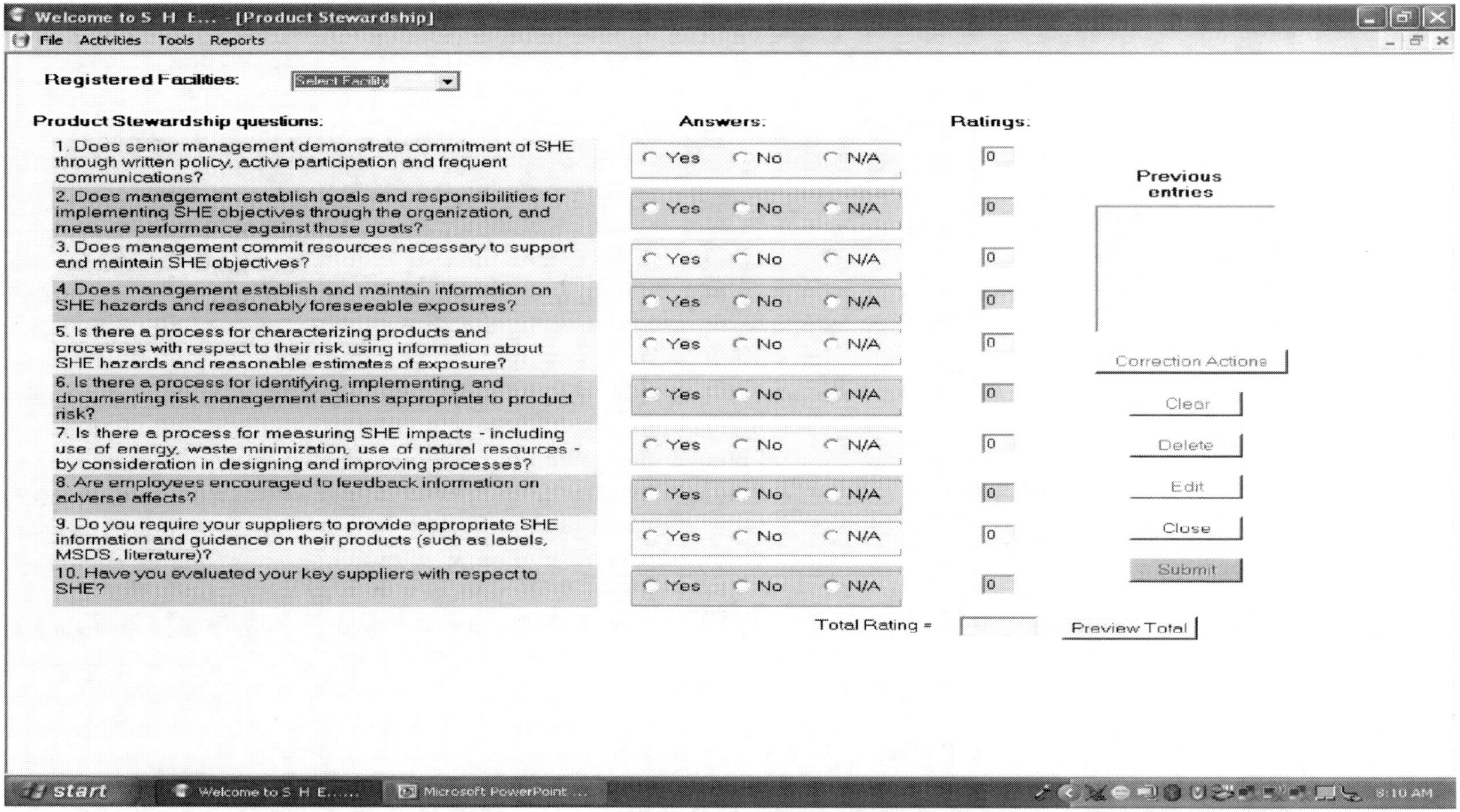

THE AUDIT FORCED THE FACILITY TO LOOK AT BETTER WAYS TO ORGANIZE AND MANAGE INFORMATION

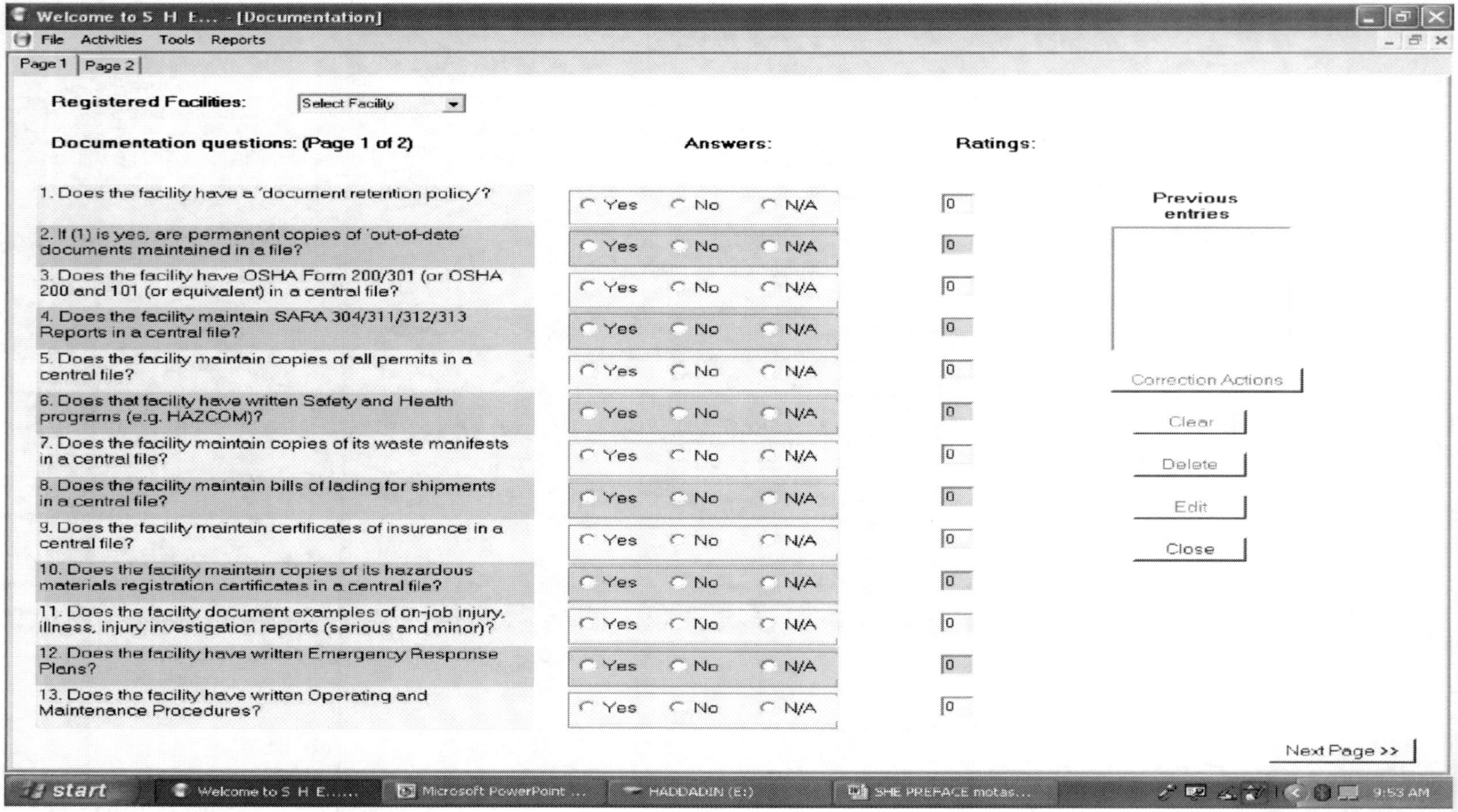

Activities

- 1. ***Environment***
  - **Air Emissions**
  - **Wastewater**
  - **Groundwater**
  - **Hazardous**
  - **Spill**
  - **Other**
- 2. ***Safety***
  - **Management Systems**
- 3. ***Safety and Health***
  - **General**
- ***Safety***
- 5. ***Process Safety Management***
- 6. ***Unloading/Loading/Storage (of Chemicals)***
- 7. ***Distribution (of Chemicals)***
- 8. ***Product Stewardship***
- 9. ***Documentation***

THE CASE STUDY FOCUSED ON PROVIDING BROAD EXPOSURE TO THESE IN ORDER FOR THE ENTERPRISE TO GAIN AN APPRECIATION OF THE VALUE OF AN EMS

THERE ARE NINE CATEGORICAL AREAS THAT THE *SHE AUDITOR* CAN PERFORM AUDITS IN, TRACK PERFORMANCE AND DEVELOP ACTION PLANS.

IN THE PILOT PROGRAM WE GOT THE TEAM TO ADOPT SIMPLE 'IT' TOOLS TO BEGIN SCHEDULING AND TRACKING CORRECTIVE ACTIONS

SHE DOES THIS WITH CALENDER AND EMAIL FUNCTIONS

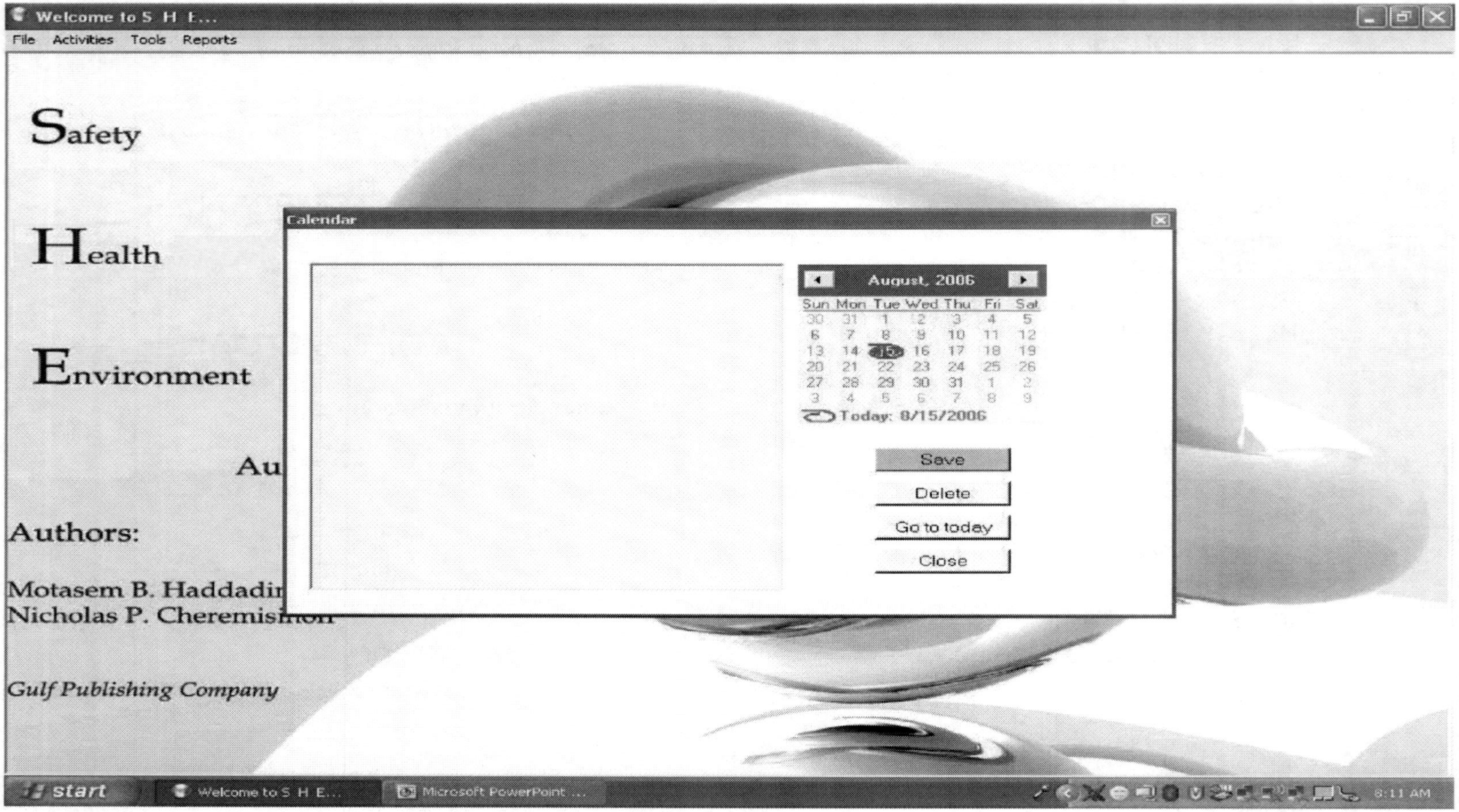

WE HELPED TO INTRODUCE A CULTURAL CHANGE THROUGH BETTER COMMUNICATION

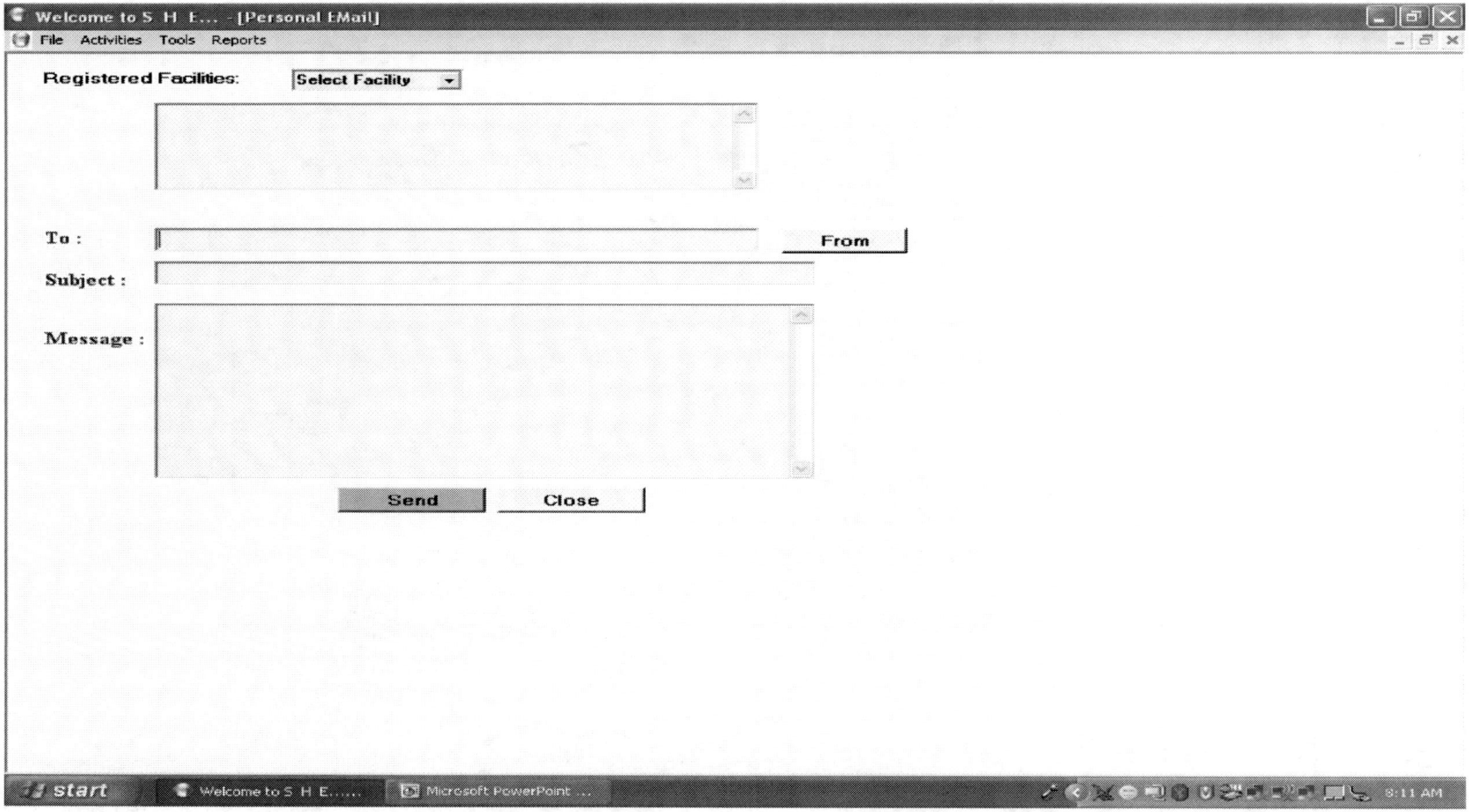

THROUGH SIMPLE METRICS (LATER ADAPTED TO 'SHE AUDITOR') WE GOT THEM TO START TRACKING ENVIRONMENTAL PERFORMANCE

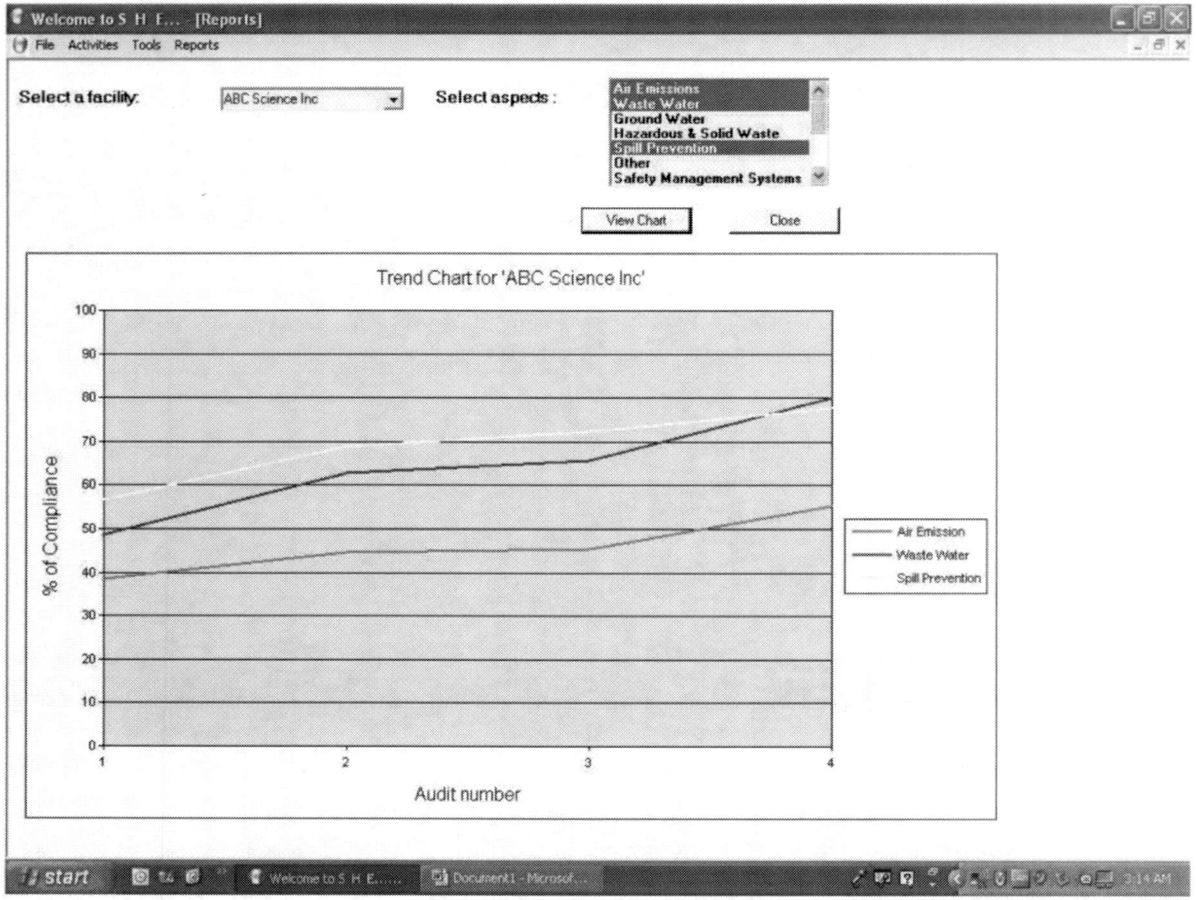

PROTOCOL APPLIED TO RANKING EAs

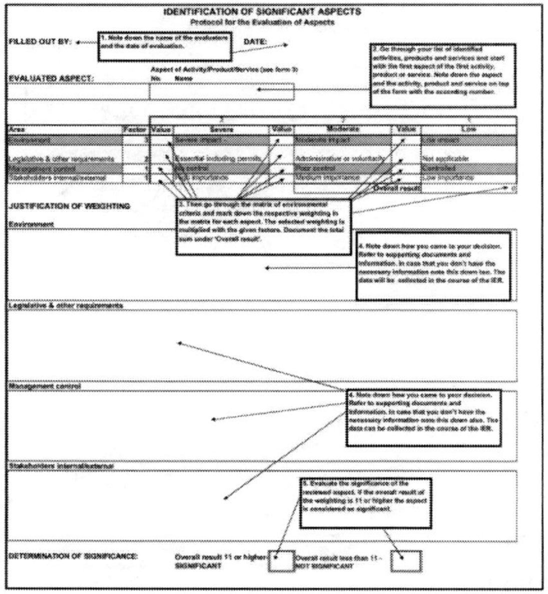

FILLED OUT BY: ← 1. Note down the name of the evaluators and the date of evaluation. → DATE:

EVALUATED ASPECT:

Aspect of Activity/Product/Service (see form 3)

No. Name

2. Go through your list of identified activities, products and services and start with the first aspect of the first activity, product or service. Note down the aspect and the activity, product and service on top of the form with the according number.

APPLIED A SIMPLE NUMERICAL RANKING SCHEME

			3		2		1
Area	**Factor**	**Value**	**Severe**	**Value**	**Moderate**	**Value**	**Low**
Environment	**3**		Severe impact -		Moderate impact		Low impact
Legislative & other requirements	**2**		Essential including permits		Administrative or voluntarily		Not applicable
Management control	**1**		No control		Poor control		Controlled
Stakeholders internal/external	**1**		High importance		Medium importance		Low importance
						Overall result	0

3. Then go through the matrix of environmental criteria and mark down the respective weighting in the matrix for each aspect. The selected weighting is multiplied with the given factors. Document the total sum under 'Overall result'.

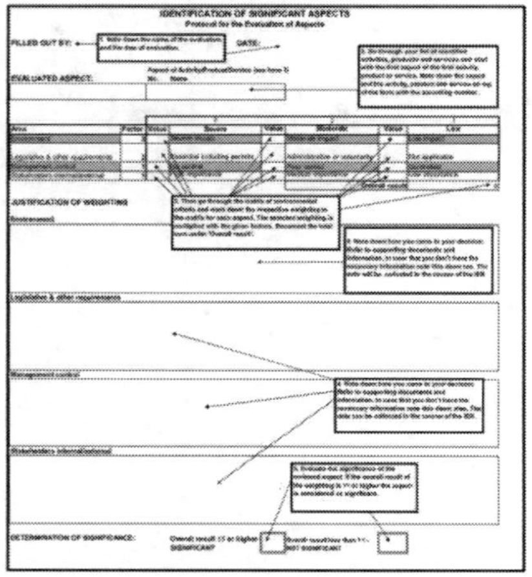

DOCUMENTED THE JUSTIFCATION FOR WEIGHTING FACTORS

JUSTIFICATION OF WEIGHTING

Environment

4. Note down how you came to your decision. Refer to supporting documents and information. In case that you don't have the necessary information note this down too. The data will be collected in the course of the IER.

Legislative & other requirements

Management control

4. Note down how you came to your decision. Refer to supporting documents and information. In case that you don't have the necessary information note this down also. The data can be
collected in the course of the IER.

Stakeholders internal/external

CALCULATED THE SCORES AND RANKED THE EAs

DETERMINATION OF SIGNIFICANCE:

Overall result 11 or higher- SIGNIFICANT

Overall result less than 11 - NOT SIGNIFICANT

5. Evaluate the significance of the reviewed aspect. If the overall result of the weighting is 11 or higher the aspect is considered as significant.

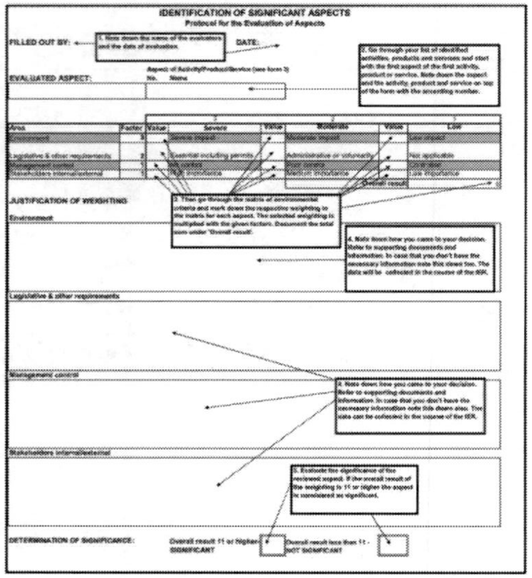

CASE STUDY

JORDAN OIL REFINERY HISTORICAL NOTES

- 1961 - REFINERY COMISSIONED
- 1970, 1973, 1982 - CONSECUTIVE EXPANSIONS AND DEVELOPMENT STAGES INCREASING DAILY PRODUCTION CAPACITY TO 8.700 THOUSAND TONS.
- 1998 – PRODUCTION CAPACITY REACHED 14 THOUSAND TONS.
- 2002 - BULK STORAGE CAPACITY INCREASED TO (1,554,605) MILLION TONS.
- 2005 - BULK STORAGE CAPACITY INCREASED TO (1,599,488) MILLION TONS

- **LOCATION: ZARQA, JORDAN**
- **PRODUCTS:**
  - LPG
  - THREE GRADES OF GASOLINE (REGULAR LEADED, SUPER LEADED, AND SUPER UNLEADED)
  - JET FUEL
  - KEROSENE
  - DIESEL
  - FUEL OIL
  - ASPHALT
  - LUBRICATING OILS
- **MARKET: 100% DOMESTIC**
- **ISO CERTIFIED: 9001 QUALITY ASSURANCE LUBES; APPLYING FOR ISO14001**
- **FEATURES:**
  - 3,493 FULL-TIME PERSONNEL
  - THREE CRUDE DISTILLATION UNITS
  - TWO VACUUM DISTILLATION UNITS
  - NAPHTHA HYDROTREATING UNIT
  - PLATFORMING UNIT
  - FLUID CATALYTIC CRACKING UNIT
  - HYDROCRACKING UNIT
  - LPG RECOVERY UNIT
  - HYDROGEN PLANT
  - TWO ASPHALT UNITS

P2 EXAMPLE 1
BOILER BLOWDOWN AUTOMATION

- MINIMIZING BLOWDOWN REDUCES
  - ENERGY LOSSES
  - MAKE-UP FEED WATER
  - CHEMICAL TREATMENT COSTS

- BLOWDOWN RATE ~ $6M^3$/HR.

- AUTOMATIC BLOWDOWN CONTROLS ELIMINATED OPERATOR ERROR
  - EXCESSIVE BLOWDOWN COST $241,000/YR
  - COST FOR SIX AUTOMATIC CONTROLLERS = $99,000
  - PAYBACK ~ 7 MONTHS

P2 EXAMPLE 2
CONTROLLING STEAM LOSSES

- STEAM LOSSES RESULTING FROM A NON-EXISTENT PMP FOR TRAPS

- STEAM LOSSES INCREASE:
  - WELL WATER CONSUMPTION
  - FUEL CONSUMPTION
  - WASTEWATER FROM RO AND ION EXCHANGERS
  - CHEMICAL USE IN R.O, ION EXCHANGERS AND BOILERS

STEAM LOSSES REPRESENT 105,120 M^3/YEAR OF WATER

RECOMMENDATION UNDERWAY

- IMPLEMENT A PMP
- INVEST IN A COMPUTERIZED STEAM TRAP MANAGEMENT SYSTEM
- COST = $19,700
- POTENTIAL SAVINGS = $632,000/YR

P2 EXAMPLE 3
REAL LOW HANGING FRUIT – REPAIRS TO CONDENSATE DRUM

- SERVICE WATER SENT DIRECTLY TO SEWER (1 M^3/HR)
  - 8,760 M^3/YEAR,
  - \$3,100/YR LOSSES

- REPAIRS TOOK 3 HOURS!
  - REPAIR STEAM TRAPS
  - REPLACE HIGH LEVEL ALARM

P2 EXAMPLE 4
RECYCLE RO FLUSH AND SAND FILTER BACKWASH

- RO FLUSHING AND SAND FILTER BACKWASH = 37,047M^3/YEAR
- WHY THROW > $13,000 PER YEAR DOWN THE SEWER?
- CAPTURE AND RECYCLE
  - CLEANING
  - FIRE WATER
- SOLUTION - *INSTALL A HOLDING TANK (120 M^3) AND TRANSFER LINES*
  - COST = $14,000

P2 EXAMPLE 5
RECYCLE ION EXCHANGE FLUSH

- REGENERATION PROCESS FOR THE EXCHANGER CONSUMES 44,125 M^3
  - (UNIT 1) 157 M^3 /CYCLE
  - (UNIT 2) 216 M^3 /CYCLE
- THERE IS 24,612 M^3 PER YR OF RINSE THROWN TO THE SEWER
- WHY THROW > $27,000 DOWN THE SEWER EVERY YEAR?
- RECOMMENDATION - REUSE THIS WATER (EXCLUDING THE FIRST 10 MINUTES) AS COOLING TOWER MAKE-UP WATER.
- PAYBACK ~ 15 DAYS

P2 EXAMPLE 6
CRUDE UNLOADING OPERATIONS

NICE LOW-COST P2 OPPORTUNITY

- RETURN OIL FROM OILY SEPARATOR TO CRUDE TANKS
- INSTALL CAPS TO STOP LOSSES FROM UNLOADING HOSES

- SAVINGS:
  - 10 TONS/DAY CRUDE LOSSES
  - 253$ /TON AND SOLD AS FUEL AT A COST OF APPROXIMATELY 112 $/TON
  - FINANCIAL LOSSES ARE 1,410 $/DAY OR 500,000 $/YEAR

- COST:
  - $42,000 (INCLUDING HOPPER TANK, TWO EXPLOSION-PROOF PUMPS, FITTINGS AND CAPS)
  - PAYBACK ~ ONE MONTH

EXAMPLES OF EAs AND CORRECTIVE ACTIONS TAKEN

LOADING OPERATIONS - TOP FILLING PROCESS RELEASE FUGITIVE EMISSIONS

RECOMMENDATION:

INVEST IN DOUBLE LOADING ARMS AND A VAPOR RECOVERY SYSTEM

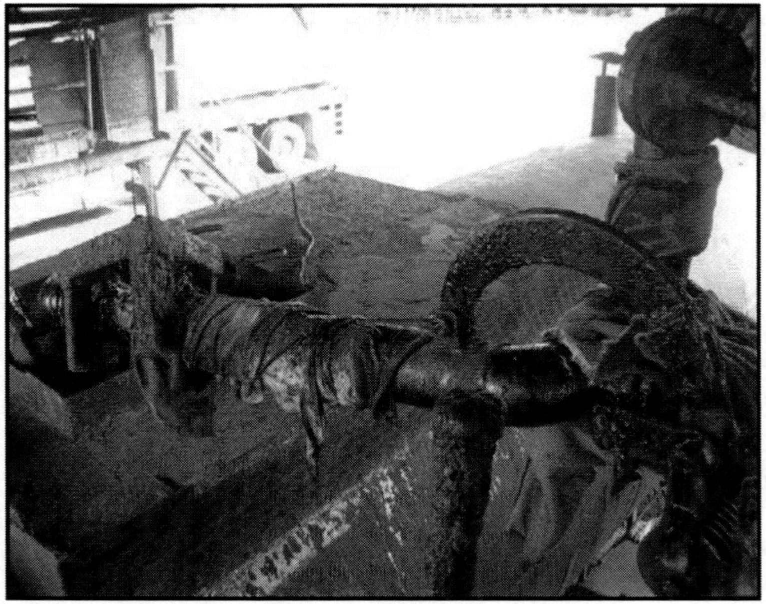

IMPLEMENT A SPILL PREVENTION, CONTINGENCY AND COUNTERMEASURES PROGRAM

EAs AND CORRECTIVE ACTIONS IN THE TRANSPORT DEPARTMENT

- **LOSSES AND ENVIRONMENTALLY UNSOUND PRACTICES**
  - 40 TONS/YR OF SPENT LUBE OIL FROM VEHICLE MAINTENANCE IS SENT TO THE REFINERY SEWER
  - OIL CONTAMINATED WATER FROM A SEPARATOR IS DISCHARGED OFF-SITE AND OIL IS SENT TO THE WWTP WITHOUT BEING RECLAIMED

UNLINED PIT CONTAINING SPENT LUBE OIL FROM VEHICLE MAINTENANCE OPERATIONS

OFF-SITE DISCHARGE OF CONTAMINATED WATER

REOMMENDED ACTIONS UNDERWAY

- RECYCLE THE SPENT LUBE OIL TO FINISHED PRODUCT FUEL TANKS
- SEGREGATE HAZARDOUS AND NON-HAZ WASTE AND BEGIN TO MANAGE
- STOP ILLEGAL DISCHARGES - SEND THE OIL TO SLOP TANKS, AND THE WASTEWATER TO THE WWTP

NEGATIVE ENVIRONMENTAL ASPECTS (EAs) IN THE OPERATIONS DEPARTMENT

- SPILLS FROM TANKER UNLOADING RESULT IN SITE CONTAMINATION AND DRAINAGE TO SEWERS OVERLOADING WWTP
- EXCESS SLUDGE (SLOP WAX) AT BOTTOM OF THE TANK NOT MANAGED
- PROCESS UNIT STEAM, CONDENSATE, WATER, CHEMICALS, CRUDE AND PRODUCTS SPILLS AND LOSSES
- SLUDGE PONDS FOR DEPOSITING WASTE GENERATED FROM ALL SECTIONS POORLY MANAGED
- SPENT SODA DRYING POND LEFT UNMANAGED
- POOR HOUSEKEEPING – RESULTING IN EXCESSIVE SITE CONTAMINATION AND UNSAFE WORK ENVIRONMENT

NEW HOUSEKEEPING RULES AND SPILL PREVENTION ARE PHASING THESE PROBLEMS OUT

THE RIAL PROJECT FOCUSED MANAGEMENT'S ATTENTION TO THE LONG-RANGE LIABILITIES

TANK BOTTOM (SLUDGE) COLLECTED IN BARRELS

A SIGNIFICANT LONG-TERM LIABILITY ISSUE THEY ARE NOW FOCUSING ON – ACCUMULATED MIXED WASTE FROM DECADES OF POOR ENVIRONMENTAL MANAGEMENT PRACTICES

THIS FACILITY NOW FACES PRIVATIZATION – THEY HAVE A MAJOR FINANCIAL INCENTIVE TO FOCUS ON LONG-TERM ENVIRONMENTAL DAMAGES

RIAL HELPED THE FACILITY TO FOCUS ON SAFETY AND TO COUPLE ENVIRONMENTAL AND SAFETY DIVISIONS UNDER A SINGLE MANAGEMENT PROGRAM

THE FACILITY NOW RECOGNIZES PROBLEMS WITH CHEMICAL AND WASTE HANDLING

- LACK OF WORKER SAFETY
- IMPROPER STORAGE AND HANDLING PRACTICES
- WASTE ACCUMULATION RECORDS NEEDED
- NO DRUM MANAGEMENT PROGRAM
- CHEMICAL SPILLS
- CHEMICAL INVENTORY CONTROLS NEEDED

ANOTHER SIMPLE CA THAT IS IN THE EAP - LUBE BARRELS STORED UNSAFELY AND PRONE TO CONTAMINATING STORM WATER RUNOFF

SOME CAs UNDERWAY AT TIME WE LEFT

- IMPLEMENT SHE AUDITS COMBINED WITH DATA MANAGEMENT SYSTEM FOR SCHEDULING AND FOLLOWING UP ON CAs
- DIKE AREA AND COLLECT SPILLAGE DURING BARREL UNLOADING AND RECYCLE PRODUCT
- AUTOMATIC TANK CLEANING AND OIL RECOVERY
- DEVELOP WRITTEN SPILL PREVENTION AND COUNTERMEASURES PROGRAM AND TRAIN EMPLOYEES
- ESTABLISH A WASTE MINIMIZATION PROGRAM WITH APPROPRIATE METRICS FOR MONITORING ENVIRONMENTAL PERFORMANCE
- ESTABLISH AN ENVIRONMENTAL ACTION PLAN AND AN ENVIRONMENTAL MANAGEMENT PLAN
- DEVISE AND IMPLEMENT A DRUM INSPECTION PROGRAM
- LABEL AND INVENTORY ALL WASTE DRUMS AND ESTABLISH A POLICY FOR REGULAR REMOVAL AND DISPOSAL. ESTABLISH GUIDELINES FOR RECYCLING AND MONITOR
- SURVEY TECHNOLOGIES FOR SLUDGE AND CONTAMINATED SOIL REMEDIATION

OTHER CORRECTIVE ACTIONS

- PROVIDE 40-HOUR HAZARD MATERIALS HANDLING TRAINING OR EQUIVALENT
- PROVIDE RIGHT-TO-KNOW TRAINING
- PREPARE WRITTEN HEALTH AND SAFETY PLAN (HASP) WITH RECOMMENDED CPC AND CHEMICAL HANDLING PROCEDURES
- CRITICALLY REVIEW AND REVISE ENERGENCY RESPONSE PROGRAM AND POLICIES
- ESTABLISH CHEMICAL INVENTORY AND RESTRICTED USE POLICY
- DEVELOP DRUM MANAGEMENT PROGRAM
- STORE CHEMICALS IN SECURED AREAS THAT ARE DIKED
- IMPLEMENT A STORMWATER MANAGEMENT PROGRAM
- IMPLEMENT REGULAR SHE (SAFETY, HEALTH & ENVIRONMENT) AUDITS
- ADOPT A MICROSOFT ACCESS DATA MANAGEMENT TOOL TO SCHEDULE AND TRACK CORRECTIVE ACTIONS
- INVENTORY, MANIFEST AND ARRANGE FOR DISPOSAL OF WASTES ON A REGULAR BASIS. COMBINE ACTIVITY WITH SHE AND WASTE MIN. PROGRAM

P2 PRACTICES THEY STARTED FOCUSING ON

MINIMIZING SOLIDS LOSSES FROM HEAT EXCHANGER CLEANING

When exchangers become fouled solids require removal. Steps to minimize solids washing to sewers include:

- INSTALLATION OF CONCRETE OVERFLOW WEIRS AROUND EXCHANGER PADS AND DRAINS TO RETAIN SOLIDS FROM TUBE BUNDLE CLEANING
- INSTALL TEMPORARY COVERS OVER SEWER DRAINS DURING CLEANING OPERATIONS
- RESTRICT CLEANING TO DESIGNATED AREAS THAT ARE DIKED OR CONTAINED FOR SOLIDS COLLECTION

LOW COST P2 PRACTICES IMPLEMENTED

MINIMIZING SOLIDS FROM VARIOUS SOURCES ENTERING SEWER

- USE OF STREET SWEEPERS ON PAVED AREAS TO REMOVE TRASH AND DEBRIS BEFORE IT CAN BE WASHED INTO SEWERS
- PAVING AND PLANTING GROUND COVER IN UNPAVED AREAS NEAR SEWERS
- INCREASED INSPECTION AND REPAIRS OF SEWER LINE BREAKS AND LINING SEWERS
- PERIODIC CLEANING OF SOLIDS FROM DITCHES AND CATCH BASINS AND VACUUMING OF SOLIDS
- USE OF CYCLONE SEPARATORS UPSTREAM OF API SEPARATPOR TO REDUCE FINES CONTRIBUTING TO SLUDGE FORMATION
- COLLECT (SWEEP AND SHOVEL) FCCU CATALYST SPILLS
- USE OF FILTERS AT SEWER DRAINS IN COKING UNITS TO KEEP COKE FINES OUT OF OILY WATER SEWER
- USE OF HYDROCLONES TO RECOVER FINES THAT ESCAPE INTO THE SEWER

LOW COST P2 PRACTICES IMPLEMENTED

MINIMIZING LEAKS, SPILLS AND OTHER LOSSES TO SEWER

- PERIODIC INSPECTION AND REPAIR OF ABOVE AND UNDERGROUND PIPING
- LEAK DETECTION AND REPAIR (LDAR) PROGRAMS FOR PUMP SEALS, VALVES, OTHER
- INSTALLATION OF TANK OVERFILL PREVENTION SYSTEMS
- INSTALLATION OF PAVEMENT IN PLACE OF BARE GROUND OR OTHER SURFACES UNDER MAJOR PIPE RACKS TO FACILITATE LEAK DETECTION
- USE OF DETECTORS TO REDUCE OIL DRAINAGE DURING TANK DRAWS AND THE USE OF AUTOMATIC WATER DRAWS ON PRODUCT AND CRUDE OIL TANKS

P2 PRACTICES THEY BEGAN FOCUSING ON

MINIMIZATION OF SAMPLE LOSSES TO SEWER SYSTEM

- **APPLY CLOSED-LOOP SAMPLING SYSTEMS SO THAT SAMPLE STREAMS RETURN TO THE PROCESS AND NOT SENT TO THE SEWER**

- **RECYCLE LAB SAMPLES OF CRUDE, REFINED AND INTERMEDIATE PRODUCT STREAMS TO OIL RECOVERY SYSTEMS AFTER LABS HAVE FINISHED ANALYSIS**

SUMMARY
RIAL SERVED AS THE CATALYST

- IN A FEW MONTHS WE FOUND
  - $1.5 MILLION DOLLARS IN SAVINGS
  - SAVED 5.8 MILLION GALLONS OF WATER
  - IDENTIFIED AND CAPTURED 226,000 TONS IN MATERIALS SAVINGS

- HELPED TO CREATE THE FIRST ENVIRONMENTAL POLICY STATEMENT

- TRANSFERRED SKILLS
  - IER/P2 AUDITS
  - SIMPLE 'IT' TOOLS TO IMPROVE DATA MANAGEMENT AND COMMUNICATIONS

- ASSISTED IN CONVINCING MANAGEMENT TO
  - FORM A DEDICATED EH&S DEPARTMENT
  - ALLOCATE PERMANENT FINANCIAL AND PEOPLE-POWER RESOURCES
  - IDENTIFY AND PRIORITIZE EAs
  - LOOK AT P2 AND NOT JUST POLLUTION CONTROL
  - BEGIN FORWARD PLANNING TO COPE WITH NEW ENVIRONMENTAL REGULATIONS
  - TRACK PERFORMANCE
  - FORMAL COMMITEMENT TO ISO-14001
    - LLOYDS OF LONDON ENGAGED

Global Warming: Who to Believe?
by Gerald T. Westbrook, TSBV Consultants

Witnesses for the Defense of the Skeptical Perspective on the Global Warming Issue.

The objective of this essay is to highlight the skeptical perspective on the global warming (GW) issue. Over the past decade the attacks on the skeptic's position have increased in frequency, shrillness and ugliness. The attacks tend to focus on the *messengers* and not on the substance. Three categories of witnesses will be cited:
(1) selected authors, specifically non scientists, but nonetheless authors with some unique perspective on this issue and who have spoken out on this issue;
(2) *distinguished veterans (DVs)* are individuals, mostly scientists, mostly retired, with incredible credentials and accomplishments to their credit; and
(3) others including brief comments on active scientists, plus the views and status of Broadcast Meteorologists and State Climatologists.

1. The Skeptics on the Global Warming Issue: Selected Authors

Four authors, and their relevant work, will be noted in this essay. These are:

- Michael Crichton, book[1]: *State of Fear (SOF)*, December 7, 2004;
- George Will, column[2]: *Global Warming? Hot Air*, December 23, 2004;
- myself, book[3]: *'Acid Rains' on Liberal Propaganda ('AR')*, December 9, 2004 and
- Alex Kozinski, essay[4]: *Gore Wars*, August 8, 2002.

Recently the Intergovernmental Panel on Climate Change (IPCC) issued an Executive Summary for Policy Makers. In spite of this release, there are still skeptics on this issue, and the debate continues. Indeed, there are many skeptics on this issue, but they get very little press coverage, and even less respect. Michael Crichton has been one exception. Because of his reputation as an entertaining and provocative writer he has received the press coverage.

His latest book, *SOF*, is a James Bond type of science fiction novel. The *warmers* scream that this is a ridiculous book, with even more ridiculous characters, and hence is not worth reading. George Will, in contrast, called this book a political broadside woven into an entertaining story.

What has infuriated the *warmers* is not really his plot, nor his characters, but the fact he repeatedly brings in current inputs on the global warming issue into his story. These take the form of *discussions* between his characters. He adds tables and graphs to this dialogue. He uses extensive footnotes, and postscripts that include *The Author's Message*—where Crichton states his convictions, conclusions and positions—and *Appendix I, on why Politicalized Science is Dangerous*.

George Will noted two major themes in *SOF*.

- Theme 1: "Crichton's subject is today's fear that global warming will cause catastrophic climate change (CCC)." This book is about CCC, but it takes some reading before one realizes that these CCC events will be *manufactured* by the radical environmental groups described in this novel. The events—ice shelf calving, flash floods, and tsunami generation—are all scheduled to occur at or near the time that a major conference on CCC is underway. In this way these groups expect to generate huge publicity, a multiplicity of new members and mega-bucks in donations.

The *warmers* talk so much about CCC, but recent examples are very hard to find. For example, one might point to the *Dust Bowl* (~1930 to ~ 1940) as a CCC. But if this were a CCC why are we not still in it? Another example could be the variation in the level of the Great Lakes. My in-laws lived on Lake Huron. My wife and I lived on this lake too. And we heard stories of unprecedented lake rise. But this turns out to be about a 15 year cycle, with about a five foot rise. Not much of a CCC.

I suspect that when *warmers* talk about a CCC, much of the time they are talking about an Extreme Weather Event (EWE). Even Crichton's CCCs are truly EWEs. Over the past years, as I write about GW, I started to build a list—those EWEs that I have experienced directly, or virtually due to the proximity to it—which is now over 100 EWEs. Some examples follow.
1900 — Galveston hurricane (HC), casualties > 6000.
1913 — Lake Huron storm took 248 lives and 71 vessels. This was the worst storm on these lakes.
1938 — LI/RI HC killed > 600. If this had moved up the Hudson casualties would have been >> 600.
1944 — Severe European winter: Battle of the Bulge.
1953 — MI tornados, killed 121, moved on to NE, killed 94. Both areas not deep in the tornado belt.
1963 — MI blizzard, April 17. Slept on church pew. Lake effect snow left seven foot drifts.
1975 — Lake Superior storm and the Edmund Fitzgerald. Largest ship to go down.
1986 — MI flood, Tittabawasee River. Flow increased from 500 to 35,000 cubic feet per second.
1997 — QC ice-storm. Hundreds of Aluminum based towers taken down.
2005 — Katrina HC, and levee collapse, casualties > 1,400.
Note that these EWEs are essentially random in date, type and site.

My vision of a CCC—and I am unaware of an official definition of a CCC—would be a change that started slowly, but lasted for a century or so. See Barbara Tuchman's description of the start of the Little Ice Age (LIA), at the start of the fourteenth century[5] .This change lasted ~ 450 years. What caused the LIA? Three ideas are noted in Chapter 10 of my book[3] namely:
(1) Changes in atmospheric circulation in Europe and North America;
(2) Changes in oceanic circulation, specifically in the thermo-haline circulation (THC). A key task[6] is to gain a better understanding of the LIA and to confirm/refute that it was due to changes in the THC.
(3) Changes in solar output, over cycles of ~ 150 to 300 years, as hinted at in the sun-spot record.

Another possible CCC would be the shut down of the Gulf Stream. Now the THC is driven by temperature and salinity differences. In contrast the Gulf Stream is a wind driven system influenced by the Earth's rotation and the lunar tides. Albert Gore Jr. has reported[6] that this THC "pump" plays a crucial role in powering this circulation, but greenhouse gas warming could disrupt it. However Dr. Richard Lindzen—The Sloan Professor of Meteorology at MIT, elected to the National Academy of Sciences in 1977, and perhaps the leading academician in this field—has noted [8] that the Gulf Stream is mostly driven by the wind. "To shut it down you'd have to stop the rotation of the Earth or shut off the wind." Lindzen went on: "And all I'm saying is that the THC is not the major driver of the Gulf Stream - - - ." Lindzen added; "we are simply saying that if you wish to issue scare remarks, you should make them accurate, according to the science."

Albert Gore Jr. has also reported[7] that 10,000 years ago the Gulf Stream [the THC] virtually stopped due to an enormous pulse of fresh water rushing out of the St. Lawrence into the Atlantic. He implies this was an unprecedented event, but there were many other ice dams and many melt water pulses down the Columbia, Mississippi, Athabaska, Nelson and the Hudson over the thousands of years de-glaciation occurred. He noted that [some] scientists are worried this could happen again. Lindzens reaction: "My read on it is that there is a certain climate of fear, to quote Mike Crichton." He went on:

we were "talking about fresh water perhaps shutting down the Gulf Stream. "But that isn't what physical oceanographers think."

The *warmers* will reject such inputs and argue how an analogous fresh water pulse could occur today. They feel that a meltdown of the Greenland ice-sheet could send a pulse of fresh water into the North Atlantic. In a recent essay[9a] I quote three references that indicate the odds of such an event are incredibly low. Indeed, the last reference, that was based on a super hyped build up of greenhouse gases, predicted—and these predictions are listed here assuming the model used was based on solid science, which was faithfully replicated in the model—a 5 % loss in this ice-sheet in about 200 years, and a complete loss in 3,000 years. This will give a few years to check out these assumptions.

To close out this theme, the more one reads about CCC, the more one is struck with the repetitious use of such words as "could" or "might" or "it is possible."

- Theme 2: "Crichton's subject is also how **conventional wisdom (CW) is manufactured**, in a credulous and media-drenched society." In '*AR*' a similar point is made namely that a *war* is going on: *warmers* versus skeptics; and as a result we live in a **tidal wave of never ending propaganda**. Today the GW issue has become conventional wisdom — it is happening, it is caused by society, it will be terrible, but we, the *warmers*, know what needs to be done to stop it.

A major contribution of Crichton's book is that it publicizes the fact that a debate on this issue is underway. One will seldom learn about such a debate from the *mass media*. Instead, the *mass media* presses ahead *manufacturing CW* about GW. In *SOF* , Crichton comes out on the skeptics side of this issue, both in the novel part and in his incorporation of much GW science in his book. As a result Crichton has been attacked vigorously by some *warmers*. However, Crichton's effort will show his readers that a debate exists, and may well change the nature of this debate.

Crichton notes that the PLM (the Politico–Legal–Media) complex, has a vested interest in keeping society in a continuous *State of Fear*, hence the title for his book. The PLM has to resort to endless fear–mongering in order to keep the populace in a constant state of fear. And climate change is a perfect subject to scare the devil out of the average citizen.

I should note that 'AR' has nothing to do with the acid rain issue. Rather this name stems from the acid satire used to describe and deflate the *Ultra Liberals, Far Lefters and Global Warmers*. In turn '*AR*' argues that these groups must use a tidal- wave of propaganda to gain and hold power. Gore's new book, *An Inconvenient Truth*[7], is a key example here . I have written elsewhere on the propaganda that I see in this book[9b]. I have also described his first book as "An excellent example of alarmism, inappropriate interpretation and outright misrepresentation [9c]." His second book can be described in the same manner.

'*AR*' is about political war, the dirty tricks involved in such war, and the propaganda vortex that any war brings. Peggy Noonan, the former speech-writer for President Reagan, commented on a TV talk show just before the 2004 election, how terrible she felt about voter fraud. She also stated, very strongly, that in her judgment, the voter fraud problem is largely a Democratic Party-driven problem. I believe the same way on the use of propaganda. While there is surely propaganda created by conservatives, it is my conviction that this is small compared to that created by the *Gorons*.

But who are the Gorons you might ask? The scientific definition is an alien from the planet Gore, the third planet in the Alpha Centauri system. And, by a very strange coincidence, one of the very first Americans they captured and converted was Albert Gore Jr. Over time he has actually become the Head Goron (HG).

In this essay the Goron name will be used as a synonym:

- for those activists/politicians who embrace simplistic science as the answer to complex problems;
- for those activists/politicians who use complex science to isolate rather than to illuminate an issue;
- for those environmentalists who exploit environmental and climate issues for their own purposes;
- for those global warmers who salute *An Inconvenient Truth*, and the like, as gospel; and
- for those activists who believe their cause is so righteous, or want their cause to be so righteous, that it justifies any means, including deceit, violent demonstrations, sabotage and terrorism.

Now some might object to the use of science fiction in a professional paper. However, I am in good company in the use of this technique. Not only does Crichton deal with science fiction, but the fourth writer, Alex Kozinski has used this approach on his 2002 commentary, entitled *Gore Wars*. Judge Kozinski sits on the U. S. Court of Appeals for the Ninth Circuit. His commentary, is in reality, a review of the book *the skeptical environmentalist (**tse**)*, by Bjørn Lomborg (BL), a Danish, environmental statistician[9d]. I don't know if Kozinski chose to couch his essay in the spirit of *Star Wars* because he felt so many of the claims and predictions of the environmental crowd come across as science fiction, but it is possible. Kozinski bemoans the long chronology of doomsday edicts that started with The Limits to Growth in the early `70s. He notes the most recent addition to this alarmist chronology is on global warming. At the end of his introduction Kozinski wondered whether anyone would ever write a book thoroughly analyzing all the environmental scares and explaining how much was legitimate and how much was hype. That book is now here, namely ***tse***. The reader may want to see the references for more inputs on the *Gore Wars*. Here I will limit myself to a few of his conclusions.

- Kozinski noted that Lomborg devotes much time to the GW issue, correctly identifying GW as the environmental-trump-card.
- Kozinski noted that Lomborg "paints a picture of a world where human welfare is dramatically improving in just about every way one might measure it." Yes, there are many ways things can be improved, but "all measurable trends point in a positive direction." - - - However, we "should not be smugly satisfied with the current state of the world. "It suggests, rather, that we must be very careful about the type of changes we bring about, lest we undermine the very values that we seek to promote." I believe GW may be the number one area that we need to be very careful on what we do.
- Kozinski noted that the *Clones* despise ***tse*** and BL and have launched a furious counter attack. Hundreds, possibly even thousands of *Clones* have challenged BL, but BL has defended his work effectively against these attacks. The volume of attacks, however, has been so large that some have compared this to the Gallileo inquisition.
- Kozinski noted that as "poverty recedes - - - the problems we fear have tended to be ameliorated or disappear, not get worse." He asks why "then do we hear so much bad news about the environment?" The answer to this question "is because environmental activists often lie, in big ways and small, in order to create the false impression that we are going from one environmental crisis to another." He notes that Lomborg "demonstrates the ways in which professional environmentalists play fast and loose with the truth."

In closing this section it is useful to stand back and take stock of the panorama of claims and predictions made by the Gorons. I would argue that never in the history of forecasting have more claims been made on forecasting capability. Never have more doom and gloom predictions been made. Not only have the Gorons predicted that we are moving into a future of hotter and more unpleasant days, but a future of terrifying CCC. Their claims have thus taken an incredible leap from *unpleasant* to *terrifying*. If that is not enough claims on their incredible clairvoyant capabilities to foretell the future, they also claim that our climate is changing so fast and this future is so imminent we have no time to contemplate, we must act. The above four writers and scientist are unlikely to buy such claims.

2. The Skeptics on *the* Global Warming Issue: the *Distinguished Veterans*

The careers and convictions of eleven *DVs* will be reported on in this section. These represent what I consider as perhaps the most interesting and compelling set of skeptics. These are individuals with incredible credentials. Many of them are retired scientists, some with the word *emeritus*, in their title. These individuals do not have to play the game of chasing after grant money. These are scientists that do not have to curry favor with the department chair-person, or other university/institute/agency *brass*. They are free to state their convictions, and to speak their mind. And the ones I will discuss are all highly skeptical on the global warming issue. These *veterans* are all agreed that the big picture on global warming, as painted by the alarmists, is seriously flawed.

Because these *DVs* have each had such grand careers one must guard against over selling them. They are not without their faults. They have their blemishes. And some have been attacked by the *warmers*. However, they are presented here as individuals that understand the sciences involved very well. And they are presented in the spirit that the odds that they will deal in hype or propaganda, or lies on this issue, is much lower than perhaps with any other group.

The breakdown of these scientists is as follows:

- hurricane experts - 4;
- experts in physics - 3 and
- agricultural/botany/food production authorities - 4.

2.1 Hurricane (HC) Specialists

Dr. Neil Frank is the Chief Meteorologist for Channel 11 in Houston, since 1987. Before he took this role, Frank served as head of the National Hurricane Center for 13 years, with a total of 26 years spent at that Center. He started his professional career as a weather officer in the Air Force.

I have had the opportunity to see him deliver the weather, be interviewed on global warming by CBS, and have heard him speak at various meetings. His conviction is that there is something wrong with the *warmers* case. A major part of this concern is in their reliance on computer models.

- He notes that the models, used in weather forecasting, can't be relied on for a three day forecast. He also notes, that for global warming predictions, we are being asked to rely on similar, but simpler models. Yet these models are applied to a far more complex task: forecasting the next century.

Dr. William Gray obtained his BSc in 1952, worked for the Air Force forecasting weather and attended U. of Chicago from 1957 to '61. He obtained an MSc in Meteorology in '59, and a PhD in Geophysical Sciences in '64. He joined Colorado State U., in '61, and has been forecasting hurricane frequency since then. Gray is a Fellow of the American Meteorological Society (AMS). He has strived to put the GW computer models into perspective.

- In a speech in Houston he noted that *climate models, while surely useful, are far from perfect. " - - - but when modelers move out onto the climate area the complexity becomes too damm much."*
- Gray recently noted [10] "While there has been warming since the middle '70s - - - this is natural, due to ocean circulation and other factors. It is not human induced." - - - "Nearly all my colleagues who have been around for 40 or 50 years are skeptical as hell about this whole GW thing."

• He noted he had "NOAA money for 30 some years, and then when - - - Gore started directing some of the environmental stuff, I was cut off. I couldn't get any NOAA money."

Dr. James O'Brien has a BSc in Chemistry from Rutgers and an MSc and a PhD in Meteorology from Texas A&M. He has worked for DuPont, the Air Force, the University of Colorado, Oregon State University and Florida State University, where he is the Distinguished Professor of Meteorology & Oceanography. He has also served as the Florida State Climatologist.
• O'Brien sees no correlation between HC intensity and GW. Rather he sees a ~15 year building & ~15 year waning cycle.
• O'Brien has also reported[11] that while the ocean may be warming in spots, it is not so in the HC formation part of the Atlantic. From 5°N to 20°N & from Africa over to near the U.S. it has cooled.

Dr. Tad Murty earned his PhD in oceanography/meteorology at the U. Of Chicago. He is considered an expert on all oceanic related phenomena. After the devastating tsunami of December 26, 2004, Murty, in six months, fielded more than 400 media interview requests. Murty spent 27 years at the Canadian Department of Fisheries and Oceans, and helped in the design of the International Tsunami Monitoring System. He has also been the director of the Australian National Tidal Facility. He has often commented on GW and HCs.
• Murty has served as editor of the Int. J. of Natural Hazards for 18 years. Out of ~2000 papers, "I cant recall one, based on actual observations, that claimed that GW has anything to do with EWEs."
• "Yes many papers on computer models tell a different story. "But after being associated with such simulations for the past 45 years, I have little faith in their predictions. "With a very slight tweaking of one single parameter in the model, forecasts can change from global warming to an ice age."
• I have examined 20 different atmospheric/oceanographic parameters associated with HCs around the globe using all historical data up to the end of 2004. "Not a single record was set after 1979."
• Since 1995 there has been an increase in the annual number of HCs in the Gulf. "However no new records have been set and nothing that cant be attributed to natural variability is happening."
• "My colleagues in India and I put together a 200 year database and found that the number of cyclones in the 20th century is about half of that in the 19th." - - -. There were three super cyclones in the 19th century and only one in the 20th."
• Murty's view about global warming: "This is the biggest scientific hoax being perpetrated on humanity[12]." There is no global warming due to human activities.

2.2 Experts in Physics

A recent news story in the local paper[13] reported on an interview with a Dr. C. Rapley, of the British Antarctic Service. He asked two rhetorical questions. "Do you agree that physics is physics? And "If carbon is increasing, how can you really deny there's going to be warming?" Rapley challenged the readers: if you really knew how physics worked, you would *stop arguing on* GW. So lets see how three *DVs* in the physics field react to the GW issue. The three *DVs* are Drs. Robert Jastrow, William Nierenberg and Frederick Seitz and are authors of a book[14] , now 15 years old, that was one of the first robust critiques of the greenhouse issue. A quick review of their credentials continues the rather incredible set of capabilities and accomplishments set by the first four distinguished veterans.

Dr. Nierenberg (1919 - 2000) was Director of the Scripps Institute of Oceanography from 1965 to 1986. He was an expert in underwater research. Nierenberg was a former member of EPA's Global Climate committee, the National Academy of Science Climate Research Board and a former chairman of the National Academy of Science CO_2 Assessment Committee. He served on several panels of the President's Science Advisory Committee.

Dr. Frederick Seitz was born in 1911, and earned his PhD from Princeton in 1934 in solid state physics. Seitz moved to the U. of Illinois in 1949, rising to the Dean of the Graduate College by 1965. From 1965 to 1968 he became the first full-time president of the National Academy of Sciences. From 1968 to 1978 he was president of Rockefeller University, a biomedical research and teaching center, and is now President Emeritus.

Dr. Robert Jastrow was born in 1925 and obtained physics degrees from Columbia. Joined NASA in '58 and formed the Goddard Institute for Space Studies in '61, which he directed for 20 years. Spent 11 years at Dartmouth as Professor of Earth Sciences, then became Chairman of the Board of Trustees at Mt. Wilson Institute. He is interested in astronomy, science fiction and religion. He is the author of *The Enchanted Loom: Mind in the Universe*; and *God and the Astronomers*.

1990 IPCC Challenge. These scientists charged the IPCC global warming range, for the next century, of 1.5 to 4.5°C, as far too pessimistic. Their analysis, based on well known observations data, but no computer model input, included:
(1) assuming the temperature increase—from pre-industrial levels to 1990—was 0.3 to 0.6°C;
(2) assuming this rise was all due to a 50% increase in greenhouse gases from pre-industrial levels;
(3) assuming a 100% increase in greenhouse gases, from pre-industrial levels, for the current century;
(4) as openers then, could see twice the warming, at the end of this century, or 0.6 to 1.2°C;
(5) assuming a correction of 0.2°C for ocean thermal lag, would give a revised range of 0.8 to 1.4°C;
(6) finally assuming an allowance of ±0.4°C for natural variability, would gave 0.4 to 1.8°C.

Based on this simple analysis, they felt the 1.5 to 4.5°C was far too pessimistic, and represented a major exaggeration of the actual physical situation. I would argue these three DVs know their physics. They don't deny there is going to be a warming but no where as big as the IPCC would like the public to believe. Note also that their range of 0.4 to 1.8°C is only 0.1 to 1.2°C above what has already occurred, a range that society can surely adapt too.

2001 IPCC Upper Value. Clearly the IPCC paid zero attention to this criticism, as the 2001 IPCC range was set at 1.4 to 5.8°C. Separate from the above analysis, the upper value deserves additional comment. This increase in the upper value from 4.5 to 5.8°C was not due to any new science, but simply a change in the scenario definition. Unfortunately the high end value gets most of the publicity. The 5.8 °C gets translated to 10.4 °F. This is rounded up to 11 °F. This value, however, is based on a hyped up scenario definition, that includes a highly unrealistic global CO_2 per capita emissions growth. In 1979 this value peaked at 1.23 tons/year per person. This ratio dropped to 1.11 by 1999. But in spite of all the attention paid to energy conservation, in spite of all the pressure to use better fuels and alternative energy, the IPCC ended up with a value of ~4.0 tons/year per person by 2100, almost four times current level. This look reinforces the conclusion that the IPCC deals in exaggeration.

2.3 Agriculture/Botany/Food Production Experts

Dr. Timothy Ball earned his PhD from the U. of London, England. His thesis was the reconstruction of climate from 1714 to 1952 using the extensive records of the Hudson Bay Company (HBC). He is the co-author of the book[15]: *Eighteenth Century Naturalists of Hudson Bay.*

Ball worked for the Canadian Air Force for eight years, then for the U. of Winnipeg for 28 years, retiring as full professor in 1996. Ball has given over 600 talks on science and the environment and has written hundreds of columns (*Weather Talk*) for the top Canadian farm magazine—Country

Guide. His interests include climate, long range weather patterns, ecosystems, air quality, silting and flooding problems, and impact of climate change on sustainable agriculture. Key comments[16]:

- Is the world getting warmer? "Yes, it warmed from1680 up to 1940, but since 1940 it's been cooling. The evidence for warming is because of distorted records." Could you summarize the evidence that suggests the world is cooling slightly, not warming up? "Yes, from 1940 until 1980, even the surface record shows cooling. The argument is that there has been warming since then, but in fact almost all of that is due to what is called the 'urban heat island effect'".- - - when you look at rural stations - if you look at Antarctica, for example - the South Pole shows cooling since 1957."
- Do we have the tools to model the climate? "We don't have the tools." - - -. "The fact is the big models don't work. The fact is we don't even understand a fraction of the mechanisms of climate."
- What is the scientific basis for the Kyoto Protocol (KP)? "There is none." - - - "So it is a policy based on ideology and economics and politics and has nothing to do with science."

Dr. Sylvan Wittwer received his BSc degree in Horticulture at Utah State and his PhD from Missouri. Wittwer, a distinguished professor of horticulture at Michigan State University (MSU), retired in 1996. He is Director Emeritus of the MSU Agricultural Experiment Station. He was Chairman of the Board on Agriculture, for the National Research Council, from '73-77 and a member of the Climate Research Board, '78-81. He has served as a consultant for all International Agricultural Research Centers, all U.S. Federal agencies relating to agriculture and environment, the United Nations Development Program, and the World Bank. He is the author of over 750 papers and five books, including: *Feeding a Billion - Frontiers in Chinese Agriculture*[17]; Food, Climate, and Carbon Dioxide; and a best selling gardening book *Greenhouse Tomatoes, Lettuce and Cucumbers*.

Wittwer conducted the original studies on CO_2 enhancement of the production of food crops, starting in 1964[18]. He notes he has "now lived through 8 decades of 'global warming'. - - - "The 'greenhouse effect' warming is non-existent - - -." He has also noted that "the evidence is that the rising levels of atmospheric carbon dioxide are very favorable for the most essential of human activities - - - namely the production of food." He writes that the effects of an enriched CO_2 on crop productivity, are positive as to the benefits for global food security. He argues that the rising level of atmospheric CO_2 is a universally free premium, gaining in magnitude with time.

He has noted[19] that the "benefits of carbon dioxide are not just limited to photosynthetic efficiency and water use efficiency. "When plants are exposed to elevated levels of atmospheric carbon dioxide, every kind of stress is alleviated to some extent that we've been able to examine." All plant stresses: water, temperature, air pollution, deficiency of nutrients, even a deficiency of light, that stress is alleviated to some extent.

Dr. Sherwood Idso obtained his BSc, MSc and PhD, in physics, all from the U. of Minnesota. In 1967, he joined the U. S. Water Conservation Lab in Phoenix as a research physicist. Idso has also served as adjunct professor of Geology, Geography and Botany at Arizona State University. In 2001 he retired from the Water Conservation Lab to become president of his own web site. Idso has published over 500 papers, plus the books *Carbon Dioxide: Friend or Foe*? and *Global Change: Earth in Transition*.

Several writings by Idso have highly influenced me. The first[20] was on a critique, in the NY Times in 1990, of Al Gore's chart, that compared temperature and atmospheric CO_2 content over the past 160,000 years. Gore was trying to make the case that the skeptics were wrong on global warming issue. Idso showed that Gore was the one who was wrong on the facts and the reasoning.

The second was his input[21] on a GW debate. His conclusion—on the impact of elevated atmospheric CO_2—tends to shatter the myth about GW: "*Rising levels of CO_2 in the atmosphere promote plant growth and at the same time reduce their demand for water. These effects should lead to a greening of the Earth and signs are that this has already happened.*" Thousands of papers—with CO_2 ranging from 350 to 650 parts per million—have verified these effects.

The third references are all output from his web site[22]: *The Center for the Study of Carbon Dioxide and Global Change.* The warmers have attacked it vigorously accusing Idso and his sons, Craig and Keith, all of being puppets of the Energy Industry. This charge surely is a great exaggeration and one of ultimate simplicity. It ignores the nature and quality of their postings. I have not seen any critiques based on the lack of quality of its work, or on the weakness of its arguments.

Dr Norman Borlaug obtained his PhD, in Plant Pathology from the U. of Minnesota in 1942 He directed the Co-op Wheat Research & Production Program in Mexico from the mid '40s to 1979. He is a distinguished professor of International Agriculture at Texas A&M University. He is credited with launching the Green Revolution and with saving perhaps a billion people via improved seeds and farming techniques. He was honored in 1970 as the winner of the Nobel Peace Prize, and was awarded the National Medal of Science in 2005.

Dr. Borlaug has expressed his concern on rabid environmentalists. He has noted[23]:
(1) the anti-biotechnology extremists will have a huge impact on humans and the environment;
(2) there is unsubstantiated scare-mongering done by opponents of genetic engineering;
(3) some scientists have jumped on the extremist environmental bandwagon in search of research $; and
(4) there are very real problems of food shortages that will emerge in the not too distant future and by 2025 this planet will have to nearly double current food production.

Dr Borlaug has been a board member of the American Council of Science and Health (ACSH). The ACSH has published a position paper[24a, 24b] on climate change entitled *Global Climate Change and Human Health*, that noted:
(1) implementation of mitigation measures—measures to stabilize the concentration of greenhouse gases in the atmosphere—would be both costly and ineffective;
(2) emission regulations would have negative effects on developing countries economies, which would therefore prevent public health improvements;
(3) if global climate change occurs gradually—as the IPCC has predicted—policymakers can safely take several decades to plan a response, and scientists will have enough time to develop cost-effective anti climate change strategies.

3. The Skeptics on *the* Global Warming Issue: Other Witnesses

3.1 The Skeptics on the Global Warming Issue: Selected Active Scientists

Other than the comments[8] by Richard Lindzen, the views of active scientists will not be covered in this essay. A review of such individuals and their work and convictions will be left for a future essay. However the lack of coverage in this essay should not be interpreted that there are no sound skeptics in that group. Indeed, there are many excellent scientists that are active in all fields of climate research, and are skeptical on this issue. There have been many surveys or petitions[25] on such scientists, over the past 15 years, some with signatures in the thousands even tens of thousands. These surveys have been attacked by the *warmers*. They argue there are no, or very few, real *climatologists* signed on and hence the survey is useless. Those who argue in this fashion conveniently ignore the large number of social scientists involved in many of the IPCC efforts, and the number of politicians and entertainers that lend their name to the *warmers* cause. No real *climatologists* in these groups. Hence, there is little doubt there are a large many of highly educated individuals that are concerned on the scope and direction of this issue, with many of these being from the hard sciences.

3.2 The Skeptics on the Global Warming Issue: Broadcast Meteorologists.

In addition to Dr Neil Frank there are many TV meteorologists that are skeptical on the GW issue. For example as I write this essay an email came in that reported on a panel discussion—held at Landerhaven in Mayfield Heights— among five north east Ohio meteorologists. Their consensus over the new IPCC report was "don't get too worked up over it." A follow up email added a sixth meteorologist - skeptic: Accuweathers Joe Bastardi.

Much of the recent discussion in this area relates to the statements by a Heidi Cullen, a self declared climate expert for the Weather Channel (TWC). This channel—a channel that relishes in climate alarmism—sponsers her weekly program *The Climate Code*. Her guests have been all the Gorons on GW, including Albert Gore jr., Ted Turner and Sir Richard Branson.

Ms. Cullen has advocated[26a] "that broadcast meteorologists be stripped of their scientific certification if they express skepticism about predictions of manmade GW." She went on: "If a meteorologist cant speak to the fundamental science of climate change, then maybe the AMS shouldn't give them a Seal of Approval." In short meteorologists better salute the CW on GW or be decertified. Muzzle the skeptics seemed to be her message. After receiving over 1000 responses, most critical of her comments, rather than an apology Cullen has commented [26b] as follows.

- "I wrote a post that has generated some pretty strong reaction and I wanted to take a moment to stop the spin."
- "I've read all your comments saying I want to silence meteorologists who are skeptical of the science of GW. That is not true. "The point of my post was never to stifle discussion. "It was to raise it to a level that doesn't confuse science and politics. "Freedom of scientific expression is essential."
- "Many of you have accused me and TWC of taking a political position on GW. "That is not our intention."
- "Our goal at TWC has always been to keep people out of harm's way." - - -"Here at TWC we have accepted that responsibility, and see it as our job to give YOU the facts on GW."

She claims the science is on her side. I doubt that Lindzen[8] would agree with her. She claims she is not confusing science and politics. I doubt that Crichton[1] would agree with her.

James Spann—Chief Meteorologist for ABC 33/40 Birmingham Alabama, and a weather entrepreneur (The Weather Company) in his spare time—commented [27]: "In Ms.Cullens world meteorologists must tout TWC's 'expert' point of view or be marginalized as a kook that has been decertified by the AMA." More by Spaan: "I have been an operational meteorologist since 1978 and I know dozens and dozens of broadcast meteorologists all over the country." - - - "I do not know of a single TV meteorologist who buys into the man-made global warming hype."

3.3 The Skeptics on the Global Warming Issue: State Climatologists.

Another interesting group of professionals in this debate are the so-called State Climatologists. At one time these offices were federally supported. Today they are supported more or less by many states, but not all. Over time a variety of individuals have been appointed to these offices—some with university positions and some with government agencies background—and have been certified by the American Association of State Climatologists (AASC). Today some of these appointees are skeptics on the GW issue and have become *inconvenient* to several governors. These governors—who at one time saw this office as one of nothing more than data collection and data storage—are now seeking to remove or silence such skeptics. In short these governors now want this office to become a political appointment.

As an example the governor of Oregon, Ted Kulongoski, wants to take the title of State Climatologist away from a Dr. George Taylor. Besides sound credentials (B.A., Mathematics; M.Sc., Meteorology; 12 years in meteorology; 19 years as State Climatologist) and serving for two years as President of AASC, Taylor rides "a bicycle to work nearly every day. "I'm a conservationist at heart. "I believe that human's should act as stewards for God's creation. "But I also believe that a balance between the environment and human welfare should occur." He reports that his favorite "balance" quote(28) is a quote from Albert Gore on this subject. That is from Albert Gore Senior, not Junior. In any event, Taylor is not a political flunky to the governor, and the governor wants a political flunky.

4. A Brief Review of the Four Key Assumptions on Global Warming

• **Earth is warming**. Most skeptics are willing to accept about 0.3 to 0.6 °C over the past century. However, some skeptics are concerned on the quality of the Surface Data Base (DB). This global DB is based on readings from thousands of surface weather stations. This **DB** shows three major trends: 1905 to 1945 - significant warming; 1945 to1975 - noticeable cooling; 1975 to 2005 - again significant warming. However, this warming for the last period may be largely due to the Urban Heat Island Effect.
• **Society is guilty**. The primary *evidence* the *warmers* cite is from the huge computer models. Skeptics have a large number of concerns on these models and their output. Also the *warmers* pay very little attention to the alternative warming cause, namely natural climate variation. Part of this variation maybe due to solar and other astrophysical phenomenon. In spite of the amazing progress being made in these astrophysical areas, one will see little mention of this progress by the *warmers*.
• **It will be catastrophic**. Consider the following input(29) from a prominent *warmer,* **Dr. Stephen Schneider**: " - - - we have to offer up some scary scenarios- - - make simplified, dramatic statements, and make little mention of any doubts we might have. "Each of us has to decide what is the right balance between being effective and being honest." One could add the Time magazine cover with *Earth* in a frying pan, as part of this approach.
• **We know how to stop it**. Consider the following inputs from two key, prominent *warmers*.

(1) **Dr. Thomas Wigley**. First of all one should know that Wigley is a highly regarded scientist, but still a *warmer*. His computer runs(30) in 1998 found a full implementation of the so-called Kyoto Protocol (KP) that called for extensive cuts in emissions, would give trivial and non-measurable results. He found the warming reduction over 2010 to 2050 was 0.07 °C; and only 0.13 °C over 2050 to 2100. The total from the KP required emission cuts was thus estimated at 0.2 °C.

If one assumed 0.2 °C reduction, per the greenhouse gas cuts that are now being discussed, and one accepted the IPCC expected global warming of 3.0 °C over this century, then one would need 15 such cuts to get to the 3.0 °C. Yet even today every nation that has signed the KP is struggling to find ways to meet the initial cut. Further the *warmers* hardly ever talk about the need for any subsequent cuts.

(2) **Dr. James Hansen**, *the father of this issue* (due to his 1988 congressional testimony), confessed ten years later[(31)]: "the forcings that drive long term climate change are not known with an accuracy sufficient to define future climate change." Here the father of this issue publicly stated we don't know. Hanson's reputation in this field, along with his *confession* should have been enough to have shut this issue down, or at least put off any KP type of policy initiative, say for 10 more years of research. It did not.

5. Conclusions

(1) The views of four non-scientist authors have been reviewed. These four skeptics are best viewed as science news reporters. When considering whether to accept the views from this group—where clearly none of the members are research scientists in this field—look at them in the role of science reporters. Such outlets as the New York Times, the Houston Chronicle, Nature magazine and Science magazine have science reporters. Further all the Non Government Organizations (NGOs) also report such news. These are the people, and these are the organizations that this group of authors is competing with as to who can best tell the public the latest scientific news. There is much food for thought in these views.

(2) The views of eleven *Distinguished Veterans* have been studied, and all are skeptical. Their lifetime publications, speeches and comments give a sample of the nature of this groups views on the global warming issue. The very best proof that a a serious, valid and vigorous debate exists on the GW issue—that it is relevant and valid, and that the skeptics may be winning—is in the speeches and papers and words of the *Distinguished Veterans*. To claim that these scientists don't understand the sciences involved is silly. To argue that these *veterans* are being conned into their positions is spurious. And to accuse them of being puppets of the energy industry is specious at best.

(3) The views of active scientists have been noted in the sense of surveys and/or petitions. The views of the TV Meteorological and the State Climatological communities have been scrutinized. There surely is no consensus in favor of the GW issue. There may be a consensus against it. The claim that "all scientists agree" is one of the more ridiculous claims made by the *warmers*.

(4) The four key assumptions on GW have been assessed. None have been definitively confirmed, at least to the satisfaction of this and to many other skeptics. Hence it is pre-mature to find CO_2 guilty. The very best one can say about the GW issue in general and the KP in particular is that it is premature. The very worst is that it is a hoax or a fraud.

References and Notes

(1) Crichton, Michael, *State of Fear*, Harper Collins, December 7, 2004
(2) Will, George, Newsweek, December 23, 2004.
(3) Westbrook, Gerald T., *'Acid Rains' on Liberal Propaganda*, iUniverse, Lincoln NE, December 9, 2004. Note that this book has nothing to say on the acid rain issue. The acid refers to the acid satire used throughout this book.
(4) Kozinski, Alex, *Gore Wars*, Michigan Law Review, August, 2002, Pages 1742 - 1767. Judge Alex Kozinski sits on the U. S. Court of Appeals for the Ninth Circuit. This 27 page essay by Alex Kozinsski, can be found by a Google search: "Alex Kozinski" "Gore Wars."
(5) See the box in Chapter 10, in reference (3), above entitled: *Life in the 14th Century*. This is from: Tuchman, B., *A Distant Mirror—The Calamitous 14th Century*, Alfred A. Knopf, 1978.

(6) Broecker, W., et al, *Possible 20 20th-Century Slowdown of Southern Ocean Deep Water Formation*, Science, November 5, 1999. Broecker is the scientist that popularized the thermo-haline circulation that includes warm surface water in the North Atlantic. As this water cools and as the salt concentration gradually increases it ultimately sinks and returns south as a deep current. The term *conveyor belt* has been used, but this is an overly simplified analogy.
(7) Gore Jr., Albert, *An Inconvenient Truth*, Rodale, Inc., Emmaus, PA, May 26, 2006
(8) *CNN: Larry King Live, A Panel Discussion: Tonight, what is global warming*. January 31, 2007. See: www.cnn.com/TRANSCRIPTS/07/01/31/lkl.01.html.
(9a) See review of *An Inconvenient Truth* at www.usaee.org. Select Publications, Under Newsletters, select. March, 2007 issue. See pages 33 - 37. Three references with inputs on Greenland are:
• Titus, James G., and Narayaman, Vijay K.,(1995), *The Probability of Sea Level Rise*, United States EPA, 230-R-95-008, 2nd Printing.
• Titus, James G., (2004), *Maps that Depict the Business-As-Usual Response to Sea Level Rise in the Decentralized United States of America*, OECD Global Forum on Sustainable Development and Climate Change, ENV/EPOC/GF/SD/RD(2004)9/Final, OECD, Paris.
• Jenkins, Geoff, et al, *Stabilizing climate to avoid dangerous climate change*, Hadley Centre, UK, January 2005.
(9b) See Amazon.com for my review of *An Inconvenient Truth*, August 31, 2006.
(9c) See Amazon.com for my review of *Earth in the Balance*, July 10, 2006.
(9d) See Amazon.com for my review of the book *the skeptical environmentalist*. January 17, 2007. The essay by Kozinski is primarily a review of the book, *the skeptical environmentalist, Measuring the Real State of the World*, by Lomborg, Bjørn, Cambridge University Press, 2001.
(10) Gray, W., et al, *Early April Forecast of Atlantic Basin Seasonal Hurricane Activity for 1997*, Dept of Atmospheric Science, Colorado State University, April 4, 1997
(11) Glassman, James, *Hurricanes and Global Warming: Interview with Dr. James O'Brien*, Tech Central Station, September 12, 2005, as reported in *Scientists Recent Comments on Global Warming and Hurricanes*, Center for Science and Public Policy, October 26, 2005..
(12) Ball, Tim, *Hardly a climate 'outlier'*, letter to the Financial Post, June 27, 2005. Ball reports on Tad Murty's quote that the climate-change scare is "the biggest scientific hoax being perpetuated on humanity." He also notes the efforts to marginalize Murty including the charge that he is practicing 'outlier science'.
(13) *Scientist rebuts global warming critics*, in an interview of C. Rapley, British Antarctic Service, Houston Chronicle, February 7,.2006.
(14) Jastrow, R., Nierenberg, W. & Seitz, F., *Scientific Perspectives on the Greenhouse Problem*, The Marshall Press, Jameson Books Inc., Ottawa, IL, 1990.
(15) Houston, Stuart; Ball, Tim; and Houston, Mary; *Eighteenth Century Naturalists of Hudson Bay*, McGill-Queens University Press, Montreal, 2003.

I first became aware of the Hudson Bay Company (HBC) as a child in Canada, through their fine department store. The HBC was, of course, far more than a department store chain. It was, in turn, a trading company; an exploration activity; a map maker; a student of, and keeper of weather data; a recorder of, and guardian of the natural resources of their region; and finally a pseudo government. No organization could have been more interested in the weather, as they had the challenge of getting their boats out of Hudson Bay *in time*. Hence they kept detailed records of weather, plus animal behavior, bird information, flower and vegetation patterns, and celestial inputs.

Ball did his PhD thesis based on these records. He has done much to organize/preserve them.
(16) Conversations from the Frontier Centre for Public Policy, Dr Tim Ball, Historical Climatologist, November 15, 2004.

(17) Wittwer, Sylvan, Yu Youtai, Sun Han, Wang Lianzheng, *Feeding a Billion: Frontiers in Chinese Agriculture*, Michigan State University Press, September 1, 1987.
(18) Wittwer, Sylvan, *The Global Environment and Food Production.* Originally published in 1997, reprinted in: www.greeningearthsociety.org/Articles/global.htm, on June 20, 2005.
(19) *Greening of Planet Earth Continues*, Panel Discussion, Bismarck, ND, November 13, 1998. Dr. Sylvan Wittwer was one of the panelists. See: www.co2andclimate.net/premier.htm.
(20) Gore Jr., Albert, *To Skeptics on Global Warming - - - You're wrong on facts and reasoning*, New York Times, April 22, 1990; Idso, Sherwood B., *Carbon Dioxide Warming Is Good for the Planet*, New York Times, May 7, 1990. This letter refuted claims made in the April 22 essay. Idso shows that Gore is the one who is wrong on the facts and the reasoning.
(21) Idso, Sherwood B., *Plant Responses to Rising Levels of Atmospheric Carbon Dioxide*, The Report of the European Science and Environment Forum, editor J. Emsley, Bourne Press Ltd, Dorset, March, 1996.
(22) See Center for the Study of Carbon Dioxide and Climate Change, www.co2science.com.
(23) *War and Social Unrest*, Center for the Study of Carbon Dioxide and Climate Change, November 16, 2005. This essay reports on a paper by Norman Borlaug, *Ending World Hunger*, Plant Physiology, October, 2000.
(24a) Shindell, Sidney, M. D., LL.B., Raso, Jack, M. S., R. D., *A Position Paper of the American Council on Science and Health*, October 1, 1997.
See www.acsh.org/publications/pubID.868/pub_detail.asp
(24b) Key quotes from Factsheet: American Council on Science and Health, ACSH, October 14, 1997. See: www.exxonsecrets.org/html/orgfactsheet.php?id=8.
(25) A wide variety of appeals, declarations, petitions, surveys and votes have been made on the GW issue over the past 15 years. This has ranged from ~ 20,000 signers for the Oregon Institute of Science and Medicine (OISM) Petition Project, down to seven voters in a *Debate on GW Science and a Surprising Conclusion on the KP*— (see Chapter 3 in reference (3) above. The seven scientists included three *warmers*, one neutral and three skeptics.)—six of the seven scientists indicated they would not sign the KP. I was at this debate, I heard them vote and I taped it. Two of the *warmers* and the neutral scientist voted with the skeptics.

The OISM Petition urged President Clinton not to sign the KP. It argued there "is no convincing scientific evidence that human release of carbon dioxide - - - or other greenhouse gases is causing or will cause catastrophic heating of the Earth's atmosphere - - -." Perhaps the key value of this petition is to counter the claim that "all scientists agree that the science is done."

Many warmers ridicule this petition. Senator Barbara Boxer[8] is one of these. They accuse the signees as being totally unfit to comment on the GW issue. However, according to the Livestock Weekly ("Popular Press Ignores Thousands Of Scientists Who Pan 'Warming'", May 14, 1998):

- of the 18,100 signatures the project has received in total so far, 16,600 have been independently verified, and 1,500 have not;
- two thirds of the signees have advanced academic degrees;
- ~2,100 physicists, geophysicists, climatologists, meteorologists, oceanographers & - - have signed;
- ~4,400 scientists are specialists in "chemistry, biochemistry, biology & other lif science degrees;
- most of the remainder are trained in related fields and possess technical knowledge "suitable for evaluation of the relevant research data";
- some of the names would appear to be plants—Perry Mason, three doctors from the MASH TV series, and one of the Spice Girls—aimed directly at discrediting the petition.

The reader is warned that the subject of GW surveys is very involved. One can be assured that any petition that emerges as a threat to the *warmers* cause will be attacked, and the signees will be presented as non-scientists, non-climatologists or puppets of the oil industry.

(26a) Marano, Mark, *Weather Channel Climate Expert Calls for Decertifying Global Warming Skeptics*, The Inhofe EPW Press Blog, January 17, 2007.
(26b) Marano, Mark, *The Weather Channel Climate Expert Refuses to Retract Call for Decertification for Global Warming Skeptics*, The Inhofe EPW Press Blog, January 18, 2007.
(27) Jones, Kieth Walters, Global Warming Weather Channel Row: Heidi Cullen vs. James Spann, January 21, 2007, as reported on The National Ledger, February 18, 2007.
(28) Taylor, George, posting on climatechangedebate.org, February 13, 2007, 2.56 PM. The quote from Albert Gore Sr. was reported as follows:
"Ecology is now a household word, but many of those who use it do not seem aware of the fact that by definition ecology is tied to economics, that man's well-being is tied to his being; that although preservation of an unsullied crystal stream, a purer atmosphere, a virgin tract of forest, or an unblemished landscape are noble goals, they are not the noblest: the noblest is to provide man with the basic stuff of existence - - - food and housing, and meaningful work."
(29) Schneider, Stephen, Interview in Discover Magazine, October, 1989.
(30) Wigley, T. M. L., *The Kyoto Protocol: CO_2, CH_4 and climate implications*, Geophysical Research Letters, July 1, 1998.
(31) Hansen, James E., et al, *Climate forcings in the Industrial era*, Proc. Natl Acad. Sci., USA, October, 1998.

Ad Hoc Operability Impact on GHG Emissions and Waste Heat Recovery

Mahmoud Bahy Noureldin [(1)], Ahmad S Aseeri [(1)] and Saleh Al-Hashimi [(2)],

[(1)] *Consulting Services Department, Saudi Aramco, Dhahran, Saudi Arabia*
nourelmm@aramco.com Tel: 9665-873-6045, Fax 9665-873-0766 aseerias@aramco.com Tel: 9665-874-6865, Fax 9665-873-0766

[(2)] *Chemical Engineering Department, The Petroleum Institute, P.O Box 2533 Abu Dhabi, U.A.E., salhashimi@pi.ac.ae, Tel: 9712-5085222, Fax 9712-5085200*

Abstract

Operability of process plants is a major process objective especially in the current oil and gas business environment. Energy supplies are tight where oil and gas companies are running their production plants close to its constraints. Process owners have become extremely conscious of their plants operability. Avoiding any process modification, even though it can result in clear benefits from energy savings point of view besides emissions reduction, which may have some impact on process operability is a word of wisdom. Operability of process plants needs availability and availability in turns needs controllability and controllability include flexibility and flexibility means switchability smoothly from one operation mode to another without losing production. Oil producers nowadays give operability a very high priority as the main process objective to the extent that heat integration of process units that may trip is normally avoided regardless. Of-course such priority does not come without price. In this paper through two industrial case studies such cost in form of more fuel consumption and more GHG emissions is introduced as an eye opener to enable oil production companies include these impacts while they are making their decisions regarding their facilities desired level of operability.

Introduction

Emissions to the atmosphere can be characterized as particulates, solid or liquid, vapor or gaseous. The majority of these species are quite small in size and diluted in concentration. Industrial emissions of major concern are particulate material of size 2.5-10 microns, particulate material of size less than 2.5 micron meter, volatile organic compounds, oxides of sulfur and nitrogen, carbon monoxide and carbon dioxide.

Environmentally, there are four phenomena associated with atmospheric emissions. They are urban smog, acid rain, ozone layer destruction and the greenhouse effect or global warming. Urban smog is commonly found in modern cities and is observed as brownish colored air. Such smog has harmful effect on living organisms. High levels of the pollutants forming such smog phenomenon can cause breathing problems and bring on asthma attacks in humans. Warm weather exacerbates this problem. Acid rain of acidic pollutants with pH values often in the range of 5 to 6 can also lead to several problems of its own. It can damage our plant life especially in forests; result in the acidification of water leading to dead lakes and streams; and increases the corrosion of buildings particularly made of marble and sandstone. The phenomenon of ozone layer destruction is further catalyzed by oxides of nitrogen and potentially increases the risk of skin cancer from sun light exposure.

The greenhouse gas effect or global warming phenomenon problem arises mainly from the burning of fossil fuels, and its main constituents are carbon dioxide and methane, of which methane has more than ten times the effect of carbon dioxide. The result is that global temperatures increase, leading to melting of the polar ice caps and thus rising sea levels, increased weather disruptions and changes to ocean currents. Atmospheric emissions are mainly formed as by-products of combustion processes. Such combustion processes, to date, remain to be the main source used for energy generation in our societies.

Operability of process plants is a major process objective especially in the current oil and gas business environment. Energy supplies are tight where oil and gas companies are running their production plants close to its constraints. Avoiding any process modification, even though it can result in clear benefits from energy savings point of view besides emissions reduction, which may have some impact on process operability is a word of wisdom. Operability of process plants needs availability and availability in turns needs controllability and controllability include flexibility and flexibility means switchability smoothly from one operation mode to another without losing production. Oil producers nowadays consider operability as a major process objective to the extent that heat integration between process streams is normally avoided regardless. Of-course such priority given to the operability concept does not come without price. In this paper an industrial case study is introduced as an eye opener to enable oil production companies include the impact of not only the waste heat cost but also extra GHG emissions in the atmosphere while they are making their decisions regarding their facilities desired level of operability. The idea that we should not integrate streams from different process areas to avoid up-sets to the operation can not be decided without the intense investigation of all pros and cons. In this paper two parts of the cons including extra waste of energy and GHG are highlighted.

Heat Integration for Pollution Reduction

In process industries, the principle sources of energy utility waste are associated with hot utilities. Furnaces, steam boilers, gas turbines and diesel engines are all producing waste by-products of combustion. If the process industries require a furnace or a boiler to provide hot utility, then any excessive use of the hot utility will automatically produce excessive utility waste and consequently excessive generation of atmospheric emissions.

Therefore, energy conservation is one way of not only saving money and natural resources, but also protecting the environment via minimizing energy-based emissions. Until late last century the preferred method of dealing with the atmospheric emissions were known to take the end-of-pipe approach where we let the problem occur and then try to treat the problem only after it happens. This approach is usually an expensive one.

This view was then thought to be myopic since it does not tackle the problem at the source where solving could and indeed has proved in many cases to be easier and more cost effective. Nowadays, energy efficiency optimization techniques have become a major tool in NO_x, SO_x, CO and CO_2 emissions minimization from combustion-based processes.

Systematic methods and tools have been developed and are currently utilized to conserve energy; protect energy-based natural resources and last but not least to minimize energy-based emissions and reduce the impact on the environment.

A major concept in the energy efficiency optimization techniques is the concept of "Heat Integration" applied in Pinch Technology to enhance waste energy recovery in industrial processes. One way of using this concept is to look first in our plants to satisfy some of their thermal energy needs through letting the hot streams (sources) that are to be cooled help the cold streams (sinks) that are to be heated and vise versa to enable both hot and cold process streams reach their heating and cooling targets.

This approach of "Integration" of heat sources and sinks in industrial facilities reduces the need for external hot thermal energy utility that mostly come from fossil materials combustion processes. This methodology in turns attacks, at the source, a main contributor to the GHG emissions by reducing the heating utility requirement. However, from the other hand such integration concept can have some impact on the process operability. In the coming two case studies the impact on the GHG emissions are calculated in case of "no integration" at all that means keeping all the degrees of freedom in the hand of the "operability" people.
As we mentioned above, such calculation will bring in the dimension of "environment" to the energy integration versus operability court for a rigorous operability consideration in oil and gas production facilities.

Industrial Case Study (I): Stabilizer-Compressor integration

The un-stabilized crude oil is pumped to the stabilizer unit via a pipeline. The stabilizer is fed with the crude oil from the inlet separator sphere that removes light ends that combines with overhead from the column to a compression system. The stabilizer is designed to fractionate, rather than strip, NGL (natural gas liquid) light ends from the crude. The fractionation capability provides a well defined split between the light ends and the crude oil. This tight split ensures maximization of stabilized crude and minimization of crude oil quality loss (API Gravity).

Heat input to the stabilizer bottom is provided through reboilers utilizing a heating media closed loop circulation system. NGL (natural gas liquid) light ends from the stabilizer overhead system is used for hot belt heater fuel; flare pilot and purge gas process blanketing. Net NGL is then compressed, condensed and pumped to the NGL (natural gas liquid) pipeline.

Feed to the stabilizer unit is composed of a blended crude oil stream from different oil fields. A 50 psig equilibrium flash is performed on the combined crude in the High Pressure Production Trap and the stream is further heated in the pipeline before feeding the stabilizer unit. The stabilized crude flows from the bottom of the stabilizer column. The crude is partially cooled by heat exchange with the stabilizer feed through the stabilizer feed pre-heaters. Final cooling of the stabilized crude is accomplished by cooling against air in stabilized crude oil cooler. The stabilized and cooled crude is then routed to the nearby storage tanks for temporary storage prior to transfer to the terminal.

The stabilizer unit is designed to process about 180,000 Barrel per day of crude oil. The stabilizing process prepares crude for atmospheric storage. The process train of stabilizing the crude consists of five main process systems:

1. Feed Preparation System
2. Stabilization System
3. Hot Belt Reboiler Loop System
4. Stabilized Crude Oil Rundown System
5. NGL Overhead Compression System

The feed preparation system performs the following three process functions:

- 3-phases separation of the crude feed
- pumping of the crude oil to the stabilizer Column
- Feed preheating

The equipment included in the Feed Preparation system is:

- Inlet Separator Sphere
- Stabilizer Feed Pumps
- Stabilizer Feed Pre-heaters

The inlet separator sphere provides: final removal of residual produced water, Surge control to smooth feed to the stabilizer and controls non-condensable components, primarily nitrogen and methane.

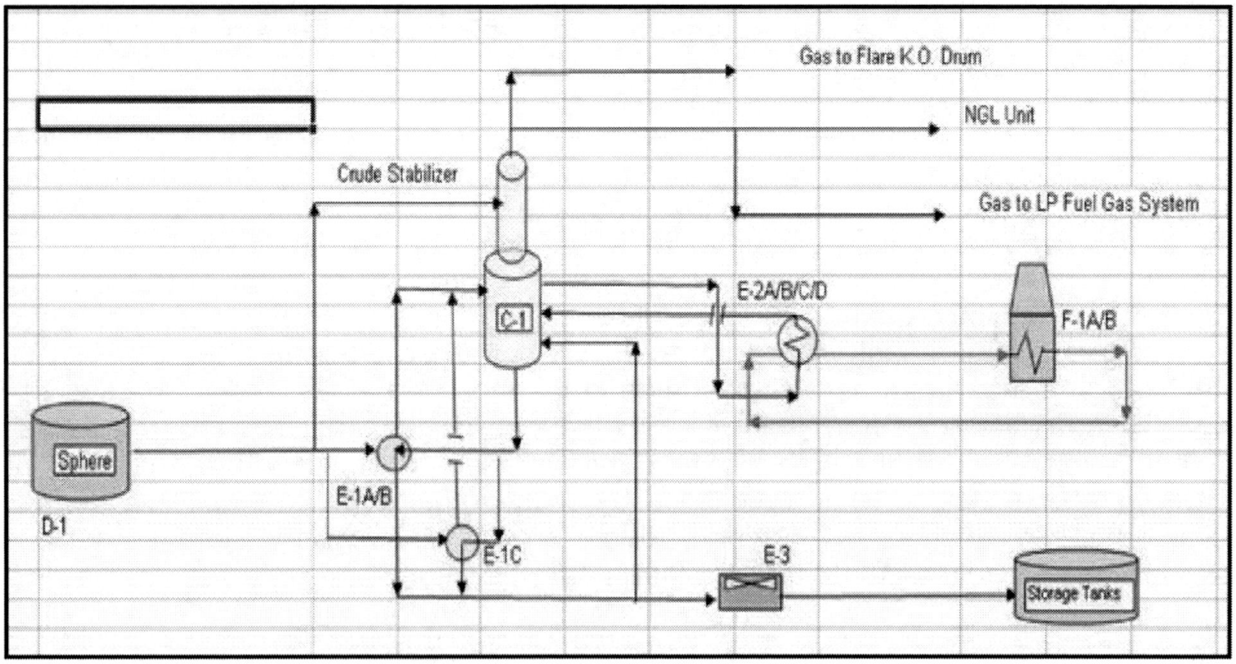

Figure 1: Stabilizer Process Flow Diagram

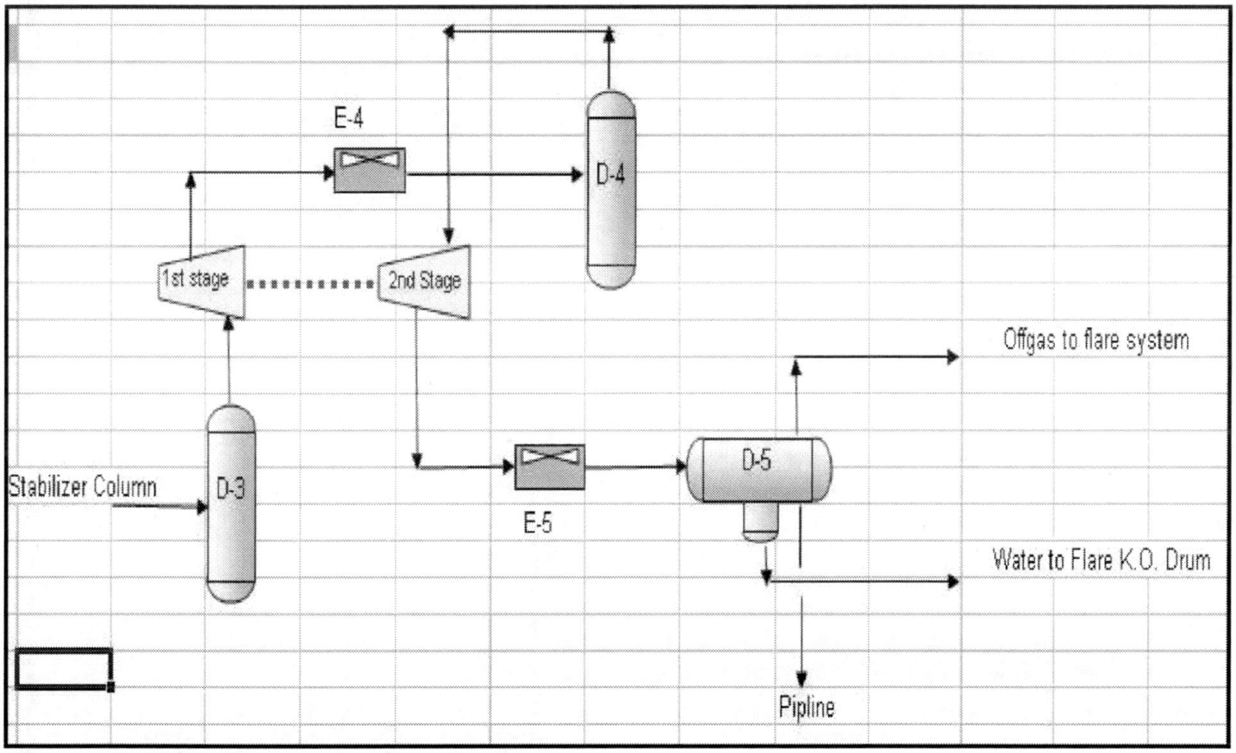

Figure 2: NGL Compression System

Methodology and Tools:

Pinch Analysis was used to evaluate the possibility of improving heat recovery in the process. To do this all hot and cold streams in the process were included to calculate the energy targets. In this case heat from the stabilizer bottom and the overhead gases can be recovered to the stabilizer feed. This can be represented in the Composite Curves of the existing process using a minimum approach temperature of 20 °F.

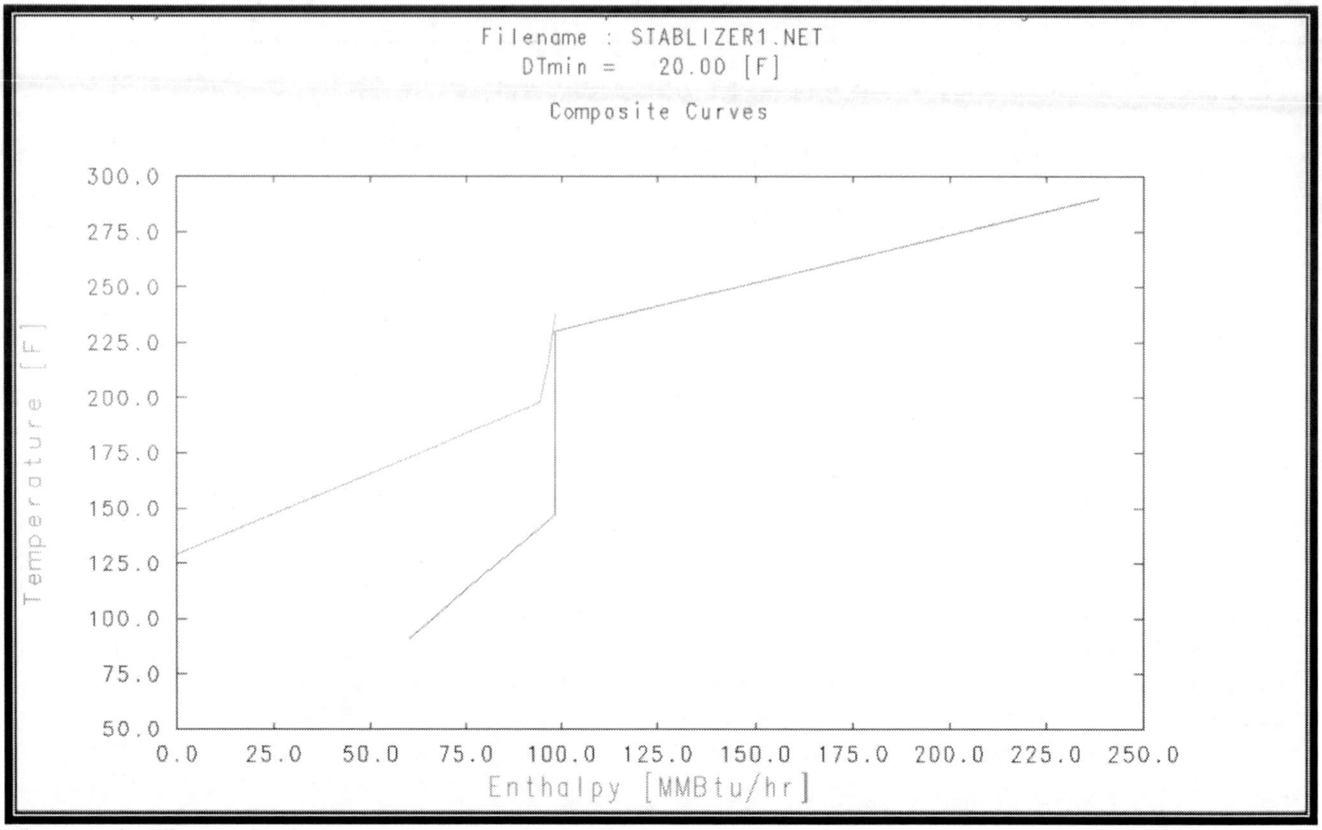

Figure 3: The existing process composite curves

The existing hot and cold process streams used in generating the Composite Curves are shown in the following table:

Table 1: Stream data for existing process

Strm	Name	TS [F]	TT [F]	DH [MMBtu/hr]	CP [MMBtu/(hr.F)]
1: 1 H	Stabilizer Bottom	198.0	129.0	85.5000957	1.23913196
2: 1 C	Stabilizer Feed	91.00004	147.0	38.10002	0.68035723
3: 1 C	Reboiled Crude	230.0	289.9994	140.499465	2.34167939
4: 1 H	1st stage gas	213.9998	129.0	2.10000279	0.247E-01
5: 1 H	2nd stage gas	230.0	129.0	4.55000568	0.04504954
6: 1 H	3rd stage gas	237.9992	129.0	5.99995581	0.05504584

The hot streams for the NGL gases have very small duties, recovering of which does not justify installing three additional heat exchangers and increasing the complexity of the plant. Thus, these streams were therefore excluded from the calculations.

The composite curves of the existing process excluding the hot streams of the NGL gases can be seen in Figure (5). The minimum approach temperature of these composite curves is 20 F°.

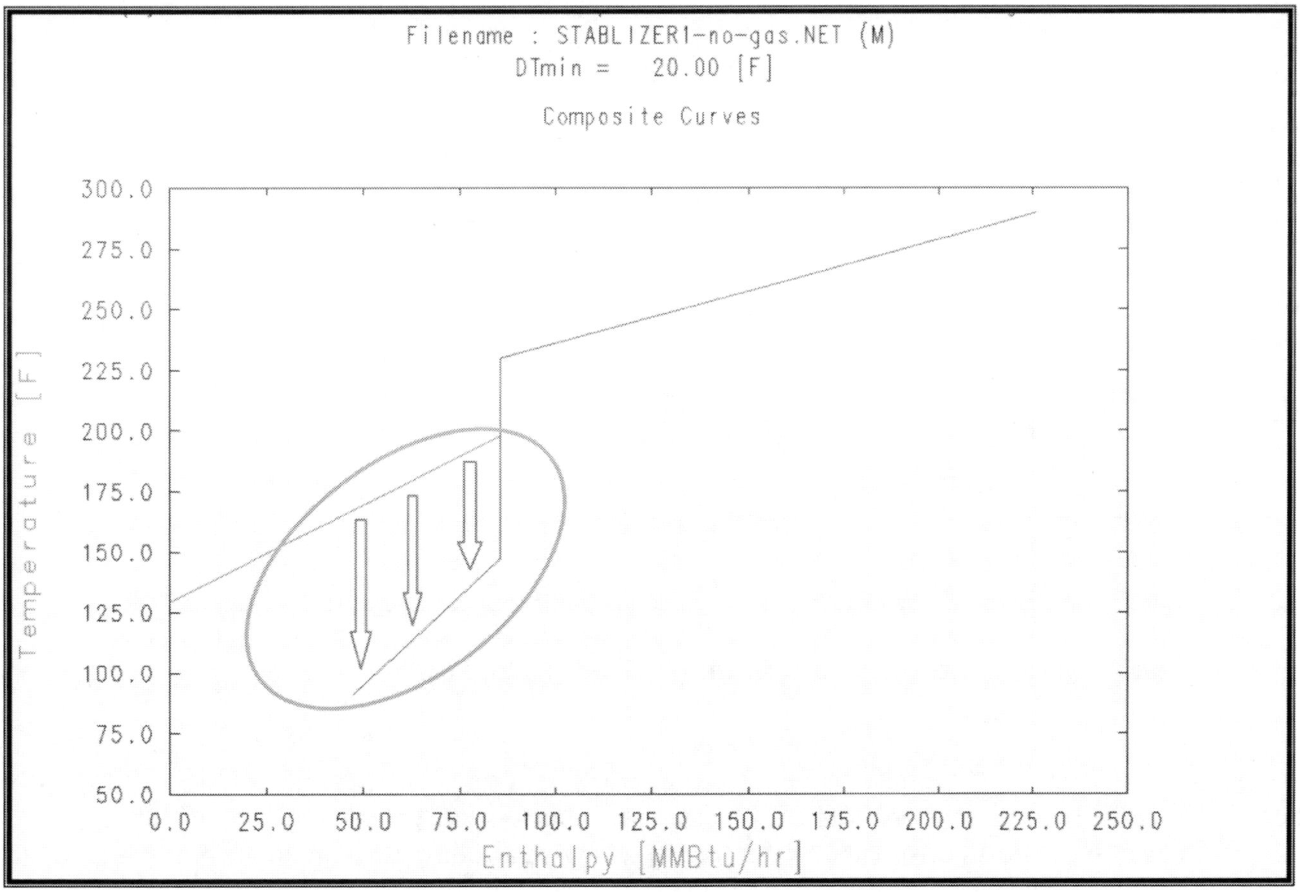

Figure (4): Composite curves excluding gas streams

Industrial Case Study (I) Results and Discussions:

The heat recovery target for the hot utility is about 38.0 MMBtu/hr. The heat recovery in this case can be increased compared to the existing process through enhancing the stabilizer feed preheating by increasing its target temperature from 147 F° to 180 F° (see table 1). This will supplement about 22.0 MMBtu/hr of the reboiler duty and hence decrease the fuel firing in the furnace.

The new composite curves are shown in Figure (7). The heat recovery duty in this case has been increased to be about 60.5 MMBtu/hr. The hot utility saving duty will be about 22.0 MMBtu/hr. Since each megawatt of heat given to the process (almost 3.4 MMBtu/hr) produces about 300 kg/hr of CO_2, assuming a very efficient furnace with 90% efficiency [1], the amount of greenhouse gas emissions then can be reduced by about 16,300 Tones/yr for 8400 hours of operation per year.

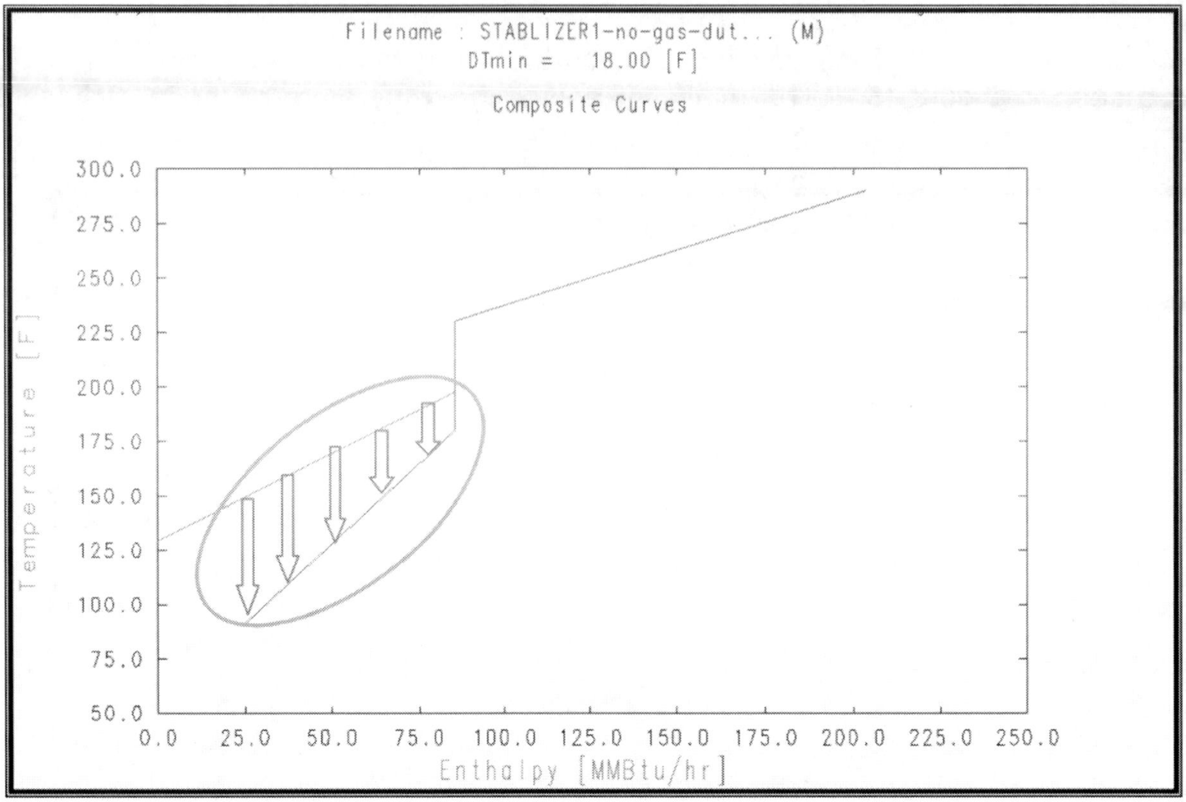

Figure 5: Composite curves after increasing feed temperature

In addition, assuming a fuel gas price of $6/MMBtu and 8400 hr/year of operation, the fuel gas saving from this heat recovery design modification is about $1.1 MM/yr. It is also important to note here that there will be another additional energy savings from this improved heat recovery system design in form of cooling utility reduction for the fin fan coolers of the stabilizer bottom product of about 22.0 MM Btu/hr.

Industrial Case Study (II):

In the second industrial case study below we introduce the price we pay from energy consumption point of view and extra GHG emissions due to the ad hoc approach to process operability mentioned earlier; that dictate no integration at all between two different process areas. In this case study a comparison is made between the no integration case, between the crude stabilization area and the gas compression area, and the cases that exhibit three different levels of heat integration.

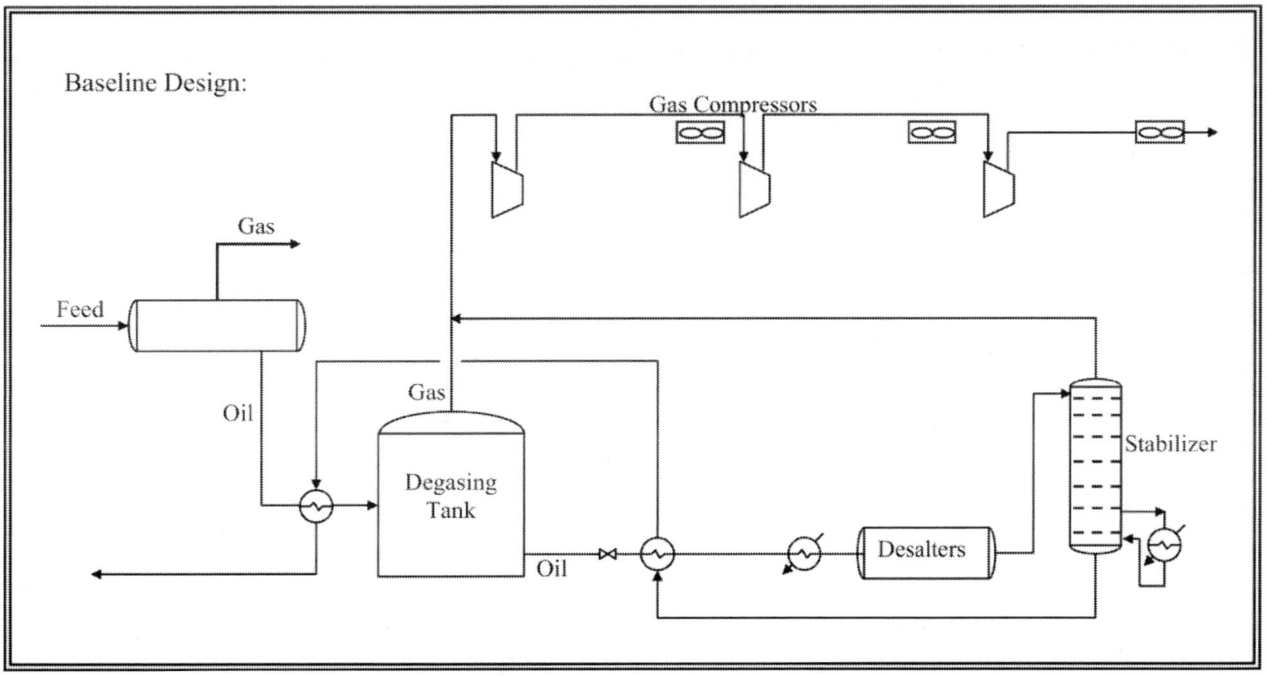

Figure 6: Process Flow Diagram without Heat Integration (preferred scheme from operability point of view)

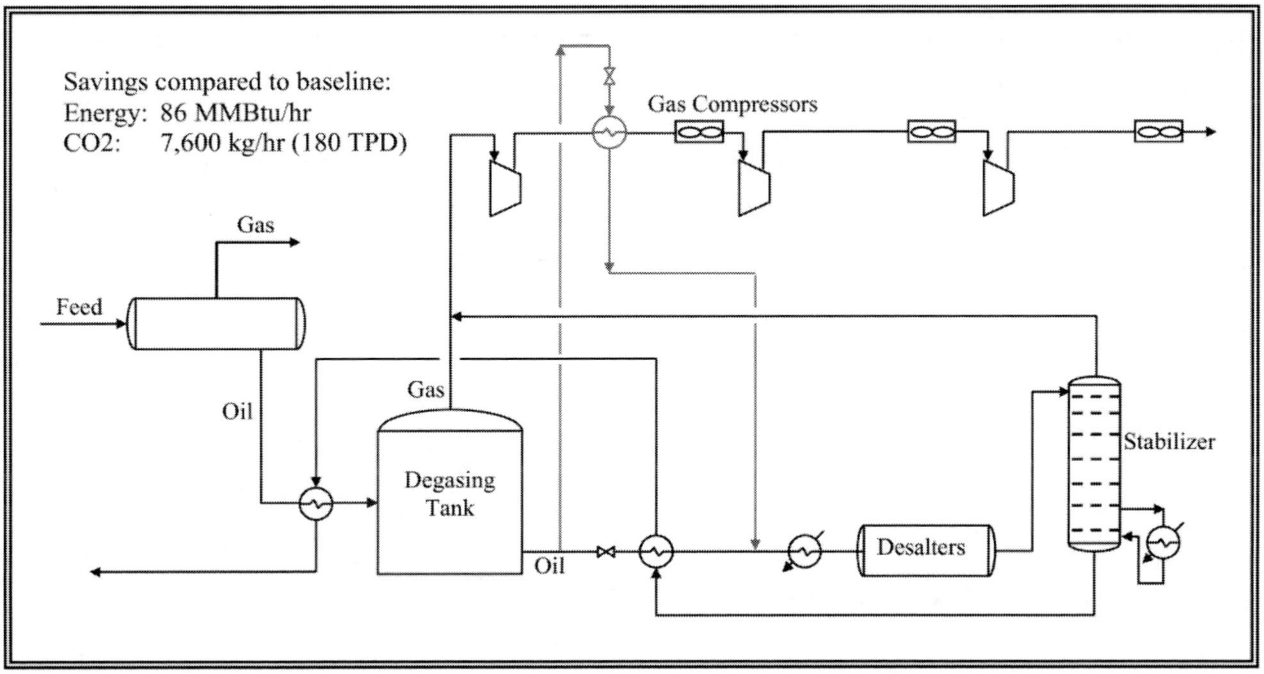

Figure 7: Process Flow Diagram with Heat Integration between gas compression process area and crude Stabilization one (one compressor)

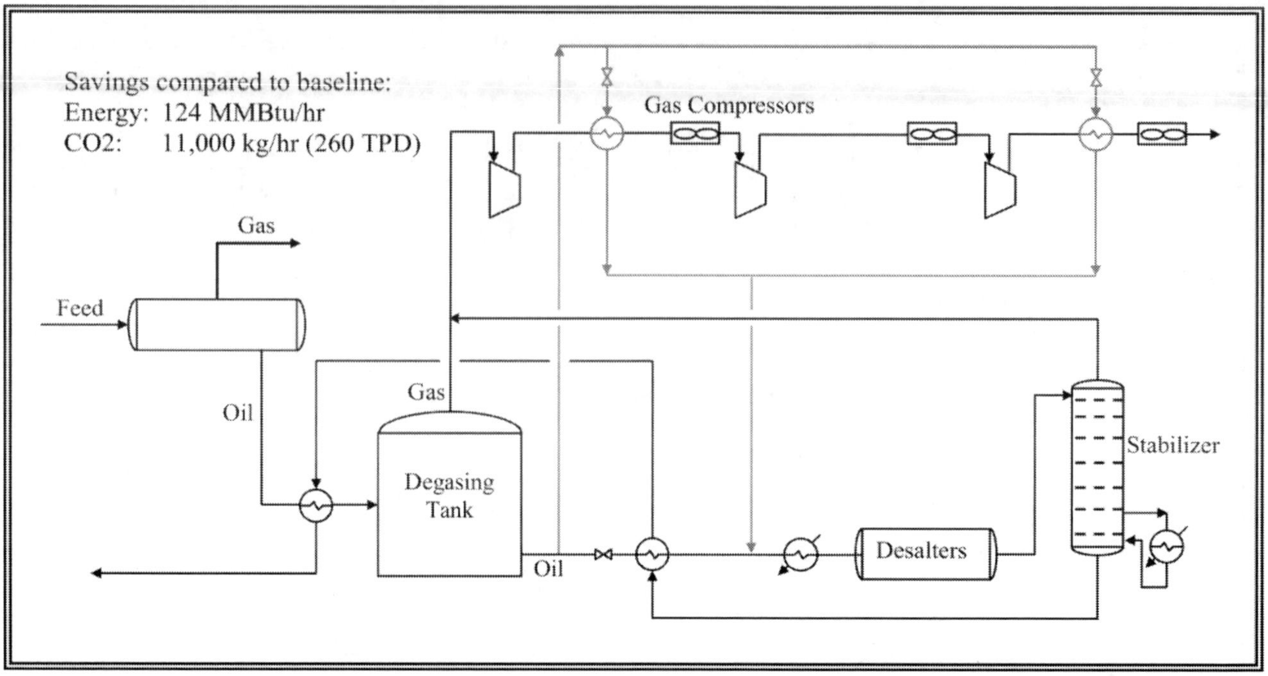

Figure 8: Process Flow Diagram with Heat Integration between gas compression process area and crude Stabilization one (two compressors)

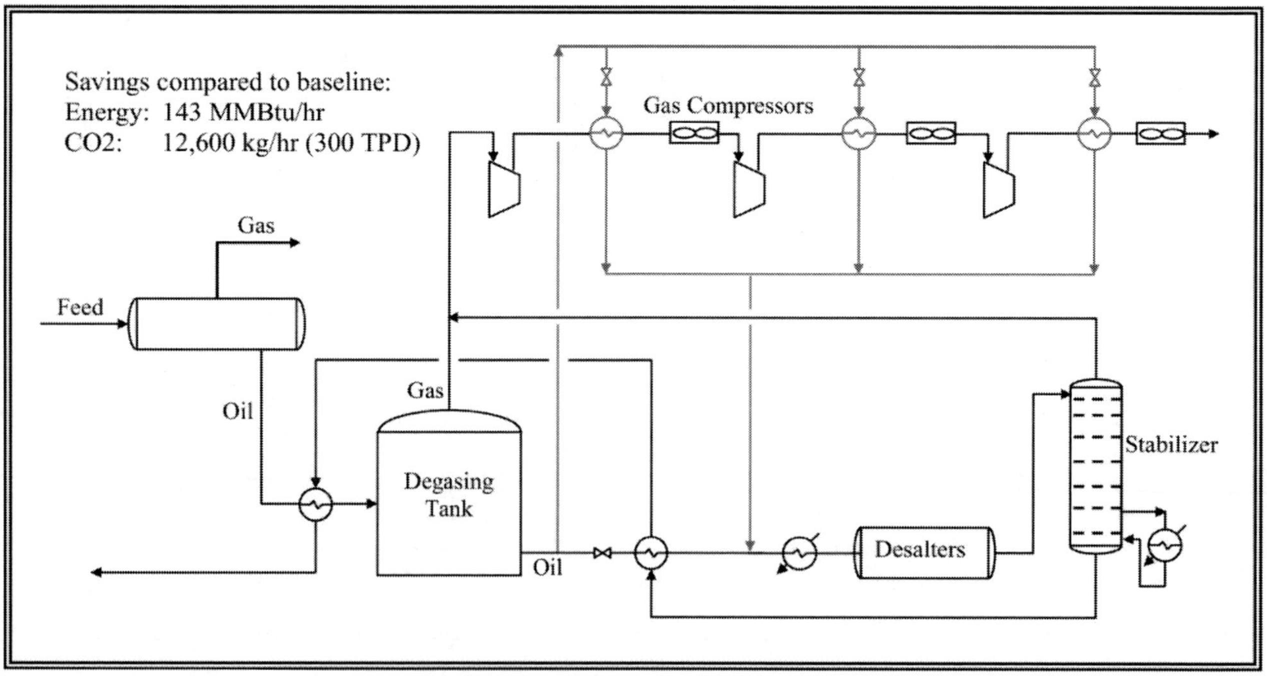

Figure 9: Process Flow Diagram with Heat Integration between gas compression process area and crude Stabilization one (three compressors)

Industrial Case Study (II) Results and Discussions:

Figures (7) , (8) and (9) above shows that the missed opportunities in the no heat integration case due to the integration of one compressor with the crude stabilization process area is about 86.0 MMBtu/hr and 1800 T/day reduction in energy consumption and in the GHG emissions. Due to the integration of two compressors of-course better results are expected as shown in figure (8).

Full integration between the compression area and crude stabilization area as shown in figure (9) leads to important savings in both energy consumption and GHG emissions where about 143 MMBtu/hr is saved and GHG emissions has been reduced by 12600 T/day due to such heat integration.

Conclusion

Atmospheric emissions such as GHG as a by-product resulting from satisfying our process heating requirements can be reduced through efficient waste heat recovery system. Heat recovery system optimization enables us not only save energy but also achieve considerable reduction in GHG atmospheric emissions. We need to remember that, roughly for every megawatt of heating utilities, obtained from boilers and furnaces in our production facilities, we can save through better trade-off between operability and optimized waste heat recovery system design and operation greenhouse gas emissions can be reduced by more than 2000 tons/yr.

References

[1] R.Smith, "Chemical Process Design" book, McGraw-Hill, Inc., 1996.

[2] M. B. Noureldin, S.H. Al Hashimi, A. K. Hasan, A.S. Aseeri and A.H. Al Qahtini, "Enhanced Waste Energy Recovery Targeting Using Real-Time Energy Integration in Pulp and Paper Plant," Proceedings of the American Society of Chemical Engineers Spring National Meeting, Orlando, Florida, April 23-27, 2006.

[3] M. B. Noureldin, A.S. Aseeri, A.H. Al Qahtini, S.H. Al Hashimi, Presentation at the AICHE Spring National Meeting on "Systematic in-Process Modification Approach for Enhanced Waste Energy Recovery in Gas Plants," American Society of Chemical Engineers Spring National Meeting Orlando, Florida, April 23-27, 2006.

[4] M.B. Noureldin, J.E. Swan, Computer-Aided design software for energy optimization through interval constraint logic propagation, Proceeding of MDP-8, Cairo University Conference on Mechanical Design and Production, Cairo, Egypt, January 2004.

[5] M.B. Noureldin, "TEM_icons™ 1.2 User's manual" Report, Department of Materials and Process Engineering, University of Waikato, Hamilton, New Zealand, 2003.

[6] M.B. Noureldin, A.K.Hasan, Global energy targets and optimal operating conditions for waste energy recovery in Bisphenol-A plant, Applied Thermal Engineering 26 (2006) 374-381

Personalizing a Greenhouse Gas Protocol: Defining the Greenhouse Gas Inventory of Your Company

B. Tod Delaney, First Environment, Inc., 91 Fulton Street, Boonton, NJ 07005 and Glenn Kramer, Amerada Hess Corporation.

GHG accounting standards, such as the WBCSD/WRI GHG Protocol and ISO 14064, as well as the development of guidance and tools, such as IPIECA's Petroleum Industry Guidelines for Reporting Greenhouse Gas Emissions and the API Compendium of Greenhouse Gas Emissions Estimation Methodologies for the Oil and Gas Industry, have established a path for oil and gas companies performing GHG inventories.

This GHG inventory path has many forks due to both organizational and operational differences in oil & gas corporations and the flexibility the previously mentioned standards, guidance, and tools maintain on many issues. For these reasons, a corporate GHG protocol customized to identify a company's specific approach or interpretation of key GHG inventory issues is important to avoid getting lost. Though development of a custom GHG Protocol requires an initial effort, such a document facilitates the performance of the GHG inventory, ensures both inventory consistency across the organization and over time, simplifies independent verification, and can enhance transparency with stakeholders, among other benefits.

This presentation will clarify what is meant by a custom GHG protocol. It will provide examples of the issues that a custom protocol would address. It will explain the uses of a custom GHG protocol and suggest the benefits it can deliver. To illustrate the effort and the value associated with developing a custom GHG protocol, the presentation will also include a case study by Amerada Hess Corporation on the development of its corporate GHG protocol.

THE NEED FOR ENVIRONMENTAL MANAGEMENT SYSTEMS IN THE WOOD PRESERVATION INDUSTRY

Nicholas P. Cheremisinoff, Ph.D.
Consultant, Charles Town, WV
(304-279-0618; Email: ncheremisi@aol.com)

ABSTRACT

This paper provides a critical review of the waste management practices of wood treating facilities which service the railroad industry. The paper examines both historical and current practices by railroad tie manufacturing wood treating plants. The work stems from both independent research conducted by the investigator, plus in-depth studies of the environmental management practices of several tie producing plants entangled in worker compensation and community toxic tort litigations. In some of the case studies presented, levels of dioxins found in attic dust and soils in communities neighboring wood treating facilities approach and may exceed the Times Beach, Missouri incident. In addition, an examination of TRI (Toxic Release Inventory) reporting for 20 wood treating plants shows omissions of priority pollutants such as benzene, dibenzofurans, anthracene, dioxins, and other toxins. The investigator points out a critical need for the industry sector to catch up with the chemical industry's Responsible Care program, and to undertake reforms on current environmental management practices through the implementation of pollution prevention programs integrated with formal Environmental Management Systems.

INTRODUCTION

The wood preservation industry is a diverse sector that provides a wide variety of innovative products that service many consumer applications, while at the same time, conserves natural resources by extending the life cycle of wood-based articles. Its origin is largely linked to the birth of the railroad industry in the 19th century. The expansion of transport rail services was so great that the demand for cross and switch ties, railroad cross panels, bridge supports, communication poles, and other wood-based articles led experts of the day to hypothesize that massive deforestation was likely within decades. This concern created a marriage between the chemical and rail industries where pesticides and various preserving formulations using creosote, pentachlorophenol, copperized chromium arsenates, and other chemical preservatives were employed and continue to be used as a means to extend the life of wood-based products. When properly treated, a cross tie has a life expectancy of 30 years. Hence, it is a fact that wood preservation is not only environmentally friendly in that it supports natural resource sustainability, but is economical since manpower, fuel, energy and pollution management associated with harvesting operations are minimized. Indeed, over many decades, responsible forest and land management have evolved into both a branch of science and industry unto themselves.

Unfortunately, many of the historical and even current practices employed by downstream wood preservation plants are not based on 'cleaner production' principles. Wood preserving facilities designed and operated to service the rail industry have historically relied upon few pollution controls and largely have been devoid of pollution prevention practices. Older plants were built with few concerns for environmental protection, worker or public safety. Because pollution controls are sunk costs, neither the railroad industry nor companies that have been in the business of supplying treated wood products to the railroads have invested in air and water pollution control technologies. Historically wood treating plants have allowed excessive levels of air emissions to be discharged in an uncontrolled fashion, allowing them to freely communicate with the environment and placing the public at risk. It has only been over the past two decades that the railroad industry has divested itself of many industrial dinosaurs, but not without leaving a legacy of enormous negative consequences to the environment, public safety, and the workers who labored over a lifetime. There are wood preserving plants today that continue to discharge significant quantities of toxic air pollutants, operate without proper pollution control, and have no pollution prevention programs.

INDUSTRY OVERVIEW

The U.S. wood preservation industry (SIC 321114) is comprised of establishments primarily engaged in (1) treating wood sawed, planed, or shaped in other establishments with creosote or other preservatives, such as chromated copper arsenate, to prevent decay and to protect against fire and insects and/or (2) sawing round wood poles, pilings, and posts and treating them with preservatives.

According to the U.S. Census Bureau's latest statistics (1997)[1] there are 451 establishments across the U.S. that employ 11,668 persons with products valued over 4.461 billion dollars annually. Compared to 1992, the number of establishments declined by 7.6%, whereas the value of shipments rose by 65.2% with employment increased by 8.2%.

Figure 1 shows the distribution of the number of establishments by state. Figure 2 reports employment figures by state for the sector. Figure 3 reports the annual value of shipments by state.

Although the value of products produced by the sector is sizable, on the average there are only about 26 employees per plant.

[1] http://www.census.gov/epcd/ec97/industry/E321114.HTM

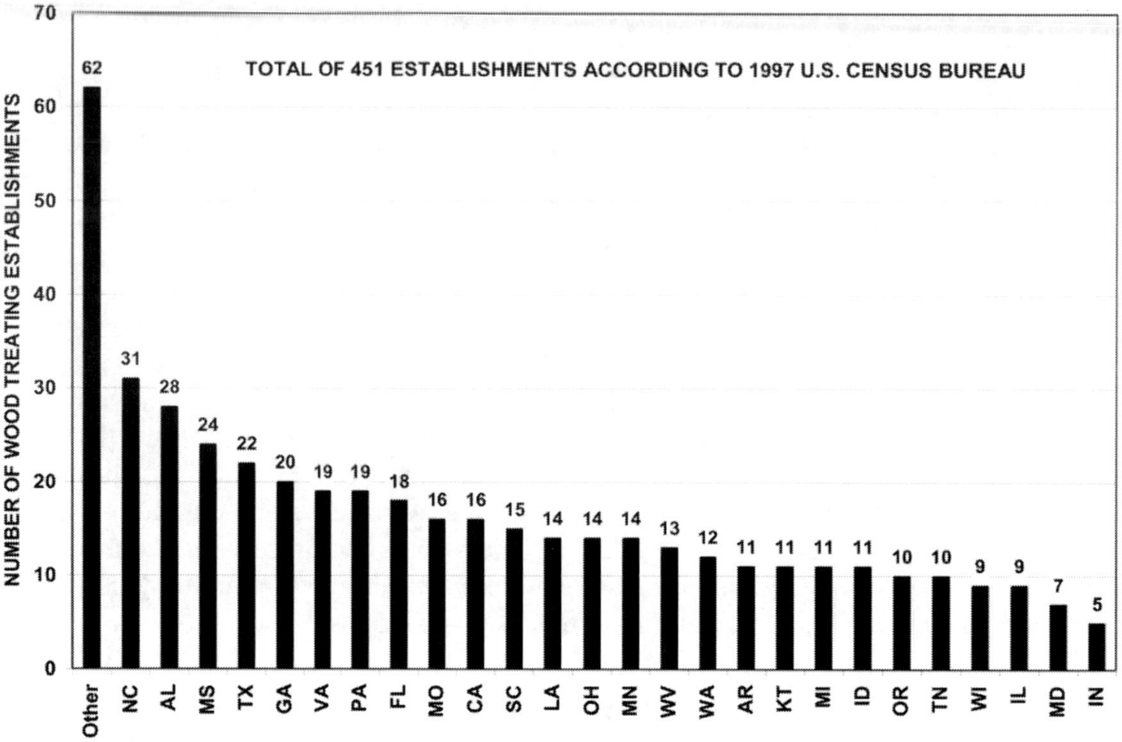

Figure 1. Number of wood treating establishments by state.

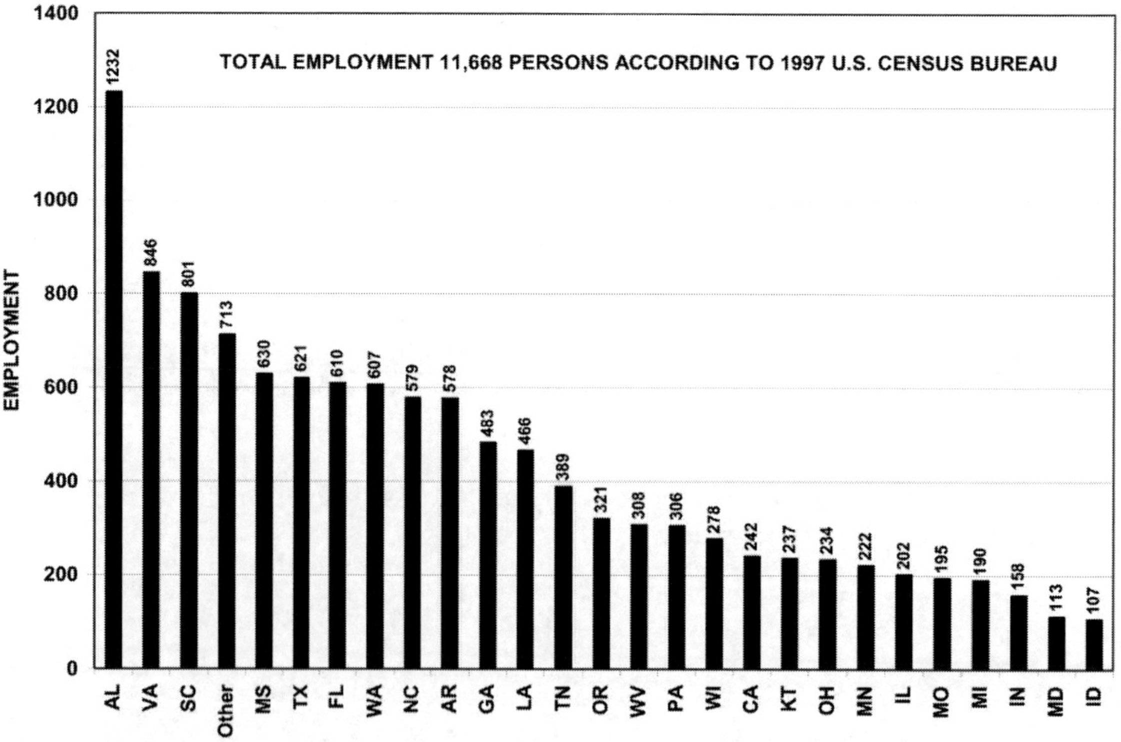

Figure 2. Employment figures for the sector.

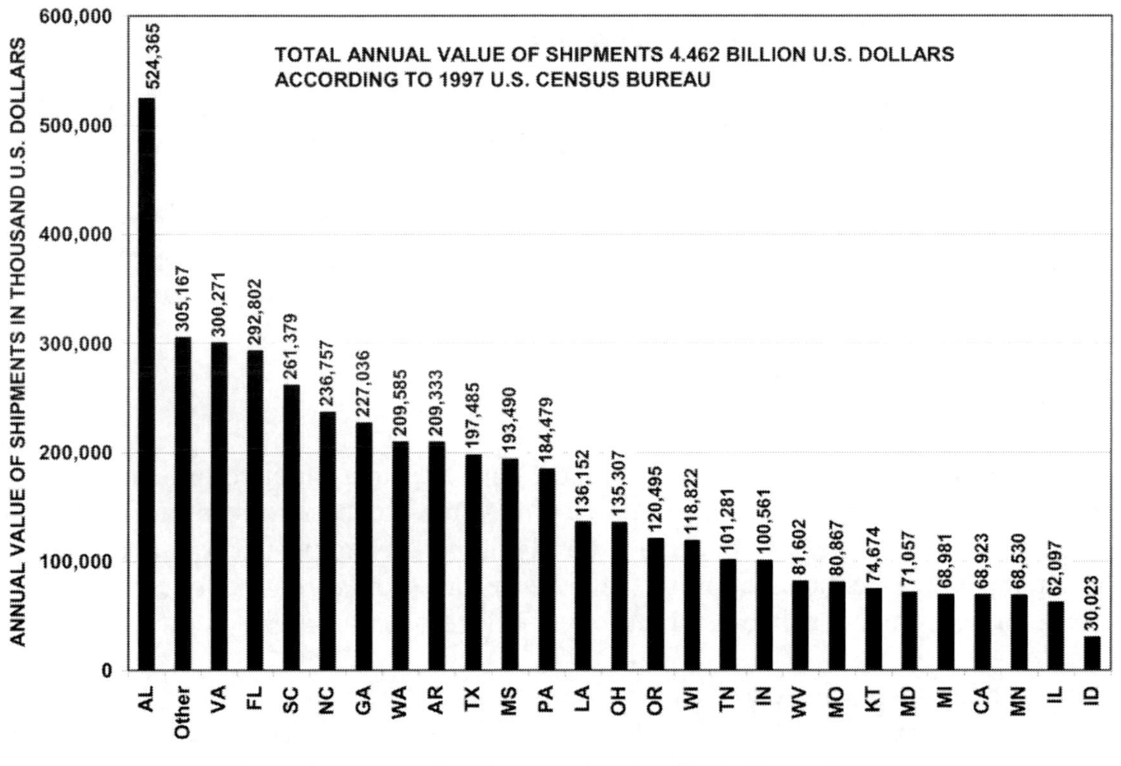

Figure 3. Annual value of shipment in thousands of U.S. dollars.

The industry sector has declined in number of establishments over the past twenty five years. A 1982 study by the U.S. EPA[2] reports that there were 524 wood preserving enterprises located throughout the country, with 65% of them employing less than twenty people. The total employment figure for 1982 was 10,900 which reflects a near stagnant employment growth (less than 0.5% growth per year from 1982 to 1997).

Due to the cost of transportation, wood preservers have tended to be located close to timber cutting operations. As was the case in 1982, nearly 75% of all facilities are situated in two regions: from east Texas to Maryland, and the northern Pacific Coast. About 23% of the sites are located in five states: Georgia, Alabama, Louisiana, Mississippi, and Florida, compared to 25 years ago where 33% of the facilities were located.

As illustrated by the case studies documented herein, wood preserving sites are plagued with past practices that have fallen below the industry standard of care prior to the regulatory era, and have been lagging in meeting compliance obligations through to today. The relatively small sizes of the enterprises as business units, an almost monopolistic control over the industry sector, uneven environmental monitoring and enforcement between states are factors which have contributed to overall poor environmental performance.

[2] U.S. EPA Sector Notebook on the Wood Preservation Industry, Washington, DC, 1982

CHEMICALS USED BY THE INDUSTRY

There are two general classes of wood preservatives: oils such as creosote dispersed in a carrier oil (e.g., No. 6 fuel oil) and petroleum solutions of pentachlorophenol; and waterborne salts that are applied as water solutions. An example of waterborne salts is chromated copper arsenate (CCA). The effectiveness of the preservatives varies greatly and depends not only on composition, but also upon the quantity impregnated into the wood, the depth of penetration, and the conditions to which the treated wood is exposed in service.

Creosote and Coal Tars

'Creosote' does not refer to a single chemical but rather to a variety of products that are mixtures of many chemicals which include wood creosote, coal tar creosote, coal tar, and coal tar pitch. Creosotes do not occur naturally in the environment. They are created by high-temperature treatment of beech and other woods (wood creosote) or coal (coal tar creosote), or from the resin of the creosote bush (creosote bush resin).

Wood preservation using coal derived creosote has a 180 year old history, beginning in Europe. The impregnation of railroad ties beginning in the middle of the 19th century established the industrial importance of this chemical. Considerable information on the chemical and toxic properties of creosote and coal tars has been known for many decades.

Coal tar creosote (C.A.S. 8001-58-9) is the most widely used wood preservative in the U.S. It is referred to by EPA simply as creosote. It is a thick, oily liquid that is typically amber to black in color, and is a distillation product of coal tar. It has a burning, caustic taste. It is used as a wood preservative and water-proofing agent for log homes, railroad ties, telephone poles, marine pilings, and fence posts. It is also a restricted-use pesticide, and is used as an animal and bird repellant, insecticide, animal dip, fungicide, and a pharmaceutical agent for the treatment of psoriasis.

There are approximately 300 chemicals that constitute the major compositional mix in coal tar creosote; however, there can be up to 10,000 different chemicals in all within a typical mixture.

Coal tar (C.A.S. 8007-45-2) and ***coal tar pitch*** (CAS #67996-93-2) are the by-products of the high-temperature treatment of coal to make coke or natural gas. These chemical products are usually thick, black or dark brown liquids or semi-solids with a naphthalene-like odor. Coal tar has a sharp, burning taste.

Coal tar creosotes, coal tar, and coal tar pitch are similar in composition with the major chemicals in them that can cause harmful health effects being polycyclic aromatic hydrocarbons (PAHs), phenol, and cresols.

Approximately 75 to 85 % or more of the coal tar mixture is comprised of polycyclic aromatic hydrocarbons (PAHs). Coal tar pitch is a residue produced during the distillation of coal tar and is distinct from coal tars and coal tar creosotes. The pitch is a shiny, dark brown to black residue containing PAHs and methyl and polymethyl derivatives along with heteronuclear components.

IARC[3] defines coal tar creosote as "the fractions or blends of fractions specifically used for timber preservation". Bedient, et.al.[4] have reported that mixture composition varies across production lots and manufacturers, but that an average composition contains about 85% PAHs and 2 to 17% phenolic compounds.

NIOSH[5] defines coal tar pitch as the tar distillation residue produced from coking. The grade of pitch can vary significantly since distillation conditions such as residence time and temperature have first-order effects on the composition of the pitch. The product is comprised mainly of condensed ring aromatics which includes 2 to 6-ring systems, along with minor components of phenolics and aromatic nitrogen bases. HSDB[6] reports that the number of chemical components in coal tar pitch is in the league of many thousands. Table 1 provides a listing of some of the chemicals identified in coal tar pitch.

Since commercial creosote mixtures are derived from different sources, which employ different distillation conditions as well as sources of coal, one may expect that the creosote components are inconsistent in type and concentration. Weyand et. al.[7] reported 2 to 20-fold differences in concentrations of several PAHs from a study of four coal tars.

Table 1. Partial list of PAH components in coal tar pitch reported by Guillen, et. al.[8]

Chemical Name	Chemical Name
11H-benz(bc)aceanthrylene or isomer	Dibenzonaphthofuran or isomer
11H-benzo(a)carbazole	Dibenzonaphthothiophene
1-methylnaphthalene	Dibenzophenanthridine or isomer
2-ethylnaphtalene	Dibenzoquinoline or isomer
2-ethylnaphthalene	Dihydrobenzofluoranthene or isomer
2-methylnaphthalene	Dihydrochrysene or isomer
2-phenylnaphthalene	Dihydroindenopyrene or isomer
4H-cyclopenta(def)chrysene or isomer	Dihydropyrene or isomer

---

[3] IARC Monographs on the Evaluation of Carcinogenic Risk to Humans, International Agency for Research on Cancer, World Health Organization, Supplement 7, Vols 1-47, 1987

[4] Bedient, P. B., A. C. Rodgers, T. C. Bouvette, et. al., Groundwater Quality at a Creosote Waste Site, Groundwater, 22: 318-329, 1984

[5] NIOSH, Criteria for Recommended Standard: Occupational Exposure to Coal Tar Products, Cincinnati, OH, SRI International for U.S. Department of Health and Human Services, National Institute for Occupational Safety and Health, NIOSH-78-107, 1977

[6] HSDB, Hazardous Substance Data Bank, National Library of Medicine, National Toxicology Information Program, Bethesda, MD, 2000

[7] Weyand, E. H., Y. Wu, S. Patel, et. Al., Urinary Excretion and DNA Binding of Caol Tay B6C3FI Mice Following Ingestion, Chem Res Toxicol 4(4): 466-473, 1991

[8] Guillen, M. D., M. J. Iglesias, A. Dominguez, et. al., Polynuclear Aromatic Hydrocarbon Retention Indices on SE-54 Stationary Phase of the Volatile Components of a Coal Tar Pitch: Relationships between Chromatographic Retention and Thermal Reactivity, J. Chromatog 591: 287-295, 1992

Chemical Name	Chemical Name
4H-cyclopenta(def)phenanthrene	Dimethylbenz(a)anthracene or isomer
4H-naphtho(1,2,3,4-def)carbazole or isomer	Dimethylbenzonaphthofuran
5H-benzo(b)carbazole	Dimethylbenzopyrene or isomer
7H-benzo(c)carbazole	Dimethylchrysene or isomer
9,10-dihydroanthracene	Dimethyldibenzonaphthofuran or isomer
Acenaphthene	Dimethyldibenzophenanthrene, -Anthracene
Acridine	Dimethylfluoranthene, -Pyrene
Anthanthrene	Dimethylnaphthalene
Azabenzopyrene or isomer	Dinaphthiophene
Azafluoranhtene, -Pyrene	Dinaphthothiophene
Benza(a)anthracene	Fluoanthene
Benzacenaphthene or isomer	Fluorine
Benzo(a)fluorine	Indenopyrene or isomer
Benzo(a)naphtha(2,3-d)thiophene	Methycarbazole
Benzo(a)naphthacene	Methyl indenoprene or isomer
Benzo(a)pyrene	Methylacenaphthene
Benzo(b)chrysene	Methylbenz(a)anthracene or isomer
Benzo(b)fluoranthene	Methylbenzacenaphthene or isomer
Benzo(b)naphtha(2,1-d)thiophene	Methylbenzo(cdf)carbazole
Benzo(b)naphtho(1,2-d)thiophene	Methylbenzofluoranthene or isomer
Benzo(b)thiophene	Methylbenzofluorene
Benzo(c)chrysene or isomer	Methylbenzonaphthifuran or isomer
Benzo(c)fluorine or isomer	Methylbenzonaphthothiophene
Benzo(c)p henanthrene	Methylbenzopyrene or isomer
Benzo(e)pyrene	Methylbiphenyl
Benzo(ghi)fluoranthene	Methylchrysene or isomer
Benzo(ghi)perylene	Methyldibenzophenanthrene, -Anthracene
Benzo(k)fluoranthene	Methylfluorene
Benzonaphthofuran or isomer	Methylphenanthrene, - Anthracene
Benzonapthofuran	Methylpyrene or isomer
Benzophenanthrothiophene	Naphthacene
Binaphthalene or isomer	Naphthonitrile
Biphenyl	Pentacene
Biquinoline	Perylene
Bisphenyl	Phenanthridine
Carbazole	Phenylphenanthrene or isomer
Chrysene	Picene
Dibenophenanthridine or isomer	Pyrene
Dibenz(ac)anthracene	Quinoline
Dibenzo(cg)phenanthrene or isomer	Tetrahydrochrysene or isomer
Dibenzoacridine or isomer	Tetramethylfluoranthene or isomer
Dibenzocarbazole	Trimethyldibenzophenanthrene or isomer
Dibenzofluorene or isomer	Trimethylfluoranthene, -Pyrene
Dibenzofuran	

The physical properties of coal tar pitch and coal tar creosote are virtually identical. Table 2 provides typical physical properties of coal tar creosote reported from several literature sources.

Table 2. Average Physical, Chemical and Fire Properties of Coal Tar Creosote

Property	Reported Value or Data	References Consulted
Molecular Weight	Not applicable because of the broad composition and large number of chemicals	
Color	Translucent brown to black; Oily colored Liquid; Yellowish to dark-greenish and brown	(Merck Index, 1989)[9]
Physical state ambient conditions	Liquid	Weiss (1986)[10]
Melting Point (°C)	20	Merck (1989); IPCS, CEC (1989)[11]
Boiling Point (°C)	194-400	Clayton and Clayton (1981)[12]
Specific Gravity	1.07-1.08	Clayton and Clayton 1981
Odor Description	Aromatic; Smokey Smell; Characteristic sharp odor	DOT (1985)[13]; Merck 1980
Odor Threshold	Naphthalene is generally the odor detected. Odor threshold for naphthalene is 0.0095 – 0.64 ppm	Agency for Toxic Substances and Disease Registry[14]
Evaporation Rate against a Reference	Slow – no reference given	Koppers Inc. MSDS[15]
Solubility in Water	Slightly soluble;	Clayton and Clayton (1981);
Solubility in Organic Solvents	Miscible with alcohol, either, fixed or volatile oils	Clayton and Clayton (1981)
Vapor Pressure, kPa @ 20°C	~ 6	Merck (1989); IPCS, CEC (1989)
Vapor Density	> 1	Koppers Inc. MSDS
Autoignition Temperature (°C)	335	Merck (1996)[16]
Flashpoint Temperature (°C)	74; >66 (closed cup)	Merck (1989); IPCS, CEC (1989)
Flammability Limits in Air	No data reported	

The International Agency for Research on Cancer (IARC) has determined that coal tar creosote is probably carcinogenic to humans. EPA has determined that coal tar creosote is probably a human carcinogen, and that coal tar pitch is a human carcinogen. EPA classified coal tar creosote as a carcinogen in the 1992 Toxic Release Inventory (TRI).

---

[9] The Merck Index: An Encyclopedia of Chemicals and Drugs, 11th ed., S. Budavari ed., Rahway, N.J., Merck and Co., Inc, 1989

[10] Weiss, G., Hazardous Chemicals Data Book, 2nd edition, Noyes Data Corp., p. 306, Park Ridge, N.J., 1986

[11] IPCS, CEC (1989) at http://www.inchem.org/documents/icsc/icsc/eics0572.htm

[12] Clayton, G. D. and F. E. Clayton (Eds.), Patty's Industrial Hygiene and Toxicology, Vol. 2B, Toxicology, 3rd revised edition, John Wiley & Sons, Inc. NY, 1981

[13] DOT, Chemical Hazard Response Information System (CHRIS) Hazard Assessment Handbook, Washington, DC, Department of Transportation, U.S. Coast Guard, 1985

[14] Agency for Toxic Substances and Disease Registry, Toxicological Profile for Creosote (update), Atlanta, U.S. Department of health and Human Services, 2002

[15] listed at http://msds.ehs.cornell.edu/msds/msdsdod/a65/m32335.htm

[16] The Merck Index: An Encyclopedia of Chemicals and Drugs, 11th ed., S. Budavari ed., Rahway, N.J., Merck and Co., Inc, 1996, pp. 435-437

Skin cancer and cancer of the scrotum have resulted from long exposure to low levels of these chemicals, especially through direct contact with skin during wood treatment or manufacture of coal tar creosote-treated products, or in coke or natural gas factories.

Cancer of the scrotum in chimney sweeps has been associated with prolonged skin exposure to soot and coal tar creosote. Eating food or drinking water contaminated with a high level of creosotes causes a burning in the mouth and throat, as well as stomach pains. Brief exposure to large amounts of coal tar creosote can result in a rash or severe irritation of the skin, chemical burns of the surfaces of the eye, convulsions, mental confusion, kidney or liver problems, unconsciousness, or even death. Longer exposure to lower levels of coal tar creosote, coal tar, or coal tar pitch by direct contact with skin or by exposure to the vapors from these mixtures can also result in sun sensitivity causing damage to skin in the forms of reddening, blistering, or peeling. Longer exposures to the vapors of the creosotes, coal tar, or coal tar pitch can also cause irritation of the respiratory tract.

NIOSH, OSHA and the ACGIH have published safe levels of exposure to limit inhalation hazards. These are referred to as 'exposure limits'. The NIOSH and ACGIH standards can be traced back to the 1950s and even earlier. NIOSH recommended exposure limits (**REL**s) are time-weighted average (**TWA**) concentrations for up to a 10-hour workday during a 40-hour workweek. A short-term exposure limit (**STEL**) is designated by "**ST**" preceding the value; unless noted otherwise, the STEL is a 15-minute TWA exposure that should not be exceeded at any time during a workday. A ceiling REL is designated by "**C**" preceding the value. Any substance that NIOSH considers to be a potential occupational carcinogen is designated by the notation "**Ca.**"

The OSHA permissible exposure limits (**PEL**) are reported in Tables Z-1, Z-2, and Z-3 of the OSHA General Industry Air Contaminants Standard (29 CFR 1910.1000). Unless noted otherwise, **PEL** are **TWA** concentrations that must not be exceeded during any 8-hour work shift of a 40-hour workweek. A **STEL** is designated by "**ST**" preceding the value and is measured over a 15-minute period unless noted otherwise. OSHA ceiling concentrations (designated by "**C**" preceding the value) must not be exceeded during any part of the workday; if instantaneous monitoring is not feasible, the ceiling must be assessed as a 15-minute TWA exposure. In addition, there are a number of substances from Table Z-2 (e.g., beryllium, ethylene dibromide, etc.) that have PEL ceiling values that must not be exceeded except for specified excursions. For example, a "5-minute maximum peak in any 2 hours" means that a 5-minute exposure above the ceiling value, but never above the maximum peak, is allowed in any 2 hours during an 8-hour workday.

Because creosote and coal tar products have such a broad composition of dangerous chemicals, a safe exposure level to the generic chemical itself is somewhat meaningless. From an inhalation exposure standpoint, the concern is for harmful PAHs. Benzo(a)pyrene [B(a)P] is an example of one of these PAHs. It is one of the most studied of these hydrocarbons, and it is a "fingerprint" chemical for detecting the presence of PAHs. The B(a)P content of coal tar is between 0.1 percent and 1 percent and it contributes to the serious potential health effects for employees exposed to emissions

Four other examples of PAH chemicals are anthracene, phenanthrene, pyrene, and carbazole; all of which are major components in creosote.

Because OSHA and EPA regulate PAHs under slightly different definitions to sources of emission, they refer to them by different names. OSHA refers to them as Coal Tar Pitch Volatiles (CTPVs) because, by definition, they come from both coke ovens and coal tar, which is derived from the coking process. EPA calls them Coke Oven Emissions (COEs), which by definition only relates to emissions from coke ovens. Both agencies, however, are referring to the same group of chemicals, including those of particular concern, the PAHs. The term CTPV (COE) is understood to mean 'benzene soluble fraction' and is therefore, a measure of the presence of Anthracene, BaP, Phenanthrene, Acridine, Chrysene, and Pyrene (i.e., the Benzene-soluble fraction is the sum of those components collected in a sample and determined to be soluble in benzene). Table 3 lists some properties of these chemicals.

CTPVs or COEs from hot industrial processes volatilize and then condense from coal or hot tar as soon as they contact normal (ambient) air temperature. CTPVs evaporate from the surface of hot coal tar during tar-processing operations. COEs (including PAHs) escape at coke ovens during coal charging (filling an oven), coking and coke pushing (removal of coke room an oven). In the case of wood treating plants, CTPVs escape from treating cylinders when the wood product is removed after the treating cycle and from fugitive emissions along the process.

The vapor pressure of CTPVs and COEs is so low at normal temperature that they behave as particulate matter. Because they are particles, they can be removed from the air the particulate filters when process temperatures are near ambient.

Table 3. Properties of some PAHs (Source of data: OSHA Web site[17])

PAH	CAS No.	Molecular Weight	Boiling Point, °C	Melting Point, °C	Color	Structure
Phenanthrene	85-01-8	178.22	340	100	white crystals	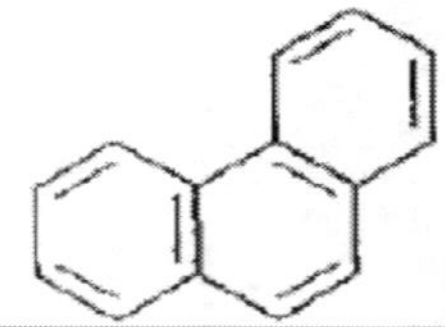
Anthracene	120-12-7	178.22	342	218	colorless crystals	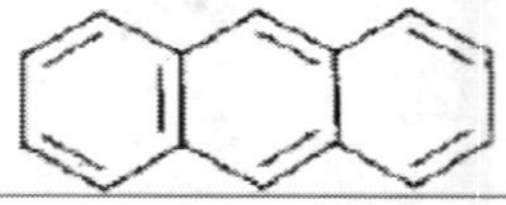
Pyrene Synonym: benzo(def)phenanthrene	129-00-0	202.24	404	156	colorless crystals	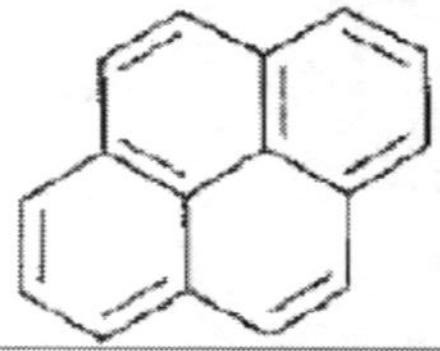
Chrysene Synonyms: 1,2-benzophenanthrene; benzo(a)phenanthrene	218-01-9	228.28	448	254	white crystals	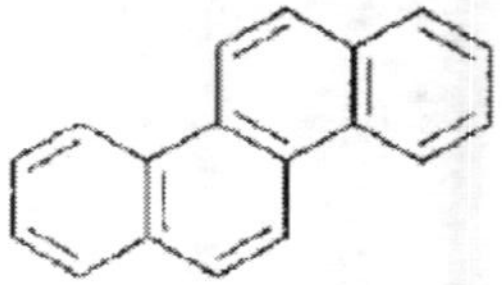
benzo(α)pyrene Synonyms: 3,4-benzopyrene; 6,7-benzopyrene	50-32-8	252.30	311	179	yellow needles	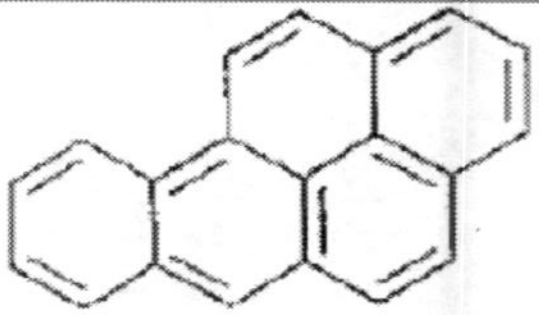

[17] OSHA Web site: http://www.osha.gov/dts/sltc/methods/organic/org058/org058.html

Standard procedures and analytical methods are provided by OSHA.[18] The following are OSHA recommended target concentrations which should not be exceeded in the workplace environment:

- 0.20 mg/m^3 for Coal Tar Pitch Volatiles (PEL)
- 0.15 mg/m^3 for Coke Oven Emissions (PEL)
- 8.88 μg/m^3 (1.22 ppm) for phenanthrene
- 0.79 μg/m^3 (0.11 ppm) for anthracene
- 9.00 μg/m^3 (1.09 ppm) for pyrene
- 3.27 μg/m^3 (0.35 ppm) for chrysene
- 2.49 μg/m^3 (0.24 ppm) for benzo(α)pyrene

Considerable understanding of community health risks from coal tar pitch-polluted air was amassed from industry-sponsored studies early on. Coal tar pitch-polluted air has been characterized at least as early as the 1950s by significant amounts of phenathrene, anthracene, pyrene, and carbozole. As noted above coal tar pitch typically contains 10% or more polycyclic hydrocarbons. A 1962 study by Sawicki, et.al.[19] reported a concentration of 1.4% B(a)P alone.

Pentachlorophenol

Pentachlorophenol (also referred to as PCP and Penta) is a chlorinated hydrocarbon. Its molecular structure is that of a phenol group with five chlorine atoms. Its chemical formula is C_6HCl_5O and as a synthetic fungicide it is part of the organochloride family. It has historically been used as a pesticide and herbicide, as well as a wood preservative chemical. Its CAS No. is 87-86-5. A major use of Penta has been as a wood preservative for power line and telephone poles, cross arms and fence posts. There are more than 130 synonyms and trade names for this chemical.

Based on 1995 production figures reported by the American Wood Preservers' Association, the largest use of pentachlorophenol and other oilborne solutions is in the treatment of utility poles. Of the 32,764,000 cubic feet of all wood treated with penta and oilborne solutions, 93 percent, or 30,617,000 cubic feet, is used in the production of utility poles. Of the 450,596,000 cubic feet of all wood treated with arsenicals, six percent, or 29,215,000 cubic feet, go into utility poles. In other words, the vast majority of pentachlorophenol is used on utility poles. Nevertheless, the treatment of utility poles are almost evenly split between penta and CCA, with 45% treated with penta, 42% with CCA and 13% with creosote.

---

[18] Go to the following Web url for detailed analytical and sampling procedures recommended by OSHA: http://www.osha.gov/dts/sltc/methods/organic/org058/org058.html

[19] Sawicki, E., T. T. Fox, E. Elbert, et. al., "Polynuclear Aromatic Hydrocarbon Composition of Air Polluted by Coal Tar Pitch Fumes," Am. Ind. Hyg. Assoc. J. 23: 482-486, 1962.

A 1983 report by NIOSH[20] reported that dioxins were present in significant quantities in commercially available grades. Table 4 reports a comparison between commercial and highly purified samples.

Table 4. Comparison between Commercial and Highly Purified forms of Pentachlorophenol Made by Dow Chemical.

Ingredient	Analytical Results	
	Commercial Brand (Dowicide 7)[a]	**Purified Product (Dowicide EC-7)[b]**
Pentachlorophenol	88.4%	89.8%
Tetrachlorophenol	4.4%	10.1%
Trichlorophenol	0.1%	0.1%
Chlorinated Phenoxyphenols	6.2%	---
Octachlorodioxins	2500ppm	15.0ppm
Heptachlorodioxins	125ppm	6.5ppm
Hexachlorodioxins	4ppm	1.0ppm
Octachlorodibenzofurans	80ppm	1.0ppm
Heptachlorodibenzofurans	80ppm	1.8ppm
Hexachlorodibenzofurans	30ppm	1.0ppm

a – Sample ID 9522 A; b – Technical grade reduced by distillation
Source: Johnson, 1978c

The presence of significant concentrations of dioxins is a serious concern for the product. Dow Chemical Co. discontinued the manufacturing of PCP in the mid to late 1970s presumably due to the increased cost associated with PCP distillation for highly purified grades and concern over a lack of consumer acceptance.

In 2001 the ACGIH[21] classified Penta as a Confirmed Animal Carcinogen with Unknown Relevance to Humans. But long before this classification Penta was understood and widely known to be a toxic and dangerous chemical linked to carcinogenicity. A 1951 study by Baader, et. al.[22] reviewed ten cases of industrial intoxications due to pentachlorophenol. Results of animal experiments, and the clinical symptomatology characterized by irritation of the mucosae and upper respiratory tract, neurological pain, generalized acne of many months duration were reported. This study shows also that well before the enforcement era of the 1970s it was recognized that Penta is readily absorbed through the skin and poses a serious inhalation hazard.

In 1987 the purchase and use of pentachlorophenol was restricted in the U.S. and the general public may no longer buy it. It has been difficult to identify sources that report

---

[20] Industrial Hygiene Surveys of Occupational Exposure to Wood Preservative Chemicals, DHHS (NIOSH Publication No. 83-106), Feb. 1983.

[21] ACGIH, Documentation of Threshold Limit Values, 7th Ed., "Pentachlorophenol," Am. Conf. Governmental Ind. Hygienists, Cincinnati, OH, 2001

[22] Baader, E. W. and H. J. Bauer, Industrial Intoxication Due to Pentachlorophenol, Industrial Medicine and Surgery, Vol. 20, No. 6, June 1951

the amount of pentachlorophenol produced in the U.S. USEPA's reported domestic production was 45 million lbs in 1983. From 1987 to 1993, according to EPA's Toxic Chemical Release Inventory (TRI), releases to land and water totaled nearly 100,000 lbs.[23]

Its earliest introduction to the wood treating industry is unclear. An early book on "Railroad Engineering" by Raymond[24] does not mention the use of pentachlorophenol among the discussions on wood treating chemical formulations of the era. A 1951 article by Badder et. al.[25] states "it was being used extensively to impregnate wood, leather, paper, paints, and glues as a protective medium, against the action of bacteria, moulds, algae, and insects." Although the general information on the WWW states that pentachlorophenol is used in the production of railroad ties, facilities studied in this review show its use has largely been restricted to the manufacture of treated utility poles. Table 5 reports some physical properties of pentachlorophenol.

Table 5. Physical, Chemical and Fire Properties of Pentachlorophenol

Property	Reported Value or Data	References Consulted
Molecular Weight	266.34	Univ. of Waterloo Web site[26]
Color	White crystals in high purity; The technical grade is a dark-grey to brown pellets/powder.	
Physical state under normal (ambient) conditions	Solid with needle like crystals	
Melting Point (°C)	191 (anhydrous), 174 (monohydrous)	UNDP[27]
Boiling Point (°C)	309 – 310	Univ. of Waterloo Web site
Specific Gravity	1.987 at 22°C	Univ. of Waterloo Web site
Odor Description		
Odor Threshold	1000-1600 ppb @ 25°C	Univ. of Waterloo Web site
Evaporation Rate against a Reference	Not Applicable under normal conditions	
Solubility in Water	3.6 x 10^{-5} mol/L at 25°C, pH 5.1. Note that there is a direct relationship between pH and solubility. That is, as the pH increases so does the solubility of PCP. Also as temperature increases, so does solubility; 20 mg/l at 30 °C.	Univ. of Waterloo Web site; UNDP
Solubility in Organic Solvents	Freely soluble in organic solvents, and alcohol, soluble in benzene, ethanol and diethyl ether; slightly soluble in cold petroleum ether;	UNDP

[23] http://www.epa.gov/safewater/contaminants/dw_contamfs/pentachl.html
[24] Raymond, W. G., Railroad Engineering, John Wiley & Sons, Inc., NY, 1923
[25] Baader, E. W. and H. J. Bauer, Industrial Intoxication Due to Pentachlorophenol," Industrial Medicine and Surgery, Vol. 20, No. 6, p. 286, June 1951
[26] http://wvlc.uwaterloo.ca/biology447/Assignments/Assignment1Submissions/On-Campus/pcp/Pentachlorophenol.html
[27] United Nations Environment Programme (Food and Agriculture Organization of the United Nations), Decision Guidance Documents: Pentachlorophenol and Its Salts and Esters, Rome - Geneva 1991; amended 1996

Property	Reported Value or Data	References Consulted
	carbon tetrachloride and paraffins.	
Vapor Pressure	5×10^{-6} mm at 19°C 3.2×10^{-4} mm at 30°C; 1.5×10^{-5} mm at 20°C	Univ. of Waterloo Web site; UNDP
Vapor Density	ND	
Autoignition Temperature (°C)	ND	
Flashpoint Temperature (°C)	Inflammable	UNDP
Flammability Limits in Air	Inflammable	UNDP

Pentachlorophenol may be released to the environment as a result of its manufacture, storage, transport, or use as an industrial wood preservative. The United Nations Development Program (UNDP) reports that pentachlorophenol has been subject to control actions primarily because of high toxicity to human and animals. Developmental and reproductive effects and liver and kidney pathology were noted in animal studies. The chemical is also highly toxic to aquatic organisms.

Pentachlorophenol contains several highly toxic dioxins which have shown carcinogenic effects in experimental animals. Those toxic impurities of greatest concern are hexachlorodibenzo-p-dioxin (HxCDD), 2,3,7,8-tetrachlorodibenzo-p-dioxin (TCDD), the chlorinated dibenzofurans and hexachlorobenzene (HCB).[28] The following is acute toxicity data reported in the literature[29]:

- Oral - LD_{50}: 27-205 mg/kg (tests with a variety of species)
- Inhalation - Rabbit (35.2%) LD_{50} 201 mg/kg
- Dermal - LD_{50}: 80-350 mg/kg (tests with a variety of species)
- Irritation - Irritating to skin, eyes and mucous membranes

The UNDP has noted that epidemiological studies on pentachlorophenol are limited. Most of the available information regarding effects of pentachlorophenol in humans comes from cases of acute over-exposure following the home use of PCP in wood preservation and herbicides and occupational exposure in agriculture and the wood treatment industry. The few available industrial surveys and epidemiological studies are limited in their usefulness because of small sample size, short follow-up periods and brief exposure periods. UNDP does however note that these studies suggest that PCP can adversely affect the liver, kidney, skin, blood, lungs and central nervous system.

---

[28] Royal Society of Chemistry, The Agrochemicals Handbook (3rd ed.), Cambridge, United Kingdom, 1991

[29] IPCS (1987). Environmental Health Criteria No. 71. International Programme on Chemical Safety, WHO, Geneva; IPCS (1989). IPCS Health and Safety Guide Series No. 19. International Programme on Chemical Safety, WHO, Geneva; U.S. Public Health Service/Environmental Protection Agency (1989). The toxicological profile for pentachlorophenol. USPHS & USEPA, Washington (Dec. 1989)

Fielders et al.[30] has reported that the extensive use of PCP to treat wood, together with its physical and fate characteristics indicate that there is likely to be widespread human exposure occurring partially through skin contact, but mainly through inhalation, which is the most dangerous route of exposure to this chemical. This is confirmed by numerous reports of its occurrence in the general environment and its presence in body-fluids, both in the general population and in exposed workers. Airborne levels of PCP production and wood-preservation facilities have ranged from several mg/m^3 to more than 300 mg/m^3 in some work areas[31].

As with creosote and coal tars, exposure may be reduced by using protective clothing. Where dermal contact is expected workers should wear gloves and other CPC. Spray applicators should wear protective clothing (e.g. overalls, jacket, gloves and boots) which is impervious to wood treatment formulations, plus a respirator, head covering and goggles when spraying. Workers should keep clothes clean or dispose of clothing. Workers should not eat, drink or smoke when contaminated from applying PCP. Automated processes and the use of closed systems, where applied, have greatly reduced worker-exposure.

Chromated Copper Arsenate

Chromated Copper Arsenate (CCA) is a mixture of chromic acid, cupric oxide and arsenic pentoxide. CCA is a wood preservative used for timber treatment. It preserves the wood from rotting due to insects and microbes. It is also known under the trade name Tanalith. Chromium acts as a bactericide, the copper acts as fungicide, and the arsenate acts as insecticide. It has had a long history in the treated wood industry and depending on end-use application was a preferred alternative to creosote, pentachlorophenol, and coal tar since the 1930s.

Like pentachlorophenol and creosote, CCA may leach from the wood into surrounding soil, resulting in concentrations higher than naturally occurring background levels. Wilkipedia[32] cites a study in *Forest Products Journal* in which up to 13% of the chromated copper arsenate leached from treated wood buried in compost during a 12-month period. Once these chemicals have leached from the wood they are likely to bind to soil particles, especially in soils with clay or soils that are more alkaline than neutral. In 1975 NIOSH concluded that chromic acid (hexavalent chrome) and inorganic arsenic were carcinogens.[33,34] Arsenic was recognized centuries earlier to be highly toxic and lethal above a threshold level.

---

[30] Fielders et al., Pentachlorophenol. In: Toxicity review, London, Health & Safety Executive, Vol., 20 pp, 1982

[31] data reported in United Nations Environment Programme, Decision Guidance Documents: Pentachlorophenol and Its Salts and Esters, Rome - Geneva 1991

[32] On-line encyclopedia: http://en.wikipedia.org/w/index.php?title=Chromated_copper_arsenate&oldid=68911387

[33] NIOSH, Criteria for Recommended Standard...Occupational Exposure to Chromium (VI), 1975

[34] NIOSH, Criteria for Recommended Standard...Occupational Exposure to Inorganic Arsenic, 1975

There are several formulations listed in the Standards of the American Wood Preservers' Association.[35,36] CCA Type C contains 47.5% chromium trioxide, 18.5% cupric oxide, and 34% arsenic pentaoxide. The preservative is usually specified on an oxide basis; and hence on an elemental basis, 1 kg of CCA contains:

- 247 gm of chromium
- 163 grams of arsenic
- 146 grams of copper

Dhamodaran and Gnanaharan[37] report that a 2% solution of CCA yields 5.0 kg/m^3 retention on a dry salt basis (DSB).

Chromium exists in several different forms: divalent, trivalent, and hexavalent. Only hexavalent chromium (CrVI) is recognized as a human carcinogen. Chromium in the CCA preservative solution exists in the hexavalent form. NIOSH considers all CrVI compounds to be potential occupational carcinogens. An increased risk of lung cancer has been demonstrated in workers exposed to CrVI compounds. Other adverse health effects associated with CrVI exposure include dermal irritation, skin ulceration, allergic contact dermatitis, occupational asthma, nasal irritation and ulceration, perforated nasal septa, rhinitis, nosebleed, respiratory irritation, nasal cancer, sinus cancer, eye irritation and damage, perforated eardrums, kidney damage, liver damage, pulmonary congestion and edema, epigastric pain, and erosion and discoloration of the teeth.[38]

Exposure to hexavalent chromium has been a recognized hazard for increased risk of lung cancer of exposed workers for more than 50 years.[39,40] Mancuso and Hueper studied the mortality experience of chromium-exposed workers employed between 1931 and 1937 at a Painesville, Ohio facility. Results of their study were published in 1951.[41] Updates on the cohorts were subsequently published by Dr. Mancuso in 1975[42]and again in 1997[43], consistently reporting an excess risk of lung cancer among exposed workers.

---

[35] American Wood Preservers' Association, Standard C1-99 All Timber Products – Preservative Treatment by Pressure Processes, In AWPA Book of Standards, Grabury, TX, pp. 23-32, 1999

[36] American Wood Preservers' Association, Standard P5-99 ALL Standard for Waterborne Preservatives, In AWPA Book of Standards, Grabury, TX, pp. 23-32, 1999

[37] Dhamodaran, T. K. and R. Gnanaharan, Optimizing the Schedule for CCA Impregnation Treatment of Rubber Wood; Holz als Roh- und Werkstoff, Springer Berlin / Heidelberg Publisher, Volume 59, Number 4 / August, 2001

[38] http://www.cdc.gov/niosh/topics/hexchrom/

[39] Machle W, Gregorius F. Cancer of the respiratory system in the United States chromate-producing industry. Public Health Rep. 1948;63:1114–1127

[40] Mancuso TF, Hueper WC. Occupational cancer and other health hazards in a chromate plant: a medical appraisal I: Lung cancers in chromate workers. Ind Med Surg. 1951;20:358–363

[41] Mancuso TF. Occupational cancer and other health hazards in a chromate plant: a medical appraisal. II. Clinical and toxicologic aspects. Ind Med Surg. 1951;20:393–407

[42] Mancuso, TF. Consideration of chromium as an industrial carcinogen. Proceedings of International Conference on Heavy Metals in the Environment: 27–31 October 1975; Toronto. pp. 343–356

[43] Mancuso TF. Chromium as an industrial carcinogen: Part I. Am J Ind Med. 1997;31:129–139. doi: 10.1002/(SICI)1097-0274(199702)31:2<129::AID-AJIM1>3.0.CO;2-V

OSHA has noted that the first occupational standard for Permissible Exposure Limit (PEL) was introduced in 1943 by the American National Standards Institute as a level adequate to prevent nasal perforations in chromium-exposed workers.[44] The 1943 recommended PEL was 52 *ug*/m^3 (0.052 milligrams per cubic meter (mg/m^3)).This limit was adopted by OSHA in 1971 when the agency was created. In 1976, OSHA announced plans to lower the Cr(VI) standard[45]. But it was not until1994 that the OSHA administrator acknowledged that "there is clear evidence that exposure...at the current PEL...can result in an excess risk of lung cancer"[46].

OSHA has revised its Permissible Exposure Limit (PEL) for CrVI to 0.005 mg/m^3 and 8-hour TWA and sets the action level at 0.0025 mg/m^3. This represents a 10-fold reduction from the previous standard. The current TLV established by the ACGIH for insoluble CrVI compounds is 0.01 mg/m^3.

Hexavalent chromium and arsenics can enter the body in two ways: by being inhaled or by being swallowed. Both can be inhaled when dust, mist, or fumes are in the air. Contaminated dust can also get on cigarettes.

OSHA defines two methods of making an exposure determination: *scheduled monitoring* and *performance-oriented option*. As a part of the scheduled monitoring, employers must conduct initial representative air monitoring to determine the 8-hour TWA exposure for each job classification and each work area per shift. After initial monitoring, follow-up monitoring is required.

TECHNOLOGY OVERVIEW

Although no two wood preserving plants are identical, they essentially the same unit processes and operate in accordance with product specifications established by the American Wood Preserver's Association. And even though production capacities can vary significantly between plants, production platform footprints are comparable between facilities with significant differences in production capacities. There are operating and process differences between plants which do make some facilities more prone to waste and pollution generation; some of which are highlighted in this review.

Figure 4 is a material process flow diagram showing the various stages of production that can be found among tie treating plants.

---

[44] Occupational Safety and Health Administration. Occupational Exposure to Hexavalent Chromium, Proposed Rule. 69 Federal Register 59313. October 4, 2004

[45] Occupational Safety and Health Administration. Occupational Exposure to Chromium, Proposed Standards. 41 Federal Register 18869. May 7, 1976

[46] Dear, JA.; Assistant Secretary of Labor for Occupational Safety and Health. Letter to Sidney M. Wolfe, Director, Public Citizen's Health Research Group, March 8, 1994. http://dockets.osha.gov/vg001/V026A/01/41/71.PDF

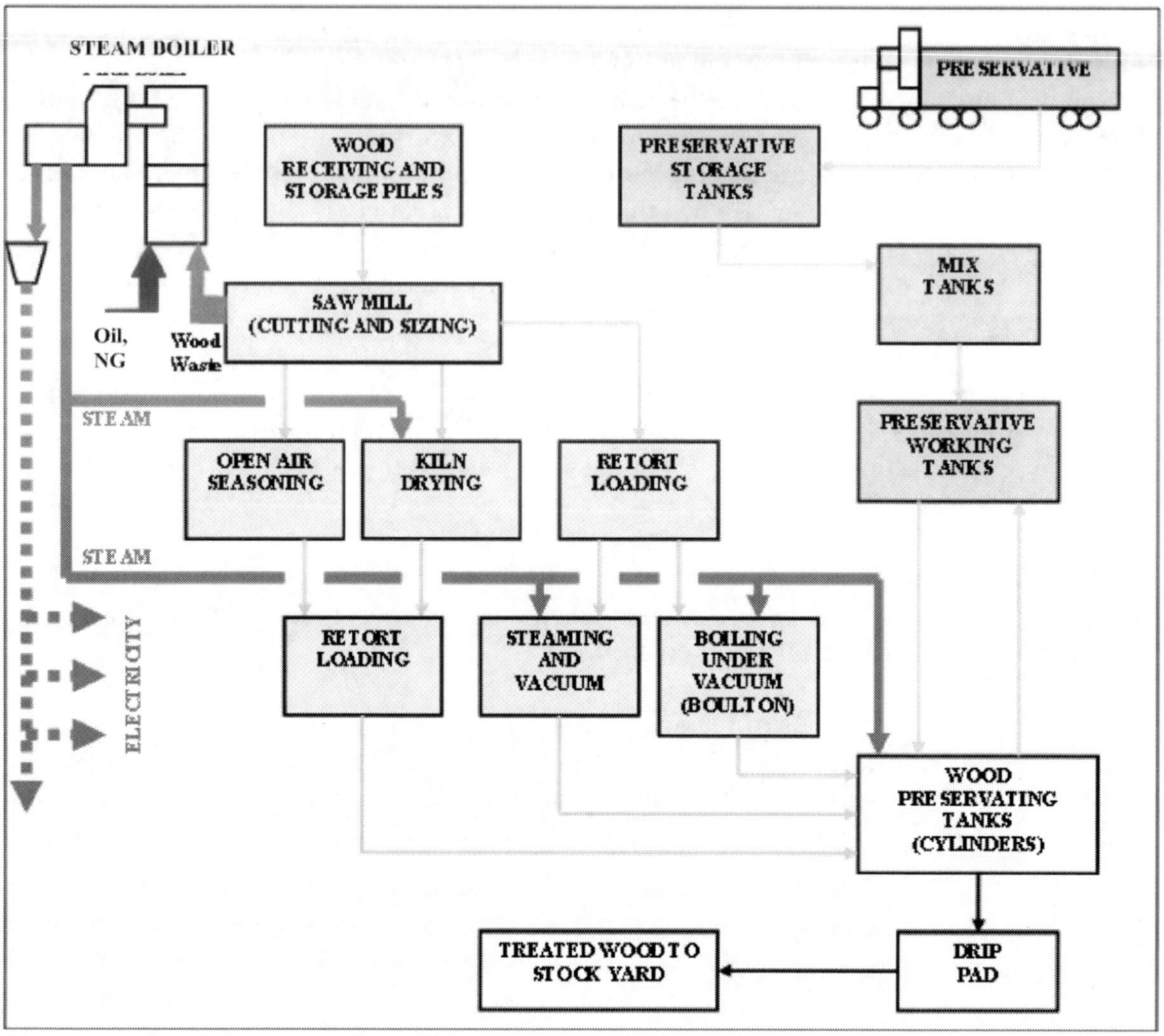

Figure 4. Material process flow diagram of a typical wood treating plant.

Conditioning: Free water entrapped in wood cell cavities slows or prevents the entrance of the preservative chemicals. Wood moisture levels must be reduced before treatment can be accomplished. Moisture reduction techniques include air seasoning and artificial conditioning.

Air seasoning involves storing untreated wood outdoors in piles. The unseasoned wood that is exposed to the open air dries slowly until it comes into approximate equilibrium with the relative humidity of the open air. The disadvantages of this technique are that large land areas are generally needed, long lead times for drying, and some wood species tend to rot before drying to required levels. See Figure 5 as an example.

Figure 5. Untreated railroad ties undergoing natural air seasoning.

Artificial conditioning includes kiln drying and the use of dehydrating agents such as naphthalene. The Somerville wood treating plant in Texas as an example relied upon the naphtha from 1953 to 1995, where wooden ties were conditioned in a cylinder using a hot solution of naphthalene. The naphthalene forces the water out of the wood pore cells; then by draining and then evacuating the cylinder under vacuum, the naphtha and water can be removed and later separated in order to recover the chemical. This technique reduces natural drying times from 12 months to under 11 hours but it is a polluting process that generates both air emissions and wastewater.

More common artificial seasoning techniques are steaming-and-vacuuming; boiling-under vacuum (known as the Boulton process); and kiln drying. These treatments remove substantial amounts of moisture from the wood but also raise the temperature of the wood closer to the treating temperature, thus saving time and energy. Steaming and Boultonizing also disinfect the wood (same as vapor drying described above). Although not true in some historical operations, most systems are segregated and conditioning is performed in dedicated cylinders that do not contain preservative.

The steaming and vacuum method has been used largely for treating southern pine poles. The process involves heating the wood with live steam for a period of time and under slight pressure. Then a vacuum is drawn.

The Boulton process was largely developed for hardwoods and Douglass fir. It is generally performed in the same cylinder that is used to treat wood. In the process, the cylinder is first charged with the wood and then the heated preservative is used to heat the wood charge for 1 to as much as 24 hours. At the end of the cycle a vacuum is drawn. The final step is to return the preservative to the work tank. The last step is referred to as "blow back" as compressed air is used to blow the preservative back into the work tank. Treatment systems may also employ pumps to withdraw the preservative from the treatment cylinder and return it to the work tank. Even though pump withdrawal systems do no actually blow back the preservative, the term blow-back is still used.

Treating: Wood preserving methods fall into two general classes: pressure processes, in which the wood is placed into treating cylinders or retorts where it is impregnated with the preservative under elevated pressure and temperature; and non-pressure processes.

Non-pressure processes include thermal methods in which heat is applied, and nonthermal methods which include brushing, spraying, dipping, and soaking. The non-pressure processes are employed only with oilborne preservatives and are not suitable for high volume productions of railroad ties and utility poles which also require high degrees of penetration.

Pressure methods are accomplished in closed cylinders. The cylinders are generally constructed of steel and are sometimes referred to as retorts. Most cylinders are standardized to 2 to 3 m (6 to 9 feet) in diameter and can be more than 46 m (150 feet) in length. They are built to withstand working pressures up to 1,720 kPa (250 psi).

As shown in Figure 6 the wood ties are stacked onto special tram cars and transported into the retort. The cylinder is closed or sealed, then filled with preservative. Steam coils generally run along the floor of the cylinders to maintain elevated temperatures. Applied pressure forces the preservative chemical into the wood until a desired penetration is achieved.

Figure 6. Shows untreated wood ties being transported on trams for loading into treating cylinders (top and middle photos). Lower photo shows close-up of a drip pad. Treated wood is allowed to cool at the pad and the drippings are collected.

There are three pressure processes: full-cell, modified full-cell, and empty-cell. The processes are distinguished by the sequence in which vacuum and pressure are applied to the cylinder. The terminology "empty" and "full" refer to the level of preservative that is retained in the wood cells. The full-cell process achieves a high-level of preservative retention in the wood cells, but less penetration is achieved compared with the empty-cell process. The empty-cell process achieves relatively deep penetration with less preservative retention than does the full-cell process.

The full-cell process is known as the Bethel process. It is used when the maximum pressure retention levels are demanded in product specifications (e.g., when treating timbers with creosote that are used for protection against marine borers). The full-cell process is intended both for creosote and waterborne preservatives.

The modified full-cell process is most often applied with waterborne preservatives. The method is similar to the full-cell process with the exception of the initial vacuum levels applied. The modified full-cell process uses less vacuum than the full-cell. Vacuum levels are determined by the wood species being treated and the desired retention levels.

The empty-cell process is capable of achieving deep penetration with a low net preservative retention level. This process is most often used with oil preservatives. The two most common processes are the Rueping process and the Lowry process.
In the Rueping process, compressed air is forced into the treating cylinder in order to fill the wood cells first with air prior to preservative injection. The pressurization times vary with wood species. With some species the pressurization period is short; on the order of minutes. More resistant wood species can require pressure periods of up to one hour. Air pressures can range from 172 to 690 kPa (25 to 100 psi) depending on the desired net preservative retention and wood resistance. Once the initial pressurization period is completed, the preservative is pumped into the cylinder, forcing the air into a Rueping tank (an equalization vessel) at a rate which maintains a constant pressure in the treating cylinder. After the treating cylinder is filled with preservative, the pressure is raised above that of the initial air pressure and maintained at an elevated level until the point when no more preservative can be forced into the wood or until a sufficient amount has been absorbed to achieve a desired retention level after the final vacuum. After the pressure period, excess preservative is removed from the cylinder and a final vacuum is applied. The final stage of vacuum can recover anywhere from 20 to 60% of the gross amount of preservative injected into the cylinder. The retort is then unloaded.

The Lowry process is an empty-cell process but without the initial air pressure. The preservative is pumped into the retort without either an initial air pressurization or vacuum stage. This process variation traps air that is already in the wood. Once the retort is filled with preservative, pressure is applied and the rest of the process is identical to the Rueping process. The Lowry process is generally thought of as being more economical than the Rueping process as it requires a lower capital investment (e.g., it does not require an air compressor, an extra tank for the preservative, or a pump to force the preservative into the cylinder against the air pressure). This also implies lower maintenance and operating costs.

SOURCES OF POLLUTION

There are numerous point and fugitive sources of emissions from wood preserving plants. Older plants are plagued with decades of poor housekeeping practices and the use of surface impoundments from historical practices predating the Resource Conservation and Recovery Act (RCRA). The use of unlined surface impoundments for waste management have resulted in significant impairment to groundwater quality in some communities due to leaching.

Wood preserving facilities have generally been devoid of even the most basic forms of pollution prevention such as good housekeeping practices. This has led to severe site contamination at some facilities which has contributed to contaminated stormwater runoff into neighboring communities.

In addition, some facilities disposed of toxic waste streams on plant property by land farming and spraying applications, whereas others fed treated wood scrap and process wastes as residual fuels to wood waste boiler operations with minimal to no pollution controls. These practices not only resulted in off-site air pollution exposures for neighboring communities, but deposition of airborne contaminants onto plant properties added cumulatively to site contamination. Cumulative deposition of toxins has likely created long-term area source emissions for some facilities since contaminated soils can be re-suspended thus creating intermittent air pollution problems. Specific examples of poor environmental management practices at various tie plants are cited under the Case Studies section.

The nature of solid and liquid wastes from wood treating facilities is that they tend to be highly concentrated and contain the hundreds of toxins that comprise creosote and or the dioxins contained in pentachlorophenol, or heavy metals as in the case of CCA. Not paying attention to and taking corrective actions has lead to long-term site contamination and highly contaminated stormwaters. See for examples the neglected housekeeping problems shown in Figures 7 through 9.

Figure 7. Drippage and leaching from treated poles resulting in site contamination and stormwater.

Figure 8. Stacked treated ties. Soil show signs of staining. The consequence is that creosote enters the storm water.

Figure 9. Drippage and leaching from treated wood ties onto soil. The contaminated areas to the left in the figure are stained soils off of the tram track drip pad.

Poor housekeeping and the lack of preventive maintenance programs on treating cylinder door seals is yet another common example where both toxic sludges and air emissions contribute to cumulative site contamination and subsequent exposures for both workers and neighboring communities. See as an example the photo shown in Figure 10.

Figure 10. Sludge and drainage from treating cylinders.

When trams of treated wood are removed from treating cylinders they are still relatively hot. Volatiles are released into the air while liquid (water and sludge containing wood lignin, oils and treating chemicals drain from cylinders.

Because the treated wood charge is pulled from cylinders at an elevated temperature, lower boiling point organic compounds are released in the form of aerosols. The emissions are quite visible at the time of product removal. See the photo in Figure 11.

Figure 11. Escaping vapor emissions from a conditioning cylinder.

The U.S. EPA reports low emission factors for these releases. An examination of the emission factors reported in U.S EPA's AP42 show that the data is nearly all reported by the industry itself based on personal air monitoring of workers working around treating cylinders. Early studies conducted by NIOSH in the 1970s have also supplemented the databank of emission factors, but a review of these original studies shows that they are plagued with analytical problems, data collected was not statistically regressed, and some personal exposure measurements were discounted for reasons that seem arbitrary. Wood treating facilities do not monitor these emissions on a regular basis, but rather rely almost entirely on calculating emissions using AP-42. The implications from this are that both workers and communities within close proximity to plants are being exposed to higher levels of PAHs (Polyaromatic Hydrocarbons) and VOCs (Volatile Organic Compounds) than are reported under the Toxic Release Inventory (TRI) program.

The primary emission point for the steaming and vacuum method of conditioning is the vacuum pump system vent. The vacuum system is comprised of condensers, which are a part of the process system and are not considered as emission controls unto themselves. In the Boulton process the emission points are the vacuum pump vent during the vacuum stage of conditioning and the work tank vent during the blow back stage of the conditioning process.

Blow back episodes also occur at the end of a treatment cycle when treating solution is returned to the work tank. Air displaced by the return solution is vented to the surroundings. Closed systems vent the displaced air back to the treating cylinder in order to fill the freeboard region that is created as the preservative is withdrawn from the cylinder. Closed systems therefore have no air emissions; however, a problem can occur when the amount of preservative that is blown back is not monitored closely. In this situation air begins to blow up through the work tank. VOCs tend to absorb in the air stream or bubble and are carried out through the tank vent. As such, tank vents should be equipped with controls such as activated carbon filters.

There are a number of point and area sources that exist after the treated wood is removed from cylinders. These include wood that are held on the drip pads prior to being moved to stock yards and the stacks of treated ties at the yard. The rate of fugitive emissions is a function of vapor pressure which in turn is driven by ambient temperature conditions. The degree of exposure to sensitive human receptors in turn depends on local atmospheric conditions such as wind speed and direction, precipitation, solar insolation and local terrain. Emission rates are greatest immediately after treated wood is removed from treating cylinders and tends to decline as the wood cools down. Large stockpiles of treated wood expose significant surface areas for evaporation of VOCs to occur which can aggravate air pollution particularly during the summer months of operations.

Other sources of emissions include wastewater treatment operations, surface impoundments and organic liquid storage tanks. Large wood treating facilities can be equipped with several million gallons of tankage for oil and treating chemicals which pose significant venting episodes from tank breathing operations. These again are aggravated during warm months of operations.

Odorous emissions from steam jet vacuum systems are another source of emissions. Single-pass water-cooled condensers tend not to condense all the organic constituents in exhaust streams. Care is needed to properly size condensers so that there is adequate cooling water to condense organic constituents in exhaust streams. Two steam jet ejectors in series with a barometric (direct contact) intercondenser between the units is an approach commonly used in oil refinery operations to reduce such emissions and seems equally applicable to wood treating plants. Another option that refineries employ is to replace steam jet ejectors with a vacuum pump and then to duct the exhaust stream to activated carbon adsorbers or to an afterburner. These approaches have proven highly effective in removing organic compounds from exhaust streams in other industry sectors but appear to have been rarely applied at wood preserving facilities.

Figure 12 provides a process step diagram which identifies the major environmental aspects of a wood preserving plant. Wood preserving facilities have a responsibility to manage every one of the aspects.

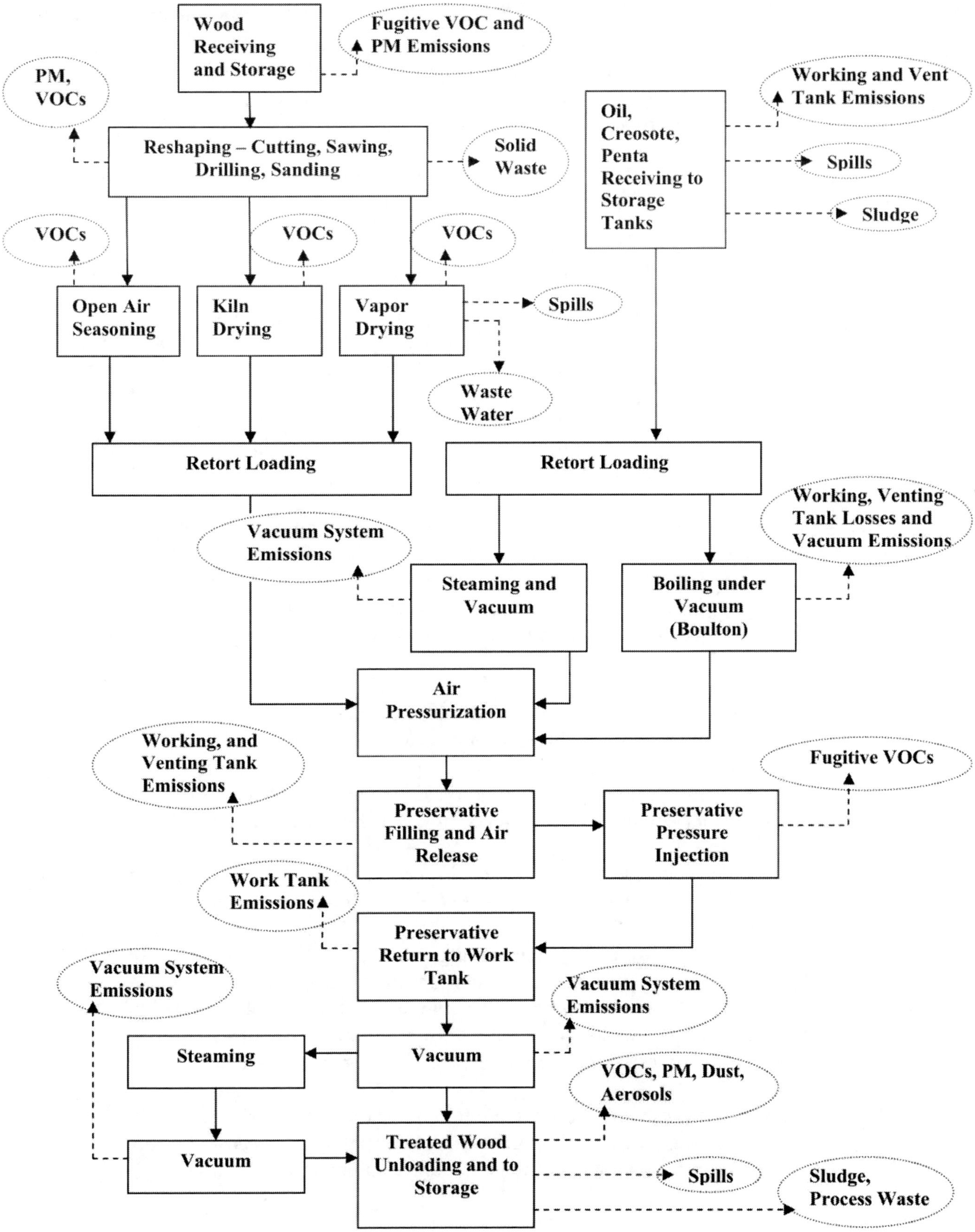

Figure 12. The major environmental aspects.

CASE STUDIES

Four tie manufacturing plants examined in some detail by the researchers are the Grenada Tie Plant, a wood treating plant located in Alexandria, LA, a wood treating plant located in Pineville, LA and the Somerville Tie Treating Plant. More information is known about the operational practices of some of the plants than others. The waste and pollution management practices of these facilities are first briefly described. A conceptual exposure model for the communities is then presented along with the results of health risk studies for the Somerville Texas plant.

Waste Management Practices at the Grenada Tie Plant

The Grenada Mississippi facility is principally a tie manufacturing plant although poles are also made. The plant was commissioned circa 1900 and is still in operation today. Both creosote and pentachlorophenol are used, but historically Copperized Chromated Zinc Chloride was used. The facility occupies a land area of approximately 172 acres, but at one point in its history it occupied a larger area which is believed to have been developed as residential. The facility has three treating cylinders that are 1.8 m (6 ft) in diameter and 36 m (120 ft) in length.

The facility lies adjacent to a residential community. The company is currently involved in a community action toxic tort which claims residents have been exposed to dioxins, PAHs, VOCs and heavy metals. Among the environmental practices that the facility relied upon that are believed to have contributed to community exposures are a land farming practice and a wood waste boiler that burned treated wood ties and process sludges.

The land farming practice involved the application of irrigating a portion of the property with wastewater containing pentachlorophenol between circa 1974 and 1988. The company contended that pentachlorophenol biodegrades and further since it is a pesticide, its application to land under controlled conditions allowed for a cost effective way to dispose of wastewater streams that it believed not to be highly contaminated.

There are several problems with this practice. Figure 13 shows a 1981 photograph of the spray field operation (there were actually three fields on the property; the largest one is shown in the photograph). The first problem is that the company did not do any active pretreatment of the wastewater prior to spraying onto land. It relied on a series of settling ponds where it has been argued that long retention times aided in the removal of settleable solids and allowed for some level of biodegradation. But as reported in Table 4, commercial grades of pentachlorophenol are known to contain dioxins and furans. These constituents were never tested for or monitored in the wastewater that was sprayed.

The second problem is obvious from the photograph shown in Figure 13; namely, the location of the field was in close proximity to a residential area. Because the company used standard garden-type spray nozzles, it created a fine mist of toxic aerosol droplets which could be readily transported throughout the residential community depending upon local environmental conditions.

To the company's credit, it conducted pilot tests in small areas in the 1970s and generated technical reports which substantiated degradation of pentachlorophenol in soil. The Mississippi Department of Environmental Quality (MDEQ) sanctioned the operation and issued a permit to operate based on these studies. The operating permit restricted spraying to days when less than 0.1 inch of rainfall occurred in order to minimize runoff.

The amount of wastewater disposed of by this method has been estimated to be 55 million gallons. In 1988 RCRA rules reclassified the spray field as a hazardous waste management unit and operations ceased. The facility then invested into a wastewater treatment plant which discharges to the municipal system. Operating records from 1988 show that spraying was conducted round the clock in order to dispose of as much wastewater as possible before the drop-dead date for the operation. This implies that residents may have been experienced near continuous exposure to toxic aerosols for several months.

Figure 13. Toxic spray field operated at the Grenada Tie plant between 1974 and 1988.

The plant also installed and continues to operate a wood waste boiler. The unit was designed to burn clean (untreated) wood waste derived from sawing and cutting operations as the primary fuel. The boiler's furnace is a mass burning bottom grate configuration.

Circa 1984 a waste to energy program was implemented in which a permit from the MDEQ was obtained to burn process wastes. The permit limited the amount of waste that could be burned based on heating value, heavy metals content, and control of combustion temperature, and further stipulated that the waste stream should be relatively clean (i.e., not contain inerts). The boiler had a multiclone which was a part of the original boiler system package. The permit issued was based on demonstration of 99.999% Destructive Reduction Efficiency (DRE) of dioxins at steady state.

In practice, the operation was not capable of maintaining high combustion efficiency. Both the composition and consistency of the wastes fed varied over wide limits which are not practical to control, thus causing fluctuations in combustion efficiency.

Evidence of this can be seen by the photograph in Figure 14, which shows sludge in a hopper. A close examination of the waste shows it to be highly inhomogeneous. This is not surprising considering the sources of the process wastes. The facility not only burned sludge derived from the cleanout of its treating cylinders, but creosote contaminated soils resulting from poor housekeeping and excessive spillage resulting from blown cylinder door seals that occurred on a frequent basis. In addition, the company shipped process wastes from 15 of its other wood preserving facilities across the country to Grenada to be burned in the wood waste boiler. More than 10 million pounds of such waste was burned between 1984 and 1992.

Figure 14. Process waste in a hopper prior to being fed to the boiler. The photo to the right shows a belt feeder in which waste was dumped in an un-metered manner.

Combustion efficiency depends on a balance between residence time, temperature and the degree of mixing or turbulence between the fuel and combustion air and gases. The ability to maintain high combustion efficiency is in part influenced by the ability to maintain uniform fuel feed. The company did not meter its waste into the boiler in a controlled fashion, nor did it have adequate quality control protocols to ensure consistent waste feed composition and consistency. Furthermore, its boiler did not have adequate sensors or controls that would allow an operator to adjust waste feed or boiler operating conditions to ensure 99.999% DRE during normal operations.

The waste to energy program devised by the company created additional negative environmental impacts. One of these additional aspects can be seen by making a closer examination of Figure 14. In the background an open top vessel emitting vapors can be seen. Wastes brought in from other facilities had wide moisture and creosote compositions. The facility had the practice of boiling off the water and recovering free oil (creosote) using a crude form of flash distillation. The practice of conducting the recovery process in an open top vessel resulted in uncontrolled air emissions of VOCs and PAHs.

Another negative impact is that the facility literally brought millions of 55 gallon drums of waste to its facility over the life of the practice. A review of more than five decades of operational practices for this facility show that poor housekeeping persisted. At no time did the owner/operator invest in waste minimization and pollution prevention programs nor did it maintain rigorous preventive maintenance practices as substantiated by records of numerous episodic releases from blown treating cylinder door seals. Since good housekeeping and generally accepted best management practices for chemical and waste handling were not applied in manufacturing, there is no reason to believe that the wastes brought to the facility were managed to any greater degree of care. Evidence of poor management of wastes can be seen by the photograph shown in Figure 15.

Figure 15. Stockpiled drums on bare ground adjacent to a drainage ditch and soil showing signs of contamination.

With the introduction of the federal Boiler Industrial Furnace (BIF) standards in 1991 and an enlightened MDEQ, process waste was reclassified as hazardous waste in Mississippi and the practice ended.

In 1994 however, the owner/operator reinvented its waste to energy program by offering to the railroad industry a means of disposing of their used railroad ties. A railroad company agreed to pay the facility to accept used railroad ties that the facility reduced in size (through chippers) and then fed to its wood waste boiler.

The railroad company and the Grenada owner/operator entered into an agreement in which the facility operator accepted all responsibility for the management of the waste stream. This agreement can be viewed as a double edge sword. On the one hand, it reflects that the railroad did not apply sufficient due diligence to determine whether its wastes were being managed responsibly, and simply passed on liabilities to the wood preserving plant. On the other side, the agreement entered into by the wood treating plant is a reflection that it had a high level of confidence that its technology and practices did offer a responsible management of the wastes.

At a glance, the practice has the earmarks of a successful recycling program. The disposal of used railroad ties constitutes a significant and growing problem in the United States. The most common form of disposal is landfilling, which is not only costly but is an inefficient use of land and has negative environmental impacts to groundwater quality. The conversion of these wastes to energy is therefore a worthwhile goal. Furthermore, as a business model, the practice makes sense. The owner/operator is being paid to accept the liability for and to manage another generator's waste, while at the same time, is receiving no-cost fuel to generate steam and electricity needed to run its plant.

It may be further argued that the use of wood as a primary fuel is less polluting than oil and thus aids in energy conservation. The operator of course did have the option of using natural gas, which is the cleanest of fuels. The owner/operator did attempt to apply for a permit to operate using spent railroad ties as the primary source of fuel but was unsuccessful.

In practice this operation also created negative environmental impacts to the community. A review of boiler operator logs, maintenance records, steam charts and internal corporate records revealed that the boiler and its feeder were plagued with operational problems that accompany poor combustion. As in the practice of sludge burning, a lack of continuous temperature and stack monitoring and real time control tools were lacking. These tools are essential for controlling boiler operations and maintaining high combustion efficiency, especially when a hazardous waste constitutes a part of the fuel mix.

Stack emissions from the boiler constitute more than 60% of the total air emissions from the plant. A review of the owner/operators 2002 Title V permit application shows that more than 1 million pounds of hazardous air pollutants in the form of VOCs, PAHs and particulates are released from the boiler stack on a yearly basis and in immediate proximity of a residential community.

Although the physico-chemical properties of PAHs in creosote vary considerably the semi-volatile property of some PAHs makes them highly mobile throughout the environment. A study by the European Commission[47] shows deposition and re-volatilization distribute PAHs between air, soil and water bodies. Both mass and chemical speciation of PAH-in-air changes following emission. PAHs containing 5 or more rings (including BaP) are predominantly in the particulate phase; whereas those containing 2 or 3 rings are almost entirely in the vapor phase. During warm summer months the 2 or 3 ring constituents tend to be more volatile and thus pose greater inhalation risks. The 4-ring compounds are particle-bound; however as noted in the EC position paper, these compounds have the greatest seasonal variability between phases. The majority of particle-bound PAHs are found on small particles that are less than 2.5 microns. These particle sizes represent the highest levels of particulate inhalation risks.

The cumulative effects of PM (Particulate Matter) deposition onto the owner/operator's property has likely also created a source of continued elevated background levels of air pollution. Since PAHs are unto themselves particulates under ambient conditions, the heavier particulates settle out closer to the source than PM10 and smaller. From years of burning treated wood and sludge the soil becomes contaminated with natural attenuation such as runoff and groundwater leaching actions removing some of the contamination. Contaminated soil becomes resuspended into the environment as dust and from attrition due to the use of heavy machinery. See as an example the photograph in Figure 16. The resuspension of contaminated soil and dust pose a significant exposure risk to workers and the community from ingestion.

---

[47] "Ambient Air Pollution by Polycylcic Aromatic Hydrocarbons (PAH). Position Paper," Office for Official Publications of the European Communities, Luxembourg, July 27, 2001

Figure 17. Forklift operation in stockyard showing a trail of dust.

Waste Management Practices at Two Louisiana Wood Treating Plants

The investigators examined the practices at two tie wood preserving plants located in Pineville and Alexandria, Louisiana. Both facilities are considerably smaller in production capacity than the Grenada plant although the footprints for the treatment portions of the facilities are comparable. These plants started up in the 1940s and 1950s, and at one point in the operational history were owned and operated by the same entity. The owner/operator is entangled in a community toxic tort in which residents claim both property value diminution and health damages from exposures to dioxins, PAHs and VOCs. Since litigation is still in early stages of discovery, only a brief overview of the environmental management practices of the two facilities is given.

Figure 18 shows satellite images of the plants. The facilities are within 3.5 km (2.2 mi) of each other and surrounded by residential communities.

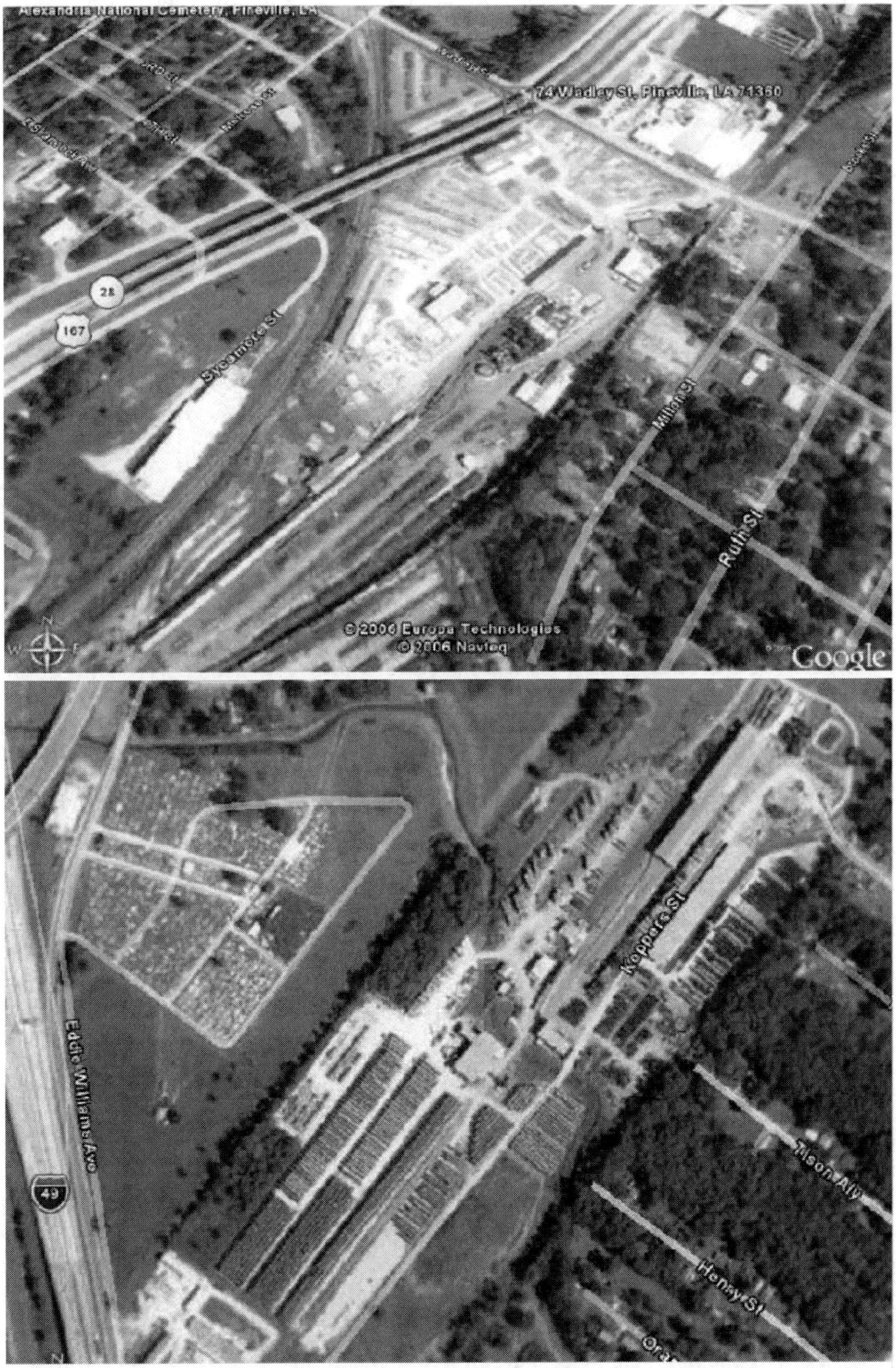

Figure 18. Satellite image of the Pineville (top) Alexandria plant (bottom). (Image downloaded from Google Earth).

The plants historically used creosote/oil mixtures, pentachlorophenol and CCA at different operational periods. Since 1987 one of these plants has had an on-site wastewater treatment unit, with the discharges from both plants sent to city sewers since the late 1980s. Among the poor environmental management practices adopted by the facilities:

- Up until circa 1970 one of the plants used an unlined pond that was a part of a natural canal as a crude oil/water settling basin to recover oil, creosote and pentachlorophenol. When a new owner took over this practice ended but the canal continued to be used to dump vacuum pump waste water until the Louisiana Department of Environmental Quality (LDEQ) discovered the practice and ordered termination in 1981.
- Up until at least the mid 1980s the facility had poor operations and maintenance practices. One explosion of a work tank caused the release of up to 12,000 gallons of creosote/oil. Since such occurrences are rare this incident is a reflection of a dangerous and non-standard operation.
- Both facilities have had a history of horrendously poor housekeeping practices. Several site visits by the LDEQ and Phase I and II site assessments from the 1980s throughout the 1990s and even early part of this decade show many parts of the property to have staining and significant levels of soil and groundwater contamination.
- Poor chemical handling practices were used for many decades and are among the contributors to poor housekeeping and site contamination. Neither facility employed chemical inventory controls; chemical unloading areas did not use secondary containment; there were no spill prevention and contingency programs that were practiced.
- Temporary treated wood storage areas were employed at one of the facilities. This practice resulted in wide site contamination from drippage and leaching of treating chemicals from the wood. The facility never placed any program into effect to address these losses and spills.
- Site inspections from the LDEQ showed chemicals to be drained from buildings to the outside. Since the facilities maintained machine and tool shops, properties became contaminated with chlorinated solvents used for parts washers and tool cleaning.
- The owner/operator had the practice of distributing treated wood scrap to the local residents at least up until 1982. This exposed local residents to dermal contact with creosote as well as inhalation risks from the burning of these wastes in home fireplaces and wood burning stoves.
- Throughout a good part of the 1980s no secondary containment of tanks was employed. Spills and runoff were allowed to freely communicate with the storm water runoff.
- At least one underground storage tank comprised of a buried railroad tank car was used for many years to store creosote. The facility never tested the tank for its integrity, leakage, spills, and appears to have closed the vessel in place without conducting a proper closure.

- The facilities have an impressive record of citations and violations that begin from the early 1980s through present times. Among the repeated violations are:
    - Areas of old spillage ignored by the owner/operator.
    - Failure to remove hazardous waste (creosote) from railroad track areas.
    - Allowing creosote wastes to remain in stagnant ditches for more than a decade.
    - Continued waste discharges to drainage ditches and to a surface impoundment that overflowed to an outfall.
    - Allowing abandoned waste pits to persist and overflow into a drainage canal.
    - Creating hazardous situations off site where bogs containing creosote were discovered and public workers came into contact with contaminated areas.
    - Allowing contaminated discharges from shop areas to drain into a canal leading off site.
    - Discharging wastes to a natural drainage course leading off site without a permit.
    - Improper storage of hazardous wastes on site with multiple violations of the RCRA.
    - Failure to submit permit applications and yearly waste reports to the LDEQ.
    - Failure to properly construct, maintain and obtain environmental samples from monitoring wells it was under compliance orders to implement.
    - Failure to notify the LDEQ of contamination discovered in groundwater in a timely fashion.
    - Failure to submit and implement corrective action plans as specified by the LDEQ for site cleanup and groundwater remediation activities.
    - Failure to devise and maintain a copy of closure plans for specific hazardous waste management units.
    - Improper sampling of monitoring wells and the adoption of environmentally unsound practices such as dumping contaminated purge water from wells onto ground.
    - Failure to collect and properly dispose of monitoring well contaminated water.
    - Failure to sample monitoring wells and report contamination.
    - Failing to maintain waste manifest records and then improperly reporting waste generation in yearly reports. In the latter citation a spot site inspection for the LDEQ revealed many tons of off-site hazardous waste manifest records despite the year end report submitted to the LDEQ reporting zero off-site shipments.
    - Releases to storm water from outfalls not reported to the LDEQ.
    - Failure to report and take corrective actions on hazardous waste discharges in ditches and releases to land.
    - Storing hazardous wastes in tanks without a permit.
    - Making false statement to regulators claiming all environmental damages were the result of previous owner operations.

- Failure to properly identify, label and contain hazardous wastes in containers stored on site.
- Being named as a Potentially Responsible Party for one of the facilities for causing severe impairment to groundwater quality.

The owner/operator has repeatedly argued with the LDEQ that it is not responsible for the cleanup of contamination off-site that has resulted from the overflow of waste pits and ponds that have drained into a natural drainage canal. It has attempted to place all the environmental liabilities onto a former owner/operator of one of the facilities, despite flagrant violations which include citations of direct discharges of hazardous wastes to drainage ditches.

Of interest is that the owner operator has attempted to sidestep environmental responsibilities for cleanup by applying legal definitions to wastes. In a most notable argument presented to the LDEQ, it articulated that it was not responsible for cleaning a contaminated portion of the property that was stained by kickback and drippage from treated wood. The operator went so far as to obtain environmental samples to demonstrate that the contamination was coal tar and not creosote, and since statutory requirements only address the cleanup of creosote, it argued that it should not be held accountable.

As noted earlier, coal tar and creosote have similar compositions and contain nearly identical toxins. The owner/operator of these facilities knows this perfectly well since they have been in the wood treating business for more than 50 years. The company's actions send a clear signal that the senior management is not committed to environmental management. Instead of attempting to sidestep its obligations, the owner/operator should be placing effort in reducing its chemical losses and spills so that it does not have to spend dollars on cleanup.

Waste Management Practices at the Somerville Tie Treating Plant

The Somerville plant was originally built in the 1890s was dismantled and reassembled in Texas where it was re-commissioned in 1904 and continues to operate. It has reportedly been the largest tie producing plant in the world.[48] A railroad originally built and operated the plant, but in 1994 sold the plant to an enterprise that continues to service it. Production records for 2005 show that more than 1.8 million railroad ties were made using creosote/oil mixes. Historically the plant used pentachlorophenol and CCA.

The facility currently occupies a 115 acre site, but at one point the property extended over an area of 300 acres. Portions of the property were sold off and developed. Figure 19 shows an aerial photograph of the facility.

---

[48] News article from Brenham Banner Press, Thurs., July 17, 1947.

Figure 19. Aerial photo taken in 1988. More recent photos show the site to be unchanged.

The plant is equipped with three treating cylinders (8 ft. diameter by 155 ft. lengths) but had as many as five cylinders (smaller in size) at various points in its history.

Up until the mid 1990s naphtha was used for the purpose of vapor drying wood. Green wood can be seasoned naturally (requiring up to 13 months) or in a high temperature/pressure vessel with an extractable solvent (in this case naphtha) which serves to remove sap and water from the wood. Using the naphtha extraction process, the time required for seasoning wood is reduced from 13 months to 5 to 13 hours. This process generates air emissions and wastewater. Roughly 15.9 kg (35 Lbs) of moisture is displaced from a single railroad tie. Hence this plant generates approximately 7.5 million gallons of wastewater per year.

A review of site features shows that there were at least 50 emission and discharge points that were either not managed or poorly managed through the mid 1990s when the facility was sold to another owner/operator.

Among the poor environmental management practices adopted by the facility were:

- Burying solid waste on the property was practiced into the late 1980s. Solid waste consisted largely of treated wood slats that were used to separate wood ties loaded onto trams in order to facilitate the flow of treating chemicals around the ties.
- Wastes streams were managed by a series of unlined pits and surface impoundments. The most prominent of these was a "waste reservoir" that occupied an area of more than 1 acre adjacent to a residential area. The cumulative volumes of the pits were close to 7 million gallons. The last pit was in operation up until circa 1991.
- There appears not to have been any storm water management program until sometime in the 1990s.
- Until sometime in the late 1980s wastewater streams were not treated except with aeration ponds.
- Wood waste boilers were employed at the facility up until circa 1995. Two boiler units installed circa 1939 were kept in operation until 1983 and thereafter were replaced with a single boiler unit. The reported stack emissions in 1981 were 200 TPY of particulates (PM). The newer boiler had PM emissions of around 60 TPY. None of the boilers were equipped with any air pollution controls. The post 1983 boiler was allowed to operate by the grandfathering of a permit. The facility began burning treated wood slats but likely burned process wastes during both boiler system histories. A material balance performed on the facility showed that cumulatively between 327 and 827 million Lbs of hazardous waste may have been burned in the boilers.
- Up until circa 1995 and possibly beyond (our analysis of waste management practices stops in 1995), the facility never used any air pollution controls. It did have a vapor recovery system for the treating cylinders and vapor drying process, but these systems were not for air pollution control purposes.
- Up until circa 1983 the facility spread ballast waste from its treating cylinders onto its property. Figure 20 shows a photograph of what is believed to be this

practice. It shows that a trail of mist was created during this practice thus forming toxic aerosols that potentially exposed workers and the community.

- The facility's first Industrial Hygiene program was introduced in 1981; however it failed to adequately identify job hazards associated with any waste handling operations. Additionally, workers were not issued any form of chemical protective clothing, neoprene rubber gloves nor given any Right-to-Know training until circa 1990. Only treating cylinder operators were issued respirators for cylinder cleaning operations beginning circa 1981.
- While the facility used CCA and pentachlorophenol intermittently, until the mid 1980s, it failed to devise any special chemical and waste handling procedures nor did it educate workers on safe handling practices.
- Under the TRI program the facility never reported dibenzofuran and anthracene air emissions. Most recently (in 2005) the facility was fined by the U.S. EPA for not reporting its benzene emissions.
- The facility did not eliminate vapor drying and convert to the Boulton process until sometime after 1995. The Boulton process has been understood for many decades to be less polluting.

Figure 20. Waste spreading was used as a basis for dust suppression at the facility.

TRI REPORTING

In addition to examining the waste handling practices of the above four plants, the investigators examined the reported air emissions from 20 wood treating plants across the country, including the case study facilities. Figure 21 shows the locations of these facilities. Not all of the plants are tie manufacturing facilities.

The TRI reporting databank revealed that most enterprises only report creosote as an air emission. A few facilities report naphthalene, dibenzofuran, anthracene, and the reporting of benzo perylene and dioxins only started in 2000. One may expect any one of the following toxins to be among the reported air emissions for wood treating plants: Anthracene, Arsenic, Benzene, Benzo(G,H,I)Perylene, Chromium, Copper, Creosote, Cumene, Dibenzofuran, Dioxin , and Dioxin-Like Compounds, Ethylbenzene, Hexachlorobenzene, Lead, Naphthalene, Pentachlorophenol, Phenol, Polycyclic Aromatic Compounds, Quinoline, and Xylene (Mixed Isomers).

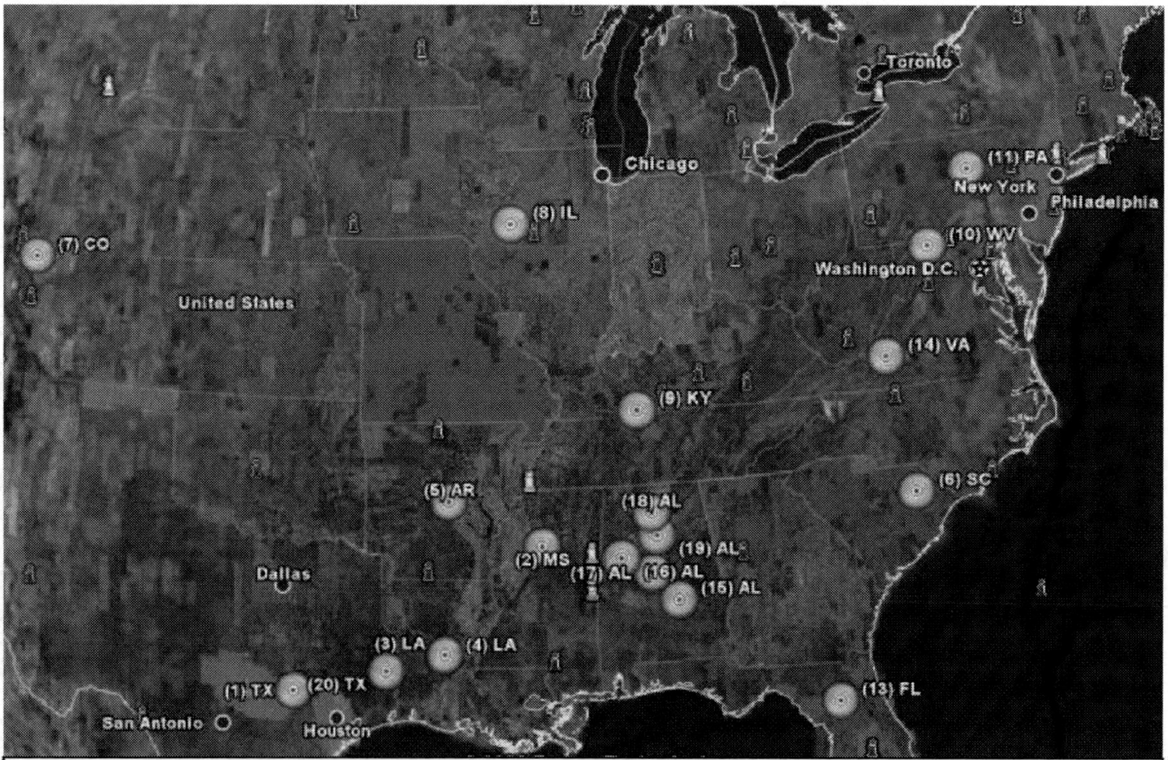

1 – Somerville, TX; 2 – Grenada, MS; 3 – Pineville, LA; 4 – Alexandria, LA; 5 – North Little Rock, AR; 6 – Florence, NC; 7 – Denver, CO; 8 – Galesburg, IL; 9 – Guthrie, KY; 10 – Green Spring, WV; 11 - Montgomery, PA; 12 – Superior, WI; 13 – Gainesville, FL; 14 – Roanoke, VA; 15 – Montgomery, AL; 16 – Northport, AL; 17 – Brierfield, AL; 18 – Falkville, AL; 19 – Warrior, AL; 20 – Jasper, TX

Figure 21. Plants examined in TRI database.

Figures 22 and 23 provide comparisons of the reported total air emissions between plants since TRI reporting began. Figure 22 shows that among the population of plants considered, the combined emissions from the two Louisiana tie treating plants are the second greatest.

Figure 22 provides a comparison between the reported total air emissions for only tie manufacturing plants considered in the study. The two Louisiana plants have the highest reported emissions. The total air emissions for the Louisiana plants in 1990 and 1991 were enormous for reasons not yet determined. See Figure 24.

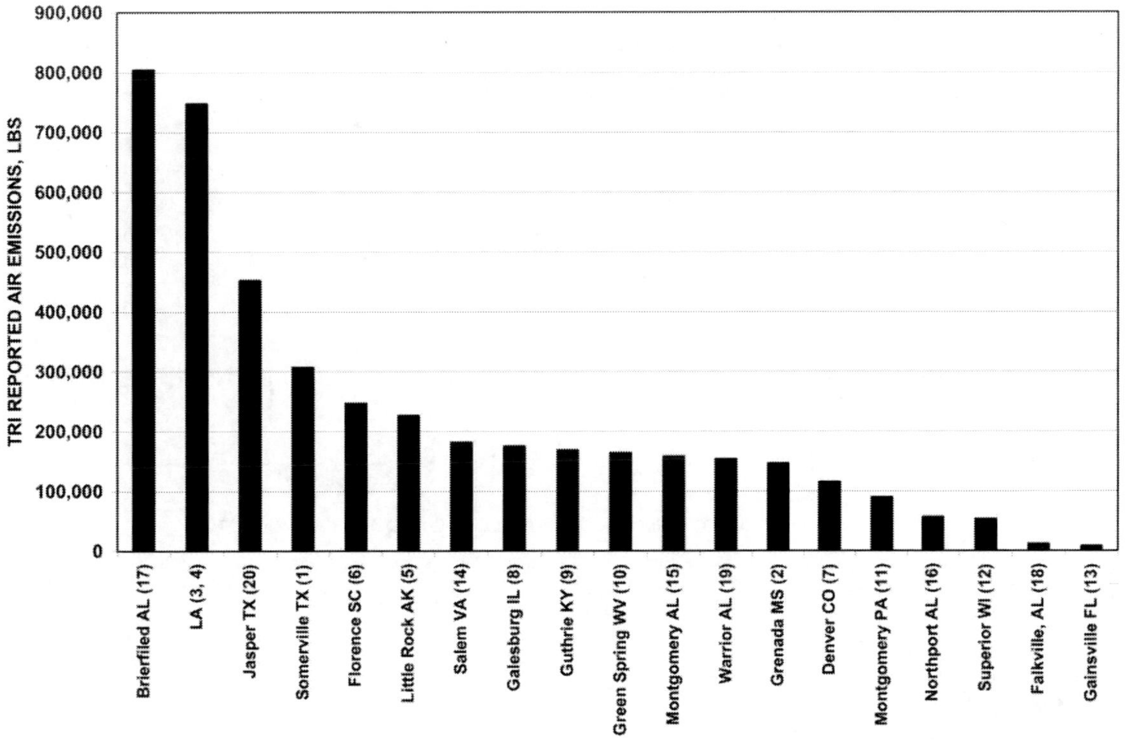

Figure 22. Comparison of total plant air emissions since TRI reporting began.

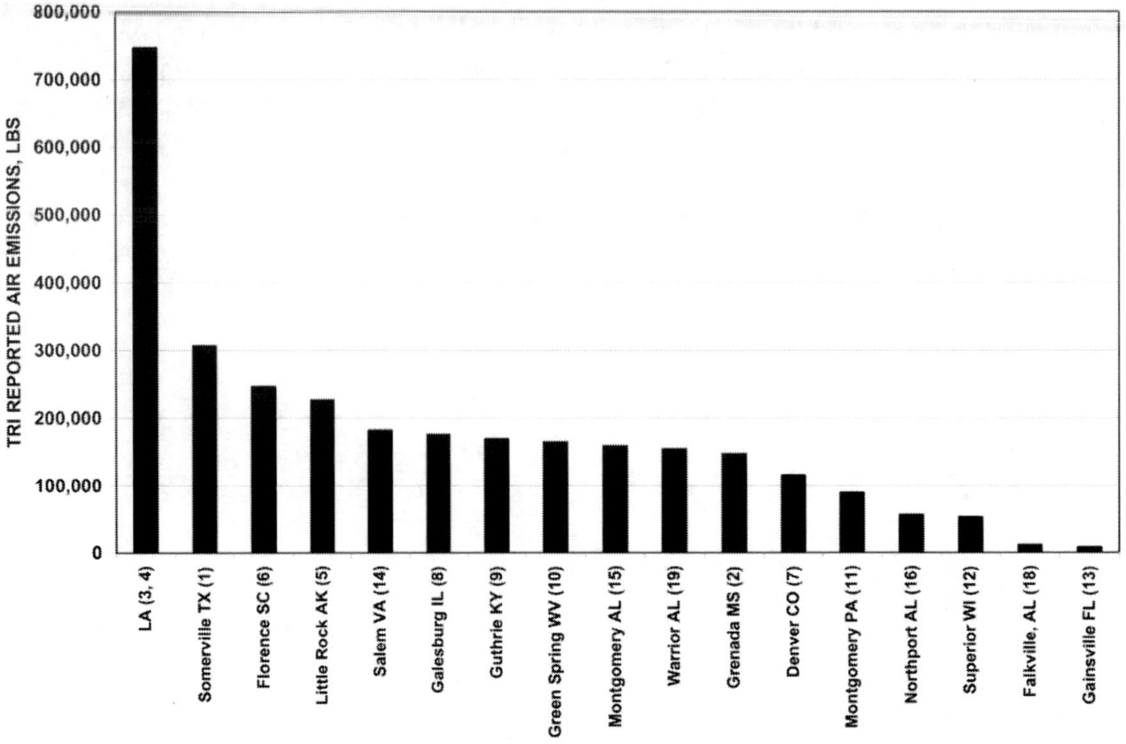

Figure 23. Comparison of total air emissions for tie plants since TRI reporting began.

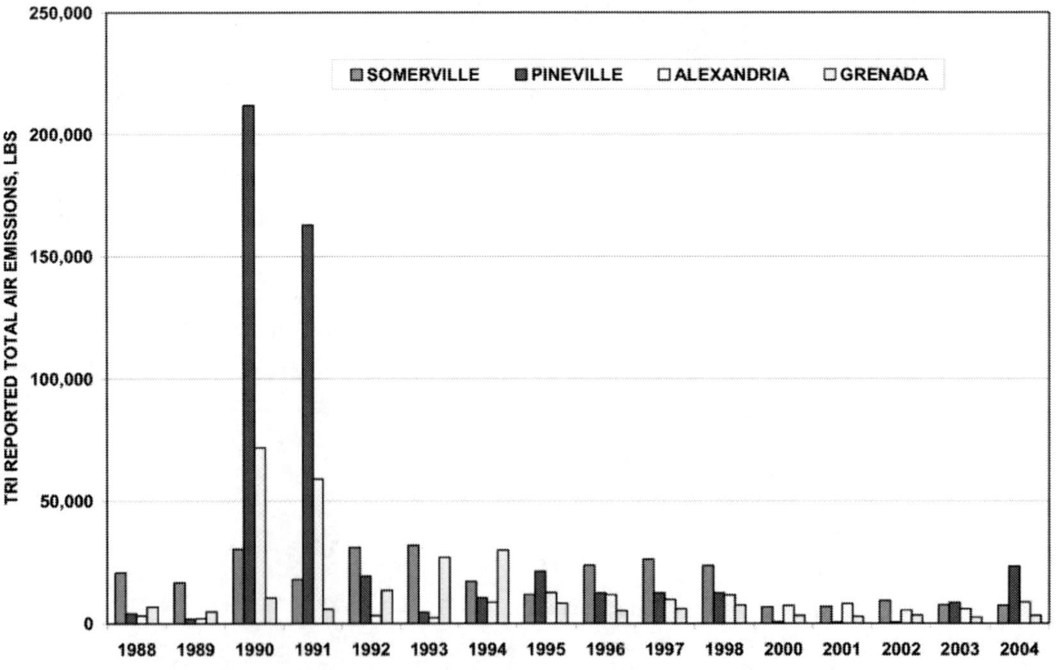

Figure 24. Comparison of Louisiana plant emissions to other facilities considered in the case studies.

The total air emissions themselves do not necessarily reflect a higher degree of risks to a community. The Grenada and Somerville facilities burned treated wood and pentachlorophenol containing wastes. These actions contributed to dioxins and furans among the mix of PICs. Very small concentrations of dioxins are a concern because of their link to carcinogenicity.

The high air emissions do; however, raise concern for both acute and chronic exposures to toxins by communities. Furthermore, many facilities reviewed in the study do not show reductions in their emissions over time. This leads to the conclusion that the population considered has not adopted P2 programs or invested into controls to reduce emissions since the TRI program began.

WASTE INCINERATION PRACTICES

Of the many poor practices identified in the case studies, the incineration of process and treated wood waste is the most troublesome. The incineration of creosote and pentachlorophenol sludges, and creosote treated wood scrap and old railroad ties generated dioxins and furans, as well as PAHs and heavy metals emitted in fly and bottom ashes. These toxins have entered into communities and have cumulatively added to exposures.

Dioxins are formed as the result of inefficient combustion. These toxins have a high melting point and tend to readily condense on to particulates such as flyash. The term 'chlorinated dioxin' refers to 75 compounds or any combination of the 75. Dibenzo-p-dioxin has been the toxin most studied and has the following chemical structure:

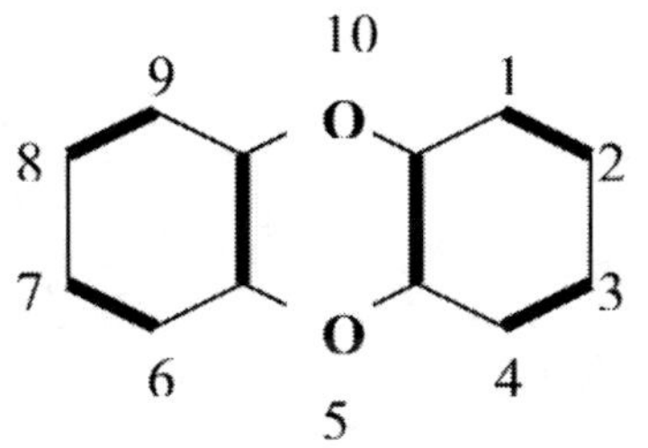

This compound is said to be chlorinated when one or more of the hydrogen atoms at the 1, 2, 3, 4, 6, 7, 8 and 9 positions is replaced by a chlorine atom. Among the 75 compound variations TCDD (2,3,7,8-tetrachlorodibenzo-p-dioxin) is considered to be the most toxic.

Since the early 1980s the scientific literature has reported that chlorinated dibenzo-p-dioxins are the result of common combustion processes and may be formed in the absence of high concentrations of chlorinated organic matter. Bumb, et.al.[49] reported the presence of dioxins from grilled meats, common refuse, paper, gasoline, fuel oil, coal and tobacco. A more recent study reported by the EPA[50] reports that relatively high levels of dioxins may be produced from the practice of barrel burning of trash, largely because of the wide composition of trash and the fact that mass burning is highly inefficient. The combustion of untreated wood will also generate dioxins and furans. In short, only trace amounts of chlorine or chlorinated compounds are needed to form dioxins under the conditions of poor combustion. But of course, the more chlorine that is present in the fuel, the more dioxins are generated when combustion is poor.

Chlorine donors and ring donors react by two known routes, and the dioxin products reach equilibrium that result in predictable isomer ratios. One mechanisms that is very relevant to the combustion of impregnated wood and contaminated sludges is the formation of dioxins through the denovo-synthesis mechanism at about 300°C (542°F) where copper and other metals in the flyash serve as catalysts in particular in the formation of chlorine radicals from the Deacon reaction. Examination of sludge compositions from the Grenada plant operations showed the waste to contain various heavy metals such as chromium, copper, arsenic, selenium and others. In the case of Somerville, CCA was used to impregnate wood and there are indications that sludge containing residues of CCA were burned in the boiler.

The overall reaction in the Deacon process is:

$$4HCL + O_2 = 2Cl_2 + 2H_2O - 116.8kJ\ (350°C)$$

As reported by Boscak[51] this reaction is catalyzed by $CuCl_2/CuCl$ and proceeds in a series of steps. The chlorine radical Cl' formed through the cleavage of the chlorine molecule is critical for chlorination of aromatic structures in dioxin formation. Also the oxidative cyclization of the aromatic structures is catalyzed by the presence of copper.

The National Academy of Sciences[52] estimated that on the average all coals contain around 1300 ppm of chlorine. There is nothing in the processes of extracting tars and refined creosote from coal that removes chlorine. Hence, the burning of creosote, process wastes contaminated with creosote, creosote and tarry sludges, and wood treated with creosote or coal tars will result in the formation of dioxins under conditions of poor combustion.

---

[49] Bumb, R. R., W. B. Crummett, S. S. Cutie, J. R. Gledhill, R. H. Hummel, R. O. Kagel, L. L. Lamparski, E. V. Luoma, D. L. Miller, T. J. Nestrick, L. A. Shadoff, R. H. Stehl, and J. S. Woods, "Trace Chemistries of Fire: A Source of Chlorinated Dioxins," Science, Vol. 210, No. 4468, Oct. 24, 1980

[50] See links to *Evaluation of Emissions from the Open Burning of Household Waste in Barrels; U.S. EPA* study published in November 1997, http://www.pca.state.mn.us/oea/res/V8_4/burnbarr.cfm

[51] Boscak, V., "Dioxins Emissions from Combustion of Impregnated Wood," http://www.bfafh.de/inst4/45/pdf/dixwood.pdf

[52] Committee on Medical and Biological Effects of Environments Pollutants, "Chlorine and Hydrogen Chloride," National Academy of Sciences, Washington, D.C., 1976.

Pentachlorophenol is a chlorinated hydrocarbon and commercial grades already do contain dioxins. Hence the combustion of any materials containing this chemical will generate dioxins under poorly controlled combustion conditions.

Bumb and co-workers have reported that combustion processes must be at least 99.9% efficient in order to ensure reduction of the concentration of dioxin from 1 part per million (ppm) to 1 part per billion (ppb). Edwards[53] reported as early as 1977 that common combustion processes are not adequate in achieving this level of reduction in stack gas emissions. And as discussed by Rezaiyan and Cheremisinoff[54], combustion is a series of complex chemical reactions that include physical, chemical and thermodynamic processes. This is especially true in the case of solid and mixed fuels with wide compositions, whereby combinations of processes including pyrolysis, oxidation, reduction and acidolysis result in the interactions of ions, electrons, free radicals, and free atoms, thus capable of producing a wide mix of PICs.

Combustion studies reported by Dow Chemical Co.[55] in the 1970s, and also by the U.S. Air Force[56] have shown that at temperatures below 800°C (1472°F) the destruction of TCDD is relatively incomplete. In the Air Force study involving the disposal of Agent Orange which contained levels of dioxin, the walls of combustion chambers were first raised to 1400 °C (2552 °F) and the waste added without any auxiliary fuel in order to bring the entire fuel mass to a uniform temperature prior to combustion. Even though the walls of the combustion chamber were maintained at an average 1273 °C (2323 °F) and a high flame temperature of 1500 °C (2732 °F) recorded, the EPA which monitored the program only reported a combustion efficiency of 99.9% for TCDD.

It has long been recognized that 99.999% destruction reduction efficiency (DRE; also known as the three 9s of destruction) is required in order to adequately demonstrate the destruction of dioxins.

Examination of the incineration practices at both the Grenada and Somerville facilities shows that the wood waste boilers used to dispose of significant amounts of treated wood waste and sludges had major operational flaws that lead to poor combustion conditions. Neither system was capable of carefully metering waste streams to the furnace sections, temperature monitors often failed or were never used, waste compositions were not carefully monitored and the practice of dumping contaminated soils into the boiler as a means of spill control was practiced. Grenada relied on nothing more than a mechanical separator (multiclone) as its air pollution control devise and Somerville operated without any air pollution controls. In neither case did boiler operators have real-time stack

---

[53] Edwards, J. B., **Combustion: The Formation and Emission of Trace Species**, Ann Arbor Science Pub., Ann Arbor, Mich, 1977.

[54] Rezaiyan, J. R. and N. P. Cheremisinoff, **Gasification Technologies: A Primer for Engineers and Scientists**, CRC Press, Boca Raton, FL, 2005.

[55] Freeman, H., Environ. Sci. Technol, 12, 1253, 1978.

[56] "Disposition of Orange Herbicide by Incineration," Department of the Air Force, Final Environmental Statement, November 1974; also "At-sea Incineration of Herbicide Orange On-board the M/T Vulcanus," Report EPA-600/2-7-086, Environmental Protection Agency, Washington, D.C., April 1978.

monitoring capabilities or were there controls provided to adjust combustion conditions enabling steady, uniform combustion which accompanies high DRE.

While dioxins can be generated from normal combustion practices entrenched among the general population, these actions cannot explain the exceptionally high levels of dioxins detected in attic dust samples obtained from homes in the subject communities. Nor can these practices explain why dioxin levels detected are highest for residences closest to facility operations. In both communities there are no other industrial sources that could help to explain the environmental samples. These factors lead to only one logical conclusion; community exposures to dioxins are a direct result of the waste incineration practices.

Both the Grenada and Somerville wood waste boiler operations relied upon conventional combustion technologies that were designed to use clean (untreated) wood as a source of either supplemental or primary fuels. Neither boiler system was ever intended by the suppliers to be operated as waste incinerators nor were they designed to achieve the industry standard of 99.999% destruction reduction efficiency (DRE; also known as the three 9s of destruction) for the burning of toxic wastes. While process wastes and treated wood scrap can be added to engineered conventional combustion systems in small quantities such that their concentrations are diluted to low levels by clean fuel, it must be done under carefully controlled conditions that ensure a steady, metered feed to the boiler's furnace. High flame temperatures as well as high turbulent mixing between combustion air and fuel, along with long residence times above 2 seconds are needed with both primary and secondary combustion chambers in order to ensure good combustion. Here also, the safeguard of pollution controls such as electrostatic precipitators, wet or dry scrubbers, high temperature fabric filters, and/or afterburners may be required.

Solid waste management constitutes a significant problem in most industry sectors. The disposal of treated wood ties alone constitutes a formidable problem for the railroad industry and one which is both costly and complex. Incineration unto itself is not necessarily an improper means of managing these wastes at the end of the product's life cycle. Compared with other disposal options such as landfilling, incineration offers both a cost effective and reasonable waste management option. Its advantages over landfilling are:

1. Up to 95% reduction in waste volume
2. Heat and power generation (i.e., conversion of waste to energy)
3. Better control of emissions and effluents (there is less potential for groundwater contamination compared to landfilling where chemicals may leach into soils and groundwater)
4. Siting flexibility
5. Up to a tenfold reduction in greenhouse gas emissions over hydrocarbon fuels

However, appropriate pollution controls and/or engineered combustion systems are critical to ensure that wastes are managed in a manner that is responsible and does not place the public in harms way. It has long been understood that good combustion is essential for the minimization of reactive aromatic species which generate precursors such as benzene, phenol and other precursors to dioxin formation. The fundamental keys to good combustion are the three T's of combustion (temperature, time, and turbulence) with the mixing of fuel and combustion air (both primary and secondary) being among the most important factors.

Our investigation reveals that two facilities were not diligent in their consideration of the negative impacts resulting from the technologies and practices used to manage treated wood ties at the end of service, and most certainly acted irresponsibly in the management of toxic wastes generated during the manufacture of these products. In the case of Grenada, although the railroad industry itself did not actively participate in waste disposal, it was a contributor to community exposures.

Part of responsible care means that enterprises must be held accountable for the entire life cycle of the products and services they provide. Companies cannot simply hand off the responsibility of disposal of their wastes to other entities without first considering whether the waste disposal facility is managing those wastes using environmentally sound technologies and practices. Facilities that invest into waste management technologies need to be diligent by considering the negative aspects of the technology and practices, and to devise safeguards to mitigate them.

CONCLUSIONS

Attic dust samples obtained from the homes of residences in close proximity to the Somerville and Grenada plants show dioxin levels to be several orders of magnitude above the World Health Organization standard for safe levels of lifetime exposure. Attic samples represent a time capsule where toxins contained in airborne dust have been captured and preserved in non-living areas of homes. In both cases dioxins detected are most probably a result of historical releases from the plants. They are a result of incineration practices associated with the wood waste boilers, re-suspension of soils that have been contaminated from years of boiler stack deposition of particulates containing dioxins, furans and other PICs (Products of Incomplete Combustion). In the case of Grenada, aerosols from the spray fields have contributed to the dioxins found. The Somerville plant also had a practice of spraying wastes onto land but with a technique that likely created a greater risk of inhalation exposure to workers. The owner/operator of the Grenada plant owns several plants, some of which are Superfund sites and were known to rely on a similar land farming practice as adopted by the Mississippi facility.

In the case of the two Louisiana wood preserving plants, only soil and groundwater samples have been reported to date. Environmental samples show significant concentrations of PAHs, dioxins and furans. Among the toxins found at levels above cleanup standards are Ancenaphthene, Acenaphthylene, Benzo(a) anthracene, Benzo (b,k) fluoranthene, Benzo(a)pyrene, Carbozole, Chrysene, Cresol, Fluoranthene, Napthalene, Phenol, Pyrene, Tetrachlorophenol, and Trichlorophenol. Dioxins have also been detected in groundwater samples. Since these are found at high levels in soil and groundwater, it stands to reason that some or all of these are emitted as air emissions. Despite this the owner/operators have never reported any of these toxins as air emissions in their TRI reporting. Soil samples both on-site and on neighboring properties to the Grenada and Somerville plants show the same contamination problems to persist along with heavy metals which include chromium, arsenic, and copper.

In addition to contaminants leaching from soil to groundwater and toxic aerosols from land farming practices, other pathways by which toxins have entered into communities include decades of contaminated storm water runoff onto neighboring properties and waste streams which flowed through drainage ditches into local water courses. The properties of the operating facilities have been badly contaminated from decades of poor housekeeping practices. Residents in some of these communities have eaten fish from local streams, used contaminated groundwater for cooking, bathing and consumption, as well as watered gardens from which fruit and vegetables were harvested and local livestock for human consumption were raised.

At some of these facilities local residents were employed by the facilities. The workers at Somerville for example were never provided Right-to-Know training, chemical protective clothing or gloves and only issued a few respirators for some job functions beginning in the 1990s. These individuals brought toxins directly into their homes and exposed their families to the chemicals handled during work days.

At all four facilities it was found that up through the early 1980s treated wood scrap was distributed to communities. Not only have some residents placed themselves at risk through dermal contact of toxins but exposed themselves and family members to inhalation risks from the burning this waste in home fireplaces and wood burning stoves.

As already noted the practice of burning treated wood and process wastes in wood-fired boilers without appropriate pollution and combustions controls are major sources of dioxins and furans.

It is logical that residential exposures to PAHs, VOCs, dioxins, furans, and heavy metals are ongoing even though some historical practices have ended more than a decade ago. While site remediation efforts have been underway for more than a decade at some of these facilities, it's important to recognize that industrial activities date back to the turn of the last century for two of these plants. Not all areas of concern have been identified. Both the Somerville and Grenada plant operations extended to land areas that were as much as two times larger than they are today. Neither facility has taken the effort to determine whether public lands and residences have been built up over areas that were used for historical waste management practices. Somerville as an example operated a creosote waste pit covering an area exceeding 1 acre in size between 1910 and 1945. Grenada operated a "creosote hole" up until the mid 1960s. Neither of these features have been fully delineated nor their locations in relationship to sensitive receptors been defined to date.

None of the facilities examined in the case studies appear to have invested into dedicated Environmental Management Systems with Pollution Prevention (P2) at the core. Additionally, of the 20 facilities surveyed under the TRI program, only a few appear to be monitoring and reporting some of the priority air pollutants, and this reporting has only been initiated within the last 2-3 years. At least one of the facilities listed under the TRI program (a pole treating plant in Jasper, Texas) invested in its first vapor recovery system in 2004; which means that its VOC and PAH emissions have been freely discharging since the 1940s when the plant was commissioned. As recently as 2004, Somerville was fined by the U.S. EPA for not reporting its benzene emissions. And the two plants in Louisiana show records of violations under CERCLA cleanup actions extending through 2002 when our analysis stops. One of these plants falls on the CERCLA National Priority List (NPL).

The above observations support that companies were not paying attention to regulatory requirements in the 1980s and they continue not to pay attention to them. Continued poor environmental performance means that these companies continue to place communities at risk as well as their own financial wellness and sustainability.

The wood preserving industry does not face greater environmental management challenges than any other industry sector that relies on hazardous chemicals to manufacture its products. The petroleum refining, chemicals manufacturing, orgnanophosphate, metals fabrication, textiles, food and beverage industries, and many others must deal with air emissions, process wastes in the form of sludge and waste water, and solid wastes. The chemical wastes and toxins in their waste streams differ, but the technologies and best practices for responsible management are universal. While there are certainly examples of poor environmental management practices and irresponsible actions among other sectors, there is clear evidence to show that overall industry sector trends have steadily increased their focus on responsible management of wastes and pollution over the past 30 years. Examples of this can be found among an almost endless collection of documented pollution prevention and waste minimization case studies for each sector on the World Wide Web, compared with very few for the wood preserving industry.

The case studies presented in this paper are examples of companies that have not and appear to continue not to make the environment an equal element in corporate strategic planning. Simply meeting regulatory requirements at the time of a compliance schedule is unacceptable.

- Companies that invested into drip pads for the first time in 1992 because they had to meet a compliance schedule were acting responsibly only minimally, because they understood and allowed decades of poor housekeeping practices, repeated spills and releases to cumulatively contaminate their properties. The consequences of these actions have been long-term negative impacts to groundwater quality, to surface runoff, and have contributed to chronic air emissions.
- Companies that apply interpretations of statutory definitions to bring wastes to their properties without considering the consequences are irresponsible.

- Companies that attempt to apply regulatory definitions of wastes in ways that allow them to avoid remediation of obvious damages caused by their operations are acting irresponsibly because they focus only on statutory issues and not elimination of negative environmental aspects that have long-term impacts on neighboring communities.
- Companies that are just now investing into air pollution controls are decades behind responsible care practices that have been in the mainstream of most industry sectors.

While meeting statutory requirements should be a focus for any corporation, acting responsibly goes beyond that. Responsible care means managing the negative environmental aspects of the manufacturing steps, regardless of whether there is a statutory requirement or not.

In the case studies we found that the management styles for environmental aspects were similar in that none of the facilities maintained trained, qualified staff. Somerville is an exception where a dedicated environmental department was created beginning sometime in the late 1990s. These facilities have largely relied upon plant superintendents, plant managers or owners, or periodic visits from corporate IH or environmental engineers responsible for managing multiple site operations to make spot inspections and recommend corrective actions which in a number of cases were never implemented. None of the facilities were monitoring accidents, releases, spills, and since records of performance or mishaps were never maintained, none of these facilities ever established baselines from which environmental action plans aimed at improving performance could be established.

The style of relying on ad-hoc and untrained personnel to manage environmental aspects dates back to the early 1960s. This type of management style reflects a critical lack of commitment towards environmental protection and worker safety. Every one of these facilities would be better off by investing the time into an Initial Environmental Review (IER) to understand their current environmental aspects and to prioritize them as a starting basis for devising structured environmental management plans. Every one of these facilities should be hiring trained personnel that are given adequate resources to manage the multiple negative environmental aspects stemming from their operations. All of these facilities should be diligently monitoring emissions, identifying the root causes for the pollution and wastes within their operations, and devising programs aimed at systematically reducing them.

Companies with contaminated properties that are now arguing that historical practices that caused off-site property and health damages were simply standard practices and the application of technologies of the day are deluding themselves. The toxic properties of creosote, pentachlorophenol and CCA were known dating back to their earliest use by the industry. Chemical suppliers and authoritative organizations such as the Chemical Manufacturers Association, the American Petroleum Institute, the American Institute of Chemical Engineers, the American Institute of Environmental Engineers, and even the American Association of Railroads and the American Wood Preservers' Association among others recommended and helped to define both safe and environmentally sound management and handling practices for these chemicals and the waste streams containing them as far back as the 1940s. Air pollution controls, wastewater treatment technologies, and best management practices for chemicals and wastes can be found among industrial and scientific literature dating back into the 1930s. It was just as irresponsible before the regulatory era of 1972 to dump creosote contaminated process wastes into a ditch leading to a public stream as it is today. The only difference between then and now is that today there are laws with stiff fines, penalties and criminal punishment for such irresponsible behavior.

The wood preserving industry has created a legacy of toxic Superfund sites. On the National Priority List (NPL) of sites identified by EPA:

- Arsenic has been found in at least 781 NPL sites;
- Penta had been found at least 314 NPL sites;
- Chromium has been found in at least 386 hazardous waste sites on the NPL;
- Copper has been found at least 210 NPL sites; and,
- Creosote has been found at least 38 of NPL sites.

It's quite possible that more wood preserving sites may be added to the NPL as neighboring communities are beginning to cope with the latent impacts resulting from decades of poor environmental management practices and continued mismanagement.

Responsible companies rely on a modern framework known as the Environmental Management System or EMS, embodied in the ISO 14001 standards. These standards are designed to focus an organization's attention and resources towards environmental issues in a way that they are managed with no less diligence than the fiscal affairs of a corporation. In fact, responsible enterprises understand that good financial performance is directly linked to good environmental performance. Through pollution prevention, investments into proper pollution controls and by applying proper metrics that monitor environmental performance along with well-defined targets for waste reduction, responsible enterprises systematically reduce emissions and wastes over time. These actions ensure the safety of workers and the public, preserve resources, and sustain business operations. Companies that simply manage environmental aspects with only a focus on meeting minimum statutory requirements may well find themselves entangled in costly class action suits as well as on-going cleanup costs that can cripple their financial wellness for many years into the future.

Improving the Process of Certifying Title V Compliance at a Refinery Marketing Terminal

Peter F. Stevenson[1], Karen N. T. Olson[1], Robert A. Morris[2], and Rebecca R. Ayala[2]
(1)Zephyr Environmental Corporation, 1515 Capital of Texas Highway S., Suite 300, Austin, TX 78746, (2) Environmental, Western Refining Company, 6501 Trowbridge Dr., North Admin. Rm. 56, El Paso, TX 79905

Introduction

Verifying and certifying compliance with Title V permitting requirements can be a labor intensive, hurried, and often inefficient process that places facility management at personal and professional risk. This is especially true for facilities that have recently changed hands or which are undergoing significant upgrades or changes, which is the case at many U.S. refineries today.

At the request of Western Refining Company's Environmental Management, Zephyr Environmental Corporation assisted Western Refining in performing a work process improvement pilot study for their Marketing Terminal facility in El Paso, Texas. The study documented work processes associated with managing and verifying compliance with the Title V permit at the Marketing Terminal, with a focus on the compliance certification and deviation reporting processes. The ultimate goal of the work process analysis was to develop recommendations to improve the efficiency and effectiveness of work processes and tools related to Title V compliance; thereby, reducing time spent managing compliance information, creating reports, and completing the Title V certification process.

The pilot study was conducted by the project team using an interactive work process analysis technique involving a four-step approach:

A) Identifying and documenting current work processes, tasks and associated responsibilities in high level work flows
B) Developing the ideal case for work processes, tasks and responsibilities in more detailed work flows
C) Identifying gaps between the current and ideal cases
D) Assessing the gaps to identify which should be addressed in order to provide the best return on investment

Although this work process analysis technique is likely a familiar tool to industrial engineers in analyzing the flow of work, it is not commonly used by environmental professionals to analyze environmental compliance management tools and methods. However, this pilot study showed that work process analysis is a very effective tool for identifying areas for improving the efficiency and effectiveness of personnel and tools involved in managing environmental compliance. Example results of this four step analysis technique are provided in this paper.

Improving the Process of Certifying Title V Compliance at a Refinery Marketing Terminal

1.0 Work Process Analysis Results

The Current Work Flows represent the actual work steps and the relative order in which they were being performed to complete Title V compliance certification for Western Refining Company's Marketing Terminal. The Ideal Work Flows represent altered work steps, relative order and associated role responsibilities needed to optimize the Title V compliance certification process for the Marketing Terminal. Both sets of work flows were developed through an intensive five day interactive process facilitated by Zephyr and conducted by the project team at the Marketing Terminal facility in El Paso, Texas, followed by 2 weeks of further analysis by Zephyr Environmental team members at the Zephyr offices in Austin, Texas. Example copies of the work flows developed using Microsoft™ Visio software are included in Appendices A and B.

1.1 Current Work Flow Process

The following current work flows were developed by the team during two and a half days of intense interaction, with intermittent updates made by the team while completing other activities. These current work flows represent the key groups of facility activities, which impact the Marketing Terminal's Title V compliance certification work process:

- Title V Certification Process
- Gather Certification Documentation
- LDAR Compliance Certification
- Loading Truck Tightness Test
- Loading Leak Detection
- Marketing Terminal Monthly Closeout
- Loading Rack De-authorization Determination
- Tank Seal Inspection
- CEMS Downtime
- Vapor Pressure Determination
- Emissions Estimation and Tracking
- Exception Reporting

The Title V Certification Process work flow represents the primary work flow or the parent activity, into which the results of all the other work flows feed. A copy of this work flow is provided in Appendix A. During the current work flow development process, the team also identified the majority of the perceived improvement opportunities or gaps, which were captured separately.

1.2 Ideal Work Flow Process

The following ideal work flows were developed over an intense one and a half day interactive session, with additional follow up work by the Zephyr team members to further expand and define the work flow steps. The fundamental strategy in developing the ideal work flows was to automate as much of the work process as possible, in order to eliminate inefficiencies associated with human data/file management. Following are the ideal work flows developed:

- Marketing Terminal Title V Certification Process

- Gather Certification Documentation
- Title V Certification Task Assignment
- LDAR Compliance Certification
- Loading
- Terminal Monthly Closeout
- Loading Rack De-authorization Determination
- Tank Seal Inspection
- Vapor Pressure Determination
- Exception Reporting
- CEMS Downtime

As with the current work flows, Title V Certification Process work flow represents the primary work flow or the parent activity, into which the results of all the other work flows feed. A copy of this work flow is provided in Appendix B.

At the core of the ideal work flows is an electronic data management system, "The System", which is actually the collection of all interconnected data management systems (PI, HMI, TMS, Hummingbird, etc) at the site. The System is used to monitor compliance dates and key data transactions at the facility. These due dates and key transactions drive the automated collection and management of data and documents or notifications to responsible individuals to perform certain actions, including the manual data and document collection and management. Key to the Title V Compliance Certification work process are electronic information storage folders, in which compliance certification related data and documents are stored and from which they are accessed during the certification process. In the flow charts, the term "cited Title V Certification folder" refers to a specific folder within this group of electronic storage folders.

2.0 Potential Gaps

During the interactive processes for the current work flow development, the team identified potential issues needing to be addressed, as well as opportunities for improving the current certification process. These potential issues and opportunities were generally addressed in the development of the ideal work flows. In addition, a comparison was performed between the current and ideal work flows, identifying further differences between the current and ideal work processes. The processes for identifying all these potential issues, opportunities for improvement and work flow differences are collectively referred to as the "Gap Analysis" for this project. The results of the "Gap Analysis" are the potential "gaps" in the Marketing Terminal's Title V Compliance Certification work process. Pending full implementation of the ideal work process as defined, the identified gaps should all be eliminated. Table 1 in Appendix C provides a compilation of the identified gaps and potential solutions by work flow category, as well as suggested priorities for implementing the proposed solutions.

Some potential regulatory gaps were also identified during the pilot study. However, the scope of the project did not include a thorough regulatory audit; therefore, the regulatory gaps identified were not considered a complete list of regulatory issues.

3. Recommendations for Work Process Improvement

The potential gaps were evaluated and prioritized to address the most significant improvement in efficiency and/or assurance of environmental compliance. The following are the most important recommendations identified for improving current Title V Compliance Certification work processes at Western Refining Company's Marketing Terminal:

- Develop (or further implement existing) interactive electronic systems for gathering data, creating documents, organizing information, tracking deadlines and sending assignments to meet those deadlines.

- Develop Management of Change (MOC) Procedures to assure that process changes and demands are adequately evaluated relative to environmental compliance before proceeding with any facility or process changes at the Marketing Terminal.

- Install loading rack instrumentation with interlocks to ensure vapor recovery unit (VRU) connection while loading and to improve the accuracy and timeliness of throughput measurements.

- Install Smart-Chip and Digi-Verify type systems to automate driver / trailer verification to reduce or eliminate potential driver fraud concerns and further assure Title V compliance.

4. Keys to Success

The process went well in large part due to the high caliber of the client staff involved in the process. The client team members were extremely knowledgeable, committed to the goals of the process, team oriented, and forthcoming in sharing the information they had concerning existing work processes.

Having access to key personnel at the operating unit was instrumental in the success of this analytical project, specifically the Environmental Compliance Specialist for the unit, the Lead Operator and Unit Environmental Coordinator, who in this case was the same person, and the Instrumentation and Electrical person.

Having large blocks of uninterrupted time with these key personnel proved very helpful, if not imperative.

The strategy for current work process development, with both the Western unit operations staff and the environmental compliance staff, worked very well. The operations staff were able to describe in detail how they performed and documented functions, while the compliance person was able to identify areas of regulatory and compliance issues that needed to be addressed. It would have been helpful to have the consulting staff very familiar with the unit's regulatory requirements to facilitate the Western staff in identifying work processes possibly perceived as unimportant by operations staff, but which could directly affect the compliance certification process.

Having a consulting team member available to transcribe the work flows as they were developed was essential to completing the interactive portion of the pilot project in one work week. In addition, transcribing the work flows into Microsoft™ Visio at the same time as they are developed on paper proved a much more efficient and effective process than having the work flow developer transcribe from draft paper copies of the work flows. The former method allowed the developer to listen to the discussion in person and obtain clarification on questions and issues immediately, instead of long after the discussion had occurred.

Having working meals brought into the facility allowed the team to maximize the limited availability of key Western team members and staff.

Meeting in the same room each day and being able to leave materials and supplies in place until the end of the process, improved the efficiency of the team tremendously and was imperative to the success of the pilot project under the compressed schedule.

4.3 Recommendations for future work process improvement projects

A. Have the same meeting room from start to finish for each interactive work session.

B. Have environmental department and unit person available for the duration of the interactive work sessions.

C. Have an IT person available to facilitate network access.

D. Have a unit person available who can help with team logistics, e.g., obtaining office supplies, printing, etc.

E. Three consecutive days, for six hours per day should be the maximum length for any interactive work session. Beyond that, team members get fatigued and are not as effective.

F. Allow adequate preparation time for the consulting team prior to coming on-site for work flow analysis. The consulting team members need to:

- Fully understand and organize the regulatory requirements, which drive the certification and deviation reporting process
- Make sure that all relevant work flows are developed to a significant level of detail during the on-site interactive process.

This assures that all Title V regulatory requirements are addressed for compliance certification.

G. Before scheduling the work flow process analysis sessions, the consulting team should meet with the environmental compliance staff and operations staff to identify the key work areas (i.e., work flows) and the key personnel or staff positions needed to be involved in the interactive sessions. Client staff time should then be planned and scheduled based on that input.

H. Provide adequate time in between on-site work flow analysis sessions to allow the consulting team to process and organize information that must be gathered in stages. This will allow consulting staff an opportunity to

- Evaluate the information,
- Put the information in a logical order and format to facilitate review and
- Identify any information gaps that need to be filled to complete the work flow.

This would also allow time for Environmental and operations staff to peer review the documentation.

I. Have two consulting technical staff involved in the interactive sessions, such that the process and regulatory issues can be analyzed and documented at the same time, without holding up the team. This will allow the analysis process to move faster and ensure process and regulatory issues are not missed in the discussion.

J. Provide the consulting team with direct access to network resources, including the internet and network printers needed to print wall-sized work flows. Access to these resources should be provided before or after normal work hours to maximize consulting team onsite time and project team interaction time.

5. Conclusions

This pilot project effectively demonstrated that work flow analysis, a non-traditional environmental compliance review method, can be used to document current compliance work processes and develop recommendations for improving those processes to make more efficient and effective use of company resources.

APPENDIX A

Current Title V Certification Work Process Flow Diagrams

Improving the Process of Certifying Title V Compliance at a Refinery Marketing Terminal

TITLE V CERTIFICATION

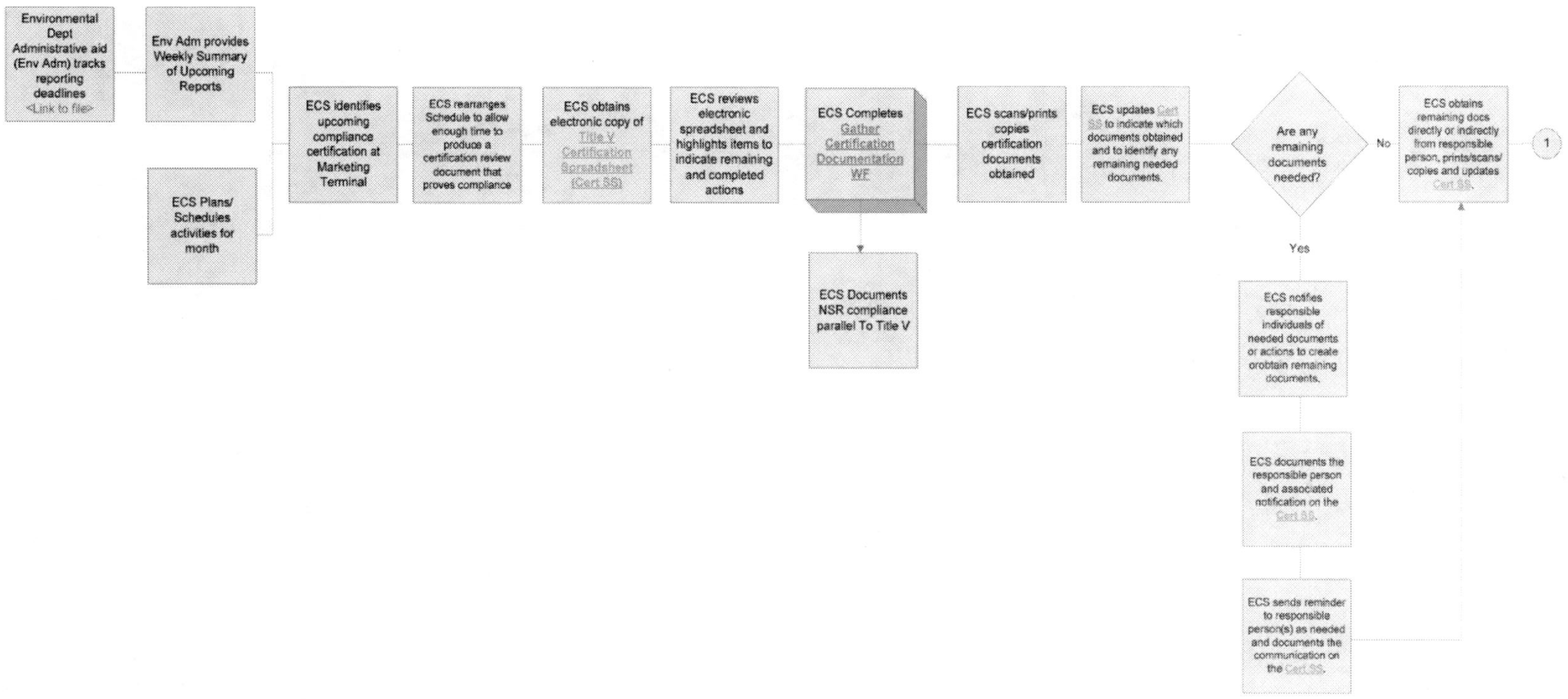

WESTERN REFINING MARKETING TERMINAL		
TITLE V CERTIFICATION CURRENT WORK PROCESS		
Designed by: CAW	Project No.: 05344.001	Revision Date: 05/19/2006
zephyr		

Improving the Process of Certifying Title V Compliance at a Refinery Marketing Terminal

TITLE V CERTIFICATION
(continued)

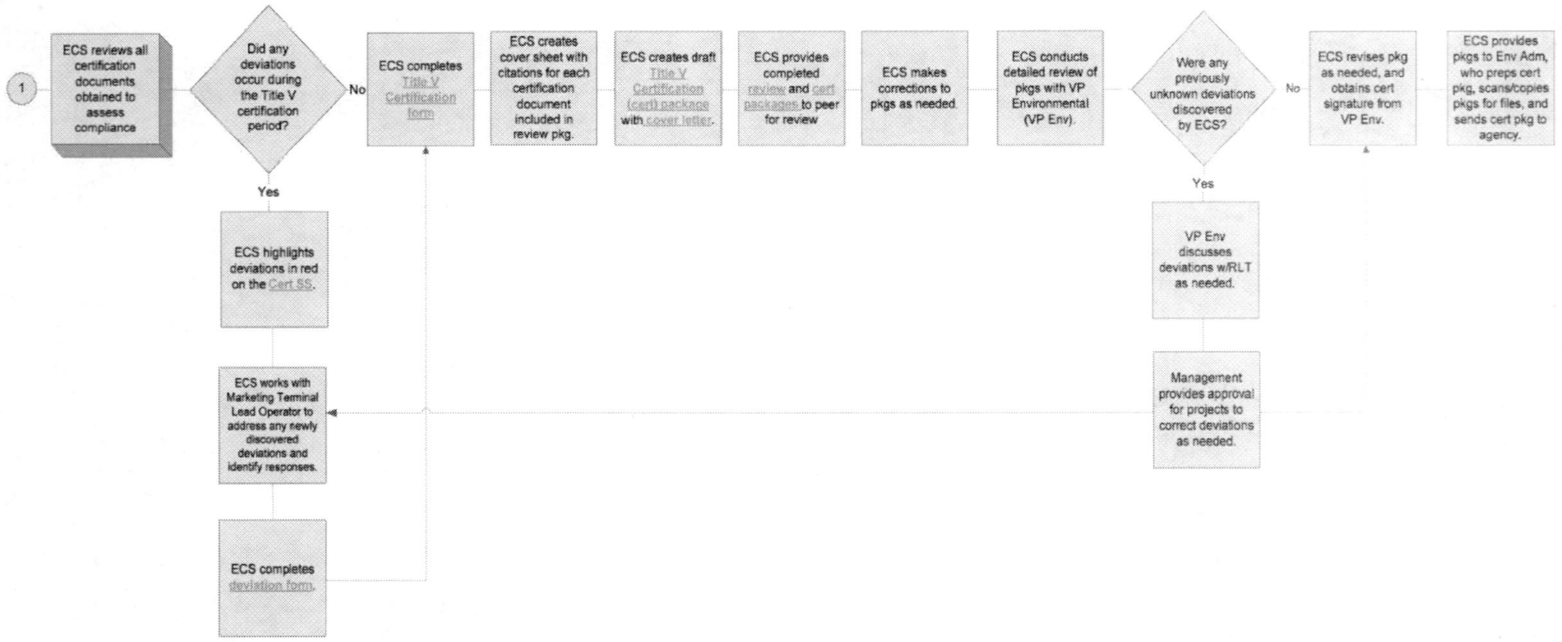

WESTERN REFINING MARKETING TERMINAL		
TITLE V CERTIFICATION CURRENT WORK PROCESS		
Designed by: CAW	Project No.: 05344.001	Revision Date: 05/19/2006
zephyr		

APPENDIX B

Ideal Title V Certification Work Process Flow Diagrams

Improving the Process of Certifying Title V Compliance at a Refinery Marketing Terminal

Marketing Terminal
Ideal Title V Certification Process

Improving the Process of Certifying Title V Compliance at a Refinery Marketing Terminal

Marketing Terminal
Ideal Title V Certification Process
(continued)

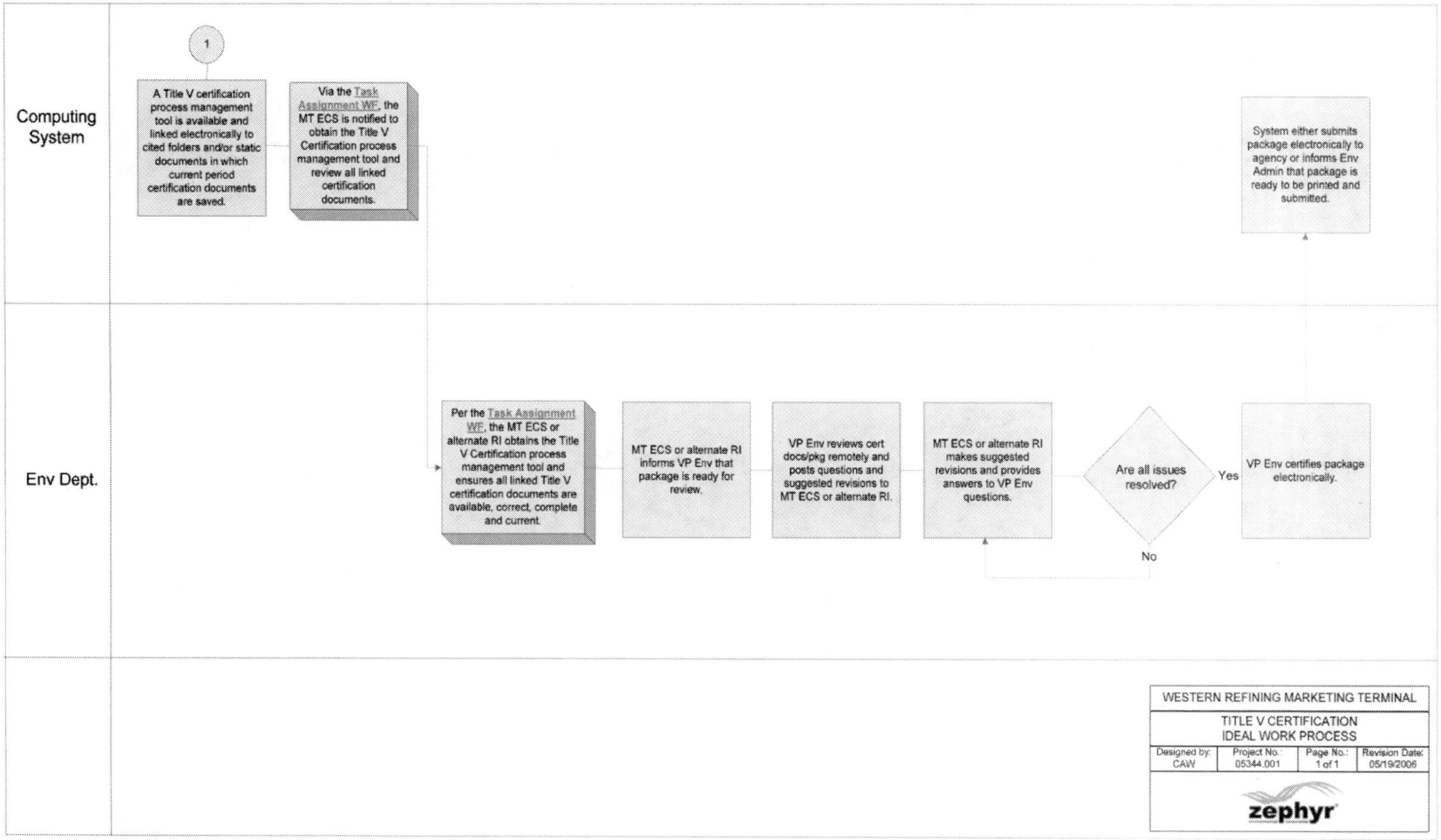

APPENDIX C

Table 1: Western Refining Marketing Terminal – Work Processes
Potential Gaps/Issues

Table 1. Western Refining Marketing Terminal – Work Processes Potential Gaps/Issues

Priority[1]	Title V Certification Gap - Potential Solution
A	Gap: Waiting until the end of the certification period to review all documentation creates too much information for the Responsible Official to review at one time. Possible Solution: All certification documentation needs to be assembled and reviewed on an ongoing basis throughout reporting period rather than at the end of the 12 month period. Possibly send each exception, deviation or other compliance certification document to Responsible Official (RO) for review and approval as it's created. RO should have capability to provide electronic certification/approval.
A	Gap: Regulation compliance deadlines and reminders for actions to meet those deadlines are currently tracked and managed manually (with tools) by the Environmental Admin. Possible solution: "System" needs to be set up to: • Send auto reminders (with lead time) (e.g., via Outlook Task) to responsible and key personnel • Created and link Electronic form templates in system • Track lab sampling and product transfers from refinery to marketing terminal and auto-send virtual Certificate of Analysis (COA) to Marketing Terminal • Setup and track (on a real time basis) limits for key values or indicators as the data or value is input (e.g., VRU shutdown log data) and send notices of any potential deviations to key individuals
A	Gap: Certification documents have to be gathered and assembled by hand. Possible Solution: Make all certification documents available electronically and link them to the certification spreadsheet. Include a reference to the regulatory driver (citation w/link if electronic) in all forms/documents/logs used for environmental compliance.
B	Gap: Manually gathering static certification documents (Drivers Loading Training Manual, John zink Operations and Maintenance Manual, etc.) during the Title V Cert process wastes valuable environmental time. Possible solution: Provide electronic copies of all static cert docs and track each for potential changes with a flag provided to Title V cert prep person.
A	Gap: The current Marketing Terminal Lead Operator is a wealth of undocumented institutional knowledge and experience concerning managing the environmental compliance at the Marketing Terminal. Potential Solution: Environmental Procedures/data gathering /review should be written for marketing terminal prior to the current lead operator's departure from the facility.

1. Priority designations indicate the relative reduction in risk per anticipated cost: A – High, B – Medium, C – Low

Kinetic Modeling of Selective Non-Catalytic Reduction (Sncr) of Nitric Oxide Using Urea-Water Solution

Rajasree Retnamma[1], Youngil Lim[1], Ho-Jae Park[1], Seong-Joon Kim[2], and Kyung-Sun Yoo[2]. (1) Chemical Engineering, Hankyong National University, Gyonggi-do, Seokjung-dong 67, 456-749 Korea, Ansung, South Korea, (2) Department of Environmental Engineering, Kwangwoon University, 447-1, Wolgye-dong, Nowon-gu, Seoul, 139-701 Korea

Selective Non-Catalytic Reduction (SNCR) is a proven post combustion technology for controlling nitric oxide from stationary sources such as power plants, incinerators, and boilers. The process operates by injecting reducing agents such as ammonia or urea into the flue gas. Use of urea (NH2CONH2) as reducing agent has been widely accepted as an alternative to ammonia due to its easier and safer handling (Muzio et al., 1997; Sun et al., 1993). In the SNCR process, injection of urea into the flue gas is often carried out as a urea-water solution because of the problems associated in delivering and storing of dry urea. Thus decomposition and oxidation of urea-water solution can play a major role in SNCR process.

Previous experimental investigations on the decomposition of urea have identified different reaction paths depending on the form of urea (Koebel and Elsener, 1992; Caton and Siebers, 1989; Itaya et al., 1997; Lentz and Wright, 1994). Gentemann and Caton (2001) extended the experimental data on the urea-water solution decomposition at temperatures from 800 to 1300 K and found that thermal decomposition dominated for temperatures below 1000 K, whereas for temperatures above 1000 K, the oxidation became important.

A detailed investigation on SNCR of nitric oxide process using urea-water solution for various inlet O2 and CO concentrations is reported by Gentemenn and Caton (2001) and for various urea/NO ratio (-ratio), and residence time by Park and Caton (2003). But the reported results on kinetic modeling of the urea decomposition showed uncertainties especially at higher temperatures (Aoki et al., 1999; Alzueta et al., 2000). A kinetic model for the decomposition of urea solution used in selective non-catalytic reduction (SNCR) is developed in our earlier work (Rajasree et al., 2007). The above model could predict a number of gases such as HNCO, NO and N2O in addition to NH3 and CO2. The mechanism has been validated using experimental data reported by Gentemann and Caton (2001).

In the present study, a reduced kinetic model has been developed for the SNCR of nitric oxide using urea-water solution. The focus was on the impact of the results by Rajasree et al. (2007) in the kinetic modeling of NOx reduction using urea-water solution. We have included those reactions in the reduced model proposed by Brower et al. (1996) for prediction of SNCR of Nitric Oxide. The mechanism has been validated using the experimental data reported by Gentemann and Caton (2001) and by Park and Caton (2003) for various inlet O2 concentration, urea/NO ratio (-ratio), and residence time at temperatures from 800 to 1300 K. There is a good agreement between the model prediction and measurements and thus confirming the accuracy and potential of the developed model. The individual rate constants are provided so that mechanism could be coupled with computational fluid dynamics (CFD) based on a turbulent reacting flow model and would play a major role for accurate modeling of SNCR process.

Acknowledgements

This study is supported by Korea Ministry of Environment as "The Eco-technopia 21 Project".

References

Alzueta, M.U., Bilbao, R., Millera, A., Oliva, M., and Ibanez, J.C., 2000, Energy & Fuels, Vol. 14, No. 2, pp 509-510.

Aoki, H., Fujiwara, T., Morozumi, Y., and Miura, T., 1999, Fifth International Conference on Technologies for Clean Environment, pp. 115-118, Lisbon, Portugal, 12-15, July 1999.

Arand, J.K., Verdes, R.P., Muzio, L.J., Nigel, L., Sotter, J.G., and Viejo, M., 1980, US Patent No. 4,208,386.

Carton, J.A., and Siebers, D.L., 1989, Combustion Science and Technology, Vol. 65, pp. 277-293.

Gentemann, A.M.G., and Caton, J.A., 2001, Second Joint Meeting of the United Sections of the Combustion Institute, 2001 Spring Technical Conference, Oakland, 25-28, March 2001.

Koebel, M., and Elsener, M., 1992, Chemical Engineering and Technology, 64, pp. 934-947.

Muzio, L.J., Arand, J.K., Teixeira, D.P., 1977, Sixteenth International Symposium on Combustion: The Combustion Institute; Pittsburg, PA, pp 199-208.

Park, Y.S., and Caton, J.A., 2003, ASME International Southwest Region X Technical Conference, Houston, Texas, 25-28, March 2003.

Rajasree, R., Lim, Youngil., Kim, Seong-Joon., Yoo, Kyung-Sun., 2007, NASCRE-2 (North American Symposium on Chemical Reaction Engineering -2), Houston, Texas, 4-8, February 2007.

Sun, W.H., Stamakis, P., Hofman, J.E., 1993, ACS Symp. Series, vol. 38, pp 734-739.

Refinery and Petrochemical Liability Protection – Audit Your Offsite Disposal, Regeneration, and Recycling Facilities

Coby A. Scher and Dennis L. Caputo. Quest Consulting, Inc., 6700 West Loop South, Suite 310, Bellaire, TX 77401

How many of you are involved in Superfund sites or other similar situations where past disposal practices are costing you a bundle of money? How many of you are still facing recent or reoccurring incidents where past offsite disposal of your wastes, regeneration of your catalysts, or recycling of your materials have led to gnashing of teeth, conferences with senior management and lawyers, and more time and money spent on addressing and fixing the problem? There is a solution – it is called auditing.

Certainly by now all of you perform compliance audits within your own facilities to ensure that you are meeting your permit conditions and other commitments. You also perform environmental due diligence assessments in compliance with ASTM and/or EPA standards and protocols to minimize your exposure when evaluating asset sale or purchase. You may even perform environmental management system audits to assess compliance with your documented system whether for certification or not. But how many of you actually audit what happens to your residuals once they leave the front gate, whether for recycling, regeneration, treatment, or disposal.

Quest Consulting, Inc. (Quest) has established a system of ranking potential environmental liability to the client (risk) as a result of using recycling, regeneration, and waste treatment/disposal facilities. In this system, the risk is ranked as low, moderate or significant, depending on the potential cost of the liability to the client and the client's risk tolerance. To reach this risk determination, the audit includes a multi-step review including the following:

• Electronic databases. • State environmental agency file review. • Agency interviews by phone or at their office to determine the results of recent inspections conducted by the agency and the facility's compliance and corrective action history. • An interview by phone with the appropriate permit engineer (s) for the facility if needed. • Interviews with agency enforcement staff as appropriate. • Facility site visit. • Interviews with facility staff.

To conduct the audit, Quest prepares an Audit Protocol that is comprehensive yet relatively brief. In addition, Quest follows good engineering and scientific principles in determining risk associated with sending waste or recyclable material to this facility. These general principles are based upon both American Society of Testing and Materials (ASTM) and Environmental Protection Agency (EPA) guidelines and protocols for performing environmental due diligence. However, not all elements of these guidelines are used, just those that contribute to an understanding and evaluation of risk. The audit protocol thus developed provides overall guidance on the type of information to be obtained and procedures to be followed as part of the audit, and is not intended as a checklist of questions to be answered. The purpose of developing the protocol instead of a checklist is to provide a "big-picture evaluation" of potential environmental liability at the audited facility.

Quest recommends that this procedure be employed at least every 3 years for low-risk facilities and more often for higher risk facilities. It should include items like catalyst regeneration, radioactive instrument rehabilitation, hazardous and non-hazardous waste disposal, scrap metal recycling, and other offsite residual management.

The paper will present and discuss the Quest Audit Protocol and its merits. The paper will also discuss how the risk ranking is ascertained and what it means. Finally, several examples of how this procedure really works with some examples of actual audits for Frontier El Dorado Refining Company (Frontier) will be presented. Using this procedure and Quest's analysis, Frontier was able to determine which facilities should be used for future recycling, regeneration, and disposal.

AUTHOR INDEX

AUTHOR INDEX

American Institute of Chemical Engineers
3 Park Avenue
New York, NY 10016-5991

ISBN: 978-1-60423-357-5